MODERN SEAMANSHIP

THIRTEENTH EDITION

AUSTIN M. KNIGHT

LATE REAR ADMIRAL

UNITED STATES NAVY

REVISED BY

CAPTAIN JOHN V. NOEL, JR., U.S. NAVY

ASSISTED BY

Captain Howard R. Prince, U.S. NAVY

Commander William C. Kincannon, U.S. NAVAL RESERVE

Lieutenant John A. Mazzolini, U.S. NAVY

Lieutenant Commander Alfred Prunski, U.S. COAST GUARD

Lieutenant Commander Laurens Dorsey, U.S. NAVY

D. VAN NOSTRAND COMPANY, INC.

PRINCETON, NEW JERSEY · TORONTO · NEW YORK · LONDON

D. VAN NOSTRAND COMPANY, INC.
120 Alexander St., Princeton, New Jersey (*Principal office*)
24 West 40 Street, New York 18, New York

D. VAN NOSTRAND COMPANY, LTD.
358, Kensington High Street, London, W.14, England

D. VAN NOSTRAND COMPANY (Canada), LTD.
25 Hollinger Road, Toronto 16, Canada

Published simultaneously in Canada by
D. VAN NOSTRAND COMPANY (Canada), LTD.

Library of Congress Catalogue Card No. 60–16932

THIRTEENTH EDITION

PRINTED IN THE UNITED STATES OF AMERICA

Preface to the Thirteenth Edition

This revision of *Knight's Modern Seamanship* has been accomplished by a small group of officers working in their off-duty hours who have found time in their busy careers to make this contribution to man's knowledge.

Lieutenant John A. Mazzolini, U.S. Navy, and Lt. Commander Laurens Dorsey, U.S. Navy, who contributed to Part I, Ships and Boats, have been on duty in the Navy Department writing training publications in the Bureau of Naval Personnel.

Captain Howard Prince, U.S. Navy, then Commander Service Squadron 4, U.S. Atlantic Fleet, revised Part II, Shiphandling. He has added new material on the hammerlock moor and has restored, after a lapse of several editions, the section on handling ships in heavy weather.

Lt. Commander Alfred Prunski, U.S. Coast Guard, revised Part III, Rules of the Road. He has succeeded the late Captain Farwell as the country's leading authority on this important aspect of seafaring.

Commander William C. Kincannon, U.S. Naval Reserve, one of the Navy's outstanding meteorologists, revised and brought up to date Part IV, Weather.

The assistance of Commander William J. Flannery, U.S. Navy, who overhauled the chapter on towing and salvage, and of Lieutenant Francis D. Furey, U.S. Navy, who reviewed the chapter on cargo handling, is gratefully acknowledged.

Commander Edward C. Hines, U.S. Navy, and Dr. Howard Johnson were of major assistance in arranging a thorough review and correction of the manuscript by the faculty of the Naval Academy and the Training Division of the Bureau of Naval Personnel, respectively.

It is hoped that those who follow the sea as a profession, students as well as seamen, will continue to find *Knight's Modern Seamanship* the useful text and ever helpful reference that it has been for nearly sixty years.

J. V. NOEL, JR.
Captain, U.S. Navy

August, 1960

Preface to the First Edition

An attempt is made, in the following pages, to cover a wider field than that covered by most of the existing works on Seamanship.

The admirable treatises of Luce, Nares, and Alston, originating in the days when Seamanship was almost wholly concerned with the fitting and handling of vessels under sail, have preserved through later editions the general characteristics which they naturally assumed in the beginning. These treatises will never be out of date until the time, still far in the future, when sails shall have been entirely driven out by steam. It will hardly be denied, however, that the Steamer has long since established its claim to consideration in Seamanship, and that there is room for a work in which this claim shall be more fully recognized than in the treatises above referred to. The excellent work of Captains Todd and Whall, *Practical Seamanship for the Merchant Service*, deals more fully than either of its predecessors with the handling of steamers; but its point of view is, as its name implies, primarily and almost exclusively that of the Merchant Service.

Shortly after the present work was begun, a circular letter was addressed to officers of the Merchant Service and extensively circulated through the Branch Hydrographic Offices at New York, Philadelphia, Baltimore and Norfolk, requesting the views of the officers addressed.

The answers received to these questions were unexpectedly numerous and complete. More than forty prominent officers of the Merchant Service replied, many of them writing out their views and describing their experiences with a fullness of detail far beyond anything that could have been anticipated.

The thanks of the author are due particularly to the following for letters or for personal interviews covering the above points: Capt. W. H. Thompson, *S.S. Belgenland;* Capt. T. Evans, *S.S. Runo;* Capt. J. Dann, *S.S. Southwark;* 1st Officer T. Anfindsen, *S.S. Southwark;* Capt. J. C. Jameson, *S.S. St. Paul;* Capt. H. E. Nickels, *S.S. Friesland;* Capt. G. J. Loveridge, *S.S. Buffalo;* Capt. F. M. Howes, *S.S. Kershaw;* Capt. T. J. Thorkildsen, *S.S. Trojan;* Capt. Otto Neilsen, *S.S. Pennland;* Capt. H. Doxrud, *S.S. Noordland;* Capt. C. O. Rockwell, Clyde S. S. Co.; Capt. S. W. Watkins, *S.S. Montana;* Capt. Anders Beer, *S.S. Nordkyn;* Capt. J. M. Johnston, *S.S. Sardinian;* Capt. A. R. Mills, *S.S. Westernland;* Capt. J. S. Garvin, *S.S. Cherokee;* Capt. Robt. B. Quick, *S.S. El Cid;* Capt. Wm. J. Roberts, *S.S. New York;* Capt. T. Richardson, *S.S. Noranmore;* Capt. E. O. Marshall, *S.S. Maryland;* 1st Officer H. S. Lane, *S.S. Mary-*

land; Capt. W. F. Bingham, *S.S. Marengo;* Capt. R. Gowing, *S.S. Greatham;* Capt. H. J. Byrne, *U.S.A.T. McPherson;* Capt. Paul Grosch, *S.S. Stuttgart;* Capt. Geo. Schrotter, *S.S. Belgravia;* Capt. F. C. Saunders, *S.S. English King;* Capt. Chas. Cabot, *S.S. Venango;* Capt. Chas. Pinkham, *S.S. Queen Wilhelmina;* Capt. A. Traue, *S.S. München;* Capt. W. Thomas, *S.S. Quernmore;* Capt. H. O. Nickerson, Fall River Line; Capt. Geo. Lane, Baltimore Steam Packet Co.

Important assistance was received from Naval Constructor W. J. Baxter, U.S. Navy, who prepared Chapters I and XVIII; and from Lieutenant E. E. Hayden, U.S. Navy, who contributed several Charts and much valuable information upon Meteorology, for Chapter XIX.

Chapter V was suggested by a paper, "Mechanical Appliances on Board Ship," by Captain Thomas Mackenzie, issued by the London Shipmasters' Society as No. 29 of their valuable series of publications.

It would be impossible to mention all the naval officers who have assisted the author with criticism and suggestions; but acknowledgment is especially due to Lieut.-Commander A. W. Grant, Lieut. John Hood, Lieut. W. R. M. Field, Lieut. John Gow, Lieut.-Commander W. F. Worthington, Commander J. E. Pillsbury, Lieut. V. S. Nelson, Lieut. Ridgely Hunt, and Chief Boatswain W. L. Hull, all of the United States Navy.

Above all, acknowledgment is due to Chief Boatswain C. F. Pierce, U.S. Navy, who not only assisted in the preparation of many parts of the text, but prepared sketches for fully one-half the illustrations of the volume.

AUSTIN M. KNIGHT.

United States Naval Academy,
April 1, 1901.

Contents

PART I—SHIPS AND BOATS

PART IV—WEATHER

APPENDICES

Part I

THE SHIP

1

Ships

Seamanship is defined by Webster as "The skill of a seaman." It is an art, not an exact science, and skill in seamanship, as in any art, can be attained only by practice and experience. There are certain principles, techniques, and descriptions that may be learned from books which will help the seaman to develop his skill faster and to learn the details of his art more readily in the only place where they can be learned and developed—aboard ship.

Ships are classified in several ways: by the materials of which their hulls are built, by their methods of propulsion, by their ownership, or by their uses. It seems handiest and most rational to classify them by their uses. Even in this method of classification we are limited by the fact that many ships are used for more than one purpose, and we must fit our grouping to the primary or principal use of each ship.

The broadest classification of ships by use distinguishes them as MEN-OF-WAR and MERCHANT SHIPS. These were originally the same; each vessel carried on its business of trade or transportation and was at the same time prepared to attack or to defend itself against any enemies encountered. As time went on, nations found it cheaper and more efficient to have certain public vessels designed and armed for warlike purposes only, and to give them the primary function of destroying hostile ships and the protection of their own unarmed merchantmen. Because of new concepts of war and of new inventions, such as the submarine and the airplane, the pendulum has now swung partly back to the point where some merchant ships are built to carry a certain amount of armament and do carry it in wartime.

Men-of-war and merchant ships together are major components of sea-power—and upon seapower rests in large part the security as well as the prosperity of the United States of America and of the Free World. Seventy percent of the earth's surface is water; in one sense we and our allies in North America and in South America are islands, surrounded by vast oceans. We must control these oceans not only to provide ourselves with the raw materials for industry, which we need, but to provide weapons and food for our allies in times of emergency. The oceans are also our shield and bulwark—in their depths are concealed nuclear submarines armed with POLARIS missiles—an invulnerable force to deter our enemies from attack.

It is well to remember that insofar as many important strategic raw materials are concerned the United States is a have-not nation. Only ships can bring us such metals as tin, cobalt, bauxite, and manganese, to mention but a few, that are so vital to our factories.

In keeping the sea trade routes open, in quelling local disorder, and in showing the Flag throughout the world, our country needs ships, both merchant and men-of-war. The Merchant Marine and the Navy are partners in maintaining American seapower.

1.1. Men-of-war. All U.S. Naval Vessels are known in the Navy as warships and fall into one of the following categories:

1. Large Combatants
2. Small Combatants
3. Submarines
4. Auxiliary Vessels
5. Amphibious Ships

Although an auxiliary ship may use her guns in her own defense, thus becoming a combatant ship, her primary mission is the logistic support of those ships whose mission it is to engage in combat.

Large Combatants

1.2. Aircraft Carriers (CVA(N), CVA, CVS) are the largest warships in the Navy, ranging from about 85,000 tons for the giant new CVA(N) to 38,000 tons for a CVS. Attack Aircraft Carriers (CVA) all have essentially the same mission: attacking all assigned targets within range of their aircraft,

FIG. 1.1 MODEL OF *U.S.S. ENTERPRISE* CVA(N) 65

providing fighter protection for the Task Force, and furnishing air support for an amphibious assault.

The *U.S.S. Enterprise,* CVA(N) 65 (Fig. 1.1), is the largest and most powerful warship afloat. She is capable of handling planes still in the design stage.

Her atomic reactors provide her with unlimited cruising range at maximum speeds. She has the latest electronic equipment. With her greatly enlarged aviation fuel tanks, the CVA(N) is able to sustain air operations continually for many days.

FIG. 1.2 *U.S.S. FORRESTAL CVA-59*

Forrestal class CVA (Fig. 1.2), presently the largest warship in the world, possesses essentially the same characteristics as the CVA(N). Her cruising range is governed by her conventional power plant, and the duration of

Legend for the following Tables of Ship Statistics

DP	Dual Purpose	*SAM*	Surface-to-Air Missiles
RF	Rapid Fire	*SSM*	Surface-to-Surface Missiles
40*MM*	Heavy Machine Gun	*SUM*	Surface to Underwater Missiles

Weapon "*A*" ASW Rocket

TABLE 1.1. AIRCRAFT CARRIER DATA

Type	Class	Approx. Full Load		Armament (exclusive of air groups)
		Tonnage	Speed	
CVA(N)	*Enterprise*	85,000	34 plus	SAM
CVA	*Forrestal*	76,000	33	5″/54DP or SAM
CVA	*Midway*	62,000	33	5″/54DP, 3″/50RF
CVA	*Essex*	42,000	33	5″/38DP, 3″/50RF
CVHE	*Santee*	24,300	19	40MM

her air strikes are in direct proportion to her much smaller aviation fuel capacity.

The *CVS* is an Essex Class carrier armed with the latest in antisubmarine aircraft. She is the key vessel in a Hunter Killer Task Force. Her primary mission is seeking out and destroying enemy submarines.

CVHE, CVE, CVU are small escort and utility carriers assigned to the Reserve Fleet. One has been converted to an LPH which is described under *Amphibious Ships*.

1.3. Cruisers (CG(N), CAG, CA, CLG, CL) are the largest of the conventionally armed warships still in commission in the Navy. Their displacement ranges from 21,000 to 8,000 tons. Many cruisers have converted to or are undergoing conversion to missile ships. The primary mission of cruisers is to provide antiaircraft defense to the Task Force. The secondary mission of cruisers, which are either partially or wholly equipped with conventional weapons, is the destruction of enemy surface units and shipping, as well as the gunfire support of amphibious operations.

FIG. 1.3 ARTIST'S CONCEPTION OF CG(N) 9, *U.S.S. LONG BEACH*

The *CG(N)* (Fig. 1.3) will be the ultimate in cruiser design. Built from the keel up as a guided missile cruiser and powered by atomic reactors, she will be capable of providing adequate antiaircraft defense for the CVA(N) and will have unlimited cruising range.

The *CAG* (Fig. 1.4) is only partially armed with missiles and still possesses 8-inch guns in her forward turrets.

The conventionally armed *CA* is now becoming an interim weapon. As more guided missile cruisers join the Fleet, she and her remaining sisters may also undergo conversion or be placed in reserve.

The *CLG* is a light cruiser converted to a missile ship in the same manner as the CAG. She still retains 6-inch guns in her forward turrets.

FIG. 1.4 *U.S.S. BOSTON* CAG-1

1.4. Frigates (DLG(N), DLG, DL) are a relatively new class of small combatants. In size as well as in mission they lie between a conventional destroyer and a cruiser. The first nuclear-powered guided missile frigate, *U.S.S. Bainbridge* DLG(N) 25, is scheduled for completion late in 1961 (Fig. 1.5).

FIG. 1.5 ARTIST'S CONCEPTION OF DLG(N)

The Guided Missile Frigate (Fig. 1.6) carries the Terrier missile, an antisubmarine rocket launcher, and conventional guns of the latest design. All guided missile frigates will have long-range sonar and antisubmarine torpedo launching systems in addition. These ships are fine offensive weapons, much more effective than older destroyers and frigates. Their size, endurance, and sea-keeping qualities enable them to operate effectively with the newest and largest aircraft carriers.

FIG. 1.6 *U.S.S. DEWEY* DLG 14

The older frigates (DL) have many of the advantages of the DLG and are superior to destroyers in size and sea-keeping qualities.

TABLE 1.2. CRUISER AND FRIGATE DATA

Type	Class	Tonnage	Speed	Armament
CG(N)	*Long Beach*	19,000	30+	SAM, SSM
CAG	*Boston*	17,350	31	SAM, 8″/55, 5″/38DP, 3″/50RF
CA	*Newport News*	21,450	30	8″/55 RF, 5″/38DP, 3″/50RF
CLG	*Galveston*	14,000	32	SAM, 6″/47, 3″/50RF
CLC	*Northampton*	17,200	32	5″/54DP, 3″/70RF
CL	*Worcester*	18,000	32	6″/47RF, 3″/50RF
DLG(N)	*Bainbridge*	7,600	30+	SAM, 5″/54DP
DLG	*Farragut*	5,700	34	SAM, 5″/54DP
DL	*Norfolk*	7,300	32	3″/70RF, Weapon "A"
DL	*Mitcher*	4,750	35	5″/54DP, 3″/70RF, Weapon "A"

SMALL COMBATANTS

1.5. Destroyers (DDG, DD, DDE, DDR) range in size from 3,000 to 4,500 tons, full load displacement. The guided missile destroyers now under construction have a Tartar missile system, 5″/54 guns and long-range sonar and antisubmarine torpedo systems. The first of this type, the *U.S.S. Charles F. Adams* (Fig. 1.7), is in commission. Other DDG are being converted from destroyers now in commission.

The postwar DD (Fig. 1.8) is equipped with the ultimate in conventional weapons and electronic equipment and is considerably more seaworthy than her predecessors. The DDR and DDE are converted and modernized Gearing

Class (long hull) destroyers. The DDR is a Radar Picket Destroyer designed primarily for the detection of enemy aircraft and the control of friendly combat air patrol. The DDE differs in that she is an Escort Destroyer designed primarily for Anti-Submarine Warfare (ASW).

FIG. 1.7 *U.S.S. CHARLES F. ADAMS DDG-2*

FIG. 1.8 *U.S.S. JONAS INGRAM DD-938*

1.6. Escort Vessels (DE, DER) are designed primarily for ASW operations, although the DER has been modified for Radar Picket duties. These ships are all approximately 1,900 tons (full load). The primary mission of the *DE* is to provide ASW protection to merchant convoys. The newest of the DE type (Fig. 1.9) is equipped with the latest in ASW equipment and embodies much finer sea-keeping qualities than her predecessors.

1.7. Minesweepers (MSO, MSC) are modern wooden vessels with non-magnetic engines and fittings throughout—equipped to sweep contact, acoustic, and magnetic type mines. The large seaworthy MSO (750 tons) (Fig. 1.10) has the primary mission of offensive sweeping in enemy waters. The smaller MSC's mission is clearing mines from our own coastal waters.

FIG. 1.9 *U.S.S. JOHN WILLIS* DE-1027

FIG. 1.10 *U.S.S. ESTEEM* MSO-438

Table 1.3. Small Combatant Ship Data

Type	Class	Tonnage (approx. full load)	Speed	Armament
DDG	*Charles F. Adams*	4,500	35	SAM, 5″/54DP
DD	*Forrest Sherman*	4,000	33	5″/54RF, 3″/50RF
DD	*Sumner*	3,320	34	5″/38DP, 3″/50RF
DD	*Fletcher*	3,050	35	5″/38DP, 3″/50RF
DDE	*Gearing* (long hull)	3,550	34	5″/38DP, 3″/50RF, Weapon "A"
DDR	*Gearing* (long hull)	3,550	34	5″/38DP, 3″/50RF
DE	*Dealy*	1,980	25	3″/50RF, Weapon "A"
DE	*Delong*	1,990	20	5″/38DP, 40MM
DER	*Harveson*	2,000	19	3″/50RF
MSO	*Dash*	750	14	40MM
MSC	*Falcon*	360	14	20MM

1.8. Submarines (SS(N), SSB(N)) range in size from 7,500 to 1,000 tons. The *SS(N)* (Fig. 1.11) possesses all the advantages of a submersible plus the added advantage of having unlimited cruising range and tremendous underwater speeds. Her primary mission is to destroy enemy shipping and sub-

FIG. 1.11 The SS(N)

marines. The SSB(N), the *U.S.S. George Washington*, Fig. 1.12 introduces a new concept in warfare. As do her sister submarines, she possesses the torpedo tubes, fire-control systems, and high speed that will make her an attack submarine. But this ship goes one step further. In addition to the basic requirements of a submarine, she will carry sixteen Intermediate Range Ballistic Missiles (POLARIS), which she will be able to launch from submerged or surfaced positions at sea. She is the first of the submarine deterrent vessels. She embodies five principles which will make her a formidable weapon carrier:

1. Practically unlimited cruising range.
2. Constantly moving and unknown launching site.
3. Precision navigation enabling her to pinpoint the target even though she is deeply submerged.
4. A constant retaliatory capability.
5. A retaliatory capability that will not draw fire to our own shores.

FIG. 1.12 *U.S.S. GEORGE WASHINGTON*

Auxiliaries

1.9. Auxiliaries (AOE, AO, AKS, AGB, etc.) support combatant ships either directly through repair, resupply, and under-way replenishment, or indirectly by gathering scientific data and testing weapons.

One of the most important developments of World War II was the ability of an under-way replenishment group to maintain the combatant ships at sea for extended periods. Today the Sixth Fleet in the Mediterranean is an example of a powerful expression of United States seapower which needs no base but is supported by its own auxiliaries.

The Fast Combat Support Ship (AOE) will be the first ship of a new type to combine the present functions of the Fleet oiler (AO) and the ammunition ship (AE). It is especially designed to accompany and support fast carrier task groups. It will have more speed than its predecessors, carry fuel oil and lubricants as well as ammunition, and will have improved rigs for transfer under way. A Combat Stores Ship (AFS) is also being planned. Designed to operate as part of an under-way replenishment group, it will combine the functions of the AF (refrigerated and dry provisions), the AKS (general stores), and the AVS (aviation stores and repairs). This ship will contain a machine accounting system for stock control and will be provided with helicopters for faster replenishment by air. See Fig. 1.13.

The AKS (Fig. 1.14) is a Fleet stores ship, a floating supply depot that provides everything from radar tubes to toothbrushes to the Fleet.

Transfer of stores is done under way. The *Altair* (AKS-32) is equipped with

FIG. 1.13 ARTIST'S CONCEPTION OF FAST COMBAT SUPPORT SHIP (AOE)

a helicopter to expedite transfer of stores and a complete electric accounting installation (IBM) to assist inventory control.

The AE carries approximately 7,000 tons of munitions and is essentially the same hull type as the AKS. Other ships of similar hull designs are the AF, which carries approximately 7,000 tons of dry and fresh food and the AK which is capable of carrying about 7,000 tons of general cargo.

FIG. 1.14 U.S.S. *ALTAIR* AKS-32 REPLENISHING CARRIER AND DESTROYER

The AGB (Fig. 1.15) is a ship in a class by herself. She is an icebreaker and has been the mainstay of all Arctic and Antarctic expeditions. The hull of an AGB is of special design, of great beam, and of relatively flat bottom. Her bow is heavily reinforced to withstand the pressure of the ice. An icebreaker does not cut through ice with her bow but rather rides up over the ice and literally

FIG. 1.15 *U.S.S. GLACIER* (AGB-4)

crushes it with her weight. Her primary mission is keeping sea lanes open and free of ice.

TABLE 1.4. AUXILIARY DATA

Type	Ship	Name	Tonnage	Speed
AO	*Neosho*	Fleet Tanker (Black oil)	38,000	18
AF	*Riegel*	Provisions Ship (Food)	15,500	20
AE	*Wrangel*	Ammunition Ship	14,400	15
AKS	*Altair*	Stores Issue Ship	12,550	17
AK	*Antares*	Cargo Ship	11,550	19
AR	*Vulcan*	Heavy Repair Ship	16,300	18
AS	*Fulton*	Submarine Tender (repair)	18,000	18
AD	*Sierra*	Destroyer Tender (repair)	18,000	18
AG	*Compass Island*	Miscellaneous Auxiliary (Scientific Test Ship for the Polaris System)	12,000	15
AGB	*Glacier*	Icebreaker	5,100 (light load)	16
AGR	*Sky Watcher*	Radar Picket Ship	12,000	15
AGS	*Requisite*	Survey Ship	13,000	16

AMPHIBIOUS SHIPS

1.10. Amphibious Ships (LPH, LPD, LSD, AKA, APA, LST) are the ships which transport and land troops with their equipment on unfriendly beaches in an amphibious assault.

The LPH is the key ship of the modern amphibious force since it brings men in helicopters to the objective area. It can combat load and land approximately 2,000 troops. The first new construction LPH, the *U.S.S. Iwo Jima,* will join the Fleet in 1962 (Fig. 1.16). In the interim, until new LPHs are available in quantity, the Navy is using a converted CVE, the *Thetis Bay,* and 2 CVS-type carriers, the *Boxer* and the *Princeton,* as LPH.

FIG. 1.16 ARTIST'S CONCEPTION OF LPH, *U.S.S. IWO JIMA*

FIG. 1.17 ARTIST'S CONCEPTION OF LPD

The LSD is equipped with floodable well-deck aft in which amphibious vehicles and landing craft are combat loaded, ready for immediate discharge. The LPD (Fig. 1.17) will be the successor to the LSD. Equipped with a partial flight deck, she will be capable of landing troops by helicopter as well as by boat.

The primary mission of the APA is the transportation and landing by boats of the troops in an amphibious operation. The AKA, with her cargo equip-

FIG. 1.18 *U.S.S. PAUL REVERE* (APA 248)

ment, carries the heavy combat equipment and most of the ammunition and other material needed to sustain the assault. The APA (Fig. 1.18) is the transport, and the AKA is the cargo ship, to be replaced by LPH and LPD.

The LST (Fig. 1.19) is the only large ship of her class. With her relatively shallow draft, she carries heavy equipment, such as tanks, trucks, bulldozers,

FIG. 1.19 *U.S.S. YORK COUNTRY* LST 1158

etc., up to the beach area. Her bow doors open, her ramp drops onto the beach, and she immediately disgorges under their own power her entire load of vehicles. The new LST are much bigger and faster than WW II models.

The AGC is the command ship from which the amphibious assault is directed.

TABLE 1.5. AMPHIBIOUS SHIP DATA

Type	Class	Tonnage	Speed
AGC	*Mt. McKinley*	13,000	15
LPH (new)	*Iwo Jima*	15,000	21
LSD	*Thomaston*	11,300	20
LPD	. . . .	11,500	21
AKA	*Tulare*	13,500	20
APA	*Bayfield*	12,900	16
APA AKA (conversion)	*Mariner*	13,500	24
LST	*York County*	7,000	16
LPH (old)	*Thetis Bay*	10,000	18

TABLE 1.6. UNITED STATES NAVY SHIP TYPES

AIRCRAFT CARRIERS
- CVA(N) . . . Nuclear Attack Aircraft Carrier
- CVA . . . . . . Attack Aircraft Carrier
- CVS . . . . ASW Support Aircraft Carrier

CRUISERS
- CG(N) . . . . Nuclear Guided Missile Cruiser
- CG . . . . Guided Missile Cruiser
- CAG . . . . . . Guided Missile Heavy Cruiser
- CA . . . . Heavy Cruiser
- CLG . . . . . . Guided Missile Light Cruiser
- CLC . . . . . . Tactical Command Ship

FRIGATES AND DESTROYERS
- DLG . . . . . . Guided Missile Frigate
- DLG(N) . . . Guided Missile Frigate, Nuclear
- DL . . . . Frigate
- DDG . . . . . . Guided Missile Destroyer
- DD . . . . Destroyer
- DDE . . . . . . Escort Destroyer
- DDR . . . . . . Radar Picket Destroyer
- DE . . . . Escort Vessel

SUBMARINES
- SSN . . . . Nuclear Submarine
- SSB(N) . . . . Submarine, Fleet Ballistic Missile
- SSG . . . . Guided Missile Submarine

AMPHIBIOUS SHIPS
- AGC . . . . . . Amphibious Force Flagship
- APA . . . . Attack Transport
- APD . . . . . . High Speed Transport
- AKA . . . . . . Attack Cargo Ship
- LPH . . . . . . Amphibian Assault Ship
- LPD . . . . . . Landing Ship, Personnel, Dock
- LST . . . . Landing Ship, Tank
- LSD . . . . Landing Ship, Dock
- LSM . . . . . . Landing Ship, Medium

MINE VESSELS
- MSO . . . . . . Minesweeper, Ocean
- MSF . . . . . . Minesweeper, Fleet
- MSC . . . . . . Minesweeper, Coastal

SERVICE SHIPS
- AO . . . . Oiler or Tanker
- AF . . . . Provision Stores Ship
- AG . . . . Auxiliary, Miscellaneous
- AE . . . . Ammunition Ship
- AGS . . . . Surveying Ship
- AKS . . . . General Stores Issue Ship
- AN . . . . Net Laying Ship
- AK . . . . Cargo Ship
- AOG . . . . . . Gasoline Tanker
- AR . . . . Repair Ship
- AKL . . . . . . Cargo Ship, Light
- AD . . . . Destroyer Tender
- ADG . . . . . . DeGaussing Vessel
- AS . . . . Submarine Tender
- AGB . . . . . . Ice Breaker
- ARS . . . . Salvage Ship
- ASR . . . . Submarine Rescue Vessel
- ATF . . . . Ocean Tug, Fleet
- ATA . . . . . . Ocean Tug, Auxiliary
- AV . . . . Seaplane Tender
- AVP . . . . Seaplane Tender, Small
- ARD . . . . . . Floating Dry Dock
- AGR . . . . . . Radar Picket Ship
- AVS . . . . Aviation Supply Ship
- AVM . . . . . . Guided Missile Ship
- AVB . . . . . . Advanced Aviation Base Ship
- ARC . . . . . . Cable Ship

PATROL CRAFT
- PC . . . . Submarine Chaser
- PCE . . . . Escort Ship
- PCER . . . . . Escort/Rescue Ship

Merchant Ships

The story of the merchant ship dates from the earliest days of recorded history after man had discovered that fish caught in the sea were good to eat. He also realized that other settlements produced certain useful articles which his settlement could use to advantage. Some of the articles or products of his own region could be exchanged for the desired imports. From these early beginnings man ventured farther and farther afield, and ships grew in size to meet the conditions of seagoing trade and the perils of wind and wave.

Merchant ships were built to carry cargo and passengers under many different conditions. The advent of the industrial age and, with it, the use of steam for motive power, changed the development.

1.11. Modern Merchant Ships. The types of vessels are:

- Mail and passenger
- Cargo and passenger
- Cargo
  - Refrigerators
  - Fruit ships
  - Tramps
  - C1, C2, C3, C4, Liberties, Victories
- Colliers
- Tankers
- Ore carriers
- Heavy lift cargo
- Coasters
- Tugs
- Fishing
- Miscellaneous
  - Channel packets
  - Whale factories
  - Icebreakers
  - Ferries
  - Sailing vessels
  - River and lake craft
- Yachts
- Barges

All of these types of merchant ships have gradually increased in cargo and passenger capacity and in speed.

The power plants of ships have changed greatly during the industrial age. Reciprocating engines were used at first (1840), but turbines were not far behind (1894). The development of the internal-combustion engine (1904) made available the diesel engine. In late years, increasing steam pressures and the use of superheat have made greater power available in smaller, lighter, turbines. Because a hull of a certain size can carry only a limited weight, the reduction in size and weight of the power plants leaves more space and weight available for cargo. Now the development of nuclear power plants has reached fruition in the N.S. *Savannah.*

The trend in ship design has been to provide greater compartmentation for safety reasons. One of the great forward steps in the development of safer merchant ships was the limiting of the total cargo allowed to be carried. This step was due largely to a British member of Parliament by the name of Samuel Plimsoll. A bill to enforce a compulsory load line was passed in 1875, and today the use of load lines is world-wide (see Chapter 3). The American Bureau of Shipping assigns load lines on U.S. ships, and the Coast Guard enforces them.

The north Atlantic has witnessed the greatest development of mail and passenger liners. The *Queen Elizabeth* is the largest merchant vessel in the world, grossing 83,673 tons. She is driven by steam turbines and has a service speed of about 31 knots. This speed has been exceeded by the *United States* which made her first crossing at an average speed of more than 35 knots.

FIG. 1.20 *U.S.S. UNITED STATES,* WORLD'S FASTEST EXPRESS PASSENGER LINER

The cargo and passenger type carries more cargo and fewer passengers than the mail and passenger liner. The tonnage and speed are reduced too. The superstructure is shorter to allow more room for cargo hatches and gear. The power plant may be steam or diesel, but the latter is beginning to predominate. The gross tonnage varies between 6,000 and 24,000, and the speed between 15 and 20 knots.

The tonnage of modern cargo ships is between 5,000 and 10,000 gross and the speed about 15 knots or better. A few passengers can be carried, perhaps

FIG. 1.21 *N.S. SAVANNAH,* WORLD'S FIRST NUCLEAR-POWERED MERCHANT SHIP

a dozen, and the cargo spaces are larger than in the cargo-passenger vessels. The engines are turbine or diesel, the latter becoming more popular. The usual speed is about 15 knots or more.

There are two well-known cargo types which are much used, the refrigerator and the fruit ships. The refrigerator ships run regularly between England and Argentina, Australia and New Zealand. These vessels have about 500,000 cubic feet of refrigerated space. Diesels are provided for propulsion, and their speeds are about 15 to 17 knots. A few passengers can be carried.

The best-known fruit ships in the United States are those of the United Fruit Line. Great Britain and Norway also have many of these vessels in use.

FIG. 1.22 S.S. *PRESIDENT MONROE,* COMBINATION PASSENGER AND CARGO LINER

Fruit cargoes are lighter than meat, and the degree of refrigeration required is not so great as may be expected. Some of the Norwegian ships are small, 3,000 to 4,000 tons gross. The refrigerater space is about 200,000 cubic feet, and the speed is about 14 to 16 knots. The United Fruit banana ships are about 7,000

FIG. 1.23 S.S. *KEYSTONE MARINER,* MODERN CARGO SHIP

gross, have 315,000 cubic feet of refrigerated space, use turbines, and make a speed of 18 knots.

The general cargo vessels have an average tonnage of 4,000 to 9,000 gross. The power plants may be reciprocating engines, turbines, or diesels. With speeds varying between 8 to 16 knots, this type of cargo vessel is built to go anywhere and carry cargoes which do not require special conditions to preserve them en route. The common name for them is "tramp." There is little superstructure amidships. In fact, the ship is designed for cargo carriage, which

FIG. 1.24 ARTIST'S CONCEPTION OF MARITIME ADMINISTRATION SEAFARER CLASS

means that cargo hatches and their accompanying king posts, booms, and gear occupy the deck space to the restriction of space for forecastle, poop, and superstructure.

FIG. 1.25 ARTIST'S CONCEPTION OF MARITIME ADMINISTRATION CLIPPER CLASS

The Maritime Administration developed three cargo carriers: C1, C2, and C3, before the war. The C1 types, two in number, were 5,100 tons gross and 412 feet long; and 6,700 tons gross and 418 feet long. Turbines were installed in one,

FIG. 1.26 ARTIST'S CONCEPTION OF LYKES LINES CARGO CLASS

diesels in the other. There were two variations in the C2: one, 8,373 tons gross and 459 feet long; and the other, 6,220 tons gross and 460 feet long. The latter had more powerful turbines which increased the speed to 15.5 knots. The C3 was turbine-driven, 490 feet long, with a speed of 16.5 knots.

The modern tanker has been designed to carry two types of oils—(a) light oils, i.e., gasoline, diesel, and kerosene, and (b) heavy oils, i.e., fuel or crude oil. Oil was carried in barrels for many years before the modern tanker was developed with its longitudinal system of framing. The present-day tanker engines are usually located aft. The forecastle is raised slightly, and the bridge island divides the long amidships section of separated tanks into two parts. The typical tanker has, therefore, three islands, usually with a connecting cat-walk and stack aft. The engineering plant may be turbine, diesel, or turbine-electrical. The tankers vary in size from small coastal types to the large, bulk

FIG. 1.27 ARTIST'S CONCEPTION OF MARITIME ADMINISTRATION PIPELINE CLASS

carriers of 110,000 D.W. tons and 18 knots' speed. Tankers also carry other liquids such as: orange juice, wine, etc.

There are many special types of vessels of which two should be mentioned: the ore carrier and the heavy-lift cargo carriers. The ore carrier looks like a tanker in that the engines and single stack are aft, the bridge is forward and there is a small, raised forecastle. The distinguishing feature is a large number of short king posts to handle the ore, and there may be a catwalk. They are about 8,500 to 9,000 tons gross, turbine- or diesel-driven, with speed varying between 13 to 15 knots.

FIG. 1.28 ARTIST'S CONCEPTION OF MARITIME ADMINISTRATION BULK CARRIER

The Sea Train Cargo Ship is equipped with railroad tracks running the length of the ship on which are secured loaded rolling stock ready for immediate discharge onto any equipped dockside sea train terminal.

1.12. Maritime Administration New Ship Design Program. The Maritime Administration, after consultation with industry and defense representatives, has developed certain preliminary designs of commercial ships to replace the units now in operation in the U.S. Merchant Marine. The basic concept of the replacement program is to provide a variety of ship designs to meet the needs of a well-balanced U.S. Merchant Marine. In this design work, the Maritime Administration has been guided by several assumptions.

1. That the dead-weight "size" of the cargo ships now in operation is adequate for the replacement program.

2. That cargo carriers which will be forming the U.S. Merchant Marine of the next twenty years must be faster than the corresponding units now in operation.

3. That, in general, a trend toward more stowage is evident, and therefore the new units must have a higher cubic displacement than those now in operation.

4. Effort must be made at this time to design cargo ships which embody the best possible cargo handling features.

Proposed Maritime Administration new ship designs are subject to modification to meet special trade route requirements.

TABLE 1.7. MERCHANT SHIP STATISTICS

Ship	Tonnage[1] (approx.)	Speed	Passengers	Crew
S.S. United States	53,000 (gross)	35	2000	1000
S.S. Brazil	22,000 (gross)	25	900	750
S.S. President Monroe	15,000 (gross)	16.5	85	54
N.S. Savannah	25,000 (gross)	20+	. . .	. . .
S.S. Keystone Mariner	12,910 (DWT)	20	12	53
S.S. America Transport	10,500 (DWT)	16.5	12	53
Seafarer Class	13,400 (DWT)	18	12	53
Clipper Class	10,800 (DWT)	18	12	53
Type (*Lykes Line*)	. . .	. . .	. . .	. . .
Pipeline Class	22,590 (DWT)	20	0	42
Bulk Carrier Class	24,000 (DWT)	16	0	42

[1] *Gross* tonnage is the internal cubic capacity of the ship expressed in tons of 100 cubic feet to the ton. DWT (dead-weight tonnage) is the total carrying capacity of the ship, expressed in tons of 2,240 pounds.

MILITARY SEA TRANSPORTATION SERVICE

1.13. Military Sea Transport Service (MSTS) ships are merchant types not fully converted to Naval use. MSTS operates approximately 400 ships of all types with civilian crews, used primarily in the supplying of this country's many bases and outposts. Although MSTS operates practically all types of ships found sailing the trade routes under various house flags, only those types which are designed for special missions and are unlike any of those types operated by private companies will be discussed here.

Roll-on-Roll-off ships (Fig. 1.29) like the MSTS *Comet* are of a new design for the handling of vehicles of all types. Ramps extend from deck to deck throughout the ship, facilitating the stowage of vehicles under their own power. The vehicles are driven aboard through the stern ramp or through side ports. Vehicles can be loaded or discharged in a matter of hours.

FIG. 1.29 MSTS *COMET* T-AK 269

Arctic Type ships like the MSTS *Alatna* (Fig. 1.30) were designed for polar operations. Her interior has the latest innovations in habitability during long and cold passages. Her hull is strengthened and heavily reinforced. She can go

FIG. 1.30 MSTS *ALATNA* T-AOG 81

virtually anywhere that icebreakers can go. A specially constructed conning station to facilitate navigation through ice has been constructed on her forward mast.

TABLE 1.8. MSTS SHIP STATISTICS

Type	Class	Tonnage	Speed	Crew
T–AK	MSTS *Comet*	16,150 (full load)	18	51
T–AOG	MSTS *Alatna*	3,200 (DWT)	12	41

2

Ship Construction

The Hull

The main body of a ship exclusive of masts, superstructure, etc., is called the *hull*. For a steel ship it is made up of plates covering a framework which in many ways is similar to the framework of a building. The hulls of various types and classes of vessels are basically similar, with certain modifications to suit the mission of the vessel.

The most important part of the frame of any vessel is the bottom centerline longitudinal, known as the *keel*. All parts of the keel are continuous from the forward end, where it joins the stem, to the after end, where it joins the stern post, or stern-post assembly if the stern post does not exist.

Radiating from the keel are a series of frames which give the hull its shape and act as supports for the shell plating, decks, etc. The frames, together with the longitudinals, constitute a framing system. The framing system is said to be longitudinal if the fore-and-aft members are continuous. Longitudinal framing system is generally used in modern merchant-ship construction because of the greater strength that can be achieved with slightly less weight. Naval surface vessels use a combination of the two systems.

Prior to the development of welding techniques, all portions of the ship's structure were joined by riveting. This practice has given way almost entirely to welding. There is a saving in weight of as much as 10 to 15 percent when welding is used. In addition, it is faster, cheaper, and gives a smoother surface that is better able to resist corrosion and to reduce skin friction in the underwater body of a ship. The use of welded construction does, however, call for a very carefully controlled assembly sequence in order to minimize locked-in stresses. Specifications for larger vessels sometimes call for a combination of welded and riveted construction, thereby using some riveted joints to provide more flexibility and to reduce local stress concentration.

Larger ships are usually fitted with inner bottoms which extend up the side of the ship to or above the water line. Large combatant vessels were often fitted with heavy armor plate surrounding the machinery spaces and other vital spaces from slightly above the water line to several feet below the water line.

The ship is further subdivided into as many small compartments as is practical, consistent with the mission of the vessel, in order to minimize leakage and flooding if the outer shell is damaged.

Bulkheads are used to subdivide the ship's interior vertically into watertight compartments for the preservation of buoyancy and stability. Oiltight bulkheads are fitted to form the necessary fuel oil tanks, and nonwatertight bulkheads are fitted to provide stowage and living spaces where watertightness is not essential.

2.1. Hull Shapes. The power characteristics, such as speed, of a ship depend to a great extent on the shape of her hull below the water line, i.e., the underwater body. No part of the underwater body is an absolutely flat surface, except in very unusual cases. The widest part of the hull is near the halfway point between the bow and stern, and the hull in this vicinity is called the middle-body section. The bottom and sides in the midship section approach plane surfaces more nearly than in any other part of the underwater body. They are joined by a curve which completes an approximate right angle and which is called the turn of the bilge. From the middle-body section the lines of the hull slope smoothly to the bow and stern in what may become a hollow or reverse curve at some point before reaching them. The narrowing part of the underwater body forward of the middle-body section is called the entrance. The corresponding part aft is called the run. A ship that has a long and tapering entrance and run and a proportionally short middle-body section is said to have fine lines. A fast man-of-war generally has fine lines. A slow cargo carrier has not. Her boxlike middle-body section is comparatively long to give her greater carrying capacity.

If the sides slope outward from near the turn of the bilge toward deck level, they are said to be flared. If they slope inward, the amount of the slope is called the tumblehome. If a deck slopes from the centerline to the side, it is said to be cambered.

There are as many refinements on this general scheme of the underwater body as there are ship designs and special uses for ships. Generally the run is somewhat longer than the entrance, and the after part of the run is narrowed for a greater distance to allow for the installation of rudder and propellers. The keel may be shortened aft, and the stern post may slant from the end of the keel to the water line for the same reasons. These characteristics are known as reduced after deadwood. Cutting away the after deadwood has an important effect on ship handling, which will be discussed in a later chapter. The lines of some ships, instead of coming to a sharp edge at the bottom of the stem, expand at that point into a rounded shape of comparatively small diameter extending from the keel to a few feet below the water line. This shape is known as a bulbous bow and enables a vessel so fitted to attain greater speeds at or near full power than a vessel not so equipped. (The stem at and a few feet on each side of the water line is a sharp edge.)

The shape of a ship's underbody explains the great difference between shipbuilding and ordinary construction. Each frame member must be fitted to the desired contour of the hull, and side plates must be bent to the proper shape before being fastened to the frames in order that the completed ship may have her designed characteristics.

2.2. Submarine Hulls. The hull of the submarine sets her apart from other ships. The entire submarine may be considered a complete hull. Actually, there are three major hull elements. The basic element is the pressure hull which contains most of the vessel's vitals—engines, living and working spaces, etc. The pressure hull is long and cylindrical. Mounted on the pressure hull is the

FIG. 2.1 SKETCH OF A POLARIS SUBMARINE FIRING ONE OF ITS MISSILES

superstructure—also called sail, which contains the conning tower. The conning tower only is watertight and contains additional vital spaces and equipment. Mounted around the pressure hull are ballast and fuel tanks. These are *exterior* to the pressure hull. The three elements are smoothly combined to give the submarine its distinctive hull shape. Figure 2.1 shows these features in the artist's conception of the hull of one of the newer missile submarines.

2.3. Decks. Decks are used primarily to provide structural strength, shelter, working spaces, and living quarters; and secondarily to subdivide the hull horizontally into greater number of watertight compartments. The decks are supported by fore-and-aft members called deck girders, and by athwartship members called deck beams. Deck beams are in turn supported by stanchions which provide the decks with support additional to that afforded by the bulkheads. The highest deck extending from stem to stern is called the main deck.

A partial deck above the main deck at the bow is called the forecastle deck; at the stern, poop deck; amidships, upper deck. The name upper deck, instead of forecastle deck, is applied to a partial deck extending from the waist to either bow or stern. A partial deck above the main, upper, forecastle, or poop deck and not extending to the side of the ship is called a superstructure deck. A complete deck below the main deck is called the second deck. Two or more complete decks below the main deck are called the second deck, third deck, fourth deck, etc. A partial deck above the lowest complete deck and below the main deck is called a half deck. A partial deck below the lowest complete deck is called a platform deck. Where there are two or more partial decks below the lowest complete deck, the one immediately below the lowest complete deck is called the first platform, the next is called the second platform, etc. Decks which for protective purposes are fitted with plating of extra strength and thickness are further defined for technical purposes as *protective* and *splinter* in addition to their regular names.

2.4. Fittings. Fittings are various structures and appliances attached to the hull to assist in handling the ship or performing the ship's work, to provide for the safety and comfort of the crew, or merely for ornamental purposes. They may be affixed solidly to the hull or may be capable of a limited amount of motion. They may be operated by hand or by power. They may be found in any part of the ship, including the underwater body, although the commonest and most useful fittings are generally encountered around the weather decks. An excellent way to identify a fitting is to decide that its absence would not affect the seaworthiness or structurel strength of the hull. The list of fittings is almost endless, but some of the commoner and more important ones will be described in order that they may be recognized and their uses appreciated (see Fig. 2.2).

Chocks are some of the most numerous and useful fittings found aboard ship. They generally take the form of castings or forgings bolted or riveted to the hull near the side along weather decks. They are used for the purpose of guiding and controlling lines led aboard, for making fast to a dock, securing anything alongside, towing, or being towed. The most common form is the open chock that has an opening on top through which the line is dropped and two curved parts called horns to hold it in. If the horns meet and the line must be led through the opening, it is called a closed chock. The heavy closed chock built at the extreme bow of destroyers and other light vessels for guiding a towline is commonly known as the bull nose. The inner surfaces of chocks are smoothed and rounded to avoid chafing the lines. Some have rollers fitted on each side for the same purpose, in which case they are known as roller chocks.

Bitts are usually found in the neighborhood of chocks and somewhat inboard of them. They are heavy vertical cylinders, usually cast in pairs and often used for making fast lines that have been led through the chocks. The upper end of a bitt is larger than the lower end or is fitted with a lip to keep lines from slipping off accidentally. As bitts are often required to take very heavy loads, extra frames are worked into their foundations to distribute the strain. Bitts are sometimes built and installed ruggedly enough so that the ship may tow or be towed by them. When built in pairs, each bitt is sometimes called a horn.

Another common fitting is the davit. Davits are set in sockets which allow them to rotate. They are made of heavy pipe or plates and are angled so that

FIG. 2.2 FITTINGS ON A MISSILE SHIP. *Official U.S. Navy Photograph*

the upper end or head of the davit will plumb some space below it at a distance from the davit's base. A tackle is rigged at the davit head so that weights can be lifted and swung as the davit is rotated. The most common use for davits is to carry lifeboats, but they are sometimes rigged to lift or lower weights over the side or out of trunks and holds.

There are numerous smaller but very useful fittings found about the weather decks. A plate with an eye attached, riveted to the deck to distribute the strain over a large area and to which a block can be hooked or shackled, is called a pad or padeye. An eyebolt serves the same purpose and may be attached to the deck with a bulkhead or a frame overhead. If the eyebolt carries a permanent ring, it is called a ringbolt. Cleats are light, double-ended fittings on which lines are made fast. Awning stanchions and lifeline stanchions are found at the ship's side on weather decks and are used to rig awnings and lifelines. These are either hinged or set in sockets so that they can be cleared away if this is necessary. Sockets are often set in the deck for special purposes, such as setting the king post for a boom and topping-lift. Deadeyes are heavy glass plates set in upper decks, and ports are fixed or hinged heavy glass plates set in the ship's side. Neither is used in the hulls of modern men-of-war in the interest of better watertightness.

Winches and cranes are the most common and most useful power fittings found aboard ship. A winch consists of a heavy frame fastened solidly to the deck and an engine or motor that turns a horizontal shaft mounted in the frame, usually with a drum fastened to each end of the shaft. If the line is revolved around the drum for several turns and is tended carefully, it will withstand a heavy strain and can be accurately controlled. Blocks rigged to padeyes or eyebolts are commonly used to give lines a fair lead to the winch drums. In case a line has a limited and repeated travel, as when cargo is being handled in and out of a hold, the line is sometimes attached permanently to a winch drum and handled by winding and unwinding it on the drum according to the direction of the shaft's rotation. Winches take the place of an enormous amount of manpower and are invaluable in speeding up operations even when sufficient manpower is available. A winch that is used primarily to handle the anchor cables, but usually having horizontal drums or a vertical capstan in addition, is called an anchor windlass.

Cranes may be regarded as large, power-driven davits. A crane may be built up as a solid structure or rigged with a boom that can be raised and lowered so as to plumb different distances from its base. The hauling part of a permanently rigged wire rope tackle for lifting weights is made fast to a drum. Motors or engines lift or lower weights by rotating the drum, rotate the crane, and lift or lower the boom. The crane is used primarily for hoisting and lowering heavy weights, such as boats and airplanes, over the side, but it can be used for many other purposes within its radius, such as for heavy vertical lifts and rotation. The winch, usually less powerful, is more flexible, and the lines leading to it can be rigged to give almost any desired result.

The location of fittings and other fixed objects on deck can be determined by the frame numbers in their vicinity. Certain frames, usually every fifth one, have their numbers cut or stamped in an accessible location near each side of the ship. Intermediate frames can be located by measuring from a marked frame in multiples of the frame spacing or the distance between frames, which is usually uniform. The number of frames in the hull and the length of the frame spacing are useful knowledge to have about any ship.

2.5. Compartment Designation for Warships. Every space in a warship (except for minor spaces, such as peacoat lockers, linen lockers, and cleaning gear lockers) is considered as a compartment and assigned an identifying letter-number symbol. This symbol is placed on a label plate secured to the door, hatch, or bulkhead of the compartment. There are two systems of numbering compartments, one for ships built prior to March 1949, the other for ships built after March 1949. Both these systems agree, however, in one respect: compartments on the port side have even numbers, those on the starboard side have odd numbers. The two systems resemble each other also in the fact that a zero precedes the deck number for all levels above the main deck. Figure 2.3 shows both systems of numbering decks. The older system uses 100, 200, 300, etc., series

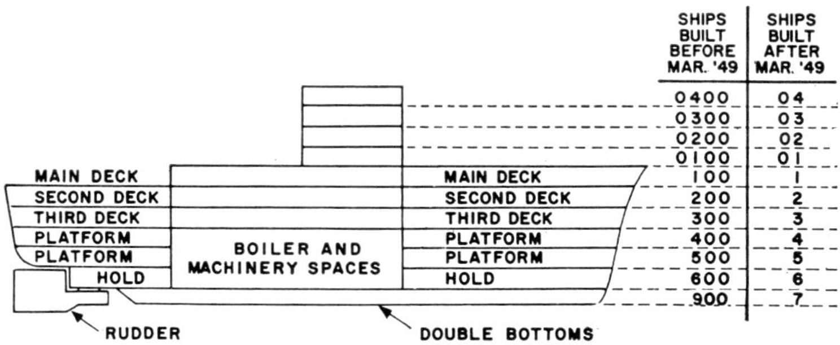

FIG. 2.3 DECK SYMBOLS FOR NAVY SHIPS

and always 900 for the double bottoms, while the new system uses 1, 2, 3, etc., and the double bottoms are given whatever number befalls them. For ships built prior to March 1949, the first letter of the identifying symbol is A, B, or C, and indicates the section of the ship in which the compartment is located. The A section extends from the bow of the ship aft to the forward bulkhead of the engineering spaces. The B section includes the engineering spaces, while the C section extends from the after bulkhead of the engineering spaces aft to the stern.

After the division letter, the deck designation comes next in the symbol. Main deck compartments are indicated by numbers such as 102, 109, 117, etc. Second deck compartments run from 201 through 299, third deck compartments form a 300 series, etc. A zero preceding the number indicates a location above the main deck. The double bottoms always form the 900 series on any ship built before March 1949, regardless of the number of decks above.

What the compartments are used for is indicated by the following letters:

A—Supply and Storage
C—Control
E—Machinery
F—Fuel
L—Living Quarters
M—Ammunition
T—Trunks and passages
V—Voids
W—Water

Here is an example of a compartment symbol on a ship built before March 1949:

C — 2 1 7 — A

- A — Supply compartment
- 1 7 — Compartment number (starboard side)
- 2 — Second deck
- C — After part of ship (C section)

2.6. Ships Built After March 1949. For ships constructed after March 1949, the compartment numbers consist of a deck number, frame number, relation to

TABLE 2.1. COMPARTMENT LETTERS FOR SHIPS BUILT AFTER MARCH 1949

Letter	Type of Compartment	Examples
A	Stowage spaces	Storerooms; issue rooms; refrigerated compartments.
AA	Cargo holds	Cargo holds and cargo refrigerated compartments.
C	Control centers for ship and fire-control operations (normally manned)	CIC room; plotting rooms; communication centers; radio, radar and sonar operating spaces; pilothouse.
E	Engineering control centers (normally manned)	Main propulsion spaces; boiler rooms; evaporator rooms; steering gear rooms; auxiliary machinery spaces; pumprooms; generator rooms; switchboard rooms; windlass rooms.
F	Oil stowage compartments (for use by ship)	Fuel-oil, diesel-oil, lubricating oil, and fog-oil compartments.
FF	Oil stowage compartments (cargo)	Compartments carrying various types of oil as cargo.
G	Gasoline stowage compartments (use by ship)	Gasoline tanks, cofferdams, trunks, and pumprooms.
GG	Gasoline stowage compartments (cargo)	Gasoline compartments for carrying gasoline as cargo.
K	Chemicals and dangerous materials (other than oil and gasoline)	Chemicals, semi-safe materials, and dangerous materials carried for ship's use or as cargo.
L	Living spaces	Berthing and messing spaces; staterooms, washrooms, heads, brigs; sickbays; hospital spaces; and passageways.
M	Ammunition space	Magazines; handling rooms; turrets; gun mounts; shell rooms; ready service rooms; clipping rooms.
Q	Miscellaneous spaces not covered by other letters	Shops; offices; laundry; galley; pantries, unmanned engineering, electrical, and electronic spaces.
T	Vertical access trunks	Escape trunks or tubes.
V	Void compartments	Cofferdam compartments (other than gasoline); void wing compartments; wiring trunks.
W	Water compartments	Drainage tanks; fresh-water tanks; peak tanks; reserve feed tanks.

centerline of ship, and letter showing use of the compartment. These are separated by dashes. The A, B, C divisional system is not used.

Deck Number. Where a compartment extends down to the bottom of the ship, the number assigned to the bottom compartment is used. The deck number becomes the first part of the compartment number.

Frame Number. The frame number at the foremost bulkhead of the enclosing boundary of a compartment is its frame location number. Where these forward boundaries are between frames, the forward frame number is used. Fractional numbers are not used. The frame number becomes the second part of the compartment number.

Relation to the Centerline of the Ship. Compartments located on the centerline carry the number 0. Compartments completely to starboard are given odd numbers, and those completely to port are given even numbers. Where two or more compartments have the same deck and frame number and are entirely starboard or entirely port of the centerline, they have consecutively higher odd or even numbers, as the case may be, numbering from the centerline outboard. In this case, the first compartment outboard of the centerline to starboard is 1; the second, 3; etc. Similarly, the first compartment outboard of the centerline to port is 2; the second, 4; etc. When the centerline of the ship passes through more than one compartment, the compartment having that portion of the forward bulkhead through which the centerline of the ship passes carries the number 0, and the others carry the numbers 01, 02, 03, etc. These numbers indicate the relation to the centerline and are the third part of the compartment number.

Compartment Usage. The fourth and last part of the compartment number is the letter which identifies the primary usage of the compartment. On dry and liquid cargo ships a double-letter identification is used to designate compartments assigned to cargo carrying. The letters are shown in Table 2.1.

The following example of a compartment number illustrates the application of these principles of compartmentation for ships built after March 1949:

3 — 7 5 — 4 — M

- M — Ammunition compartment
- 4 — Second compartment outboard of the centerline to port
- 75 — Forward boundary is on or immediately aft of frame 75
- 3 — Third deck

Hull Systems

Piping built into the hull which carries a liquid or a gas is known as a hull system. The principal hull systems are:

- Firemain
- Damage-control pumping
- Flushing
- Flooding
- Fresh water
- Drainage
- Compressed air
- Fuel-oil filling and transfer
- Aircraft fuel

2.7. Firemain System. The firemain system in large ships forms a horizontal loop around the machinery spaces, and is ordinarily served through cross-connections by the fire-and-bilge pumps and fire-and-flushing pumps in the main machinery and boiler rooms. In newer ships, there are secondary loops, or single smaller mains, forward and aft, extending through the magazine areas. In late aircraft carrier designs, a secondary loop parallels the main loop on a high level; and in late cruiser designs, a vertical loop is added to the main loop.

In smaller ships, such as destroyers, the firemain is a straight main, leading fore and aft throughout the machinery spaces, with extensions forward to the forward living spaces, and aft to the steering gear room, below the water line. Vertical risers lead from the firemain to spaces above decks and serve superstructure fireplugs. In a typical destroyer the firemain system is divided into two groups, one forward and one aft, each supplied by pumps in one engine room and one fire room.

2.8. Sprinkling System. The magazines of a warship are divided into groups, according to location. Each group is supplied by a separate sprinkling system connection leading from the firemain at a convenient location and controlled by a group control valve. The group control valves are operated from remote control stations, hydraulically, pneumatically, electrically, or mechanically. Sprinklers may also be actuated automatically by a thermostat when there is a temperature rise in the magazine.

2.9. Damage Control Pumping System. In large warships a damage-control pumping system is installed to provide for rapid flooding of counter-floodable voids after damage, and for rapid transfer of ballast sea water. A large central main runs fore and aft below the protective deck, with transverse leads to spaces served.

In a typical installation, the system is supplied by two large-capacity centrifugal pumps, electric, or diesel-driven. By making appropriate valve connections, this system can be used for rapid removal of water from flooded spaces, after the rate of entrance of flooding water has been checked.

2.10. Flushing System. For sanitary spaces, flushing water at pressures around 30 psi is supplied. In older ships of large size, a separate flushing loop was installed, receiving pressure directly from flushing pumps, or from the firemain through reducing valves.

Present practice is to provide branches from the firemain, via stop valves and reducing valves, wherever flushing services are required. A typical cruiser has five such branches.

The flushing system as well as the firemain system will become fouled with a foreign growth which accumulates within the piping. This fouling decreases the supply of water to firemain risers and flushing outlets. To restrict fouling, a plastic interior coating has been developed which is applied to the inner surface of the piping in naval shipyards. Fresh-water flushing of these systems will kill the marine growth.

2.11. Flooding System. In large ships, remote-operated, sea-flood valves were formerly installed to permit flooding of magazines in the event of fires in the vicinity. Modern efficient magazine sprinkling systems in new ships are sufficiently flexible to be operated even after damage, and such ships have little need for sea-flooding systems.

In large ships, remote-operated, hydraulically controlled flood valves are installed in counter-floodable voids. The latest practice provides a single flood valve to service several such voids in order to minimize the number of openings in the hull. These valves permit rapid counter-flooding even when power is temporarily lost.

2.12. Ballasting System. It is often necessary to ballast fuel oil stowage tanks with salt water after the fuel oil is burned, in order to maintain proper list, trim, draft, torpedo protection, and stability. Fuel oil ballast tanks in such ships are flooded with salt water from the sea or from the firemain through a drainage manifold. Removal of ballast water from the tanks is accomplished through pumps. Fuel oil service tanks should never be ballasted with water.

2.13. Fresh-Water System. Fresh or "drinkable" water, called potable water, is usually stored aboard ship in special tanks, low in the ship. From these tanks, it is delivered to necessary outlets, such as scuttlebutts, washbowls, galley sinks, and the like, through the fresh-water system. This consists of a pressure tank and continuously operating, constant-pressure centrifugal pumps, which maintain pressure in the system. These pumps are usually located near the fresh-water tanks, and frequently in engineering spaces.

The fresh-water system affords a source of possible progressive flooding if damaged. Cutout valves are installed to permit segregation.

2.14. Drainage System. Each ship has some means provided for removing water from within its hull. Systems of piping, with or without pumping facilities installed for this purpose, are termed *drainage systems.*

Drainage systems are divided, on most ships, as follows:

1. Main drainage system
2. Secondary drainage systems
3. Plumbing and deck drains
4. Weather-deck drains
5. Feed drains in machinery spaces

In addition to the above systems, the following portable pumps are used to drain flooded areas not provided with drainage facilities:

1. Electric submersible pumps
2. Gasoline handy billies, the "P-500-type" pump, gas turbine pumps
3. Jet pumps (eductors)

The main drainage system runs throughout the main machinery compartments. However, on some ships it extends well into the bow and stern. On smaller ships the main drain consists of a single pipe running fore and aft, usually amidships. On larger ships it is a loop system, extending along both sides of the engineering compartments and joined at the ends. Main drainage systems may be used on many later type ships to drain "floodable" voids used in counter-flooding, after such voids have been flooded, and to empty fuel oil tanks which have been ballasted with sea water.

Types of pumps installed in the main drainage systems are: steam-driven reciprocating pumps, turbine or motor-driven centrifugal pumps, and jet pumps (eductors).

Secondary drainage systems serve to drain spaces forward and aft of the main machinery compartments. The piping is smaller in size than that used in main drainage systems. It may be a continuation of the main drainage system, but in many instances they are not connected.

Plumbing and deck drains are provided for draining fixtures and compartments within the ship by gravity. Gravity drainage piping is installed most extensively in compartments above the water line. On large ships, some compartments near or below the water line may be drained to compartments lower in the ship, where the water can be pumped overboard. These lower compartments are bilges and bilge wells, shaft-alley sumps, drain tanks, or sanitary drain tanks.

2.15. Compressed Air System. Compressed air systems aboard ship have a wide variety of uses. The principal importance of compressed air to the stability of the hull and to the safety of many surface ships is in the starting of the emergency Diesel electric power generators.

When some boiler casualties occur, all steam or electrically driven equipment on board stops operating until steam can be raised again. The ship loses propulsion and steering, goes dead in the water and, except for a few battery-operated lanterns, everything below decks is in darkness. To raise steam quickly, it is necessary to use an electrically driven fuel-oil pump. Electric power for this must come from the emergency power supply—the emergency diesel.

On the submarine, the compressed air system achieves major importance. Two large compressors provide the air. Use of this air in the various ballast, buoyancy, and safety tanks permits routine diving and surfacing. Air is also used for torpedo ejection in submarines and in surface ships.

2.16. Fuel-Oil System. The fuel-oil pumping system in large ships consists of a loop serving all fuel-oil tanks and permitting transfer of fuel from storage tanks to service tanks and thence to fuel-oil service pumps. The latter pump the fuel oil to the fuel-oil heaters and thence to the burners in the boilers.

Included in the system are topside fuel-oil-filling connections, which lead down to the loop.

This system is used for *transfer* of liquid for correction of list and trim, or improvement of stability or reserve buoyancy after damage, and furnishes a possible avenue for progressive flooding.

2.17. Aircraft Fuel System. Ships carrying or tending aircraft carry gasoline and jet aircraft fuel. Gasoline systems are designed to minimize the inherent fire and explosion hazard.

The majority of the gasoline systems installed are of the salt-water displacement type in which salt water enters the tanks as gasoline is withdrawn. This keeps the tanks full and eliminates dangerous pockets of gasoline vapor. The salt water enters the bottom of the tanks from the firemain, through reducing valves, and forces the gasoline up and out through the fueling piping. Delivery to high spaces, such as catapults or flight decks, is accomplished by boosting the discharge pressure of the gasoline with water turbine-driven pumps.

The gasoline tank compartment is fitted with a fixed CO2 system which can be instantly discharged when danger threatens, flooding the entire space with CO2. In new ships, a blanket of inert gas surrounds the tanks. In addition, later designs provide a "saddle-tank" arrangement, in which each gasoline tank is also always directly surrounded by a tank of salt water.

3

Damage Control and Hull Preservation

The original stability of a ship and its ability to withstand hull damage and flooding is the business of her designers and builders, but the basic principles of stability and damage control are important to the seaman for his own safety and that of his ship.

3.1. Stability. A ship has two principal kinds of static stability, longitudinal and transverse; the first tends to keep a ship from rolling end over end; the second tends to keep her from capsizing. Longitudinal stability is always enough to avoid danger, although poor longitudinal stability characteristics may cause discomfort by excessive pitching and make a very wet ship. The knowledge of the transverse stability of any ship, on the other hand, is important to the seaman in order to gauge the amount of roll allowable without danger of capsizing.

Unless acted upon by some external force, a ship that is properly designed and loaded remains upon an even keel. A righting moment which develops when the vessel rolls tends to return the ship to an even keel. The righting moment is the product of the righting arm (defined later) and the force of buoyancy. The force of buoyancy is the sum of the vertical components of the hydrostatic pressure on the underwater body. It is also equal to the weight of the water displaced by the underwater body. The force of buoyancy also keeps the ship afloat, but it may be overcome and the ship sunk if too much weight is introduced, as is the case when too many holds or compartments are flooded. The righting moment tending to bring the ship back on an even keel may be overcome and the ship capsized if too much weight is introduced on one side of the centerline, as when all compartments on one side are flooded. A ship may also capsize when it is stranded and the righting arm becomes negative due to reduction of the underwater volume.

The foregoing examples are static forces tending to capsize a ship. The effects of wind and waves are dynamic forces. A smaller dynamic force than static force may be sufficient to overcome the righting arm and to capsize a ship because the effect of a dynamic force depends on the speed with which it is applied as well as on its magnitude. A common example is that a boat under sail may be capsized by a sudden squall when it might weather perfectly a wind of the same force that came up gradually.

An elementary idea of the forces tending to prevent a ship from capsizing can be obtained from (a) and (b), Fig. 3.1. In (a), G represents the center of gravity of the ship. It is the point where the sum of the moments of all the weights of a ship with reference to any axis through this point are equal to zero. In other words, the ship acts as though all its weight were concentrated. Because the loading is usually the same on both sides of the vertical centerline plane, G usually lies in this plane and, in ships of conventional form, near the water line. The center of gravity remains in the same place if no weights are added or subtracted, or if weights are added or subtracted whose algebraic sum

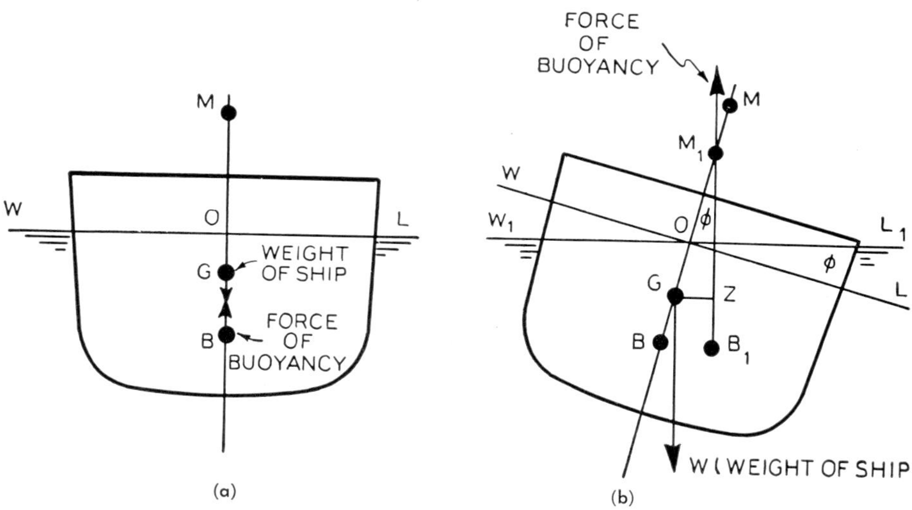

FIG. 3.1 FORCES ACTING ON HULL

lies in the same horizontal plane as the center of gravity. It is raised if the sum of the weights added lie above this plane, or the sum of the weights subtracted lie below it. Conversely, it is lowered if the sum of the weights subtracted lie above it. The pertinent formula is:

$$KG_1 = \frac{WKG \pm w_1Kg_1 \pm w_2Kg_2 \cdots \pm w_nKg_n}{W \pm w \pm w_2 \cdots \pm w_n}$$

where W = original displacement of ship
KG = original height of center of gravity
w = weight added or removed
Kg = height of w above keel

The symbol B is used to indicate the ship's center of buoyancy which may be considered as that point at which the component of all upward forces are considered to act and which lies in the geometric center of the underwater form of the vessel. When the ship is on an even keel, B is in the vertical longitudinal centerline plane, and the upward force of buoyancy is directly under the point

G where all the weight of the vessel is considered to be concentrated. When the ship lists or rolls, B moves to the lower side, and a vertical line drawn from B cuts the vertical plane of the centerline. The point M, where this line cuts for an infinitesimal angle of inclination, is called the metacenter, and distance GM is called the metacentric height. BM is called the metacentric radius and is mathematically equal to I/V, where I is the moment of inertia of the water-line plane and V is volume of displacement.

For any given condition of loading, G may be considered as fixed, but both B and M move as the vessel rolls. In (b) the ship has rolled to the right an angle of ϕ degrees, and B has moved to B_1, the center of gravity of the volume of liquid displaced in this new position. A vertical line from B_1 cuts the centerline at M_1, the new position to which the metacenter has moved. A perpendicular line drawn from G to the intersection of line B_1M_1 (point Z) is the righting arm and is indicated by the symbol GZ. As can be seen, it is the distance between the lines of action of the force of buoyancy B acting up and the force of gravity acting down. The righting moment is equal to the value of the couple set up by W and B. Its value is $W \times GX$ or $B \times GZ$, where W is the weight of the ship.

As the ship heels over further, M_1 moves down the centerline, and the distance GM_1 decreases by an amount varying with the characteristics of the individual ship. Theoretically, M_1 could move below G and the righting arm GZ become negative, but in practice other factors could usually capsize the ship before this point is reached. The angle at which GZ becomes negative is the upper limit of the range of stability.

As the angle of inclination increases, the length of the righting arm increases for a time until it reaches a maximum, after which its length decreases due to the movement of M_1 down the centerline. The righting moment varies directly as the righting arm, since W remains constant; hence it, too, increases until it reaches a maximum after which it decreases sharply. This is called the point of maximum righting arm.

The preceding paragraphs give a very rough idea of the forces tending to prevent a ship from capsizing. All the factors involved are treated thoroughly under the science of naval architecture. Cross curves of stability show the ship's static stability characteristics (i.e., righting arms, range of stability, etc.) for different displacements at an assumed position of center of gravity. Curves of form show geometric characteristics for different drafts (such as underwater volume plotted as $35W$ and $36W$, and KM). GM may then be calculated from the relationship $GM = KM - KG$, provided the weight effects have been evaluated. Even more practical, a stability diagram may be solved. It is emphasized, however, that KG and KM must *both* be known, before an appraisal of ship stability is accurate.

A stiff ship may not have as great a range of stability as a lively ship. She is hard to roll initially because of her greater metacentric height but may have a much smaller range of stability than a lively ship. For instance, an aircraft carrier in a very steady platform and has several times as great a metacentric height as a destroyer, but her range of stability is much less, and consequently she cannot roll to nearly so great an angle with safety. Because of the great displacement W of a carrier, the righting moment $W \times GZ$ is also large and would never be exceeded under any conditions, except after extensive damage to the watertight shell.

The addition of weight above the center of gravity decreases the metacentric height and consequently the range of stability. The same effect is caused by removal of weight below the center of gravity. The most common example of this is in warships when the fuel tanks are emptied without admitting water ballast for compensation.

Wind and waves are usually dynamic forces. Sudden strong gusts of wind or heavy seas, especially in shallow water, may build up a dangerous roll when the same force applied as a steady pressure would cause no trouble. A rough method of keeping out of trouble is to watch the period or time required for a complete roll from side to side. The period should remain approximately the same regardless of the magnitude of the angle or roll. Should the period increase appreciably, or the ship appear to hesitate at the end of the roll before coming back, she is probably approaching or past the position of maximum righting effect, and immediate steps should be taken to decrease the roll by changing course or speed or both.

Certain artificial methods are sometimes used to reduce a ship's roll. Rolling keels or bilge keels are built-up structures of roughly triangular cross section attached outside the hull near the turn of the bilge and extending part of the length of the ship. These false keels have the function of damping the roll of the ship. More effective devices are antirolling fins (*Sperry Gyro-Fin* or *Denny-Brown Stabilizer*) which are adjusted by machinery inside the hull.

To sum up this discussion, the seaworthiness of a vessel is dependent on three things:

1. Its initial and overall stability. Initial is the resistance of a ship to initial heeling when on an even keel. Overall stability is the resistance of a ship to heeling caused by static forces throughout her range of stability.

2. Its range of stability. Range of stability is the total angle through which the righting arm is positive. It is the angle of heel either to port or starboard through which a vessel tends to return to an upright position.

3. Its dynamic stability. Dynamic stability is the righting energy available to resist heeling through an angle not greater than the range of stability.

Initial stability is measured by the transverse metacentric height in feet which is the distance from the center of gravity up to the metacenter. The center of gravity remains fixed for any particular condition of loading, but the metacenter may change. Any change in this can be quickly estimated by the formula:

$$GM = \frac{0.44B^2}{T^2}$$

where GM = the transverse metacentric height in feet
B = ship's beam in feet
T = time in seconds of a complete roll; e.g., port to starboard to port

NOTE: The constant 0.44 represents an average for various hull forms.

Thus if the period of the roll is doubled, the GM is quartered, and in all probability the ship is in danger. The decision to abandon a ship or to attempt to save her is based greatly on this new calculated GM.

This GM can be increased by lowering the center of gravity by completely flooding some of the lowest tanks and by casting overboard top hamper, boats, torpedoes, spars and mast. The removal of any free surfaces by the complete flooding or pumping of fresh water or fuel-oil tanks, and the removal of water in the bilges will increase the GM, especially if the tanks run athwartship.

A ship with a long, easy roll makes a good platform and a good passenger ship, but the very fact that she has an easy roll is a sign of a low metacentric height or narrow beam. A ship with a large GM or a large beam will have a quick, jerky roll which is uncomfortable. Beam also has a great effect on initial stability (GM) since the moment of inertia of the water-line plane is a geometric function of the beam.

Static stability for any angle is measured by the righting arm at that angle of inclination. As a ship is inclined, this arm increases until it becomes a maximum approximately when the main deck waterways are awash. Further inclining decreases this arm until, at about twice the angle at which the arm was a maximum, it becomes zero and the vessel will no longer right itself.

This last angle gives the range of stability of a ship and totals approximately twice the angle at which the main deck waterways are awash. Therefore, a high freeboard as compared to the beam is desirable. Between two different types of ships, the relative range of stability is measured by the ratio of freeboard and beam; whereas between two ships with the same ratio of freeboard to beam, the GM is the measure of the relative range of stability. Merchant ships with their high freeboard compared to their beam are very seaworthy, even though they have a low value for their GM.

The work utilized in the inclining of a ship is the measure of its dynamic stability. If the force of the sea acting on a ship becomes great enough to heel her over until the righting arm becomes zero, the ship will not be able to right itself. The apparent force of the sea is greatest when the period of the ship and that of the waves are in synchronism, thereby building up a much deeper roll. This force can be controlled by the ship changing course or speed or both, which alters the apparent period of the waves relative to the ship.

3.2. Tonnage of Ships. As there is sometimes confusion as to the difference between displacement and the several kinds of tonnage, the following definitions are given:

Displacement. The weight of the volume of water displaced by a ship is called her "displacement" and is normally expressed in tons. A cubic foot of sea water weighs 64 pounds, and of fresh water 62.5 pounds; therefore, a ton is equal to 35 cubic feet of sea water or 35.9 cubic feet of fresh water. Displacement tons are always 2,240 lb.

Tonnage, Gross Register. The total enclosed space or internal capacity of a ship expressed in tons of 100 cubic feet each is the gross register tonnage. The unit of volume is that figure which was used originally in "Moorsom's System" of measuring ships, and this system has, with slight variations in application, been adopted by most of the nations of the civilized world.

Gross register tonnage is used for calculating net register tonnage in the United States as a basis for drydock charges for steamers.

Tonnage, Net Register. The actual earning power of a ship is expressed by the net register tonnage, and this figure is secured by deducting from the gross tonnage such spaces as may have no earning capacity: for instance, the engine, boiler and shaft alley spaces, coal bunkers, spaces used in steering and working the ship, and such spaces as may be necessary for the accommodation of the crew. The laws of the several nations vary with reference to the various deductible spaces. Net register tonnage is generally used in charging harbor dues, canal tolls, and other similar charges to which merchant ships are liable.

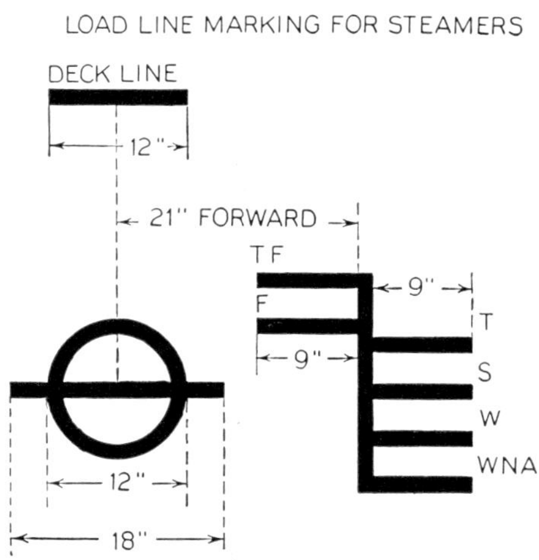

FIG. 3.2 EXPLANATION OF SYMBOLS OF THE LOAD LINE MARK FOR CARGO STEAMERS AND TANKERS

TF—Tropical Fresh Water Load Line
F—Fresh Water Load Line
T—Tropical Load Line
S—Summer Load Line
W—Winter Load Line
WNA—Winter North Atlantic Load Line

3.3. Load Line Markings.[1] In accordance with the International Load Line Convention, 1930, and an Act of Congress passed in 1929, load lines were estab-

[1] The earliest load line markings, still found on all British merchant ships, are called Plimsoll marks after Samuel Plimsoll, by whose efforts the act of Parliament to prevent overloading was procured.

lished for merchant vessels of 150 gross tons or more proceeding to sea on an international voyage. (Special load lines for vessels engaged in voyages on the Great Lakes and in Coastwise Voyages by Sea are in conformity with the Coastwise Load Line Act, 1935, as amended in 1936, and apply also to all merchant vessels of 150 gross tons and upwards.) These load lines indicate the drafts at which, for various conditions and types or classes of vessels, there will still be left a sufficient percentage of reserve buoyancy to ensure the safety of the vessel. On it are indicated the maximum safe drafts for fresh and salt water, for winter and summer, and for certain oceans.

As provided in the Load Line Act of 1929, the American Bureau of Shipping assigns load lines and issues load-line certificates. The authority by whom the load lines are assigned may be indicated by letters marked alongside the disc and above the centerline.

3.4. Pressure. Hydrostatic pressure on a submerged body, such as the hull of a ship, is proportional to depth and acts at right angles to the surface of the hull. Each square foot of surface of the hull is subject to a pressure of $\frac{1}{35}$ ton for every foot of depth (or 64 pounds per foot of depth). This water pressure is applied to the hull and transmitted through the frames, decks, and bulkheads to the various parts of the ship. Although the horizontal pressures of water exerted on each side of the ship cancel each other, the force still acts on the hull. The decks and transverse framing and bulkheads prevent lateral crushing of the hull.

If the skin of the ship is ruptured, the hydrostatic pressures formerly exerted on the plating are now imposed on the bulkheads of the flooded compartments. This is why all bulkheads require stiffeners to prevent them from bulging, and why bulkheads well below the water line are thicker, require more stiffeners, and are subjected to higher test pressures. Flooding water will exert a considerable *upward pressure* against the overhead deck of a flooded compartment if the deck in question is some distance below the water line.

Hull Damage Control

The stability principles considered in this chapter are useful to the seaman to the extent that they explain in general why the undamaged, properly loaded ship remains upright in heavy weather. They also reveal the range and type of response that may be expected from the undamaged ship. If damage occurs it is necessary to apply these principles quickly to save the ship. In the application of these principles it is important to realize how damage may reduce margins of buoyancy and stability. Damage sustained by vessels in peacetime accidents is often similar to the effect of wartime enemy action; a peacetime collision with possible attendant fires and explosions may be just as serious as enemy hits.

After serious damage or in heavy weather the captain must be guided by three factors if he is to save his ship: He must maintain *power*, *buoyancy*, and *stability*. The damaged ship sometimes may be able to survive in calm waters by maintaining only buoyancy and stability. But it may often be impossible to do this without power. Flooded engineering compartments or water reaching main switchboards (through flooding or through ventilators) are important causes for power loss.

3.5. Overall Consequences of Hull Damage. War experience has shown that whenever a ship suffers damage involving serious flooding, either the damage is so extensive that the vessel never stops listing or settling in the water, going down within a few minutes, or the vessel stops heeling, changing trim, and settling in the water shortly after initial damage.

Experience also shows that vessels which survive several hours after damage and then sink, suffer PROGRESSIVE flooding whose control requires proper training of the crew. This flooding comes about in the following manner:

As a result of a hit or collision a large hole is opened in the side. Several bulkheads and decks may be carried away just inside this hole. Immediate flooding occurs through these large holes, giving the ship her initial list, trim, and reduced stability.

In addition to the large holes, there may be a certain amount of subsidiary damage, with riddled or warped bulkheads and decks, opened seams, leaking doors and hatches, etc. These permit slow leakage and progressive flooding past the boundaries of the damage. The slow flooding is aggravated if personnel escaping from the damaged area leave doors or scuttles open behind them.

3.6. Action Before and After Damage. The preparedness before any damage is sustained will often determine whether or not efforts to save the ship afterward can be successful. As a rule, men-of-war are better able to cope with damage. They are usually more fully compartmented than merchant ships and have more men and equipment to rally against the damage. There has been an increasing tendency both to design and to operate merchant ships with a higher capacity to resist damage. And on both naval and merchant ships there has been a growing understanding that the captain or master has a vital primary role in ensuring such preparedness against damage at sea. This preparedness consists generally of:

1. Utilization of designed hull safety features at all times when at sea (such as ensuring that watertight boundaries are faithfully maintained.)
2. Ensuring that the ship is not overloaded.
3. Ensuring that deck loads are not exceeded.
4. Ensuring proper amount and distribution of liquid and other cargo and ballast.
5. Ensuring that crew members are trained to localize damage insofar as facilities of the ship permit.

The loss of the *Andrea Doria* is well known by mariners as an example of how improper liquid loading can doom a ship to sinking. When one of her huge, off-center, empty fuel tanks ruptured and filled, the *Andrea Doria* suddenly increased her list nearly 20 degrees.

There exists, in any situation where the vessel does not sink immediately, an excellent chance of saving her if slow leaks can be patched and plugged. Bulkheads that have not collapsed under the blast and onrush of water from the hit are not likely to collapse under hydrostatic pressure.

Immediately after serious damage TWO IMPORTANT DECISIONS must be made: (1) Whether all hands should remain aboard, all but the salvage party should be evacuated, or all hands should abandon ship; and (2) what corrective measures will improve the situation instead of making it worse. The first of these decisions is made by the captain, but his conclusions must be based on information that he receives from the engineer. The second decision is frequently the problem of the engineer unless it involves ship handling or loss of military efficiency (as through jettisoning ammunition).

3.7. The Enemies of Stability. If a ship's tank or void is only partially full, the liquid contents may "slosh" back and forth with the motion of the ship. This effect is known as *free surface.* A similar effect is noted if a compartment is partially flooded.

If the hull is ruptured so that one or more compartments are open to the sea, *free communication with the sea* results.

Free surface and free communication with the sea are, when combined, the deadly enemies of stability.

There is little excuse for free surface; tanks partially full should generally be ballasted. The captain who takes his ship into heavy weather with free surface in his tanks is foolhardy. The ship with initial free surface which is damaged so that it also acquires loose water and free communication with the sea will almost surely be in danger. Free surface should be avoided, since it always causes a reduction in GM and overall stability. (It should be remembered that a reduction in GM reduces a ship's stability.) Free communication with the sea not only reduces GM, *but* GZ *as well.* Thus, some initial stability is not only lost but, since GZ decreases, there is a decrease in the ship's righting moment.

Care and Preservation. Without proper care, both wood and metal hulls would rapidly deteriorate to the point where they would be unsafe and unusable. Wooden hulls are subject not only to rot and deterioration when not properly protected from the atmosphere, but also to damage from marine animals and marine growths that attack the underwater body. Metal hulls must be protected from both corrosion and erosion in order to maintain their seaworthiness.

Marine animals and marine growths which damage the hull of a ship or retard its speed are natives of salt water and, generally speaking, are more common and damaging in tropical waters than in colder climates. The greatest hazard to wooden hulls is the teredo or marine borer, a wormlike mollusk which eats its way through wood and riddles planks and timbers with small holes. Like all marine animals and growths, the teredo is poisoned by copper and its derivatives, and the standard protection for wooden hulls is to cover them with a copper base or similarly poisonous paint. Originally a tropical pest, the teredo has spread until it is now found in virtually all salt-water harbors of the United States.

The underwater pest which causes the most trouble on steel hulls is the barnacle, although mussels and marine grasses of various sorts are also found attached to the underwater body. The barnacle is a univalve mollusk which attaches itself tightly to the skin of the ship and forms a shell that is roughly cone-shaped. It propagates freely and if undisturbed will soon build up a layer several inches deep. Barnacles and other marine growths apparently do little damage to a steel hull, except to destroy the protective paint and allow sea water to promote corrosion, but they may reduce the speed of a ship as much as several knots by increasing skin friction between the water and the hull. Modern plastic paints, applied hot in drydock, protect a hull from barnacles for several years.

The greatest problem in the care and preservation of metals is corrosion. Corrosion is the gradual disintegration of a metal due to chemical or electrochemical attack by atmosphere, moisture, or other agents. There are many different types of corrosion. A technical discussion of the various types may be found in any standard reference work on the subject.

Although a ship does not encounter any corrosion problems which may not be found anywhere else, it does encounter virtually all types of corrosion. A bronze propeller secured to a steel shaft and turning in an electrolyte (sea water) introduces the possibility of electrochemical attack on the shaft and the ship's hull. The propeller itself may suffer from cavitation and corrosion-erosion attack. Pumps handling sea water are subject to similar conditions. Propeller and pump shafts are also subject to corrosion fatigue.

The hull and superstructure of the vessel are usually of plain carbon steel, and any seaman knows the amount of chipping and painting necessary to combat the severe rust caused by a marine atmosphere. Some superstructures have been built of aluminum alloys which resist atmospheric corrosion.

The various means for preventing or minimizing corrosion may be classified under four headings:

1. *The use of alloys that resist attack by the particular environment.* The resistance to atmospheric attack of such materials as the copper base alloys, stainless steels, and Monel metal are well known, but there is no single alloy that is immune to all corroding media. However, the addition of alloying elements in even small amounts may greatly improve the resistance of a particular material to a given environment. About 0.25 percent copper added to a carbon steel doubles its resistance to atmosphere corrosion. Copper-nickel is an excellent material for condenser tubes, but its resistance to sea water is greatly improved by the addition of 0.2 to 0.6 percent iron. Depending on environment, the corrosion resistance of stainless steels may be improved by small additions of manganese, silicon, columbium, titanium, molybdenum, or nitrogen.

2. The use of galvanic protection. The electrochemical mechanism may be utilized for protecting a structural metal in contact with an electrolyte. This galvanic protection consists of attaching to the structure a metal, anodic to the one to be protected, thus sacrificing the added metal and protecting the structure. Zinc is commonly used to protect steel, cast iron, brass, and bronze. The zinc must be in good, electrical contact with the metal to be protected and the electrolyte. Galvanic protection is used for propeller shafts, rudders, and hull plates by attaching zinc plates near the propellers.

3. The control of environment. Metals have very low corrosion rates in dry gases and in pure water free from air. An outstanding example of control of environment is the dehumidification done in the moth-balling program carried out by the armed forces since the end of World War II. The sealed interiors of ships were kept dry with air-conditioning machines; guns, tanks, planes, etc., were completely sealed in moistureproof covers and provided with moisture-absorbing agents.

4. The use of protective coatings. These may be divided into three classes: (a) chemical coatings, (b) nonmetallic coatings, and (c) metallic coatings.

Chemical coatings are those that are formed by chemical reaction between the metal surface and an appropriate solution. The application of phosphate coatings to the ferrous metals is representative of this group. The part to be protected is dipped in or sprayed with a solution of metallic phosphates, and the coating that forms is not only resistant to atmospheric corrosion but serves as an excellent base for paints. Some magnesium and aluminum alloys are similarly protected with chromate coatings. Oxide and silicate coatings may also be chemically produced on certain metals.

Nonmetallic coating materials include paint, varnish, plastics, natural and synthetic rubbers, bituminous, and petroleum products. In general, these materials form a mechanical film to exclude air and moisture from the metal surface. Their effectiveness depends on the initial cleanliness of the metal and on the film thickness. All of them may be satisfactory for prolonged periods of time, depending on environment and whether or not the surface is subjected to abrasion.

Metallic coating processes serve to cover a corrodible metal with a thin layer of another metal which is more resistant to attack. Sometimes the coating also provides galvanic protection, as is the case with zinc or cadmium coatings on the ferrous metals. Metallic coatings may be applied by electroplating, hot dipping, or metallizing.

Electroplating consists of making the clean base metal the cathode in an electrolytic cell containing a water solution of some salt of the metal being deposited. Close control of the process is essential to good nonporous plating. Copper, nickel, chromium, cadmium, and zinc are electrodeposited on iron and steel as corrosion preventives.

Hot dipping consists of immersing the cleaned base metal in a molten bath of the coating metal. The process is suitable for those metals that will wet each other. Zinc and tin coatings are commonly applied by hot dipping.

Metallizing is a metal spraying process in which a wire or powder of the coating metal is melted by an oxyacetylene flame in the presence of an air jet which atomizes the metal and sprays it onto the base metal. The base metal must have a roughened surface since the bonding is purely mechanical. Virtually all metals can be sprayed, and the process is used to build up worn parts and provide hard surfaces as well as for corrosion resistance.

3.8. Drydocking. The protective measures just described for the underwater hull cannot be used at discretion, as is done in the case of surfaces above water. The submerged hull is virtually inaccessible while the ship is waterborne. No protective coating that will last indefinitely under water has ever been devised and it is unlikely that one ever will be. Damage to underwater fittings may occur that will remain undetected as long as the fittings remain submerged. The only practical method for preserving the underwater hull is to remove it from the water at intervals, make needed repairs, and give it the best protective coating available before refloating. This routine is generally known as drydocking.

There are several methods of drydocking. The oldest and simplest is known as beaching and careening. This process consists of putting a ship on a shelving beach at high tide, working on alternate sides during periods of low tide by heeling her over at an angle with tackles made fast from some solid object to her spars, and finally hauling her off during a subsequent period of high water. This method will never be seen in modern practice, except in the case of very small vessels or possibly for temporary repairs in a case of accident or emergency. For the ordinary steel vessel the usual methods of drydocking are by the use of marine railways, floating drydocks, and graving docks.

A *marine railway* (Fig. 3.3) is an inclined shipway by means of which a vessel of moderate size, resting in a cradle, can be drawn out of the water above the reach of the tide. The cradle is on wheels that run on rails and is moved by means of a windlass and endless chain. Few marine railways will handle a ship larger than a moderate-sized destroyer.

The usual *floating drydock* (Fig. 3.4) is made up of rectangular, open-ended sections that can be fastened firmly together. Tanks are flooded with water to sink the dock far enough for a ship to enter the cradle and then are pumped out so as to raise the ship and the inner part of the dock clear of the water. By using enough sections of suitable dimensions, all but the largest types of ships can be docked. Sections of the dock also can be used to dock each other. One of the great advantages of floating drydocks is that they can be towed to the localities where they are most needed. Mobility of floating drydocks has been increased greatly in the types of construction developed by the U.S. Navy in the past few years. Floating drydocks for the smaller types of vessels are built in one piece with the usual ship type bow and are equipped with a steering mechanism. A section with a ship type bow for ease in towing may also be provided for larger sectional floating drydocks. Most floating drydock sections are self-contained. They have pumps for emptying their ballast tanks, power plants and other facilities for the storage and distribution of such

FIG. 3.3 MARINE RAILWAY. *Official U.S. Navy Photograph*

FIG. 3.4 FLOATING DRYDOCK. *Official U.S. Navy Photograph*

services as oil, compressed air, steam, and electric current; and may be equipped with machine shops and with quarters for their crews.

A *graving dock* (Fig. 3.5) is a permanent installation in a shipyard. It is a narrow basin having walls and a floor, usually built of reinforced concrete, into which vessels may be floated and from which the water may be pumped out, leaving the vessels dry and supported on blocks. It is used for repairing and cleaning the underwater hull of ships and in some cases for building ships.

FIG. 3.5 GRAVING DOCK. *Official U.S. Navy Photograph*

Graving docks are built at an angle to the shore line, and one end must be open to navigable water in order to allow the entrance of ships. This open end is usually closed by a caisson, either floating or sliding, although in some docks a double swinging gate is used. The most usual type, the floating caisson, is a self-pumping hollow gate which is floated and towed clear when the dock is flooded. After a ship enters the dock the caisson is moved into place and flooded with water, which causes its ends to sink into wedge-shaped grooves cut in the walls of the dock while its base fits against and is supported by the sill, which extends across the entrance of the dock and is raised somewhat above its floor. The depth of water over the sill at high tide determines the greatest draft of any ship that can be docked. In a naval shipyard dock especially, it is desirable that this depth be greater than the ordinary draft of any ship it is expected to accommodate in order to allow for the increased draft due to battle damage.

The walls of a graving dock are usually built in steps which are known as altars. Power capstans and fixed bollards are installed to control the lines used in hauling the ship in and out of the dock. Blocks are heavy wooden structures used to build up the cradle in which the ship rests while in dock. The keel blocks, upon which the ship's keel is supported, are large cube-shaped, semi-fixed structures built up of concrete, hardwood, and soft pine caps. Other blocks, called bilge blocks, can be hauled in and out to the desired position on tracks set at right angles to the center line. Docking keel blocks are the blocks upon which a vessel's docking keels rest. Should a vessel have no docking keels and insufficient flat bottom for the installation of enough blocks to keep her upright, wale shores may be used. These are spars extending from the ship's side to the side of the dock and wedged in place as she settles upon the keel blocks. In case several small vessels are to be docked at once, blocks may be set up for them on appropriate sections of the dock floor.

3.9. Preparations for Docking. In preparing a dock to receive a vessel, the dockmaster or docking officer first refers to the ship's docking plan. This furnishes necessary information concerning the underwater hull for docking purposes. For United States men-of-war a copy will usually be found in the files of the ship's home yard, but every ship should carry its own docking plan. To place the blocks accurately and to build them up to the proper height and bevel, the docking plan must give the following information:

1. Full extent of keel with flat and rising portions accurately delineated.
2. Peculiarities of stern post and rudder.
3. Sections, amidships and elsewhere, to show proper height and bevel of bilge blocks, if these are necessary.
4. Shape and location of keels, docking keels, struts, propellers, underwater fittings, and projections of all kinds.

Further, since it is often necessary to provide more blocks than usual under heavy weights, the docking plan should show:

5. Location of turrets, boilers, engines, and other unusual weights.

To assist in locating sighting battens the plan should show:

6. The length on the load water line.

Finally, in order to enable the dockmaster to determine whether or not the dock can take a ship, or to place her accurately in the dock, the plan must show:

7. The length overall, the beam, and all projections, such as blisters, increasing the normal beam.

3.10. Docking. The dockmaster of any particular dock, knowing the ship's draft, the maximum depth over the sill, and the current and tidal variations in the vicinity, decides on the time the ship should enter the dock and so informs the commanding officer. The commanding officer makes the necessary arrangements to ensure that at the time specified the ship is without any list to either side and with as nearly zero trim as possible.

The dock being already prepared, water is admitted, the caisson is floated and removed, and the ship is brought to the dock entrance, usually with the assistance of tugs. When the bow of the ship has safely crossed the sill of the dock, the responsibility for her safety rests upon the dockmaster, who hauls her into the dock, replaces and sinks the caisson, centers the ship, starts dock pumps, and proceeds with the docking until the ship is safely landed on the blocks and the dock is pumped dry.

Should any extraordinary conditions exist, such as those due to accident or battle damage to the ship, the special precautions taken in docking her must depend upon the judgment of the dockmaster.

No change of any kind in her weights during the period when the ship is in dock should be made without the knowledge and consent of the dockmaster. Improper changes in weights may cause her to do serious damage to herself or the dock when she is floated, due to sudden changes in list or trim.

3.11. Routine Work in Drydock. The following routine drydock work is done in addition to any special underwater repairs or alterations decided upon:

1. Clean bottom, including scaling or wirebrushing of badly corroded parts.
2. Cut out and re-drive all loose or badly corroded rivets.
3. Caulk leaky seams and rivets.
4. Overhaul underwater valves.
5. Repack underwater stuffing boxes.
6. Renew zinc and mild steel protectors as necessary.
7. Take propeller shaft clearances and re-wood stern and strut bearings as necessary.
8. Check pitch of propellers, clean and polish them.
9. Examine rudder pintles and gudgeons and rudder shaft packing. Take rudder bearing clearances.
10. Paint bottom, paint draft marks, paint boot-topping.

3.12. Undocking. After all underwater repairs are completed and the bottom is painted, a time for flooding the dock is agreed upon between the commanding officer of the ship and the dockmaster. The former stations men at the outboard valves and elsewhere as he deems necessary, to ensure that water does not enter the ship. The latter stations men at the various shores and lines to prevent as far as possible any injury to the ship or dock from a change of weights or any unexpected alteration in tide or wind.

Water is admitted to the dock under the dockmaster's control. When all sea-valve openings are covered, flooding is usually checked until their water-tightness can be reported. When the water has risen to a sufficient height the ship lifts from the keel blocks. When the ship is safely afloat flooding is continued until the water level within the dock is the same as that outside. The caisson is floated as quickly as possible, then removed, and the ship is hauled out of the drydock.

4

Steering and Propulsion

A knowledge of seamanship would not be complete without an understanding of the means available to maneuver a ship: the power plant and the steering mechanism. A ship's captain or deck officer must know the minimum and maximum responses of his ship to backing, accelerating, and rudder angle under all conditions. With this knowledge the expert shiphandler will always avoid putting his ship into such a situation that judicious use of his engine and rudder cannot extricate him. Regardless of the apparent complexity, any propulsion device is based on the fundamental principles of energy conversion. The energy available in a fuel, whether conventional fuel such as oil, or nuclear, is converted to mechanical energy, which is then used to give the ship kinetic energy or energy of motion. This chapter outlines the devices used to accomplish these energy conversions.

Boilers and Combustion

Naval boilers supplying steam for propulsion are now mainly water tube boilers. In this type the water is inside tubes within the boilers, and the products of combustion change the water to steam by the transfer of heat through the tube walls. Fire tube boilers, in which the products of combustion are within the tubes and heat is transferred outward to water surrounding the bank of tubes, are now seldom seen in naval service except as auxiliary boilers supplying steam for heating or for evaporators. Water tube boilers, which have replaced the old fire tube boilers, may be of the large tube or small tube (express) type. The latter is found in most modern steam-propelled ships. The small tubes give a greater heat-transfer surface and became feasible when feed-water treatment had progressed far enough that deposits would not block the path of water in the small tubes. Modern boilers can be safely lit off cold and brought up to full pressure in 1 hour, which is quite an improvement over the approximately 8 to 12 hours required by the old fire tube boilers.

4.1. Nuclear (Atomic) Reactors. Heat for turbine propulsion and turbogenerators on the nuclear-powered ship comes not from a boiler but from a *reactor*. It might be easiest to understand the shipboard function of a reactor by considering that it is a radically new type of boiler which consumes a unique fuel. As a matter of fact, the water is converted into steam through an arrangement roughly similar to the old fire tube boiler. Once the steam is made, it is used in the geared turbine-drive plant.

There is one major difference between boilers and reactors. In the boiler, the water in the tubes is heated directly by the boiler fires. This is not the case in the reactor. Here the heat is initiated and controlled by moving a neutron absorber in or out of the fuel area. The resultant controlled nuclear fission heats water under pressure (known as a *coolant* because it removes heat from the reactor). The coolant then goes through a heat exchanger known as the steam generator.

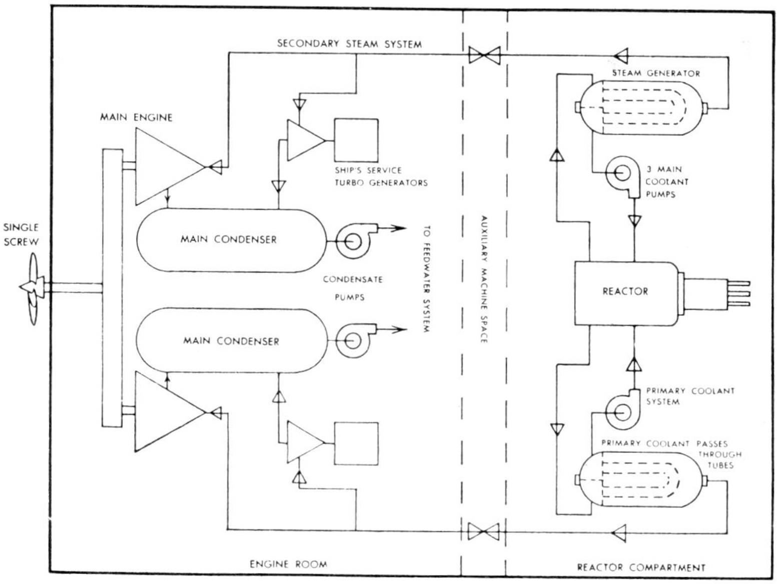

FIG. 4.1 POWER PLANT OF THE NUCLEAR SUBMARINE

The following description of the engineering plant of the *U.S.S. Skipjack* has been provided by her builder, the Electric Boat Division of General Dynamics Corporation.

Primary or Main Coolant System. The reactor compartment equipment includes one reactor, and two primary loops.

The reactor gives up heat to the main coolant, water, which then is forced through the boiler heat-exchanger tubes where it gives up heat to form steam on the shell or secondary side of the boiler. The main coolant is then pumped back into the reactor where it is again heated up.

The water from the starboard and port loops mixes together in the reactor, but the rest of the plant is divided into identical units port and starboard. The main coolant water is kept pressurized to ensure that boiling will not take place in the reactor.

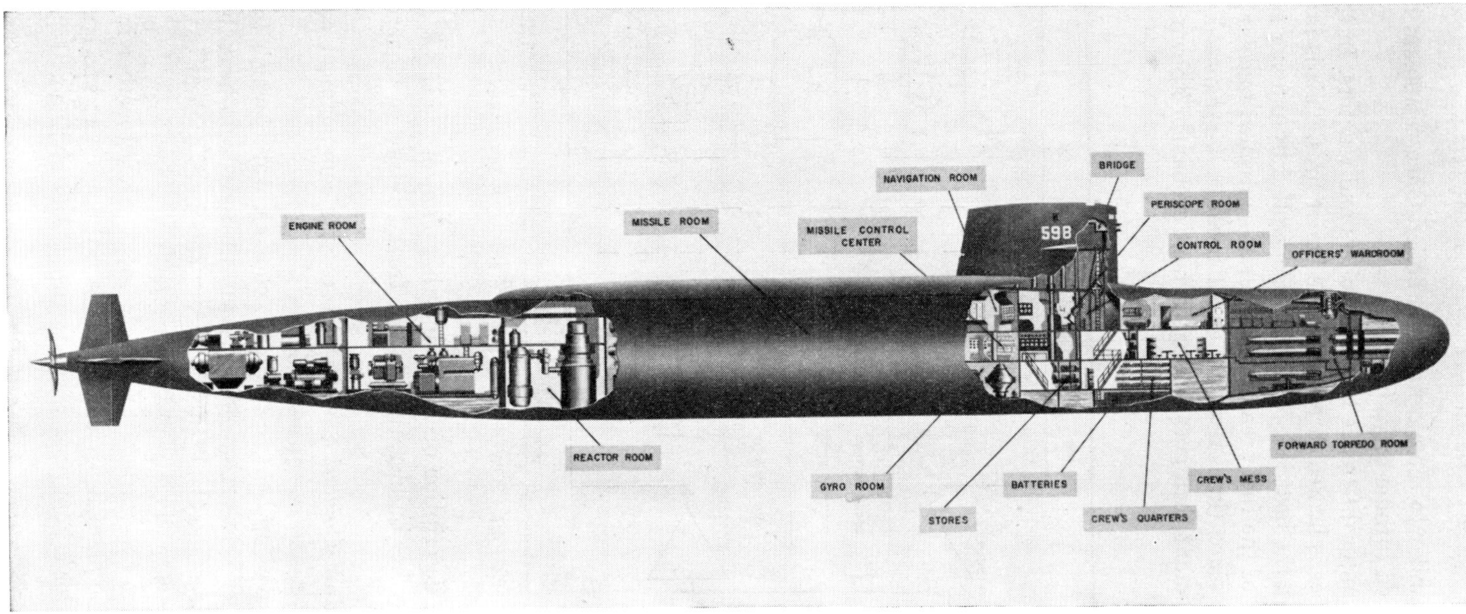

FIG. 4.2 FLEET BALLISTIC MISSILE SUBMARINE SHOWING NUCLEAR POWER PLANT. Electric Boat Division of General Dynamics Corporation, Groton, Conn.

Secondary or Main Steam System. The secondary system is the steam system. It is completely isolated from the primary system since the primary water goes through the tubes of the boiler while the secondary water, which is boiling to make steam, is on the shell side of the boiler.

Steam rises from the boiler to the steam drum where the water is separated from the steam. The dry saturated steam then flows back to the engine room where it drives ship service turbogenerator sets (SSTG), coolant turbogenerator sets (CTG), and the main propulsion turbines.

Provision is made in some ships for declutching the propulsion turbines and reduction gears from the propeller shafts so that the ship can be driven through the water by electric motors mounted integrally on the propeller shafts. The electric motors can receive power from the battery, from diesel engines, or from AC-DC motor generator sets.

Radiation. When the reactor is in operation, the lower level of the reactor compartment is kept isolated and personnel cannot enter this space. Within a few minutes after shutdown the reactor compartment lower level can be entered to perform maintenance work.

The shield of the *Skipjack* reactor reduces the radiation to a level such that during a cruise lasting the life of the reactor, the average crew member will receive less radiation than he would during a lifetime from x-rays and cosmic rays and natural radioactivity in the sea, air, drinking water, and ground. In one year of operation the average crew member received less than the Bureau of Standards allowable radiation dosage for one week.

Startup. A typical schedule for startup from a cold condition follows:

Four hours before getting under way—one man starts a precritical checkoff, which is a thorough check of all reactor control equipment. The in-port watch in the engine room and reactor compartment checks systems lined up for operation.

Two hours before under-way time—engineering duty section stations the watch. Commence pulling rods.

One and one-half hours before under-way time—reactor startup completed—warming up primary loop and steam lines.

Thirty minutes before under-way time—warm up turbines. Put turbogenerator sets in operation.

Fifteen minutes before under-way time—ready to answer bells.

4.2. How the Deck Officer Can Promote Efficient Boiler Operation. Some deck officers seem to feel that the engineer's only reason for blowing tubes is to spread fine carbon dust over the clean topside, and consequently they refuse to allow tubes to be blown on their watch. An understanding of the reason for blowing tubes may temper their reluctance to grant the necessary permission. The oil burned in boilers leaves a layer of soot on the outside of the small water tubes previously discussed. This soot is undesirable for the following reasons: (1) The soot acts as an insulator and slows heat transfer to the water within the tubes. (2) If the soot remains in a boiler when fires are secured, it absorbs moisture from the air; the moisture activates the sulfuric acid in the soot, and this acid in turn attacks the metal of the tubes and boiler drum. (3) If allowed to remain too long, the soot packs into a solid mass and can be removed only by tedious hand cleaning. To maintain maximum boiler efficiency, tubes should be blown, while under way, once every 4-hour watch and, while in port, twice a day. When "tubes are blown," high-pressure steam is admitted to perforated tubes (known as soot blower elements) which are permanently installed within the boiler. The elements are rotated so that steam jets from the perforations play on all the tube surfaces within the boiler, cleaning them thoroughly. Tubes should be blown when the wind is abeam. The officer of the deck should also be aware that maximum efficiency usually is served by steaming with a light brown haze

4.3. Combustion In Internal-Combustion Engines. The fuel for an internal-combustion engine is burned within the engine and the products of combustion pass directly through the engine, resulting in the transformation of

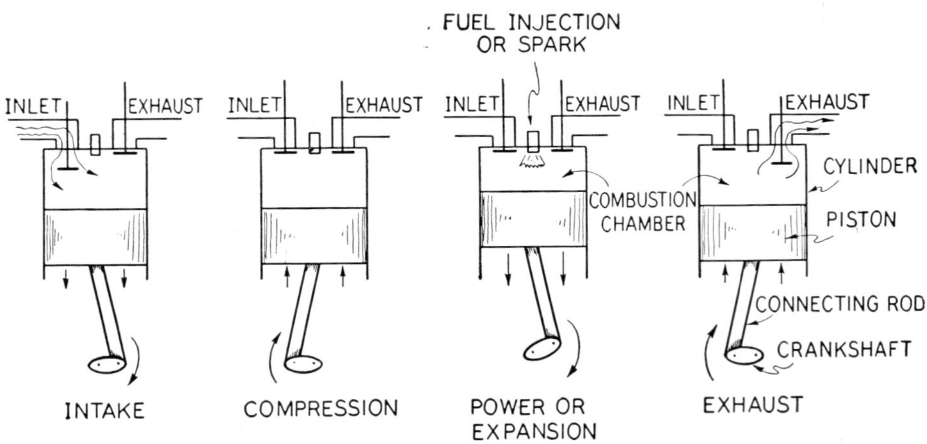

FIG. 4.3 ONE CYLINDER OF A SPARK IGNITION OR COMPRESSION IGNITION ENGINE (FOUR STROKE CYCLE) SHOWING THE FOUR PHASES IN ONE COMPLETE FIRING CYCLE

heat into mechanical energy. There are basic differences in the three common types of combustion chambers in internal-combustion engines. In the spark ignition, commonly referred to as the gasoline engine, combustion chambers are located in each cylinder and comprise the space between the top of the piston and the top of the cylinder. Fuel is admitted periodically and is ignited by an electrical spark. The fuel and air are mixed in the carburetor and enter the combustion chamber together. The gasoline engine is seldom seen in naval vessels or boats now because of its extreme fuel fire hazard.

The compression ignition or diesel engine is a reciprocating engine very similar to the spark ignition engine shown in Fig. 4.3. The main differences are that the fuel is ignited in the diesel engine by the high temperature of the highly compressed air in the combustion chamber; air only is taken in on the intake stroke, and the fuel is injected in a fine mist at the start of the power stroke. The diesel engine is used extensively in the Naval Service for main propulsion on ships and boats and for auxiliary or emergency electric generators even in steam-propelled ships.

The gas turbine engine combustion chamber is simple in construction and is merely the container in which air is mixed with a burning fuel. The air and fuel flow are continuous (in contrast to the spark and compression ignition engines), and once the fuel is ignited no further outside spark or high temperature is required to maintain burning. The combustion chamber, although simple, must be designed to provide cooling, proper air and fuel mixing, and freedom from flame "blow-out." A more complete discussion of the internal-combustion engines and their naval applications will be given later in this chapter.

We have seen that energy is extracted from fuels and becomes available for further conversion in the form of a heated working substance, either steam or the actual products of combustion. There will follow a discussion of how the heated working substance is used to produce mechanical energy.

4.4. The Steam Turbine. Steam enters a turbine through nozzles which direct the steam on to moving blades mounted solidly on a revolving drum. The drum is enclosed in a casing and supported on bearings at each end of the casing. When the steam has passed through the first row of moving blades, making them spin, it enters a row of stationary blades attached to the casing which in turn direct the steam against a second row of moving blades attached to the same revolving drum. Alternate rows of fixed and moving blades are located along the length of the turbine. The steam flows through the turbine because of the pressure difference between the point of entry and the point of exit. As it passes through each set of fixed and moving blades (known as turbine stages), the steam pressure and temperature drop as some of the energy is extracted to make the turbine drum revolve. Theoretically the stages necessary to extract all the energy available between the entering and final steam pressures could be in one casing, but actually such a turbine would be too long, the shaft would tend to sag in the middle, and unequal expansion could take place. Actually, the stages are divided between two or three turbines known as the cruising, high-pressure, and low-pressure turbines. Large ships have only a high-pressure and a low-pressure turbine for each shaft. Destroyers usually have the third cruising turbine as well. A disadvantage of the turbine is that it is not reversible, and you cannot supply steam at the normal outlet or low-pressure end of the turbine and have it run in reverse. To provide backing power another turbine is required, designed to turn the shaft in the reverse direction. This backing turbine could be housed in a separate casing, but because of its small size it is more economical to mount it in one of the existing turbine casings, usually the low-pressure turbine. The backing turbine can be small because naval ships are not required to go as fast astern as they are ahead; hence fewer stages are required in the backing turbine.

Turbines are most efficient when running at high speed (3,000 to 4,000 rpm), whereas most propellers[1] are more efficient at slow speeds (up to about 200 rpm). To allow both to run at their most efficient speed, a reduction gear is used between the turbine and the propeller shaft. The reduction ratio varies but is in the neighborhood of 20 to 1 for large ships.

4.5. What the Deck Officer Should Know about His Turbines. When the officer of the deck gives permission to the engineers to start warming up the main plant he should know that the engines are slowly turned by the jacking gear. The jacking gear consists of an electric motor and a hand-operated gear clutch which can be connected to the reduction gear. The clutch is connected, and the electric motor is started, which turns over the turbines slowly and the propellers even more slowly. While no way is put on the ship by this, the OOD should make sure that nothing can foul the propellers in their slow revolving before he grants permission to warm up.

Approximately 15 minutes before the time set for getting under way, the engineer officer of the watch will request permission to "Spin main engines with steam and to continue spinning at 5-minute intervals." The officer of the deck should realize that it is only the engines that are spun and not the propellers. A quick puff of steam sufficient to start the turbines rolling is admitted from the ahead throttle, and, as soon as the turbines start, the ahead throttle is closed and the astern throttle is opened, stopping the spin. The propeller may turn only $\frac{1}{20}$ of a revolution, but no way should be put on the ship from this spinning.

The reason for spinning the turbines at 5-minute intervals is to prevent uneven heating of the turbine rotor with the possibility of its developing a sag. With the close fit of the rotor in the turbine casing a slight sag would cause the blades to scrape on the casing and possibly snap off. Before spinning the main engines with steam, the jacking gear must be disconnected. The purpose of jacking is the same as for spinning with steam. Whenever the engines are stopped, they should be spun every 5 minutes. When word is given to secure the plant, the jacking gear is again engaged, and the turbines are jacked over until cooled.

4.6. Locking a Shaft. If one engine or propeller is damaged while a ship is under way, and the ship cannot be stopped, the damaged engine and shaft must be locked to prevent worse damage. If only the steam is shut off, the shaft will continue to turn as the water acts on the dragging propeller. To lock a shaft under way the astern throttle is opened enough (steam is admitted to the backing turbine) to stop the shaft from turning; then the jacking gear is engaged and a friction brake on the jacking gear is tightened. The other shaft (or shafts on a 4-shaft vessel) are usually limited in their allowable revolutions when one or two shafts are locked. The limitation is required to avoid overloading the turbines driving the undamaged shafts. The limiting rpm's for naval vessels are given in the Bureau of Ships Manual which is provided to all ships.

[1] Propellers will be discussed in more detail in a later section of this chapter.

INTERNAL-COMBUSTION ENGINES

4.7. Gasoline Engines. As previously mentioned, the gasoline engine is seldom used in the Navy for propulsion because of the fire hazard. However, because of the light weight of a gasoline engine for the power it develops, it is used for boats where speed and light weight are most important.

4.8. Diesel Engines. The diesel engine is the most common internal-combustion engine now found in boats. For horsepower up to approximately 4,000 to 6,000, it has definite weight, space, and efficiency advantages over a steam plant. The diesel engine can be found as the main propelling machinery in landing ships and craft, small auxiliary tankers and cargo vessels, patrol vessels, tugs, and submarines. The diesel can be either connected directly to the propelling shaft or connected through a reduction gear to the propelling shaft. Another method is to use the diesels to drive electric generators which supply electricity to drive motors which turn the propeller shafts. This operation will be discussed further at the end of this section.

The direct-connected diesel must be of slow speed in order to allow the propeller to operate at an efficient speed. However, to reduce weight and improve its own efficiency, the diesel should run at high speed. The reduction gear is the logical answer and allows the diesel to turn at its high efficient speed while the propeller turns at its slow efficient speed. Most diesels have a flexible coupling between the engine and the reduction gear of the propeller shaft to prevent engine vibrations from being transmitted to the gears or the shaft. The coupling can be hydraulic—similar in style but larger than those now used in the automobile fluid drive—spring packs, or an electromagnetic device.

4.9. Gas Turbines. The gas turbine holds promise as a lightweight, compact plant which can be warmed up and loaded in a matter of minutes. One of the main problems involved is to find a material for the turbine blades which will withstand the high temperatures of the products of combustion. Figure 4.4 shows a schematic diagram of a simple gas turbine plant. The engine is started

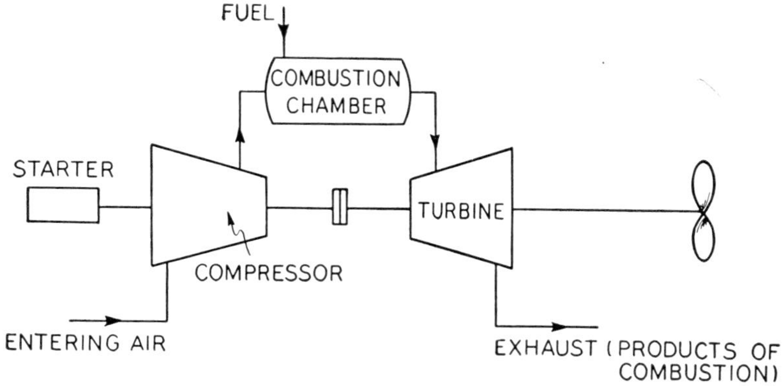

FIG. 4.4 SCHEMATIC DIAGRAM OF A SIMPLE GAS TURBINE PROPULSION PLANT

by an electric starting motor; air enters the compressor where its pressure is raised before going to the combustion chamber where fuel is added and ignited. The products of combustion enter the turbine and cause it to turn. The starter is then disconnected and the turbine continues to run, driving the compressor and the propeller. In order for the turbine to develop enough power to drive the compressor and the propeller, the turbine inlet temperature must be high, but the high temperature tends to burn out the turbine and reduce its useful life. The gas turbine is most efficient when running at its designed speed and does not necessarily lend itself to variable speed operation. It may have an application as a lightweight, relatively short-lived auxiliary plant for a steam-turbine-driven vessel. The steam turbine plant should be used for normal cruising and the gas turbine plant cut in only when the highest speed was necessary.

4.10. Electric Drive. Any of the propulsion engines mentioned so far could be used with an electric drive. In the electric drive the engines already mentioned drive an electric generator at a constant speed. The generator or generators in turn supply electricity to electric propulsion motors which drive the propeller either directly or through a reduction gear. Reversing the propellers is accomplished by electrical switches which reverse the driving motors. The engine driving the generator can always turn in one direction. The electric switches can be located on the bridge and hence give the officer of the deck complete control of his speed and direction of travel. Diesel-electric tugboats are frequently fitted with this type of control.

Engine Control Instruments

The bridge is the primary station for controlling the engines as well as for steering and navigating the ship. In large ships various auxiliary control stations are provided with part or all of the bridge control instruments duplicated and with sound-powered telephone communication with the bridge and the engine rooms. These auxiliary control stations are usually in the same places as the auxiliary steering stations. In merchant ships an auxiliary control and steering station is generally located on or above the main deck, in the after part of the ship. In men-of-war, where the chance of battle damage is great, such controls may be found in the damage control central, control engine room, steering engine room, and possibly in other locations.

The standard and usual instrument for engine control is the engine telegraph. This instrument consists of a dial with a pointer operated by a handle or lever, by means of which orders are transmitted to a similar dial in the engine room by either mechanical or electrical connections. When the order is understood and obeyed, the handle on the dial in the engine room is moved and the pointer on telltale on the bridge dial moves to the proper position to show acknowledgment of the order. A separate engine telegraph is usually provided at the control station for each engine, but sometimes two are combined in one stand. The telegraph for a merchant ship can usually indicate the following orders: stop, slow speed, half speed, and full speed ahead; slow speed, half speed, and full speed astern. In men-of-war the dial indicates: stop, one-third speed, two-thirds speed, standard speed, full speed, and flank speed ahead; one-third speed, two-thirds speed, and full speed astern. These speeds call for a number of revolutions previously agreed upon for standard and full speeds, and corresponding fractions of standard speed for one third and two thirds.

A revolution per minute indicator, actuated by the engine or propeller shaft, is also located at the control station so that the speed and direction of rotation of the engines can be checked directly.

There are various auxiliary methods for giving orders to the engine room in case of mechanical or electrical failure of the engine telegraph. The commonest one is the sound-powered telephone.

All methods of engine control must be operated and checked carefully a short time before getting under way. This test is provided for in the ship's routine. Further tests should be given when the instruments have not been operated for long periods under way or when circumstances require them.

4.11. Propellers. Now that the energy of a fuel has been converted to heat and in turn the heat converted to mechanical energy, a way must be found to use the mechanical energy to drive the ship. The propeller is the device used to drive all large modern vessels. The propeller in a fluid, such as a marine propeller, is a device to obtain a reactive thrust by increasing the velocity of the fluid through its disc. It thus changes the momentum of a mass and provides a propulsive force, or reactive thrust. One method of describing a propeller is by the number of blades it has. The usual number is three or four, although propellers are in use having two, five, and six blades. Normally a propeller with more blades will have a smaller diameter for the same "pushing" power than one with fewer blades.

Another method of describing a propeller is by the direction it turns when driving the ship ahead. Obviously we must specify the side from which we view the propeller in determining its direction of motion. By convention, a right-handed propeller is defined as a propeller that turns in a clockwise direction, viewed from astern, when driving the ship ahead. Similarly a left-handed propeller turns counterclockwise viewed from astern, when driving the ship ahead. Vessels may have one propeller and be designated as single-screw ships; two-twin screw vessels may employ three or four propellers. It is most common for ships to be equipped with one, two, or four screws. The blades of a propeller may be fastened to the hub with bolts or may be cast with the hub in one piece.

There is one type of propeller which can reverse the direction of a ship without requiring a change of direction of the drive shaft. This type of propeller is known as a controllable pitch propeller. The blades are mounted so that they can each rotate on a shaft which is mounted in the hub as shown in Fig. 4.5. It is an excellent propeller for making a ship very maneuverable. With the increased application of gas turbine main propulsion, the controllable propeller

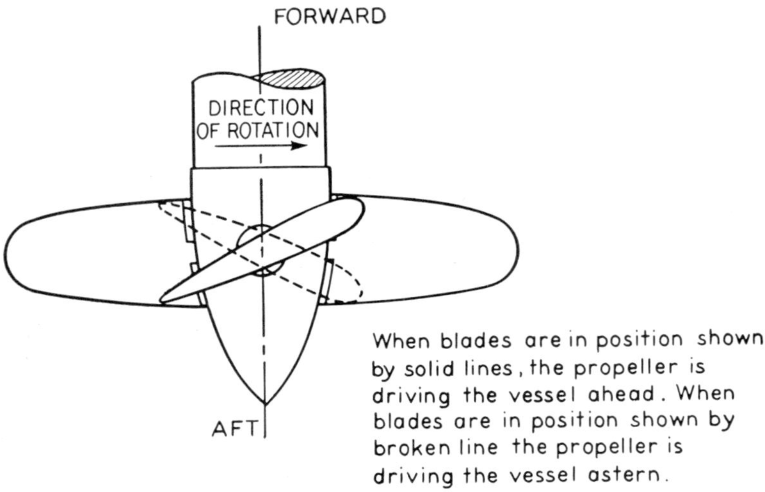

FIG. 4.5 SCHEMATIC DIAGRAM OF A CONTROLLABLE PITCH PROPELLER

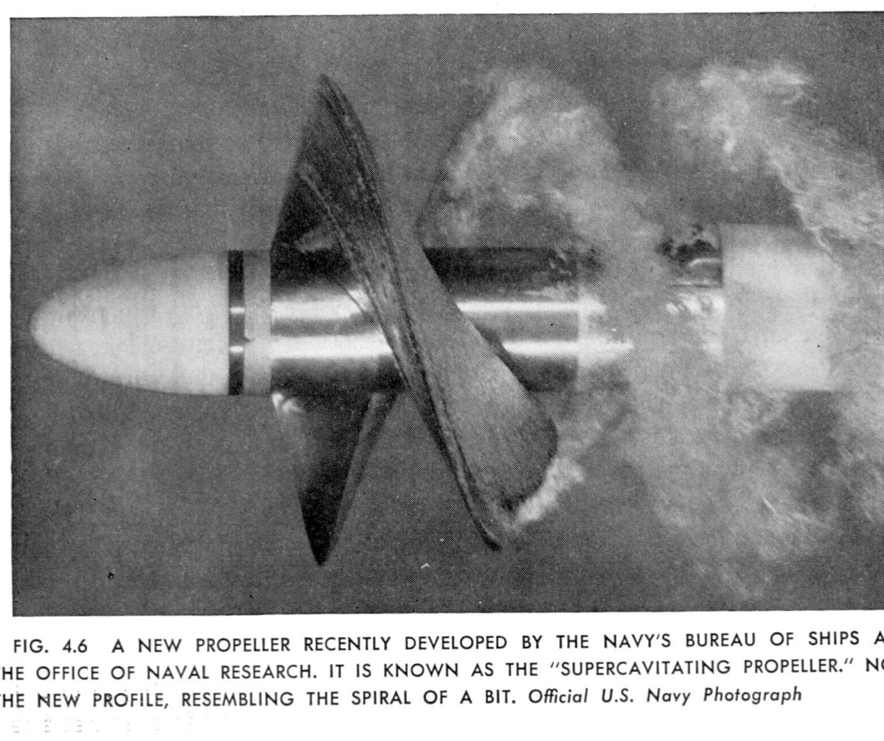

FIG. 4.6 A NEW PROPELLER RECENTLY DEVELOPED BY THE NAVY'S BUREAU OF SHIPS AND THE OFFICE OF NAVAL RESEARCH. IT IS KNOWN AS THE "SUPERCAVITATING PROPELLER." NOTE THE NEW PROFILE, RESEMBLING THE SPIRAL OF A BIT. *Official U.S. Navy Photograph*

is a necessary adjunct. Its primary advantages are reversibility and maximum or peak efficiency over wider range of rpm resulting in better economy.

A new propeller, using an old design, the water screw, promises revolutionary advances in ships speed with no increase in power. The propeller, shown in Fig. 4.6, was developed by the Navy and is known as a supercavitating propeller. It greatly reduces the *effect* of cavitation, if not the cavitation itself, as can be

FIG. 4.7 A CONVENTIONAL BLADED PROPELLER AS IT REVOLVES. NOTE THE TURBULENCE OR CAVITATION WHICH LIMITS ITS SPEED AND EFFICIENCY AND HOW IT COMPARES WITH FIG. 4.6. *Official U.S. Navy Photograph*

seen by comparing it with a conventional propeller, Fig. 4.7. Cavitation is the partial vacuum in a fluid about a rapidly revolving propeller. It results in the formation of vapor pockets that stream behind the blades and consume much energy. The supercavitating propeller, operating at high speed, thus eliminating reduction gears in some installations, creates a vapor cavity along the trailing edge of the new blade that results in a minimum of power loss.

Steering

The usual method of changing the course of a ship under way is by putting the rudder over to one side or the other. The action of the water on the rudder forces the stern of the vessel sideways, and the vessel changes course. The larger the rudder, the faster the ship will turn, because there is a larger area on which the water can act. Similarly, two rudders will make a ship turn faster for a given rudder angle than one rudder which has a smaller area. We can see then that, if a ship is to be very maneuverable and have a small turning circle, the rudder should be as large as possible or two rudders should be provided. Modern destroyers are fitted with twin rudders so that they can make quick turns when attacking a submarine.

Besides the area of the rudder, the speed of the water past the rudder also effects the response of a ship to putting the rudder over; the faster the water is traveling, the better the response. Because the water is traveling faster directly

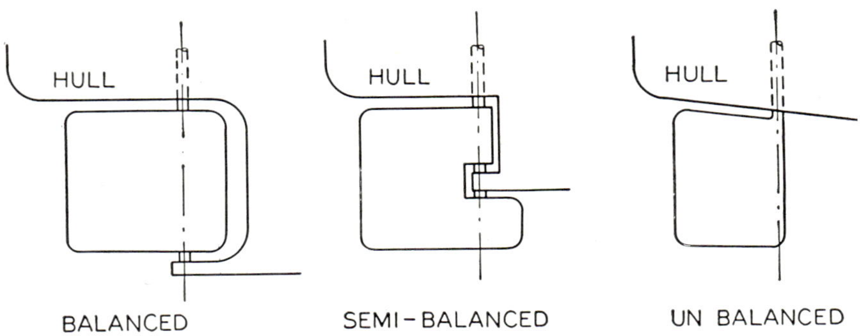

FIG. 4.8 COMMON TYPES OF RUDDERS

astern of the propellers, it is good practice to put the rudders directly abaft the propellers. In a twin-screw ship, if twin rudders are used directly behind the screws, their combined area can be equal to the area of a single rudder placed amidship, and the steering characteristics of the ship will be better than if it had the single rudder.

Rudders are of three general types: balanced, semibalanced, and unbalanced, as shown in Fig. 4.8. When part of the rudder area is located forward of the rudder stock, the rudder is easier to turn because the action of the water on that part tends to help the rudder turn. The unbalanced rudder is the hardest to turn. The choice of rudder is determined by the shape of a vessel's stern, the number of propellers, and the speed the vessel will develop. To turn the rudder a steering gear is required.

4.12. The Steering Engines. The electrohydraulic type of steering gear (Fig. 4.9) is used on all modern naval vessels. The hydraulic power is furnished by a variable stroke hydraulic pump, such as the Hele-Shaw pump or the Waterbury pump driven by a continuously running constant-speed electric motor. There are many variations of the number and arrangement of the hydraulic rams, pumps, and driving motors, and of the method of transmitting the motion of the rams to the rudder cross-head, but the schematic drawing shown in Fig. 4.9 illustrates the essential principles of this type of steering gear. As shown in this plan, movement of the control mechanism *E* in one direction causes the pump *S* to take suction from one pipe leading to it and put pressure on the other. Movement of the control mechanism in the opposite direction causes the direction of pumping to be reversed. When the control mechanism is in the neutral position no fluid is pumped and no pressure is exerted on either line. The rate of pumping in either direction depends on the amount of movement of the control mechanism.

The electrical self-synchronous type of control is used by all modern naval vessels. Briefly, the system consists of a selsyn transmitter controlled by the motion of the steering wheel, suitable electric leads, and a selsyn receiver connected to the control mechanism at the steering gear. Selsyn transmitters and receivers are alternating-current electric motors so designed that the rotor of

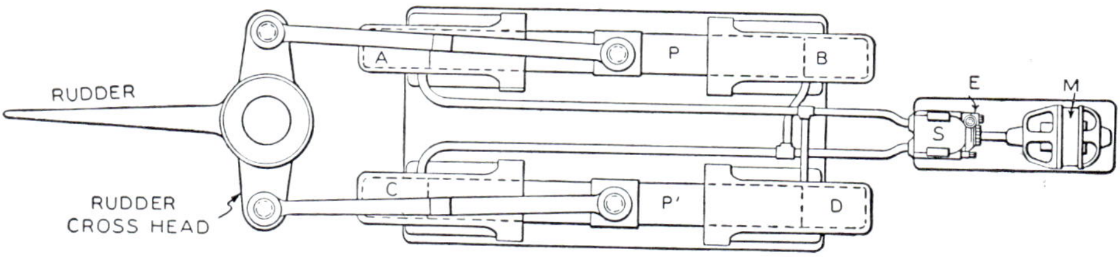

FIG. 4.9 DIAGRAMMATIC PLAN OF ELECTROHYDRAULIC STEERING ENGINE

the receiver follows exactly, in speed and amount of angular displacement, the motion of the transmitter rotor. Motion of the steering wheel, which is carried on an extension of the shaft of the selsyn transmitter rotor, is, therefore, transmitted directly to the control mechanism which acts to cause the steering gear to produce the desired rudder angle.

With all types of steering gear control, provision is made on all naval vessels and on most other ships for quickly shifting control from station to station. The practice of making this shift of steering control part of the daily routine offers valuable training for handling a casualty in any ship control station.

When twin rudders are provided, they both turn the same amount in the same direction if steering control is on the bridge. However, on some ships, when control is shifted to the steering engine room (large ships have separate steering engines for each rudder), each room can independently control the motion of its own rudder. In a ship where this is true, the old adage that a ship has no brake is proved false. While the design was not made expressly to permit this, both the rudders can be turned outward, thus slowing the ship just as landing flaps slow an airplane. Even though it is not recommended that officers of the deck make a practice of slowing down by using the rudders in this manner, and it involves using sound-powered telephones to give orders to the steering engine room, the OOD should know if such a method of slowing is available in his ship for use in unusual situations. For example, a ship may have way on and power to the steering engines but cannot back down.

4.13. New Devices for Steering. There are now being developed and used a variety of new rudder and propeller mechanisms that give certain vessels improved handling qualities. Some new Navy minesweepers have an active rud-

FIG. 4.10 THE DRAVO PIONEER, a new tug fitted with the KORT nozzle or shrouded propeller.

der that has the propeller with its own power source installed with it as an integral unit. Another example (Fig. 4.10) is a tug whose improved steering control, particularly when backing, is due to its Kort nozzle or shrouded propeller.

5

Navigational Instruments

Seamanship has been defined as the art of handling a ship or a boat under all conditions of weather, tide, current, or other influences affecting its immediate movement or safety. It is with this movement and the major navigational instruments that this chapter will deal.

In order to conduct a ship safely to her destination, it is necessary to know her course and speed as accurately as possible at all times. In addition, the depth of water must be known, or at least the fact that it is deep enough for safety. It is often necessary to know the bearing, or direction, and the distance to fixed objects and other vessels sighted. The instruments and appliances used to obtain this information are sometimes called navigational instruments because they are also used by the navigator to fix the geographical position of the ship and to establish a safe course to the desired destination.

5.1. Compasses, General. The course or direction in which the ship is moving is determined primarily by the compass, which may be of either the magnetic or gyroscopic type. Sailing vessels and small craft usually carry only the magnetic compass, but large ships have both.

The earth may be considered to be a huge magnet, the poles of which are slightly offset from the earth's geographic poles. A magnetic needle freely suspended in the magnetic field surrounding the earth would lie with its axis parallel to the lines of force of this field. These lines of force (the horizontal components of which are called magnetic meridians) are not great circles terminating at the magnetic poles. They are irregular because of the uniformly changing strength and direction of the magnetic field at each location. The lines of force are parallel to the earth's surface at the magnetic equator and make an angle with the horizontal at all other places. Because the magnetic and geographic meridians do not coincide, the needle will lie parallel to the lines of force and, over most of the earth's surface, to the east or west of the geographic meridian and at an angle to the horizontal. A magnetic compass is equipped with two or four so-called permanent magnets which are constrained to rotate in a horizontal plane and hence can move only east or west of north. Being influenced by the horizontal component of the earth's magnetic field, they lie parallel to the magnetic meridian at their position on earth.

Variation is the angle between the magnetic meridian and the true meridian. A compass unaffected by any local magnetic force will point to the east or west of true north by an amount equal to this angle. It does not point exactly to the magnetic north pole but only in that general direction. The compass magnets are affected, of course, by a local magnetic force, especially on a steel ship.

The angle between the magnetic meridian and the north to south (0 to 180 degrees) axis of the compass card is the deviation. Compasses are adjusted by inserting other magnets below the compass card and by securing iron spheres or other shapes about the compass card. They are adjusted, i.e., corrected, so that the deviation is neutralized and the compass needle points along the magnetic meridian. Deviation varies with the ship's magnetic heading and with its position on the earth.

The *compass error* is the algebraic sum of the variation and deviation.

5.2. The Navy Standard 7½-Inch Liquid Compass (Fig. 5.1). This is a magnetic compass consisting of a skeleton compass card 7½ inches in diameter

FIG. 5.1 COMPASS CARD AND BOWL. *Official U.S. Navy Photograph*

made of tinned brass, on the underside of which are fixed the magnets. The whole rests on a pivot in a liquid composed of Varsol or of 45 percent alcohol and 55 percent distilled water to resist freezing.

The compass itself is mounted in a stand called the "binnacle" (Fig. 5.2). The binnacle, however, is more than a mere support for the compass. It contains all the various correctors used to adjust the magnetic compass and reduce deviations.

Since the earth's magnetic field is not uniform throughout, and since the ship is seldom on a level plane and the majority of ships are not made of wood or nonmagnetic material, the compass is acted upon by a series of outside magnetic forces which tend to reduce its accuracy. The various correctors contained in or on the compass binnacle act to reduce or neutralize these outside forces.

5.3. Compass Compensation. With the advent of the magnetic mine and torpedo which were attracted or detonated by the ship's magnetic field, a compensating practice known as *degaussing* was introduced.

A series of coils are run both athwartships and fore and aft on a ship at various locations, depending upon her size and type. These coils are then energized, setting up a magnetic field nearly equal to, but directly opposite to, that set up by the ship itself. This counter field in effect neutralizes the ship's own magnetic field to a large degree and thus provides the ship with some degree of immunity to magnetic devices. However, this counter field also affects the ship's magnetic compass. Since it is not practical to adjust the compass each time degaussing is energized or de-energized, another system is used. This system of neutralizing the effects of the magnetic field set up by the degaussing coils in the vicinity of the compass is called *compensating* the compass. Compensating coils (Fig. 5.2), installed around the head of the binnacle, tend to neutralize the counter magnetic field created by the degaussing coils.

FIG. 5.2 COMPASS BINNACLE WITH DEGAUSSING COMPENSATING COILS

5.4. Boxing the Compass. To *box the compass* is to name the points and fractional parts of points from north, around clockwise, back to north. Old style compass cards were graduated only in points and fractional points, and courses and bearings were so designated. The points, half points, and quarter points are listed in Table 5.1 for all four quadrants.

5.5. The Gyrocompass. The gyroscopic compass has two distinct advantages over the magnetic compass: first, it is entirely independent of the earth's magnetic field and all the technical tables, instruments, and measurements associated with it; second, it can be made to indicate the true geographic north instead of the magnetic north. Unfortunately, it is a complicated mechanism, and in case of casualty it requires expert technical attention. Because of this,

TABLE 5.1. BOXING THE COMPASS.

NORTH TO EAST	Points	Angular measure ° ′ ″	EAST TO SOUTH	Points	Angular measure ° ′ ″
North:			East	8	90 00 00
N¼E	¼	2 48 45	E¼S	8¼	92 48 45
N½E	½	5 37 30	E½S	8½	95 37 30
N¾E	¾	8 26 15	E¾S	8¾	98 26 15
N by E	1	11 15 00	E by S	9	101 15 00
N by E¼E	1¼	14 03 45	ESE¾E	9¼	104 03 45
N by E½E	1½	16 52 30	ESE½E	9½	106 52 30
N by E¾E	1¾	19 41 15	ESE¼E	9¾	109 41 15
NNE	2	22 30 00	ESE	10	112 30 00
NNE¼E	2¼	25 18 45	SE by E¾E	10¼	115 18 45
NNE½E	2½	28 07 30	SE by E½E	10½	118 07 30
NNE¾E	2¾	30 56 15	SE by E¼E	10¾	120 56 15
NE by N	3	33 45 00	SE by E	11	123 45 00
NE¾N	3¼	36 33 45	SE¾E	11¼	126 33 45
NE½N	3½	39 22 30	SE½E	11½	129 22 30
NE¼N	3¾	42 11 15	SE¼E	11¾	132 11 15
NE	4	45 00 00	SE	12	135 00 00
NE¼E	4¼	47 48 45	SE¼S	12¼	137 48 45
NE½E	4½	50 37 30	SE½S	12½	140 37 30
NE¾E	4¾	53 26 15	SE¾S	12¾	143 26 15
NE by E	5	56 15 00	SE by S	13	146 15 00
NE by E¼E	5¼	59 03 45	SSE¾E	13¼	149 03 45
NE by E½E	5½	61 52 30	SSE½E	13½	151 52 30
NE by E¾E	5¾	64 41 15	SSE¼E	13¾	154 41 15
ENE	6	67 30 00	SSE	14	157 30 00
ENE¼E	6¼	70 18 45	S by E¾E	14¼	160 18 45
ENE½E	6½	73 07 30	S by E½E	14½	163 07 30
ENE¾E	6¾	75 56 15	S by E¼E	14¾	165 56 15
E by N	7	78 45 00	S by E	15	168 45 00
E¾N	7¼	81 33 45	S¾E	15¼	171 33 45
E½N	7½	84 22 30	S½E	15½	174 22 30
E¼N	7¾	87 11 15	S¼E	15¾	177 11 15

wherever gyrocompasses are used, magnetic compasses are still retained as auxiliaries.

The heart of the gyrocompass is the gyroscope which can be looked upon as a mass concentrated in a plane and spinning about an axis perpendicular to that plane. When placed in gimbal rings, it has three properties; first, three degrees of freedom; second, rigidity of plane; third, precession. The first property refers to the ability of a gyroscope to be rotated about any one of three mutually independent axes independently of the other two. These three axes are termed the spinning, the horizontal, and the vertical axes. When a gyrocompass is operating normally, the spinning axis is parallel to the earth's surface and the meridian; the horizontal is parallel to the earth's surface and the lines of latitude; the vertical points to the zenith. In the gyrocompass, by means of mechanisms controlled by the force of gravity, the freedom of motion about the horizontal axis is almost entirely suppressed in order to keep the spinning axis parallel to the earth's surface and the meridian.

Conversion of Points to Degrees

South to West	Points	Angular measure	West to North	Points	Angular measure
South	16	180 00 00	West	24	270 00 00
S¼W	16¼	182 48 45	W¼N	24¼	272 48 45
S½W	16½	185 37 30	W½N	24½	275 37 30
S¾W	16¾	188 26 15	W¾N	24¾	278 26 15
S by W	17	191 15 00	W by N	25	281 15 00
S by W¼W	17¼	194 03 45	WNW¾W	25¼	284 03 45
S by W½W	17½	196 52 30	WNW½W	25½	286 52 30
S by W¾W	17¾	199 41 15	WNW¼W	25¾	289 41 15
SSW	18	202 30 00	WNW	26	292 30 00
SSW¼W	18¼	205 18 45	NW by W¾W	26¼	295 18 45
SSW½W	18½	208 07 30	NW by W½W	26½	298 07 30
SSW¾W	18¾	210 56 15	NW by W¼W	26¾	300 56 15
SW by S	19	213 45 00	NW by W	27	303 45 00
SW¾S	19¼	216 33 45	NW¾W	27¼	306 33 45
SW½S	19½	219 22 30	NW½W	27½	309 22 30
SW¼S	19¾	222 11 15	NW¼W	27¾	312 11 15
SW	20	225 00 00	NW	28	315 00 00
SW¼W	20¼	227 48 45	NW¼N	28¼	317 48 45
SW½W	20½	230 37 30	NW½N	28½	320 37 30
SW¾W	20¾	233 26 15	NW¾N	28¾	323 26 15
SW by W	21	236 15 00	NW by N	29	326 15 00
SW by W¼W	21¼	239 03 45	NNW¾W	29¼	329 03 45
SW by W½W	21½	241 52 30	NNW½W	29½	331 52 30
SW by W¾W	21¾	244 41 15	NNW¼W	29¾	334 41 15
WSW	22	247 30 00	NNW	30	337 30 00
WSW¼W	22¼	250 18 45	N by W¾W	30¼	340 18 45
WSW½W	22½	253 07 30	N by W½W	30½	343 07 30
WSW¾W	22¾	255 56 15	N by W¼W	30¾	345 56 15
W by S	23	258 45 00	N by W	31	348 45 00
W¾S	23¼	261 33 45	N¾W	31¼	351 33 45
W½S	23½	264 22 30	N½W	31½	354 22 30
W¼S	23¾	267 11 15	N¼W	31¾	357 11 15
			North	32	360 00 00

By systems somewhat similar to those used in gunnery fire control, headings of the master compass are synchronously transmitted to repeaters which, in external appearance, look not unlike the magnetic compass and are placed at all points on the ship where, at some time or other, the knowledge of the ship's heading is likely to be a necessity. On the largest naval ships there may be dozens of these.

The main equipment of a gyrocompass installation consists of the master compass and its supports; batteries in case of failure of ship's supply voltage; one or two small motor-generator sets for supplying other than ship's voltages to the compass; a panel for instruments, switches, rheostats, fuses, and alarm relays; and a panel for fuses and switches for the repeaters. The large naval ships have two master compasses, one forward and one aft, both placed below protected decks; the smaller ships have one (see Fig. 5.3).

The gyrocompass should be started at least 4 hours before the ship gets under way. At the time the compass is started the gyro electrician should be advised as to the ship's head (true) in order that the gyrocompass may be started pointing north. While under way, corrections for latitude and speed must be kept mechanically set on the master compass to the nearest degree or knot permitted by the graduations on the latitude and speed correctors. Correction for course is automatically cared for within the compass itself.

Before getting under way, the repeaters should be adjusted to read the same as the master compass. While under way, repeaters should be checked with the magnetic and the master compasses every half hour and the results recorded in a book provided for this purpose. The officer of the deck should always re-

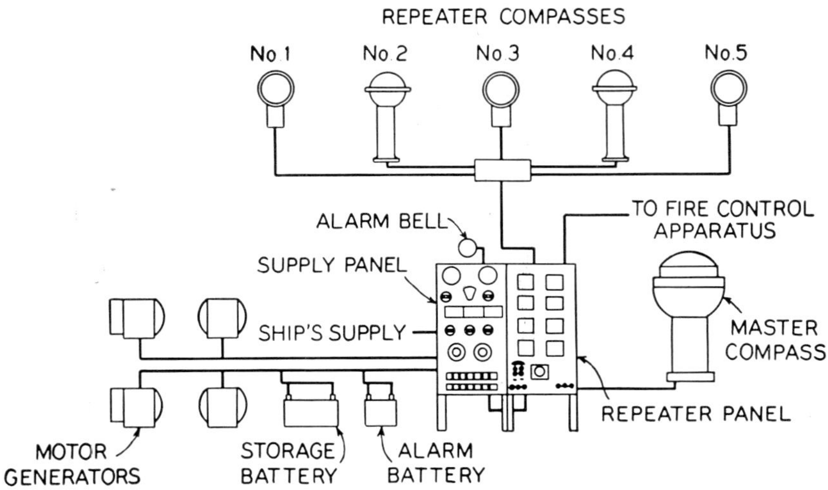

FIG. 5.3 DIAGRAM OF GYROCOMPASS SYSTEM

ceive an immediate report of the results of the compass check and should watch carefully for any indication of possible compass disarrangements. The record of compass checks may prove invaluable in a latitude other than that for which the magnetic compass deviation table was made when there is no opportunity for actual calibration. The series of compass checks in the new latitude may be used to make up an accurate magnetic compass deviation table.

5.6. Automatic Steering Device. One of the gyrocompass remote units is the automatic steering device or "gyropilot," as the Sperry Gyroscope Company unit is called. Many of the larger Navy auxiliaries and merchant ships are equipped with this unit. The entire assembly consists of the gyropilot control unit (Fig. 5.4), the master gyrocompass, the drive motor and various other minor electrical accessories. The drive motor is coupled to the steering engine. It receives indications from the gyropilot and, in turn, operates the rudder through the ship's standard steering engine, bypassing the normal telemotor system.

The gyropilot contains three settings:

(1) Gyro position. In this position the desired course is set remotely and

when the gyro indicates the ship is on the proper heading, the gyro position lever is locked in place. The ship then, in effect, steers itself continually, maintaining the desired course. There is also a rudder and weather adjustment which compensates to a high degree for rough seas and changing rudder angles normally required to keep the ship on course.

(2) Hand position. In this position the ship is steered manually by the wheel on the gyropilot control unit.

(3) Off position. In this position the automatic steering device is disengaged and the ship is steered through her own telemotor system using the normal ship's wheel.

During extended voyages where frequent course changes are not required, the gyropilot has proved more accurate than the steersman. In the event of an emergency where radical course changes are required, the gyropilot may be quickly disengaged and normal hand steering resumed.

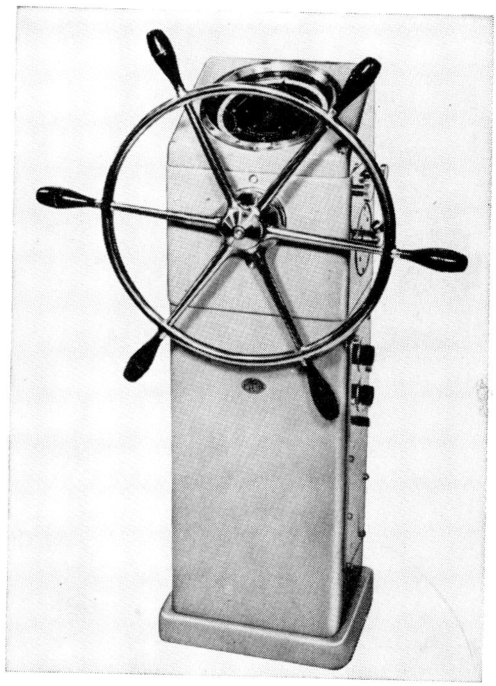

FIG. 5.4 GYROPILOT STEERING STAND

5.7. Dead Reckoning Equipment. The majority of Navy combatant ships and many of the larger auxiliaries are equipped with dead reckoning equipment. This equipment indicates on dials the ship's position in respect to longitude and latitude and also provides a graphic record of the ship's position with respect to a fixed starting point. There are three components to the dead reckoning equipment:

(1) The dead reckoning *analyzer* (DRA) provides north-south and east-west components of the ship's travel. There are counters which indicate how far the ship has run in each one of those directions, and a counter which indicates the total distance the ship has traveled.

(2) The dead reckoning *indicator* (DRI) provides a continuous dial indication of the ship's position in longitude and latitude, and

(3) The dead reckoning *tracer* (DRT) provides a graphic record of the distance and direction traveled by the ship.

The indicator and tracer are normally mounted in a single unit known as the DRT navigational plotting table (Fig. 5.5) and the analyzer is mounted separately on a bulkhead, usually near the plotting table.

FIG. 5.5 DRT NAVIGATIONAL PLOTTING TABLE

5.8. New Developments. The dawn of the ballistic missile era brought with it a need for a more exact method of obtaining navigational positions.

Today such a system of advanced navigation is a reality and is instilled in surface ships and submarines. One of the ships which is evaluating the various components and modifications is the *U.S.S. Compass Island* EAG-153 (Fig. 5.6).

The heart of this system consists of the following three major components:

1. SINS (Ships Inertial Navigation System)
2. NAVDAC (Navigation Data Assimilation Computer)
3. STAR TRACKER

5.9. SINS. This system provides the ship with an accurate and continuous position at all times. It is based on Newton's Laws of Motion, and mechanically senses, by means of ultrasensitive devices which may be called accelerometers, the accelerations (changes of course and changes of speed) which the ship undergoes. These accelerations, when combined with time, can be used to deduce continuously the ships position in latitude and longitude. The SINS, once set up and provided with an appropriate source of power, requires no information from outside the ship. Because the SINS utilize inertial forces, it is sometimes termed an inertial navigator.

5.10. NAVDAC. The computer combines, evaluates, and stores data received from other elements of the SINS system. In effect, it is a memory bank of highly accurate navigational data capable of rejecting navigational solutions of poor quality and accepting only those of a high degree of accuracy.

5.11. Star Tracker. The tracker is an extremely sensitive device with radio or infrared components that calculates elevation (altitude) and azimuth data of celestial bodies. The use of the star tracker is not limited by poor visibility but may be used effectively in the heaviest of overcasts with a high degree of accuracy, since it can detect radio emanations of celestial bodies.

FIG. 5.6 *U.S.S. COMPASS ISLAND EAG-153. Official U.S. Navy Photograph*

Today this new concept of navigation is used in the firing of missiles from ships and submarines and in the underwater navigation of nuclear submarines. The future may well see this system, with its components modified and simplified, installed as the principal system or method of navigation on all types of ships.

The Navy has released some details of a previously classified navigation system that enables submerged submarines to determine their position from the stars. The system is called SCAR—Submarine Celestial Altitude Recorder.

The value of SCAR was demonstrated by the record-breaking submerged voyages of the nuclear-powered submarines *Nautilus, Seawolf,* and *Skate.* Fitted into the periscope, SCAR made it possible for these submarines to use the time-tested system of navigating by the sun and the stars without having to surface.

A celestial fix can be taken with SCAR from periscope depth. When the star, moon, or sun is sighted, a switch on the scope is pressed and the exact altitude of the celestial body is computed automatically, giving the angle of sighting in degrees and minutes as well as the time in hours, minutes, and seconds. This information is printed instantaneously on a slip of paper that looks very much like the bill at a supermarket checkout counter. The timing device used in SCAR is accurate to within a second a day.

After two or more stars have been sighted through the periscope, all the navigation officer has to do is consult the Navy almanac and work out a fix on a chart. The point where the lines of the star fixes intersect indicates the position of the ship.

BEARINGS

5.12. The Azimuth Circle (Fig. 5.7). An azimuth circle is an instrument which is used to take bearings; i.e., to note the directions of either terrestrial or celestial bodies. The azimuth circle is a ring which fits snugly over the top of the compass bowl but can be moved easily around the edge.

On the upper surface of the ring are secured a pair of sight vanes, a set of sun azimuth observing appliances, 2 spirit levels, and 2 lugs. One of the sight vanes consists of a metal leaf with a vertical slit in the center; and the other is a frame which holds a thin wire vertically and has a movable, shiny black leaf outboard of it. One of the vanes is located at each end of a diameter of the ring; and, because the ring rotates concentrically with the compass bowl, this arrangement causes the line of sight to pass through the vertical axis of the compass at all times. The vane which contains the vertical wire has a system of prisms and mirrors beneath it by which the bearing figures by compass are brought into the field of view of the observer. By sighting through the vertical slit and lining it up with the wire and the terrestrial object being observed, a correct compass bearing may be read. Bearings of stars are sometimes taken by sighting across their reflection in the shiny black leaf.

FIG. 5.7 BEARING CIRCLE AND AZIMUTH CIRCLE. *Official U.S. Navy Photograph*

The inner lip of the azimuth circle is graduated counterclockwise in degrees from 0 to 360, thereby permitting the angular offset of the observed object from the ship's centerline, i.e., the relative bearing in degrees of the object, to be read at the lubber's line.

The set of sun azimuth observing appliances consists of a concave mirror which is capable of being moved on its horizontal axis, and a boxlike apparatus opposite it, which contains mirrors and prisms to reflect a ray of light, entering a vertical slit in the face of the box, on to the compass card. One of these observing appliances is secured at each end of another diameter of the ring which is at right angles to the line of sight of the sight vanes, the horizontal axis of the concave mirror being placed at right angles to the diameter on which it is mounted. This set of observing appliances uses the brightness of the sun to take bearings, and these bearings are called *azimuths*. By means of the mirror, a narrow ray of light is reflected into the slit in the face of the box; from there it is reflected down to the compass card as a pencil of light which shines on the compass scale that indicates the direction to the sun. The azimuth is read, then, on the compass card, where this pencil of light shines.

Two split levels are provided because the azimuth circle must be in a horizontal plane before a correct bearing can be taken. The lugs facilitate the moving of the ring.

5.13. The Bearing Circle. The bearing circle is similar to the azimuth circle except that it is not fitted with the set of sun azimuth observing appliances. The sight vane which contains the vertical wire may have an oversized system of prisms and mirrors for greater ease in reading the bearing.

5.14. The Pelorus. A remote device for taking bearings is known as a pelorus. One type consists of a gyro repeater compass on a stand, usually located on the wings of a ship bridge. Above the gyro repeater is a bearing circle, azimuth circle, or other bearing measuring device.

5.15. Bearings. The bearing or direction of an object from the ship may be stated in one of the following ways:

1. Compass bearing. Bearings taken with any compass are compass bearings. Record of such a bearing should include identification of compass used.
2. True bearing. Bearings of an object taken with a compass which is pointing true north will be true bearings. Bearings taken with a gyrocompass are true bearings provided the compass has no error. Bearings taken with a magnetic compass must be corrected for the compass error (variation and deviation) to correct them to true bearings.
3. Relative bearing. The bearing of an object with respect to the fore-and-aft line of the ship is a relative bearing.
   a. In degrees. The relative bearing of an object is expressed in degrees from the ship's head around to the right through 360 degrees.
   b. By "points." There are 32 points in a circle; hence, 1 point equals $11\frac{1}{4}$ degrees.

Radio and Electronic Aids to Navigation

5.16. The Radio Direction Finder. The radio direction finder was first used during World War I to locate enemy radio stations. It has since found extensive use, particularly in marine navigation.

The radio direction finder is a valuable instrument developed for receiving radio signals and at the same time determining the line of direction from which the signals are being received. In other words, it determines the bearing of the origin of the received radio waves. Its operation depends on the directional characteristics of the loop antenna, and it consists essentially of a coil of wire (loop antenna) for receiving the signals. The wire is wound around a rectangular or circular form and the ends of the coil are connected with suitable apparatus for making the signals audible through headphones or for indicating signal intensity by visual means. The coil is mounted on a vertical axis above a horizontal circle graduated in degrees, and a pointer connected with the coil indicates the angle at which the coil stands with reference to the zero line of the circle.

On land the horizontal graduated circle has its zero line fixed at true north; on board ship the zero line is fixed in the ship's head or is replaced by a gyrocompass repeater and possibly an automatic deviation compensating device giving true bearings. Radio signals impinging upon the receiving coil become audible in the headphones of the operator, the intensity of the sound varying with the position of the coil relative to the line of direction of the incoming radio waves. If the plane of the coil is at right angles to this direction, minimum sound is heard; if the plane of the coil is parallel to the line of direction, maximum sound is heard. It is found in practice that more accurate determination of direction is possible by the disappearance of sound or by minimum intensity than by its maximum intensity; and the pointer is accordingly placed to point at right angles to the plane of the coil.

In practical operation, it is found that maximum signal when using the loop and the vertical antenna together is considerably stronger over a fairly wide arc in one half of the circle of rotation than it is in the other half. This tells the observer that the source of signal lies in the half of the circle in which the signal is stronger, and the auxiliary pointer thus indicates the general direction of the source. It is then only necessary to shift over to the loop alone and, utilizing its bilateral characteristics, determine by minimum intensity of sound the accurate single direction or the bearing of the signal source.

There are various ways of applying the radio direction finder to navigational purposes. Installed on board ship, it may be used for taking bearings on any source of signals, as, for example, another ship or a radio beacon. The bearing corrected for the deviation of the radio direction finder for that particular bearing relative to the ship's head may then be used as is any other navigational bearing. Also the radio direction finder may be installed at a shore station to give any ship requesting it, its true bearing from that station. When several such stations are connected together in a group by land wire and centrally controlled, they are able to give a requesting ship its geographical position.

5.17. Measurement of Distance. At some lightships and radio beacon stations on shore, sound signals are synchronized with radio beacon signals. When within audible range of the sound signal, navigators on vessels with radio receivers capable of receiving radio beacon signals may readily determine their approximate distance from the station by observing the time in seconds which elapses between hearing any part of the distinctive group of radio dashes and the corresponding part of a group of sound dashes. The interval should be multiplied by 0.18 or divided by 5.5. The answer is in nautical miles. Since the speed of sound in air is subject to marked variation, and may not follow a direct path from source to ship, distance measured in this way should be used with caution.

5.18. Measurement of Bearing and Distance. Radar (derived from Radio Detection and Ranging) was developed between the two World Wars. It underwent intensive research and development during World War II, and a large number of sets were manufactured and installed on planes and ships. The basic equipment consists of:

1. A transmitter to send out high-frequency (3,000 to 30,000 megacycles) short radio waves.
2. A modulator to send out these waves in short bursts or pulses.
3. A rotating directional antenna which is used for transmitting and receiving.
4. A receiver to amplify the returning echo.
5. An electronic switch to disconnect the receiver from the antenna during transmission.
6. A scope or face of a cathode-ray tube which receives the stream of electrons set up by the returning "echoes."

Two general types of scopes are used: the A scope and the PPI (Plan Position Indicator).

The A scope shows a line with two or more notches in it. The left-hand notch appears when the signal is transmitted; the right-hand notch, when the echo is received. Because radio waves travel at a nearly constant speed, about 186,218 statute miles per second, an unchanging distance scale can be used to measure the distance between the pips or notches. The line itself is drawn during the duration of the transmitted signal.

The PPI scope presents a distorted picture (Fig. 5.8, which was taken from a cruiser entering Tokyo Bay). The antenna rotates and the returning echoes from objects encountered cause a part of the screen to glow. The glow persists. A compass rose gives the bearing of any particular spot, and the range scale is centered at the center of the scope.

The A scope is more accurate than the PPI, but the latter has navigational advantages when used in connection with a chart or in the open sea.

Radar (1) is often more accurate than other methods of navigation, (2) can be used at night or in low visibility, (3) can be used at greater distances from land than some methods of piloting, and (4) is rapid. It must be used with a thorough understanding of its capabilities and limitations, and accuracy varies with the type of gear and the skill of the operator. A radar signal is directional, narrow, and fan-shaped. The image on the scope is wider than reality, which makes the center of the image the correct bearing. Two nearby objects may form one image, and a number of objects along but off shore may appear as a continuous shore line. The length of the pulse or emission affects the depth of the image. The shorter the pulse, the more accurate the depth of the image. Two objects on the same bearing could appear as one, and an object near the

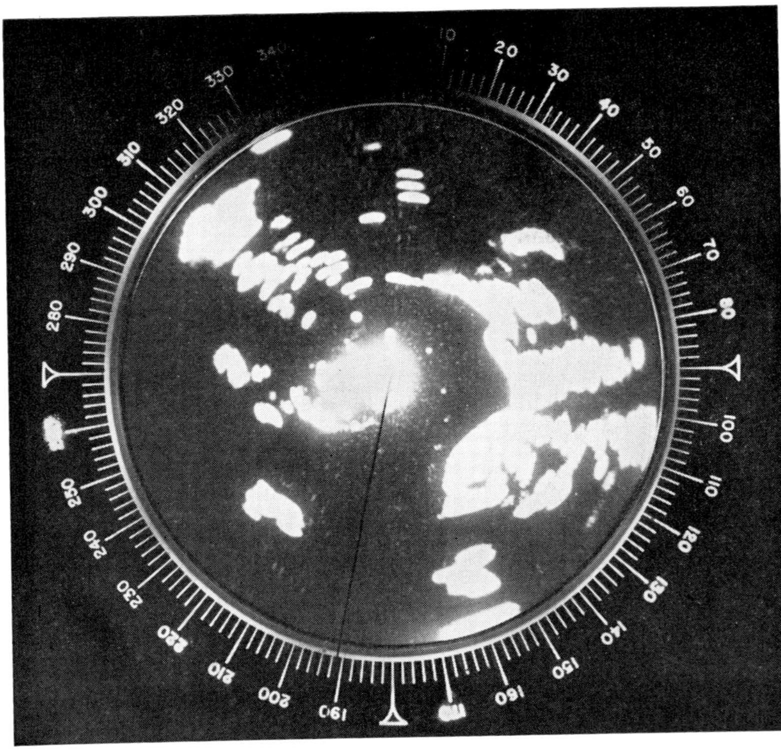

FIG. 5.8 PPI SCOPE. *U.S.S. CUMBERLAND* ENTERING TOKYO BAY. *Official U.S. Navy Photograph*

shore might appear to be a part of the shore. A low beach might not appear, but a high bluff some distance inland would appear as the shore line. High objects may shield and thus eliminate other objects behind them. The minimum range is about 50 yards. The maximum range is limited by the curvature of the earth for small objects and by the line of sight for tall ones.

The VPR scope (Virtual PPI Reflectoscope) shows the usual PPI image, and by mirrors the image is thrown on a chart of the area to the same scale as the PPI scope.

When using the radar for fog navigation, particularly when the vessel is in piloting waters and when she is in company with or in the proximity of other vessels, several people may need to see the bridge PPI scope more or less simultaneously. At night when the hood is not required on the scope, they may all gather around the scope just as a group might watch television, each watching the situation and all being able to maintain a continuous mental status board plot of the overall situation. However, during daylight in a fog the master and mate (Commanding Officer and Officer of the Deck) must each take turns at viewing the scope and each must take time to dark adapt and reorient himself to the current picture.

A radar data computer which would provide automatic evaluation of potential collision courses for radar equipped ships is being developed under the sponsorship of Maritime Administration research.

5.19. Loran (LOng RAnge Navigation) is a cathode-tube, electronic navigation aid which measures the time interval between the reception of synchronized signals from two stations as shown in Fig. 5.9.

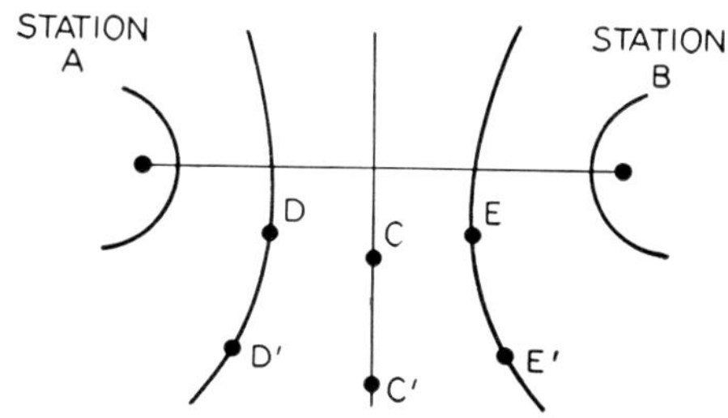

FIG. 5.9

A ship at C on the perpendicular bisector of AB would receive both signals at the same time if sent simultaneously. In like manner, a ship at C' would also receive the two signals at the same time. A ship at D would receive signal A before signal B. The difference in time of receipt at D would be repeated at D'. In fact, a ship anywhere along a hyperbola would have the same time difference as any other ship on the same hyperbola. A number of such hyperbolas could be drawn with foci at A and B. Each curve would have its own unique time difference, if the signals, instead of being simultaneous (two symmetrical curves per time difference), were properly phased in time.

A ship obtains a line of position, therefore, from the signals of two stations (Fig. 5.10). If a fix is desired, at least a second line of position must be obtained from two other loran stations, a celestial sight, or other means.

The loran waves are received by ground wave or by one or more reflections (or refractions) from the ionosphere layers. Ground waves are more accurate but are absorbed more quickly than sky waves. The ship does not transmit when using loran navigation. The service is available 24 hours in the day in any weather, and coverage extends from the United States to Europe, Alaska, and the Far East. But there still remain many parts of the seven seas that are not covered by loran. The system is unusable near the base line extended because it is inaccurate. Consider, for example, how the hyperbolas flare out in this area. The range of any two stations is about 700 miles by day and 1,400 by night.

Special loran charts for plotting loran lines of position are published by the Hydrographic Office. Coordinates of lines of position may also be determined by computation, or by special tables, also published by the Hydrographic Office.

FIG. 5.10 SKETCH OF LORAN RECEPTION. *Official U.S. Navy Photograph*

5.20. The Measurement of Speed. The speed of a ship may be measured in any of a variety of ways, among which are the following:

1. By a log, of which there are many types (Fig. 5.11).
2. By the revolutions of the propellers.
3. By running over a measured course or between bearings of charted points on shore.

Whatever method is used, due allowance must be made for current and wind and the distinction recognized between speed through the water and speed over the ground.

5.21. The Pitometer Log (Fig. 5.12). Several instruments for indicating continuously the speed of a vessel through the water and registering the distance traveled may be encountered. The pitometer log manufactured by the Pitometer Log Corporation is an example of these instruments. The rodmeter transmits the sea-water pressure through piping to a bellows in the rotary distance transmitter. This pressure is balanced by one from an automatically controlled pump. The bellows controls the speed of a motor whose speed is therefore proportional to the ship's speed. Both speed and distance readings are reproduced electrically in different parts of the ship by means of repeaters.

5.22. Revolutions of the Propellers. The revolutions of the propellers afford a very convenient indication of the speed of the ship but are subject to certain corrections, nearly all of which are matters of estimate rather than of exact determination. Revolutions are calibrated with reference to speed under standard conditions: draft and trim normal, the bottom clean, and weather moderate. Under these conditions runs are made at different speeds over a

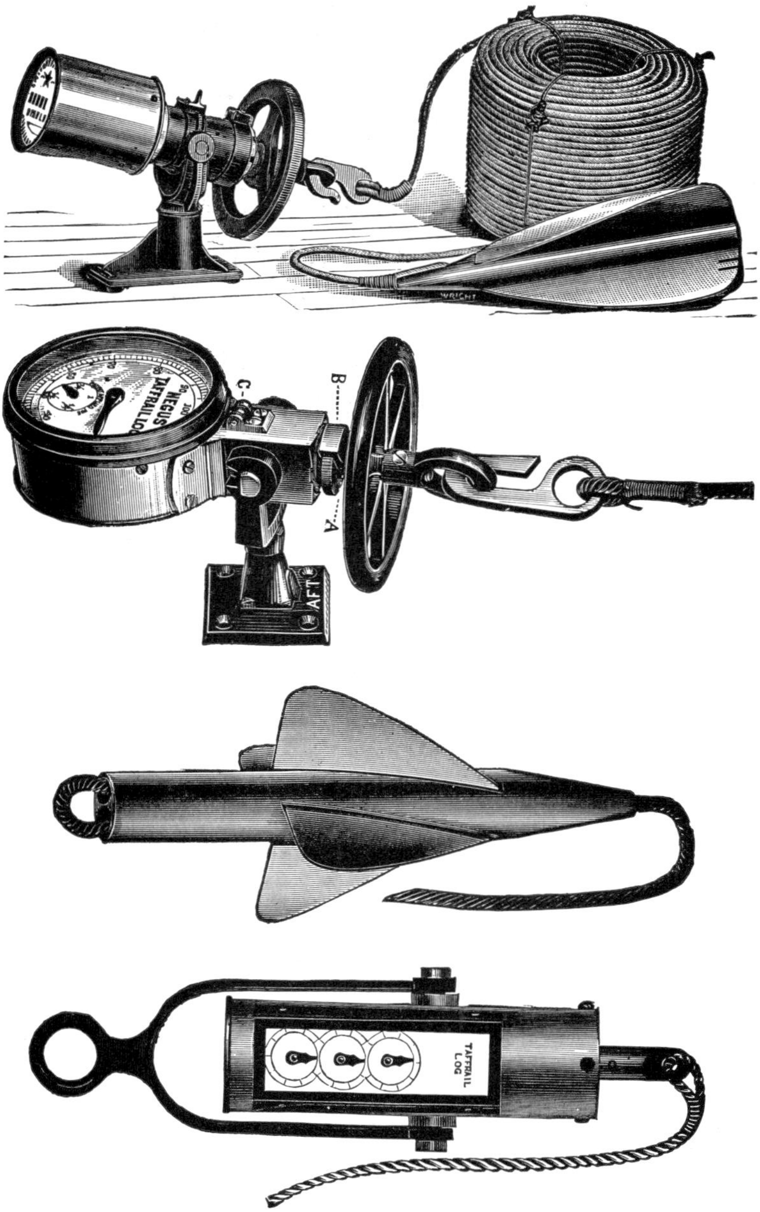

FIG. 5.11 PATENT LOGS. (*Reproduced by courtesy of Messrs. Negus and Sons, and John Bliss and Co.*)

measured course. The average of several runs at each speed in both directions is used to eliminate the effect of wind and current. The averages for the different speeds are then plotted to obtain a calibration curve, and from this curve a table of revolutions required for any speed is made up for the given

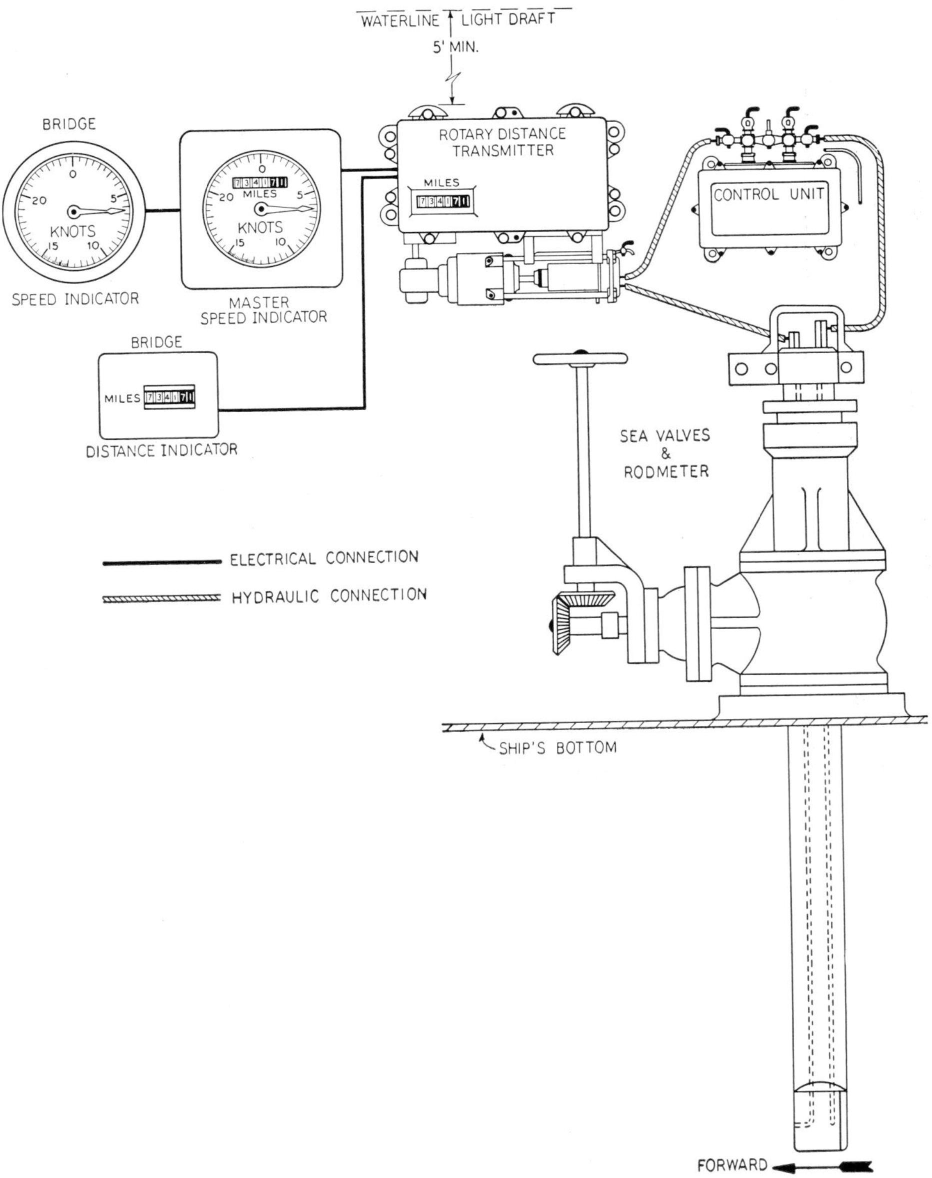

FIG. 5.12 PITOMETER LOG

conditions of the calibration runs. Over a period of time data may be collected for varying conditions of draft, trim, and foulness of bottom.

If the length of time out of dock were an accurate measure of the degree of fouling, the condition of the bottom would always be approximately known, but this is not always the case. This element in the problem must be estimated from what is known of the facts. As experience is gained with any particular type of ship, very close approximation can be made of the correct revolutions to be used for the desired speed. There are numerous types of mechanical and electrical revolution indicators installed, and descriptive information will always be available for the type found on any given ship.

In high-speed ships, the determination of speed by means of revolution of the propellers is generally preferred to other methods, provided that prevailing conditions of weather and currents are taken into consideration.

Equipment to Determine Depth

5.23. The Measurement of Depth. Sounding is the act of determining the depth of the water. Soundings may be taken with the lead, sounding machine, fathometer, or sonic depth finder.

The hand lead (Figs. 5.13 and 5.14). Soundings to ascertain the depth of the water on entering or leaving port or in any case where there is supposed to be less than 15 fathoms of water are taken by the hand lead. Soundings are taken while the vessel has head way on; the leadsman heaves the lead forward and gets the depth as the vessel passes when the line is nearly perpendicular. A good leadsman in sufficiently high chains should get accurate soundings in 7 fathoms at 12 knots, 10 fathoms at 10 knots, and 15 fathoms at 7 knots; in depths greater than this the speed must be reduced or other means used.

The lead line is made up of about 25 fathoms of three-quarter inch braided cotton twine marked in fathoms with a wooden handle or toggle lashed into the line about 2 fathoms from the lead. Braided cotton twine best meets the requirement that the lead line should neither stretch nor shrink excessively as it is alternately wet and dry. The lead is securely attached to the line and weighs either 7 or 14 pounds.

Before marking, the line is well soaked and then measured carefully and marked as follows from the lead:

2 fathoms........	2 strips of leather
3 fathoms........	3 strips of leather
5 fathoms........	white cotton rag
7 fathoms........	red flannel rag
10 fathoms........	piece of leather with a hole in it
13 fathoms........	as at 3
15 fathoms........	as at 5
17 fathoms........	as at 7
20 fathoms........	2 knots
25 fathoms........	1 knot
30 fathoms........	3 knots
35 fathoms........	1 knot

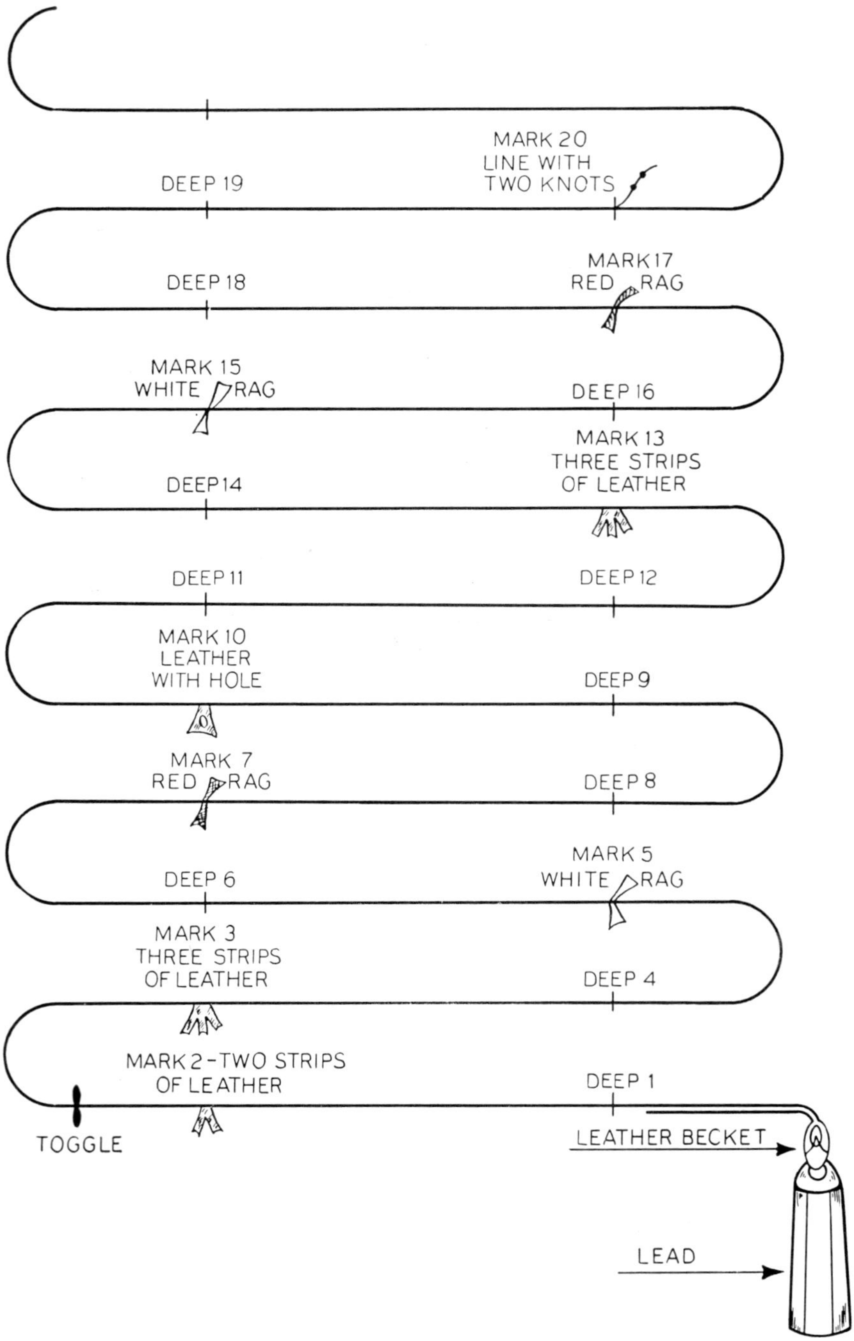

FIG. 5.13 THE HAND LEAD

FIG. 5.14 SAILOR IN CHAINS WITH HAND LEAD. *Official U.S. Navy Photograph*

These are known as "marks," and the numbers omitted are the "deeps." Soundings should be sung out in the exact phraseology indicated here:

Report:	*Depth Corresponding with:*
"By the mark five!"	Five-fathom mark.
"And a quarter five!"	Greater than, but not one-half fathom greater than 5-fathom mark.
"And a half five!"	One-half fathom greater than 5-fathom mark.
"Quarter less five!"	Less than, but not one-half fathom less than 5-fathom mark.
"By the deep four!"	Estimated deep between the marks.
"No bottom at fifteen!"	No bottom at 15-fathom mark.

At night the leadsman should know the distance from his hand to the water's edge and at each cast should subtract this distance from the mark in his hand for the true sounding; or a special lead line may be made up on which the distance from the water's edge to the leadsman's waist is laid off on the line before beginning to measure for the marks. Then the actual depth is indicated by the mark in the leadsman's hand.

Lead lines for use in accurate sounding work, such as hydrographic surveys, are usually marked in feet. The hand lead has been largely replaced by the echo sounder where the area has been accurately surveyed and charted.

The Deep-Sea Lead. The deep-sea lead, pronounced "dipsey," is briefly described, although its purpose is now better served by the sounding machine and sonic depth finder. The "dipsey" is still required in the merchant marine. The lead weighs from 30 to 100 pounds and is attached to a line correspondingly heavy, marked every five fathoms—1 knot at 5, 15, 25, 35, etc., fathoms; 2 knots at 20; 3 knots at 30; 4 knots at 40; etc. The bottom of the lead is hollowed out, and this hollow is filled with tallow before dropping. By thus "arming the lead" a sample of the bottom is brought up for comparison with the kind of bottom indicated on the chart. Knowing the depth and kind of bottom aids a navigator in fixing the ship's position.

In obtaining a sounding, the lead is taken to the forecastle and the line passed aft on the weather side, outboard of everything. Men are stationed at short intervals along the rail, each with a small coil of the line in his hand. The engines are stopped, and, when the ship has nearly lost her way, the forecastle boatswain's mate sings out, "Watch-ho, watch," and heaves the lead over the side, well clear. Each man in succession holds onto the line until he feels the weight of the lead; then he sings out, "Watch-ho, watch," and lets go the bight. The man who obtains bottom while the line is in his hand reads it off as he would a hand lead line. The line is then carried to a small snatch block, hauled in, and the character of the bottom recorded.

5.24. The Sounding Machine. The original sounding machine was invented years ago by the great Scottish scientist Lord Kelvin (then Sir William Thomson), and all later machines are based upon the principle which he first applied: the principle of determining the depth of water by the pressure of the water at the depth in question, this pressure being registered on the inside of a chemically sensitive or frosted-glass tube open at one end and closed at the other. The tube with closed end up is lowered in the water. The pressure of the water acts upon the air in the tube to a height which depends upon the pressure and therefore upon the depth. How high the water rose in the tube is indicated by the height of color of its sensitive interior, or by the level of clear (wet) frosted glass.

5.25. Continuous Soundings. There are several instruments that may be installed for the purpose of taking continuous soundings and for keeping a record of the depths encountered in the track over which the ship has passed. The best known of these instruments are the fathometer and the sonic depth finder.

The *fathometer* is operated on the principle of echo depth finding. This consists of producing a sound of such intensity that it will travel to the ocean floor and be reflected back. With the speed of sound known, the time required for this sound to reach the bottom and return is used to measure the depth of water below the ship's keel at any instant.

The Submarine Signal Company's fathometer is often found in U.S. men-of-war and is designed for continuous soundings. The latest oscillator and receiver are both located in a single unit which can be lowered or raised through the bottom of the ship. In older installations, two water-filled tanks are mounted against the skin of the ship inside the hull near the keel. One tank contains the sound producer, an electromagnetic submarine oscillator in a watertight case. The other tank contains the hydrophone echo receiver. The echo received in the hydrophones is amplified and transmitted to the depth indicator mounted in some convenient location near the bridge. The depth of the water is shown on a dial graduated in fathoms by means of a beam of red light flashing about twenty-four times a minute. This is the rate at which the oscillator sends out its signals. A recorder is connected mechanically to the fathometer indicator. It records the soundings taken as a solid line on a moving strip of paper graduated for depths and corresponding times of soundings.

The *sonic depth finder* has an indicating and recording apparatus very similar to that of the fathometer. With this instrument depth is measured by the time required for a supersonic signal to echo from the bottom. It will measure great depths quite accurately and has been found very valuable in running lines of deep-sea soundings when kept in continuous operation during long voyages. By this means, several unsuspected deeps and many other previously unknown features of the ocean's floor have been discovered.

Caution should be exercised in relying on the indications of echo sounding instruments in shallow water, and their readings should always be checked by the use of the hand lead. The reason for this is that the results given by an echo from a soft or muddy bottom may vary from the true depth by an amount which may be critical in these circumstances. Echo sounders used in surveying are calibrated at frequent intervals against a hydrographic lead line.

5.26. The Drift Lead. The drift lead should be put over the side when there is any possibility of the ship's dragging anchor because of a swift tideway, bad weather, or the poor condition of the bottom for holding an anchor. A drift lead consists of a heavy lead, 30 to 50 pounds, secured to the end of a line. Enough hand leads to make up the necessary weight may be used if a heavy lead is not available. The drift lead line is not marked.

The best location for the drift lead is near the bow where it will be less affected by swinging and yawing. After the lead has been placed on the bottom close to the side of the ship, the line should be secured to the ship at a point directly over the lead but with a fathom or so of slack in the line. With this method a tautening of the line will give a visible indication that the ship may be dragging. The drift lead should be taken up before a swing and replaced immediately after the swing. A ship is particularly likely to drag as the current starts to run after a swing because the swing may loosen the anchor, and the growing current then breaks it out.

When there is any likelihood of dragging, the riding chain should also be observed for the characteristic jumps caused by a dragging anchor, and in addition a range should be watched. A range should always be checked from the same point on the ship and consists of any two well-defined objects on shore that may be in line with the observer. With two ranges, one near either beam, the difference between yawing or swinging and actual dragging is easily distinguished. This is not necessarily true of the drift lead, although the drift lead is a most useful additional precaution. An accurate bearing or bearings may be used in the same manner as ranges; and of course *fixes,* if available, may be used to detect a ship's movement.

6

Boats

The term boats, as used in this chapter, is defined as: small craft which can be hoisted aboard a large vessel. This definition is a rule of thumb and is considered accurate in only a restricted sense. Actually a large vessel carrying small craft can in turn be carried by an even larger vessel, as is the case in several types of landing ships. Another general rule states that the term boats refers to small vessels that are restricted in use by their size, and that are unsuited to make regular independent voyages of any length in unsheltered waters.

6.1. Uses of Boats. Ship's boats are used to carry out ship's business, such as saving life, making landings in water too shallow for the ship, transporting passengers and stores between ship and shore when in harbor, or transporting troops and equipment between ship and beach during the assault on an unfriendly shore line in time of war. The number and the type of boats carried vary with the type of ship and the uses for which the boats are intended. Before discussing the various types of boats in use, it will be best to describe briefly boat and sail nomenclature, as well as boat construction.

6.2. Construction of Wooden Boats. Boats are built of either wood, metal, or plastic. The majority of the boats presently used by the United States Merchant Marine are made of metal, and those used by the Navy are of wood and plastic. Larger types of landing craft, however, are made of metal. The majority of boats used by the Coast Guard are made of wood.

There are four types of wooden boat construction:

Carvel (Fig. 6.1A). In carvel building, the planks lie alongside one another without overlapping, and the seams are calked. Where the construction is too tight to admit calking, a narrow batten or ribband is run along the seams inside. The calking in this case is limited to the garboard seams and to the ends of the planks.

Where heavy boats are built on this system, a second layer of planking is sometimes used, and the two layers are separated by canvas. Seams of inner planking are staggered from those of the outer planking. Frequently on the larger-sized boats additional planking called "ceiling" is installed inside the frames.

Clinker (Fig. 6.1B). In clinker building, the planks overlap at their edges like the clapboarding on a house and are fastened to one another as well as to the frames. As the planks thus support each other, this system has greater strength for a given weight than the carvel building, and the frames can be placed farther apart. On the other hand, the planks are likely to split along the line of fastening, and repairs are made with some difficulty because of these seam fastenings and added care required in fitting new planks. The seams are not calked. The swelling of the planks causes them to bind tightly upon one another. To keep them tight the boat should be put in the water frequently or,

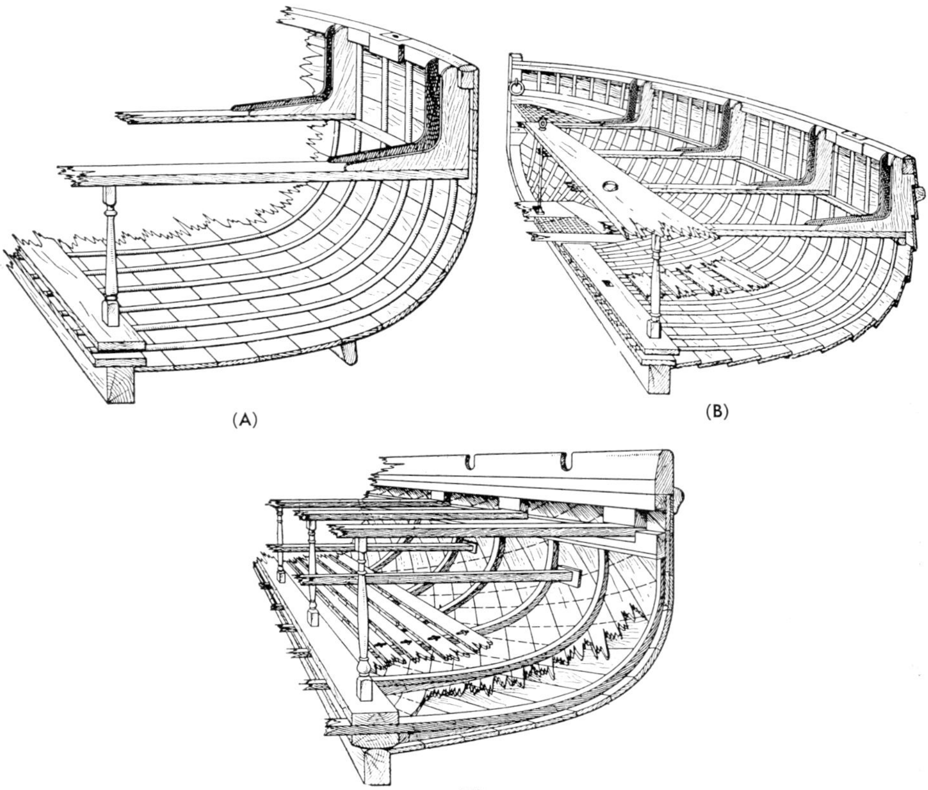

(A) (B) (C)

FIG. 6.1 (A) CARVEL. (B) CLINKER. (C) DIAGONAL CONSTRUCTION

if that is impractical, well wetted with a hose from time to time. The same care must be taken of tightly built carvel boats.

Diagonal (Fig. 6.1C). In diagonal building, the planks run diagonally at an angle of 45 degrees from the keel to the gunwale, and two thicknesses of planking are used at right angles to each other. This offers a very strong type of construction but a more costly one. As a rule it is used only for large boats which normally carry heavy weights.

The carvel and diagonal systems are sometimes combined, two layers of planking being used, one carvel laid, the other diagonal.

Plywood. Marine plywood has become an important material for boat construction and is used in several types of smaller landing craft. Sheet plywood is used for boats of straight or nearly straight section, affording an economical boat requiring a minimum of maintenance. Some boats are built of molded plywood in which thin strips, impregnated with phenol glue, are diagonally bent over a mandrel or mold to the shape of the hull desired, building up a skin of several layers thickness. The mold is then placed in an autoclave where, at increased temperatures, the strips are curved and formed into a one-piece hull that possesses many of the characteristics of plastic construction.

6.3. Metal and Plastic Boats. The plating of metal boats is of galvanized sheet steel or aluminum. The gunwales are sometimes of wood, but more often they are steel angle bars or aluminum. The keel, stem, and sternpost are usually of steel.

Special precautions are taken in the design to make all parts accessible for inspection, drying, cleaning, and painting in order to prevent damage from corrosion. The danger of such damage is the principal objection against this kind of boat. Another objection to this type of construction is that, unless buoyancy tanks are provided, the boat will sink if the hull is punctured or if she swamps or capsizes. By marked contrast, a wooden boat under these conditions will not sink and will provide a limited degree of support for her crew and passengers.

Plastic boats are of a relatively new type of construction. Currently the Navy uses five designs, and a number of other designs are in the experimental stage. The hulls of all designs, in service and experimental, are made as a single piece from fibrous, glass-reinforced plastic resin.

The key to the construction is in the master mold. Layers of glass cloth or mat are laid in the mold. Each consecutive layer is impregnated with liquid resin and the completed hull cured before removal from the mold.

Plastic boats do not have seams, and those of 26 feet or less generally do not require ribs. The high resilience inherent in the plastic shell eliminates the need for ribs. Longitudinal strength is provided by plastic members built up inside the shell.

The construction cost of plastic boats is more than that of wooden boats of equal size but the upkeep is far less; they do not deteriorate from dry rot while in stowage; they are not subject to rot while in service; water absorption is negligible; plastic has greater impact resistance due to its resilience; and repair is considerably easier, cheaper, and faster than wooden boats. They are impervious to teredos (sea worms), and barnacles (marine shell animals) are easier to remove than from wood. Although plastic boats are somewhat lighter than wooden boats of the same size, their handling characteristics are the same.

6.4. Boat Nomenclature. In order to learn the fundamentals of boat operation and to understand more readily boats and their differences, it is first necessary to learn the words used in connection with boats. See Fig. 6.2.

6.5. Sail. The use of sail is fast disappearing today except in pleasure craft and a few coastwise lumber and fishing vessels.

Figure 6.3 shows the names of the parts of the four-sided fore-and-aft *lug* sail used in the 26-foot Navy whaleboat and in most merchant lifeboats. The

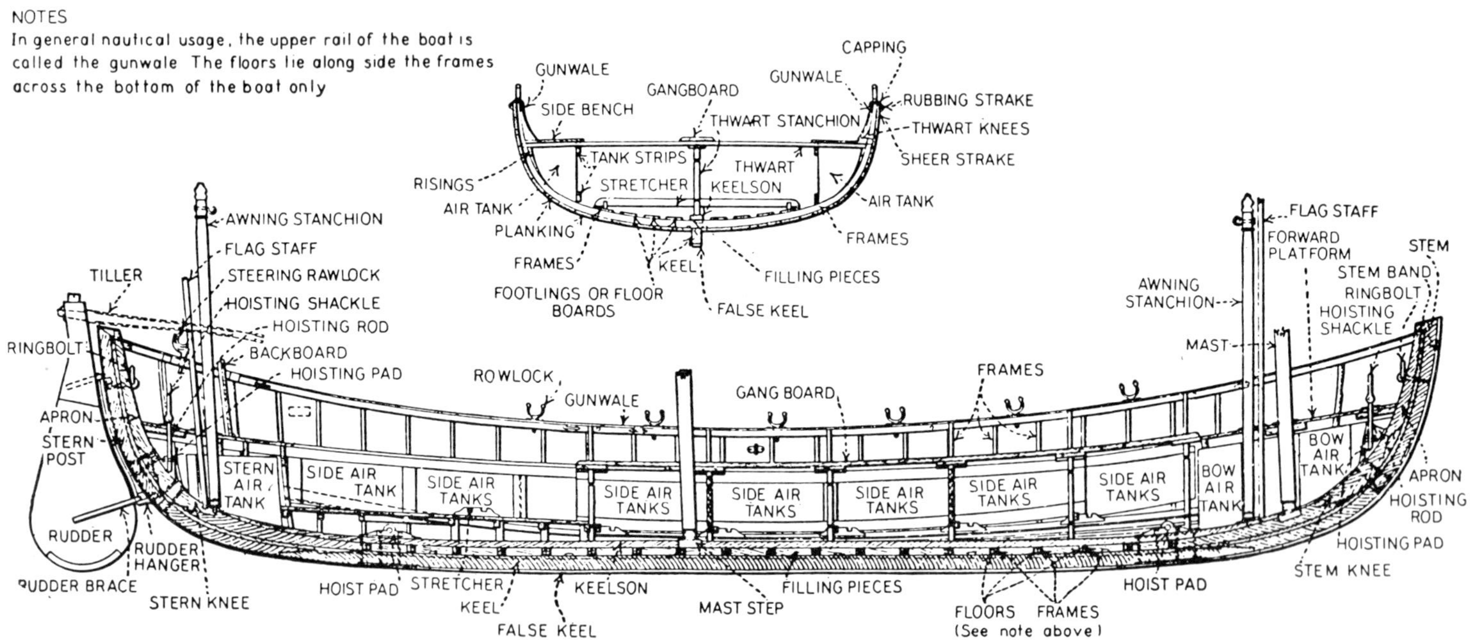

FIG. 6.2 PARTS OF A BOAT

only difference in nomenclature between this and a three-sided sail, which rises to a sharp point, is that the head of the latter is the sharp upper corner and a three-sided sail has no throat.

Sails are reinforced at the corners with extra patches and are strengthened by *bolt ropes* sewed to their edges all the way around. Part way up a large sail there is usually a long patch called a *reef band* attached to either side. Each reef band carries a row of short lines called *reef points,* which are brought under the sail and bent together to *shorten* or *reef* the sail. A sail is reefed when

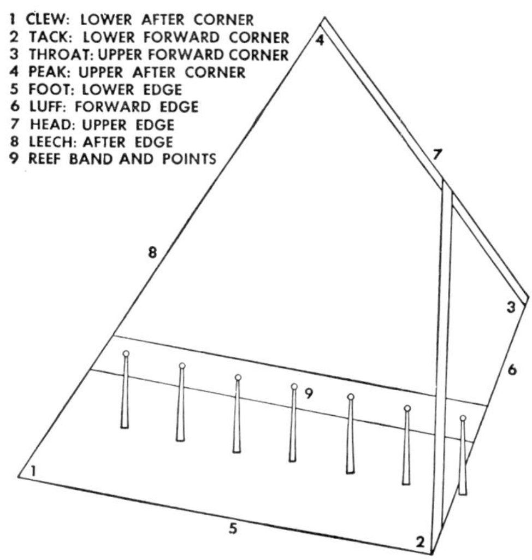

FIG. 6.3 NOMENCLATURE OF THE FORE-AND-AFT LUG SAIL

the wind is too strong for its full area. If there are two sets of reef bands on a sail, reefing by the upper one is called *close* or *double* reefing. Sails are named for the masts on which they are, such as *foresail, mainsail,* etc. Sails forward of the foremast are *headsails* or *jibs.*

The mast or masts which support fore-and-aft sails are a permanent feature in large sailing vessels. Navy boats and merchant lifeboats are provided with means for setting up or removing their masts, and the process is called *stepping* or *unstepping* the mast. The foot of the mast is set in a socket in the keelson called a *step* and is held there by a *kingpin.* The mast is steadied upright by a hinged collar on one of the thwarts. The part of the collar which opens on the hinge is called a *gate.* Fore-and-aft support of a mast is by lines or wires called *forestays* and *backstays.* Athwartship support is afforded by *shrouds.*

The foot of a fore-and-aft sail may be bent to a boom, or the sail may be free-footed (without a boom). The line or tackle which controls the thwartships motion of the boom of a sail, or of the foot of a free-footed sail, is called a *sheet.* Sheets are named for the sail they control, as foresheet, mainsheet, jib-sheet, etc.

Sails are hoisted by *halyards,* which are named for sail and the part of the sail to which they are attached, as *foresail throat halyards, mainsail peak halyards,* etc. Halyards which hoist a jib are *jib halyards.* Jibs are triangle *headsails,* bent to foremast fore stays and running from the upper part of the mast to the bow, or to a spar attached to the bow called the *bowsprit.*

6.6. Buoyancy of Boats. Boats, whether of wood, metal, or plastic, usually are fitted with air tanks for reserve buoyancy. Such tanks are made of copper or galvanized sheet steel. Some boats, such as those used by the Coast Guard, obtain their reserve buoyancy through fitted-in, waterproof balsa or styrene-plastic buoyancy blocks.

In a self-bailing boat a watertight deck runs throughout the length of the boat slightly above the load water line. The space below is thus converted into a watertight compartment. Water from this deck runs overboard through freeing ports or scuppers in the side of the boat. Any water which may enter the compartments below the deck from leaks in the side or bottom of the boat is pumped out by a bilge pump and discharged onto the deck where it flows off through the freeing ports or scuppers.

If the buoyancy of a boat can be carried high, the boat will tend to right itself if it is capsized. This tendency will increase if the keel is made as heavy as practicable. Self-righting boats are accordingly built in such a shape as to permit carrying the bow and the stern air tanks very high and are fitted with a heavy keel of iron or lead. The self-bailing and self-righting features are sometimes combined, as in the large motor lifeboats used by the Coast Guard.

6.7. Boats Under Oars. Pulling boats are no longer assigned to naval vessels for general service; therefore, the occasion for commanding a Navy boat

FIG. 6.4 TEN OAR DOUBLE-BANKED WHALEBOAT

under oars seldom occurs. However the basic boat used by the Navy, Coast Guard, and Merchant Marine in the training of their seamen recruits and their officers as well in the rudiments of boat handling is the whaleboat.

The pulling whaleboat is the light, fast, seagoing boat developed by whalers for riding out a "Nantucket sleighride," that is, being towed by a harpooned whale. She is double ended (her bow and stern have the same shape). Her lines curve from a low freeboard to a relatively high bow and stern, and she is capable of riding out some of the roughest seas if properly handled. She may be pulled with either 5 or 6 oars single-banked, or 12 oars double-banked. She may also be sailed by use of a standing lug rig or other type of sail dependent upon her size. Although this boat is not common today, it is still used by the Coast Guard for types of rescue operations, and her steel counterparts are used on merchant ships as lifeboats. Figure 6.4 shows a 10-oar double-banked whaleboat underway.

Boats of the United States Navy

A large part of the Navy's work is carried out by means of its small boats. Their design is determined largely by the following considerations; mission or purpose of the boat, seaworthiness, strength, cargo or personnel capacity, speed, minimum maintenance, and ease in stowage on board ship when required. Wooden boats are the most common today, but more and more of the newer replacements and specialized types are now being made of metal or plastic.

6.8. Label Plate. Every boat in the naval service is fitted with a label plate which provides data concerning its design, manufacture, and maximum capacity. The latter is calculated in terms of carrying capacity of men, approximately 10 cubic feet of internal volume per man, based on an average weight per man of 165 pounds, fully clothed and wearing a life jacket.

The data on a sample label plate follows.

26-Foot Motor Whaleboat
Boat registry No. 13681; maximum capacity, 22 men. (Maximum capacity includes boat crew and passengers and assumes all passengers in cockpits and seated in so far as possible.)
Naval Shipyard, Norfolk, January 1959, Buships Plan No. 248 628

In addition to appearing on the label plate, the registry number of a standard boat is cut in the keelson immediately below and at the after end of the stern sheet grating. If there is no keelson, the number is cut into the covering board on top of the keel. In the case of motorboats, the registry number is cut into the covering board or into the sternpost extension.

6.9. Boats. *Dinghies* are small boats of about 9 feet overall length, carried aboard larger craft, such as aircraft rescue boats, landing craft, and ships, to provide transportation for the crew and as a general tender. Equipped with sail, a dinghy is also used for recreational purposes. It is normally equipped with a pair of oars.

Punts are general-purpose workboats, square ended, and usually used for work along the ship's water line. They are equipped with a pair of oars, although usually they are propelled by sculling. Three sizes of punts are used in the Navy: 10, 12, and 14 feet long.

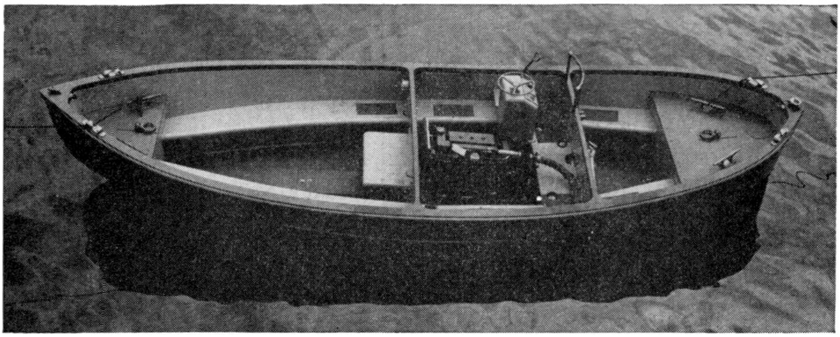

FIG. 6.5 26-FOOT MOTOR WHALEBOAT MK 5 (PLASTIC)

Wherries are larger versions of the dinghies used for the same general purpose. They are normally propelled by two pairs of oars.

Motor whaleboat (Fig. 6.5) is built along the same lines as the pulling whaleboat. It is, however, necessarily heavier, and the weight of the engine makes it considerably less seaworthy. The motor whaleboat, originally made of wood,

FIG. 6.6 28-FOOT PERSONNEL BOAT MK 2

is now made of plastic and differs from its predecessor in that it is of greater beam and is equipped with a wheel and rudder in lieu of a tiller. It is used to transport personnel and as a lifeboat.

Personnel boats (Fig. 6.6) are built along the lines of commercial cabin cruisers. They are used to transport enlisted personnel and officers. These boats are replacing the motorboats presently being used as officer personnel carriers. They range in size from 28 to 40 feet.

Motorboats (Fig. 6.7) are used to transport officer personnel. Motorboats used by commanding officers are known as *"gigs"* and those by flag officers as *"barges."*

FIG. 6.7 35-FOOT MOTORBOAT (ADMIRAL'S BARGE)

Utility boats (Fig. 6.8) are used to transport personnel or cargo. In the past they were assigned primarily to boat pools, but they are now replacing the motor launches aboard ship.

FIG. 6.8 40-FOOT UTILITY BOAT

15 Person CO_2 inflatable lifeboat—MK III (Fig. 6.9) are carried aboard most Navy ships. They have replaced the various types of life rafts which were carried in the past. The inflatable lifeboat, which when inflated is 15 feet, 8

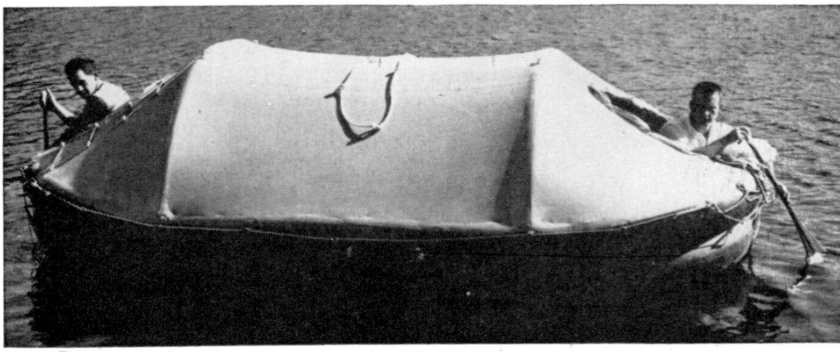

FIG. 6.9 INFLATABLE LIFEBOAT MK III

FIG. 6.10 36-FOOT LANDING CRAFT LCP(L) MK 4 (STEEL)

inches long with a beam of 7 feet, 4 inches, is carried deflated in specially constructed racks. Its size before inflating is 5 feet, 0 inches long; 2 feet, 2 inches wide; and 2 feet, 9 inches high.

FIG. 6.11 LANDING CRAFT MECHANIZED (LCM 6)

6.10. Amphibious Craft. The Landing Craft Vehicle or Personnel (LCVP) is the most common of all the landing craft. Many thousands of them still exist, although they are being phased out for use in amphibious operations. Today they are being used primarily as utility boats, and large numbers are assigned to boat pools and carried by ships.

FIG. 6.12 LANDING VEHICLE, TRACKED, HOWITZER LVT(H)-6

The *Landing Craft* (LCPL) (Fig. 6.10) is one of the most important of the amphibious craft. It is used for the control of other landing craft in an amphibious landing and is equipped with communication facilities and radar.

The *Landing Craft Mechanized* (LCM 8) is the largest boat used in the Navy. It is similar in design to the LCM(6) (Fig. 6.11) but has an overall

Table 6.1. Data on Navy Boats

Type of Boat	Hoist WT of boat fully equipped exclusive of crew (pounds)	Maximum displacement of boat fully equipped inclusive of fuel, crew, and full load (pounds)	Maximum capacity (number of men), including crew	Cruising range at full power in nautical miles	Speed at full load
A. Standard Boats:					
9′ Dinghy w/sail	160	N.A.	2	N.A.	N.A.
12′ Punt	685	N.A.	N.A.	N.A.	N.A.
16′ Wherry	400	N.A.	10	N.A.	N.A.
22′ Utility boat (plastic)	5,500	7,900	16	67	8
26′ Utility boat	7,300	11,425	25	100	9
26′ Motor whaleboat	5,550 *	9,180	22	110	7
26′ Motor whaleboat (plastic) MK 5	5,326	8,626	22	110	7
26′ Motor launch	7,300	11,425	25	180	6.5
28′ Personnel boat MK 2	9,300	12,375	22	120	16
28′ Personnel boat MK 4	13,500	16,800	22	95	15
28′ Personnel boat MK 5	13,000	16,300	10	95	15
28′ Personnel boat MK 6	11,700	14,700	20	95	15
30′ Motor launch	8,900	15,500	40	160	6
33′ Utility boat (wood or plastic)	10,960	17,460	45	140	10
35′ Motorboat	13,500	17,955	27	115	10
36′ Motor launch	11,750	23,960	70	290	6.5
40′ Utility boat (wood or plastic)	17,550	27,900	75	110	11
40′ Motor launch	16,300	31,150	90	150	8
40′ Motorboat	14,700	20,805	37	115	10
40′ Personnel boat MK 1	16,500	23,595	43	210	12
40′ Personnel boat MK 2	16,500	23,595	43	210	12
40′ Rescue boat MK 1	24,500	25,985	13	140	20
45′ Rescue boat	30,000	32,000	14	125	33
50′ Utility boat (wood or plastic)	23,300	48,050	150	150	11
50′ Motor launch	23,000	49,160	175	150	8
63′ Rescue boat MK 3	62,800	64,000	14	450	28
7 Person CO2 inflatable lifeboat	120 (stowed)	223	7	N.A.	N.A.
15 Person CO2 inflatable lifeboat MK III	267 (stowed)	370	15	N.A.	N.A.
B. Special Boats:					
17′ Line handling boat	2,900	3,890	6	70	11
24′ Plane personnel boat	6,100	7,900	14	110	13
27′ Motor work boat	6,500	9,776	26	127	8
33′ Plane rearming boat	10,000	18,600	4	55	10
35′ Plane rearming boat	10,000	18,660	4	70	10
42′ Aircraft refueling boat	57,552	57,552	21,025 †	160	6
45′ Picket boat MK 5	33,200	39,320	5	200	13
52′ Sounding boat	67,200	73,748	6,448 ‡	600	10.5
72′ Torpedo retriever	N.A.	106,814	24,000 ‡	180	18.5
73′ Noise measuring boat	N.A.	111,843	N.A.	220	17.5
C. Landing Craft:					
35′ LCVP (obsolete) §	18,500	26,600	36	110	9
36′ LCP(L) MK 4	18,500	22,500	8	100	15
56′ LCM(6)	56,000	124,000	80	130	9.5
73′ LCM(8)	134,000	254,000	120,000 ‡	190	9

* 6,450 pounds with 8 man lifesaving crew.
† Fuel.
‡ Pounds.
§ There are several thousand LCVP's still operational; however, they no longer meet the requirements of amphibious operations and are generally regulated to utility craft duties.

length of 73 feet. It is anticipated that this craft, although smaller than the Landing Craft Utility (LCU), will eventually replace it as the carrier of tanks and heavy equipment. It has a capacity of 120,000 pounds.

The *Landing Vehicle, Tracked* (Personnel and Howitzer), LC T(P) and LV T(H)-6, are armored amphibious assault vehicles. They are normally carried, combat loaded, in the well deck of an LSD or in the tank deck of an LST, ready for discharge well out from the beach. The LVT(H)-6 (Fig. 6.12) equipped with an 155-mm howitzer will give close gunfire support to assault troops until conventional tanks and artillery can be brought ashore. The LVT(P)-5 lands the troops she carries through a bow ramp. Since both of these vehicles are armored and amphibious, they can carry troops ashore protected from small-arms fire and discharge them only in areas which provide adequate cover.

6.11. Weights and Capacity of Navy Boats. Table 6.1 gives the approximate scale weights of standard, special boats and small landing craft in pounds and the maximum carrying capacity including crew (based on 165 pounds per man).

The hoist weights in Table 6.1 are based on either the reported scale weights in the case of the smaller boats or on detailed weight calculations in the case of the longer boats.

Frequent repainting, repairs, alterations, and water absorption in wooden boats result in increases in boat weights of as much as 10 percent or more.

In designing booms, cranes, etc., for single-point suspension, the design working loads should be used on the weights given in the table plus an added 10 percent.

In designing davits for two-point suspension, the design working loads should be based on the weight of the heavier end of the boat appearing on the sling plan for the particular boat under consideration plus 10 percent.

When carrying liberty parties, the designated carrying capacity should never be exceeded; in carrying stores the load in pounds, including men and stores, should never exceed the maximum allowable cargo load. In motorboats, the practice of carrying passengers, stores, or baggage on the top sides should be prohibited. When it is necessary to carry stores or baggage, a corresponding reduction in maximum number of passengers should be made.

In connection with the rated capacity designated on the label plate, it should be borne in mind that this represents the maximum capacity under normal weather conditions in enclosed waters. Reduction is always necessary under extreme weather conditions or in the open sea. Frequently, conditions will be such as to reduce greatly this rated capacity.

Except for small landing craft which are not fitted for stowage of life jackets for passengers, life jacket allowance in boats is established as 50 percent of maximum boat capacity. One life jacket for each member of the crew must be aboard before embarking. If conditions warrant, i.e., rough seas, extra life jackets may be loaded aboard before the boat leaves the ship on her next run and all passengers embarking may be issued life jackets before the boat departs.

6.12. Blocks and Falls. Table 6.2, a table of the sizes of blocks and manila falls for boat davits, is a general guide. Boats which do not appear on this list

TABLE 6.2

Boat	Block	Falls
12-foot wherry	4-inch double	$1\frac{3}{4}$-inch circ.
20-foot dinghy	4-inch double	$1\frac{3}{4}$-inch
24-foot whaleboat	8-inch double	3-inch
30-foot whaleboat	12-inch double	4-inch
35-foot motorboat	14-inch treble	5-inch
36-foot motorlaunch	14-inch treble	5-inch

are those which, when ship-borne, are hoisted with slings only, except for 40-foot motor launches and landing craft. The latter are frequently handled on gravity davits utilizing wire rope falls.

The block assignment in the table contemplates the practice of fitting metal upper fall blocks at the davit heads. If an arrangement of sheaves in the head of the davit is provided in lieu of the upper falls block, the number of sheaves should be equal to the number of sheaves required on the block specified in the table.

The lower falls blocks should have wood cheeks, and they should be fitted with standard disk-bearing hooks suitable for the block assigned by the table.

6.13. Releasing Gear. The standard boat-releasing gear used by the Navy (Fig. 6.13) is known commercially as the Raymond releasing hook. This gear consists of a disk-bearing swivel hook with a tripping device hinged at the bill of the hook. The tripper is so weighted at its outer end that when the boat is water-borne and the load is removed from the hook, it automatically tumbles and throws the boat shackle out of the hook and releases the boat. When the boat is not water-borne, the load on the hook prevents the tumbling of the tripper. To facilitate hooking on prior to lifting the boat, the weighted end of the hook is fitted with a lanyard which is passed through the shackle and held taut in order to prevent tumbling of the tripper.

BOATS OF THE UNITED STATES COAST GUARD

The boats described below are used at the Life Boat Stations of the Coast Guard along both coasts, the Inland Waterways, and on the Great Lakes. These boats are designed for specific tasks, and those who man them are considered among the world's finest small boat handlers.

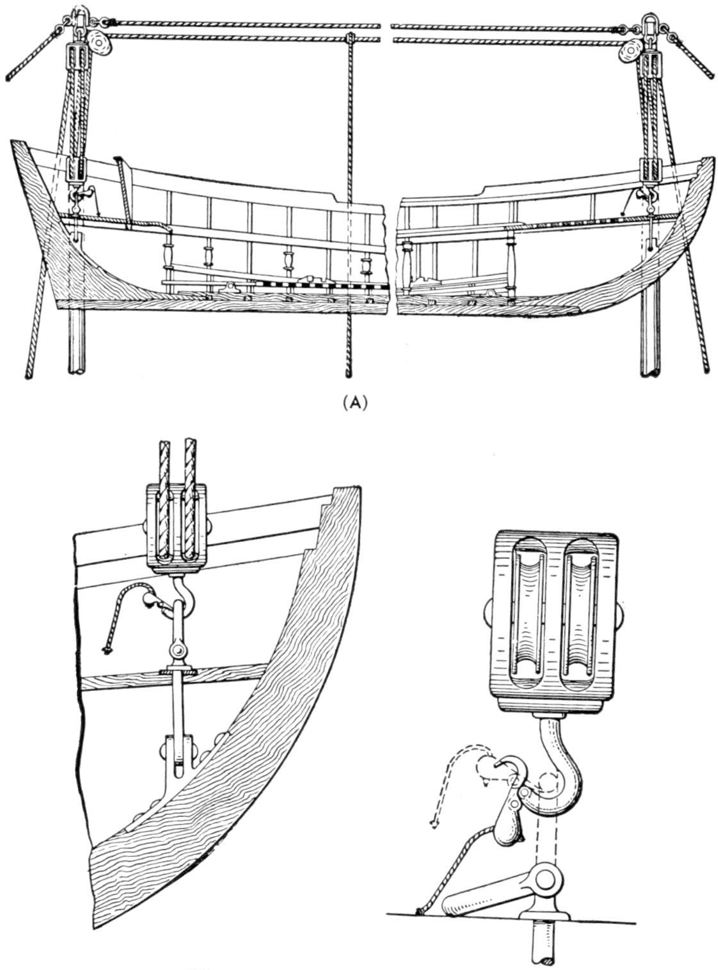

FIG. 6.13 RAYMOND RELEASING HOOK

6.14. Lifeboats. *52-Foot Motor Lifeboat (Fig. 6.14).* This boat is of entirely new design with emphasis on survival in rough seas. Although the conventional double-ended hull form has been retained, the new motor lifeboat is constructed of steel, carries twin screws, and is powered by diesel engines. The superstructure and interior trim are made of aluminum alloy. Advanced features include fire and salvage pumps of 500 gallons per minute capacity driven by the main diesel engines; improved visibility and protection at inside and outside steering stations; power driven windlass and towing hawser capstan; forced ventilation and electric heating for compartments; complete galley, messing, and living accommodations; fathometer; radio communication equipment; radio direction finder.

This new craft with a cruising range of 500 miles represents the first major change in Coast Guard lifeboat design in more than twenty years. It will be assigned to lifeboat stations for rescue work.

FIG. 6.14 52-FOOT MOTOR LIFEBOAT. *U.S. Coast Guard Official Photograph*

40-Foot Utility Boat. This boat is a general-purpose, twin-screw, diesel-powered boat designed to meet service requirements for light rescue, security, and off-shore duty.

36-Foot Motor Lifeboat (Fig. 6.15). This is a heavy-duty, steel-hulled, self-righting, nonsinkable lifeboat. Coast Guard lifeboat stations with equipment

FIG. 6.15 36-FOOT SELF-RIGHTING, NONSINKABLE MOTOR LIFEBOAT. *U.S. Coast Guard Official Photograph*

such as this, up and down the coasts of the United States, save many lives each year. This boat is capable of weathering severe storms at sea.

30-Foot Utility Boat (Fig. 6.16). This boat is similar in design to the 40-footer and has many of its best features. The 30-footer is a diesel-powered, single-screw, general-purpose boat designed for light rescue, security, and offshore duty. It has a mahogany reinforced wooden hull and is capable of 22 knots in seas up to 4 feet.

FIG. 6.16 30-FOOT UTILITY BOAT. *U.S. Coast Guard Official Photograph*

26-Foot Self-Bailing Motor Surf Boat. This boat is carvel built, powered by a gasoline or diesel engine. It is also equipped with oars and can be launched and landed on a beach. This boat has good maneuverability and has been used

FIG. 6.17 16-FOOT (PLASTIC) OUTBOARD MOTORBOAT. *U.S. Coast Guard Official Photograph*

for flood operations and general work in shallow waters. It is extremely seaworthy even in rough seas.

26-Foot Pulling Self-Bailing Surf Boat. This boat has the same hull as the motor surf boat but is propelled by oars and steered by a sweep. The boat is light enough to be moved by trailer along the beach, launched through surf and then landed on the beach.

16-Foot (Plastic) Outboard Motorboat (Fig. 6.17). This boat is powered by a 35-horsepower outboard motor and is used by United States Coast Guard Boarding Team to move around in inland waterways to check on the safe operation of pleasure boats.

6.15. Merchant Marine Lifeboats. The boats of the Merchant Marine are lifeboats and are seldom, if ever, used for other purposes. The modern lifeboat

FIG. 6.18 24-FOOT MERCHANT MARINE MOTOR LIFEBOAT. *U.S. Coast Guard Official Photograph*

is a double ender of the whaleboat type. The hulls are normally made of metal, although a new plastic type is being introduced. They are sturdily built and are able to be used under severe weather conditions at sea. They are intended to be propelled by oars or sails, hand gear, or motors.

All passenger vessels in ocean service, and all cargo vessels, including tank ships of 1,600 gross tons and over on an international voyage, are required to carry a motor-propelled or hand-propelled lifeboat. Figure 6.18 shows a 24-foot motor lifeboat with associated equipment stowed on board.

TABLE 6.3. REQUIRED EQUIPMENT FOR LIFEBOATS

Letter identification	Item	Ocean and coastwise	Great Lakes	Lakes, bays, and sounds; and rivers
a	Bailer	1	1	None
b	Bilge pump	1 [1]	None	None
c	Boathooks	2	1	1
d	Bucket	2	1	1
e	Compass and mounting	1	None	None
f	Ditty bag	1	None	None
g	Drinking cups	1	None	None
h	Fire extinguishers (motor-propelled lifeboats only)	2	2	2
i	First aid kit	1	None	None
j	Flashlight	1	1 [2]	None
k	Hatchets	2	2	1
l	Heaving line	2	None	None
m	Jackknife	1	None	None
n	Ladder, lifeboat gunwale	1 [3]	None	None
o	Lantern	1	1	1
p	Life line	1	1	1
q	Life preservers	2	2	2
r	Locker	1	1	None
s	Mast and sail (oar-propelled lifeboats only)	1	None	None
t	Matches (boxes)	2	1	1
u	Milk, condensed (pounds per person)	1	None	None
v	Mirrors, signaling	2	None	None
w	Oars	1 unit [4]	1 unit [4]	1 unit [4]
x	Oil, illuminating (quarts)	1	1	None
y	Oil, storm (gallons)	1	1	None
z	Painter	2	1	1
aa	Plugs	1	1	1
bb	Provisions (pounds per person)	2	None	None
cc	Radio installation	1 [5]	None	None
dd	Rowlocks	1 unit [4]	1 unit [4]	1 unit [4]
ee	Rudder and tiller	1	1	None
ff	Sea anchor	1	1	None
gg	Searchlight	1 [5]	None	None
hh	Signals, distress, floating orange smoke	2	None	None
ii	Signals, distress, red hand flare	1 unit [4]	1 unit [4,6]	None
jj	Signals, distress, red parachute flare	1 unit [4,7]	1 unit [2,4]	None
kk	Tool kit (motor-propelled lifeboats only)	1 unit [4]	1 unit [4]	1 unit [4]
ll	Water (quarts per person)	3	None	None

[1] Motor-propelled lifeboats, certified for 100 or more persons, shall be fitted with an additional hand bilge pump of an approved type or a power bilge pump.
[2] Optional—see footnote 6.
[3] Not required on lifeboats of less than 60 persons capacity.
[4] A unit of oars is a quantity equal to the number of oarlocks in the boat plus spares.
[5] Required only on motor-propelled lifeboats fitted with radio cabin.
[6] An approved flashlight, item (j), or 12 approved parachute red flare distress signals, item (jj), may be substituted for 6 of the required 12 hand red flare distress signals.
[7] Vessels in coastwise service need only carry 1 unit for each 5 lifeboats or fraction thereof.

TABLE 6.4. MASTS AND SAILS REQUIRED FOR LIFEBOATS

Length of lifeboat (feet)		Standing lug sail							Mast [1]		Yard [1]	
Over	Not over	Area (square feet)	Luff and head length	Leach length	Foot length	Clew to throat	Ounces per square yard	Commercial designation number	Length (feet and inches)	Diameter (inches)	Length (feet and inches)	Diameter (inches)
..	17	58	5–11	12–1	8–10	10–10	14.35	10	11–2	3	6–11	2
17	19	74	6–8	13–8	10–0	12–2	14.35	10	12–6	3	7–8	2
19	21	93	7–5	15–1	11–2	13–8	14.35	10	13–10	3½	8–5	2½
21	23	113	8–3	16–11	12–4	15–1	14.35	10	15–2	3½	9–3	2½
23	25	135	9–0	18–6	13–6	16–6	14.35	10	16–6	4	10–0	3
25	27	158	9–9	20–0	14–7	17–10	17.50	8	17–10	4	10–9	3
27	29	181	10–5	21–5	15–7	19–1	17.50	8	19–2	4½	11–5	3¼
29	31	203	11–0	22–8	16–6	20–3	20.74	6	20–6	4½	12–0	3¼
31	..[2]	...[2]	....[2]	....[2]	....[2]	....[2]	.....[2]	..[2]	....[2]	...[2]	....[2]	...[2]

[1] Mast lengths measured from heel to center of upper halyard sheave. Mast diameters measured at thwart. Mast and yard shall be of clear-grained spruce, fir, or equivalent.
[2] Subject to special consideration.

Every lifeboat must be equipped with sufficient number of air tanks of suitable size and appropriately placed to keep it afloat when filled with water.

All lifeboats are equipped with a lifeline, properly secured the length of each side, festooned in bights not longer than 3 feet, with a seine float in each bight, and hanging within 12 inches of the water. Lifeboats on vessels in ocean and coastwise service are required to be equipped with grab rails with three ½-inch diameter manila grab lines extending from gunwale to gunwale under the keel, to enable persons to cling to or climb upon the lifeboat when it is capsized.

Every lifeboat must be equipped with an automatic drain plug with a cap or plug attached to the lifeboat by a suitable chain.

When built, a plate on every lifeboat shows, among other things, the cubical contents and the number of persons the boat may carry. Every lifeboat must also have its cubical contents and the number of persons it is allowed to carry plainly marked or painted on each bow in letters and figures 1½ inches high. In addition, it must have the number of persons allowed to be carried plainly marked or painted on top of at least two of the thwarts in letters and figures not less than 3 inches high.

All lifeboats must have the vessel's name and the number of the boat plainly marked or painted on each bow in figures not less than 3 inches high. Where boats are carried on both sides of a vessel, lifeboat No. 1 is forward on the starboard side; lifeboat No. 2, forward on the port side; lifeboat No. 3 next abaft lifeboat No. 1 on the starboard side, and so on. The odd-numbered boats are on the starboard side and the even-numbered boats are on the port side. When

TABLE 6.5. REQUIREMENTS FOR NUMBER OF LIFEBOATS

Registered length of ship in feet	Minimum number of sets of davits	Minimum number of open boats of the first class	Minimum capacity of life boats (cubic feet)
100 and less than 120	2	2	980
120 and less than 140	2	2	1,220
140 and less than 160	2	2	1,550
160 and less than 175	3	3	1,880
175 and less than 190	3	3	2,390
190 and less than 205	4	4	2,740
205 and less than 220	4	4	3,330
220 and less than 230	5	4	3,900
230 and less than 245	5	4	4,560
245 and less than 255	6	5	5,100
255 and less than 270	6	5	5,640
270 and less than 285	7	5	6,190
285 and less than 300	7	5	6,930
300 and less than 315	8	6	7,550
315 and less than 330	8	6	8,290
330 and less than 350	9	7	9,000
350 and less than 370	9	7	9,630
370 and less than 390	10	7	10,650
390 and less than 410	10	7	11,700
410 and less than 435	12	9	13,060
435 and less than 460	12	9	14,430
460 and less than 490	14	10	15,920
490 and less than 520	14	10	17,310
520 and less than 550	16	12	18,720
550 and less than 580	16	12	20,350
580 and less than 610	18	13	21,900
610 and less than 640	18	13	23,700
640 and less than 670	20	14	25,350
670 and less than 700	20	14	27,050
700 and less than 730	22	15	28,560
730 and less than 760	22	15	30,180
760 and less than 790	24	17	32,100
790 and less than 820	24	17	34,350
820 and less than 855	26	18	36,450
855 and less than 890	26	18	38,750
890 and less than 925	28	19	41,000
925 and less than 960	28	19	43,880
960 and less than 995	30	20	46,350
995 and less than 1,030	30	20	48,750

lifeboats are nested, the lifeboat under lifeboat No. 1 is numbered 1-A; the lifeboat under lifeboat No. 2 is numbered 2-A, etc.

One of the lifeboats on each side of a passenger vessel is fitted for emergency work at sea. Each of these boats is provided with at least four lines fitted to a span between the davit heads of sufficient length to reach the water at the vessel's lightest seagoing draft, with the vessel listed 15 degrees either way.

All other boats on passenger vessels and all boats on cargo vessels, in ocean or coastwise service, are provided with two such lifelines. A releasing gear of the type which may be unhooked under tension is used in the emergency lifeboat and a sea painter is passed along forward on the vessel when at sea and in the lifeboat a long eye, strap, and toggle is fitted.

All vessels must have one approved ladder for each set of davits to enable passengers and crew to descend to lifeboats and rafts. They must be kept ready for use on the boat deck. On all passenger vessels and on all cargo vessels, and all vessels where the boat deck is more than 30 feet from the lightest seagoing water line, illumination must be furnished from the vessel for the lifeboats when being launched or while alongside.

6.16. Boat Handling Equipment. Boat davits are usually of the following types: *gravity, sheath screw, quadrantal,* and *radial* or *round-bar.*

With the gravity type davit (Fig. 6.19a), the boat is carried in two cradles, mounted on rollers which move over two parallel tracks at right angles to the

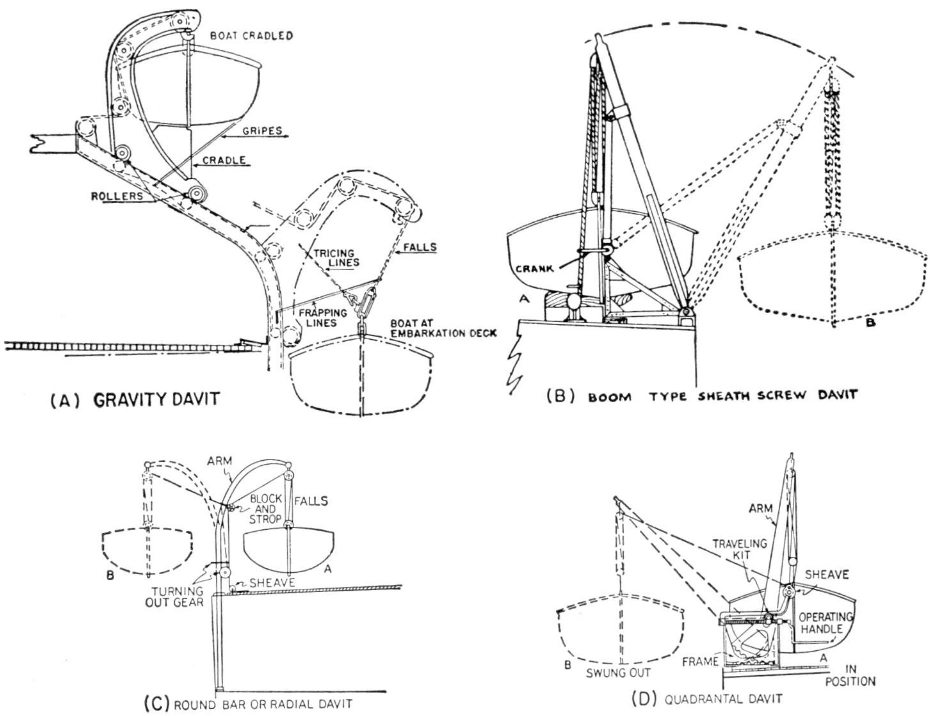

FIG. 6.19 TYPES OF BOAT DAVITS

vessel's side. After the gripes are released, a brake liner is raised which permits the lifeboat and the entire assembly to roll down the tracks by gravity, stopping with the lifeboat suspended over the side and at the embarkation deck. Tricing lines swing the boat into the ship's side and hold it in position until the frapping lines are passed around the falls and secured, frapping the boat in position to receive personnel. After this the tricing lines are cast adrift by tripping the pelican hooks before the boat is loaded. Raising the liner again permits the boat to continue down until it reaches the water.

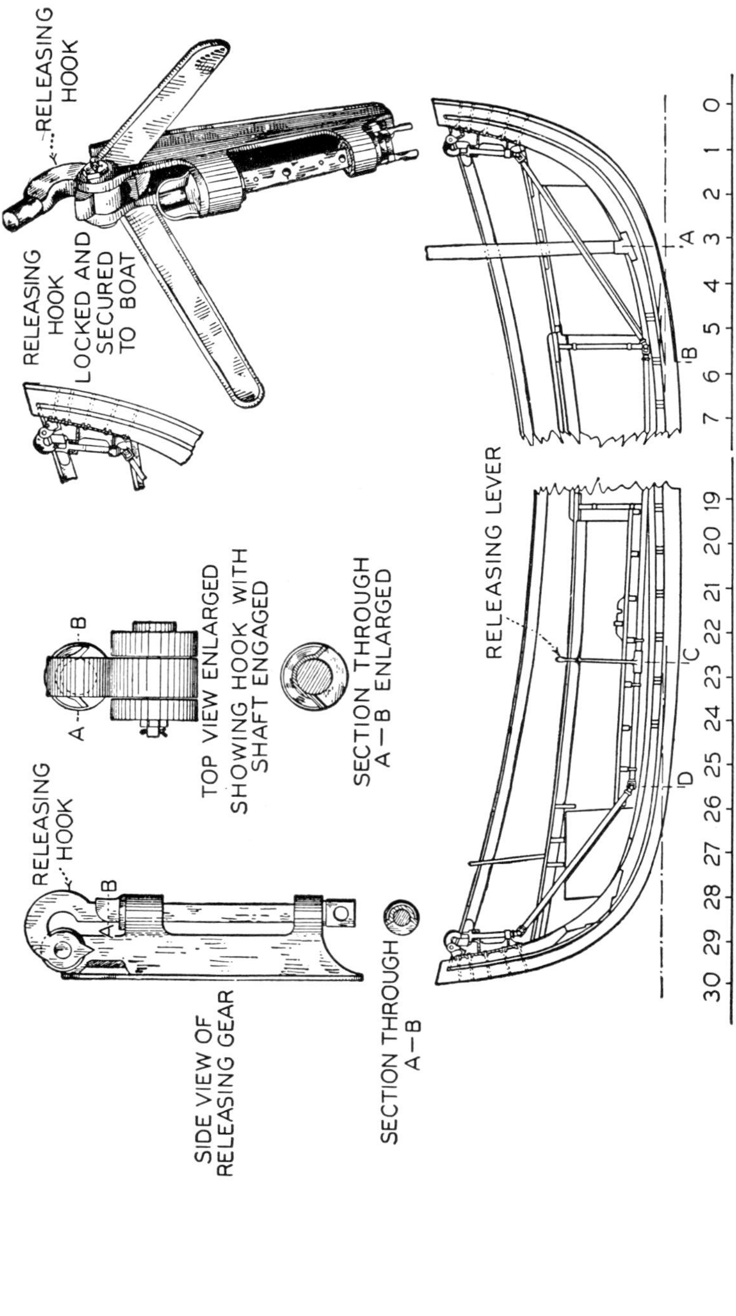

FIG. 6.20 ROTTMER RELEASING GEAR

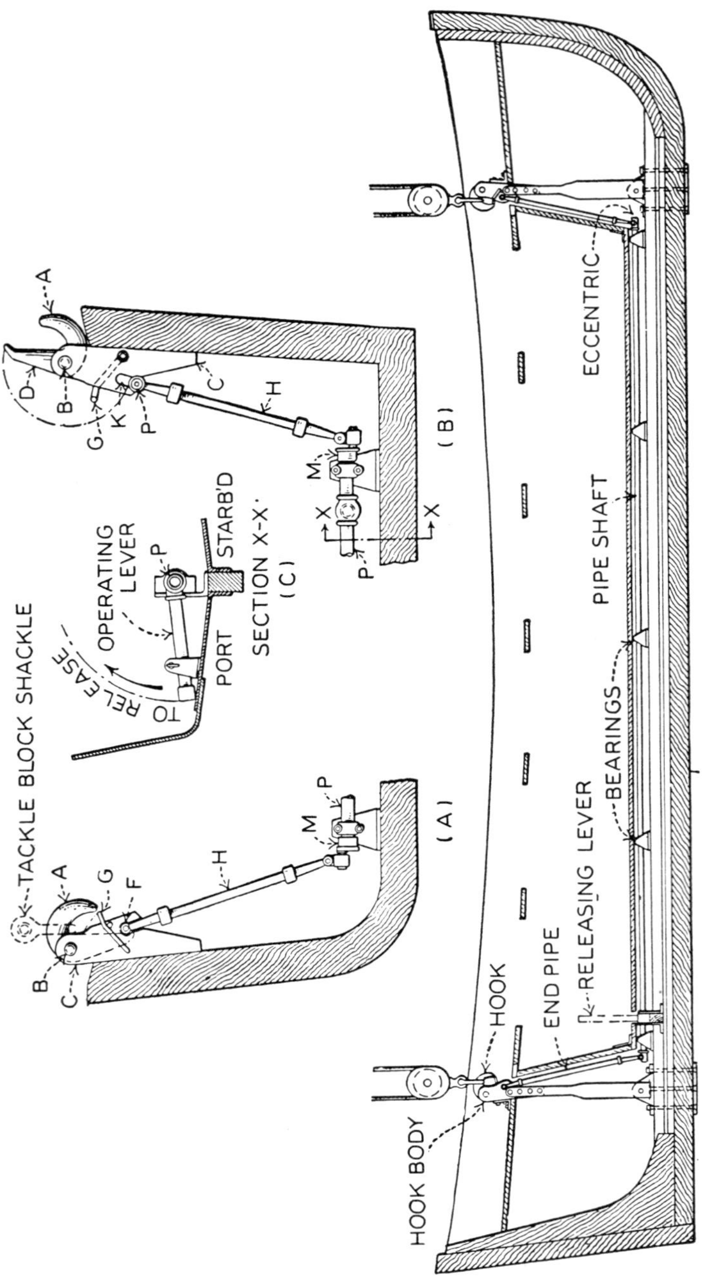

FIG. 6.21 STEWARD RELEASING GEAR

With the sheath screw type davits (Fig. 6.19b), the lifeboat is carried either on chocks under the davits or is cradled between the davits. In operation, the davit, which is pivoted near the foot so that it will turn in an arc at right angles to the vessel's side, is rotated outboard by a crank operating a sheath screw, so that the boat is suspended over the side and frapped in to the embarkation deck. It is then lowered to the water in the usual manner by its falls.

With the quadrantal type davits (Fig. 6.19c), the lifeboat is carried on chocks under the davits. The davits themselves stand upright with the tops curved in toward each other so that the ends come directly above the hoisting hooks of the lifeboat. In operation the davit, which rotates so that it will turn in an arc at right angles to the vessel's side, is turned outboard by a crank operating a worm gear, so that the boat is suspended over the side and frapped into the embarkation deck. It is then lowered to the water in the usual manner by its falls.

FIG. 6.22 BOAT EQUIPPED WITH HYDROFOILS

With the radial or round-bar type davits (Fig. 6.19d), the lifeboat is carried on chocks, under davits which may be rotated. In operation, the davits are both turned inboard until the bow of the boat clears the forward davit. The forward davit is then turned out so that the bow swings out, and the same procedure is followed with the after davit, so that the boat is suspended over the side where it is frapped in to the embarkation deck. It is then lowered to the water by its falls. This type of davit must always be guyed.

Releasing gear is required on lifeboats. Its purpose is to permit releasing both falls from the boat either before the boat is waterborne or immediately thereafter while under a towing strain. Types in use are the *Rottmer* (Fig. 6.20) and *Steward* (Fig. 6.21). These consist of two releasing hooks in each end of the boat to which the falls are attached. These releasing hooks are hinged on pins and held engaged by a locking device. A jointed shaft running the length of the boat is connected to both locking devices. Upon throwing a lever attached to this shaft, the hooks are capsized and the boat released.

Another common type is the *mills* chain type. The gear release handle is conveniently located on deck at the after end of the boat. When this handle is pulled, trigger hooks with ball weights to which the boat falls are made fast drop, and the boat is free fore and aft. The boat must be waterborne before the releasing gear can work. When being lowered, the weight of the boat and its load is, for safety's sake, directly opposed to releasing.

6.17. Experimental Craft. A new development is the *hydrofoil* boat. Although in the experimental stage at present, continuing evaluation is proving the versatility of this type of boat. The hull is similar to other types of existing conventional boats. However, both forward and aft, hydrofoils are fitted to the hull. As the speed of the boat increases and commences to plane, the boat rises up and rides on the foils. The foils, being relatively small in comparison with the bottom of the hull, decrease water resistance and thus the boat is capable of high speeds with no increase in size of the power plant.

7

Communications

Communications is a subject of vital interest to the deck officer and particularly the ship's commanding officer or master. No officer on the bridge of a ship today can be considered an expert in seamanship if he cannot employ communications effectively. Communications takes place within the ship—bridge to main engine control, bridge to radio central, etc.—and communications takes place to and from the ship. Every ship has various internal or *interior communication* systems, such as the telephone, public address, engine order telegraph, etc. In addition the ship has postal, radio, visual, or sound *exterior communication* facilities.

The term for all rapid communications *at a distance*, of which radio is only one example, is *telecommunications*. *Communications* is a term applied to all methods of communications—rapid as well as nonrapid. The term *communications* is used customarily when the term *telecommunications* would sometimes be more accurate.

7.1. Telecommunications. The three major classes of telecommunications are electrical, visual, and sound.

Electrical Telecommunications:

a. Radiotelegraph
b. Teletypewriter
c. Radio-teletypewriter
d. Radiotelephone
e. Television
f. Facsimile

Visual Telecommunications:

a. Flaghoist
b. Flashing light
c. Semaphore
d. Pyrotechnics
e. Colored lights

Sound Telecommunications

7.2. Electrical Telecommunications. Radiotelegraph (often called CW for "continuous wave") is a system for transmitting messages by a radio wave which an operator separates into the dashes and dots of the Morse code by opening and closing a hand key. Radiotelegraph was in use by the Navy as early as 1903; and even today, in spite of the development of faster and more convenient methods of electronic communications, it is one of the most reliable and trustworthy systems.

The mental and manual actions performed by an operator in converting letters to Morse code (and vice versa) are replaced in teletypewriter by electrical and mechanical actions. To transmit a message the operator types on a keyboard similar to that on a typewriter. As each key is pressed, a sequence of signals is transmitted. At receiving stations the signals are fed into receiving machines, which type the message automatically.

Teletypewriter signals may be sent either by landline or by radio. Landline teletypewriter communication is used both by the military services and by commercial communication companies, such as Western Union. Radio-teletypewriter (RATT) is primarily intended to furnish high-speed automatic communication over ocean areas. Today the primary shipboard use of RATT is for receiving fleet broadcast schedules, for which it is very well suited. Radio-teletypewriters can clear traffic at a rate up to 100 words per minute as compared to the 18 to 25 WPM speed of the CW fleet broadcasts. Since the shipboard operator is freed from manual copying, and hundreds of vessels may be receiving a single broadcast, the total saving in trained manpower is considerable.

Radiotelephone (sometimes called voice radio) is one of the most useful communication methods. Because of its directness, convenience, and ease of operation, radiotelephone is used by ships and aircraft for short-range tactical communication. Its direct transmission of voice makes it possible for a conning officer to have in his hands a means of personal communication with the Officer in Tactical Command (OTC) and with other ships. There is little delay while a message is prepared for transmission, and acknowledgments can be made instantly. Radiotelephone equipment is usually operated on frequencies that are high enough to have line-of-sight characteristics—i.e., the waves will not follow the curvature of the earth. This limits the usual range of radiotelephone from 20 to 25 miles, thus giving a certain degree of security. Radiotelephone procedure can be learned fairly easily by persons with no training in communications.

A new technique in radiotelephone is called Single Side Band (SSB). Here, voice radio is not limited in range to the line of sight. Ships and aircraft many hundreds of miles apart can "talk" to each other.

During World War II, equipments were produced for transmitting voice communications without using radio waves or emissions. The transmitting medium was infrared light (invisible to the unaided eye).

Through a ship's radiotelephone, the nearest commercial shore "marine operator" can link voice radio communications of a ship with any "party" ashore having a conventional telephone. With the advantages of radiotelephone go some disadvantages. Transmissions may be unreadable because of static, enemy interference, and a high local noise level caused by shouts, gunfire, and bomb or shell bursts. Wave propagation characteristics of radiotelephone frequencies are sometimes freakish, and transmissions may be heard at great distances, but not by the addressee. Most radiotelephone messages are in plain language, and if information is to be kept private, users must keep their messages short, stick to proper procedures, and use care in what they say.

Military television is still in the developmental stage. There is little such equipment in the Fleet today, but television promises to be useful for purposes requiring rapid and continuous transmission of pictorial data from a distance. It seems especially suitable for:

1. Remote guidance of missiles.
2. Reception of reconnaissance data from aircraft.
3. Remote inspection of underwater salvage operations.
4. Simultaneous briefing of many commanding officers or aviators of a task force when the tactical situation is too urgent to permit duplication of weather data, charts, or other pictoral information.

Facsimile (FAX) resembles television in that it is a process for transmission of pictures. It is unlike TV in that (1) facsimile gives the receiving station a permanent record of the transmission while television does not; and (2) facsimile requires several minutes to transmit a picture twice the size of this page, while television sends a continuous stream of 30 pictures per second.

FAX is very useful for transmitting such matter as photographs and weather charts. The image to be sent is scanned by a photoelectric cell, and variations in the cell output due to the character of the picture are used to modulate a radio wave. At the receiver the signal operates a recorder which reproduces the picture.

7.3. Visual Telecommunications. Visual communication systems have been in use since men first sailed the seas and are still the best means for communicating at short range. In reliability and convenience they are the equal of radio and are more secure.

The most important visual systems are flaghoist, flashing light, and semaphore. Pyrotechnics, colored lights, and sound have important wartime uses.

Flaghoist is a method whereby various combinations of brightly colored flags and pennants are hoisted to send messages. It is the primary means for transmitting brief tactical and informational signals to other ships. Signals are repeated by addressees, thus providing a sure check on the accuracy of reception. Texts of messages which may be sent are limited to those found in signal books.

Directional flashing light is a visual telegraphic system in which an operator opens and closes the shutter of a searchlight to form the dashes and dots of the Morse code. The light may be pointed and trained to be seen only from the viewpoint of the receiver.

Nondirectional flashing light is sent out from a lamp on a yardarm. Dots and dashes are made by switching the lamp on and off. Since the light is visible in every direction, this method is well suited for messages which are for several addressees.

In wartime, flashing light communications that must be carried on after dark are usually conducted by means of infrared beams, which are not visible unless viewed through a special receiver.

Semaphore is a communication method in which an operator signals with two hand flags, moving his arms through various positions to represent letters, numerals, and other special signs. It is especially suitable for long administrative messages because of its speed. It is not readable much farther than 2 miles, even on a clear day.

Pyrotechnics have a wide variety of uses. Often their use can be a matter of life or death. Examples are identification of own ship or attracting attention when in distress. Coastal lifesaving stations also employ pyrotechnic flares to signal to vessels in distress.

Pyrotechnics used for signaling are, for the most part, of the fireworks variety. The term "pyrotechnic light" includes all types which provide a temporary source of light as opposed to pyrotechnic smoke. Common sources of pyrotechnic light are *Very* pistol flares, colored shell bursts (parachute flares), aircraft parachute flares, Roman candles, and float type flares. Common types of pyrotechnic smokes are also *Very* pistol bursts, colored smoke from shell burst, aircraft parachute smoke pots, float type smoke pots, smoke generators, and smoke puffs from surface ship smoke stacks. The meaning of a pyrotechnic signal is dependent on the color rather than on the type of pyrotechnic used.

The following limitations which are inherent in pyrotechnic signals can be considered guides in using them:

Signals consisting of a succession of pyrotechnics or a combination of colors should not be used, because there is the danger that an observer may not have seen the whole of the signal and may consequently misinterpret it.

The standard colors red, white (or yellow), and green are the only colors which give satisfaction under varying conditions of visibility. Under certain atmospheric conditions, white signals may appear as yellow; these two colors are therefore considered synonymous. Under certain conditions of humidity a white pyrotechnic is liable to be mistaken for green.

Tracer is particularly liable to be confused with red pyrotechnics. At a distance it is difficult to identify the exact position from which a pyrotechnic signal was fired; consequently, a single pyrotechnic fired by each of two separate originators may appear to an observer as two pyrotechnics fired simultaneously or in succession from one originator.

The range of visibility for "pyro" is largely dependent on weather conditions.

Colored lights (beyond their use as navigational lights) can be used at night to maneuver ships at sea. Colored lights could be used for purposes of identification. Colored lights used for these purposes are usually shown for short periods only. Colored lights should not be confused with pyrotechnic lights which are also of short duration. The meanings of colored lights generally tend to be different from the meanings of the same colors of pyrotechnic lights. Remember colored lights are normally electric light type illumination, whereas pyrotechnic lights are produced by some chemical combustion (burning) or explosion of the "fireworks" variety.

7.4. Sound Telecommunications. Sound signals are used routinely according to the Rules of the Road. Sound signals are also made by navigational aids, such as buoys, beacons, lighthouses, lightships, etc., as well as fog-bound ships. Sound may also be used in the maneuvering of ships (either as a supplement to visual signals or as a separate signal).

It is important that signals in use for one purpose not be misinterpreted as being made for another purpose. Sound signals incorrectly interpreted as Rules of the Road signals can cause serious confusion.

When signaling with sound in fog or listening to such signals, bear in mind:

Fog signals are heard at greatly varying distances. Under certain conditions of atmosphere, when an air fog signal is a combination of high and low tones, one of the tones may be inaudible. Occasionally a fog signal may be wholly inaudible in certain areas.

A fog bank may exist a short distance from a fog signaling station or ship and not be seen; therefore, the signal may not be sounded. Some fog signaling apparatus cannot be started immediately after signs of fog have been observed. Fog signals can sometimes be heard by fog lookouts in the eyes of the ship or aloft but may not be heard on deck or on the bridge.

Shipboard Communication Facilities

7.5. Communication Coordination Facilities. For rapid exterior communications, ships have two main stations from which messages and signals are sent and received. These stations are known as the radio station and the visual station. Coordination of both stations in large warships takes place in the communication center under the supervision of the communication watch officer (CWO). The communications center (Fig. 7.1) is usually located in the vicinity of the main radio room or radio central. Both the visual station (signal bridge) and the radio station consist of a number of outlying shipboard facilities.

At the visual station there are semaphore flags, portable signal lights, and a loud hailer ("bull horn" or electric megaphone). Signal flag hoists and flags are in the vicinity. A ship's bell and whistle are nearby. Running lights and other navigational and anchor light fixtures are located on the masts, superstructure, and above the main deck; controls are near the visual station—usually in the pilothouse. Signal searchlights are in the superstructure in the vicinity of the visual station; blinker lights are installed on the yardarms with keying controls at the visual station. A fog gong is located on the main deck. Pyrotechnics for signaling are located near the signal bridge.

The radio station facilities are concentrated mainly at radio central. Radio antennas are located throughout the superstructure. Radio transmitters and receivers are found not only in radio central but in special transmitter rooms and separate radio rooms. Portable and emergency radio equipment is located in a number of places. Remote operated radiotelephone units—send/receive facilities—are found on the bridge, in the pilothouses, CIC, and in radio spaces.

Radioteletype and radiotelegraph facilities are normally operated at the main radio station.

A number of interior communications systems are employed to link the bridge and pilothouse with the key shipboard communication facilities. Regular telephones as well as sound powered telephones are used. Voice tubes are also used, and on some ships pneumatic tubes are installed to send messages quickly.

FIG. 7.1 THE SHIPBOARD COMMUNICATIONS CENTER. NOTE THE CW RADIO OPERATORS. *Official U.S. Navy Photograph*

Amplifier type announcing systems (often referred to as "squawk boxes" or "intercom units") are widely employed also.

7.6. Semaphore Facilities. Semaphore requires little equipment—the two hand flags attached to staffs are all that is needed (Fig. 7.2). The standard semaphore flags are usually 15 or 18 inches square, and each staff is long enough to enable the sender to grasp it firmly. The flags are similar to the *oscar* alphabet flag. The *papa* flag is sometimes substituted. Most semaphore flags issued to the fleet today are fluorescent and are made of sharkskin cloth.

When fluorescent flags are used, background is not too important. When cotton flags are used, it is imperative that a good background be selected. Otherwise the other ship(s) may not be able to see the flags clearly.

7.7. Searchlights. Searchlights may be classified as incandescent and arc. In the 12-inch searchlight the source of light is a specially designed incandescent lamp. Control of the light is by means of a shutter. This is illustrated in Fig. 7.3. The front and rear doors are hinged to the searchlight case to permit access to the interior for relamping and cleaning. There is usually a handle on the rear of the case to elevate and depress the light or turn it in azimuth.

The Navy standard 12-inch searchlight is simple to operate. Anyone can learn in a short time the proper procedure for turning the lamp on and off, positioning it horizontally or vertically, and operating the shutters.

FIG. 7.2 SEMAPHORE IN USE. *Official U.S. Navy Photograph*

The carbon-arc 24-inch searchlight is intended primarily for signaling and secondarily for navigational use. It consists of a stationary pedestal which is secured to the searchlight platform, a turntable which carries the two trunnion arms and rotates on the pedestal, and a drum which is pivoted on the trunnion arms to allow it to be elevated and depressed.

The drum contains an automatic high-intensity carbon-arc lamp which is operated from a direct current power supply. At the rear of the drum is a metal reflector. At the front of the drum is a dome glass, an iris shutter for shutting off the searchlight beam completely, and a high-speed sector vane shutter which may be operated locally or remotely.

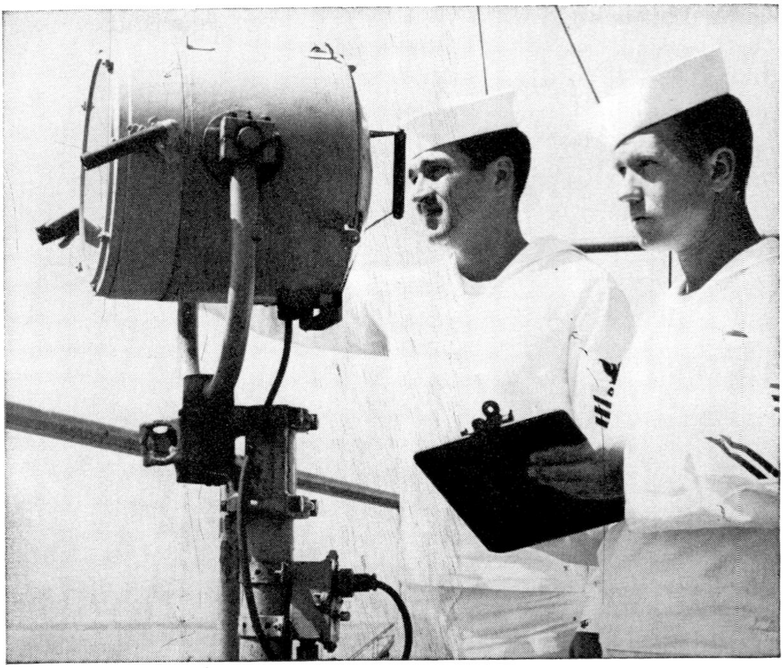

FIG. 7.3 THE 12-INCH SEARCHLIGHT IN USE. *Official U.S. Navy Photograph*

Handles at the rear of the drum provide a means of swinging the searchlight in train and elevation to direct the beam of light.

7.8. Nancy Facilities. *Nancy* is a system of visual communications which uses a special light that is visible only with the aid of equipment designed for this purpose (Fig. 7.4).

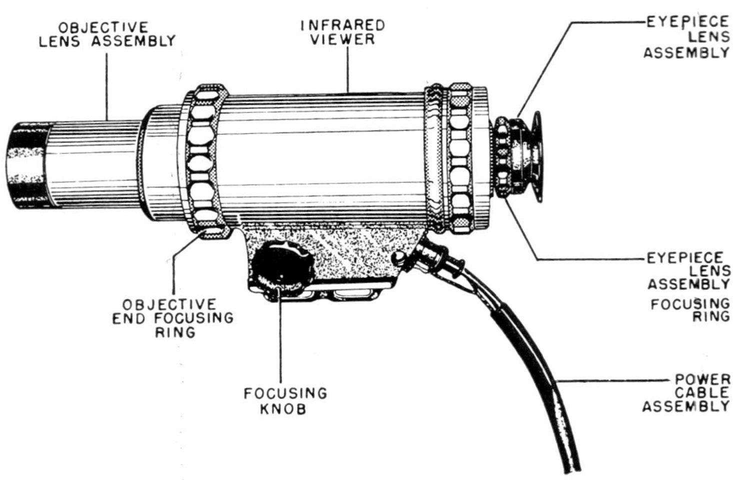

FIG. 7.4 A NANCY RECEIVER

During World War II the most widely used nancy signaling gear was the type with a filter lens and hood for mounting on the standard Navy 12-inch searchlight. With this gear attached, the light is operated in the same manner as an ordinary communication searchlight. Under average conditions it has an effective range of 10,000 to 15,000 yards, depending on the type of receiver used.

In addition to this equipment there are various other types of nancy beacons which have been designed for recognition and reconnaissance as well as communication.

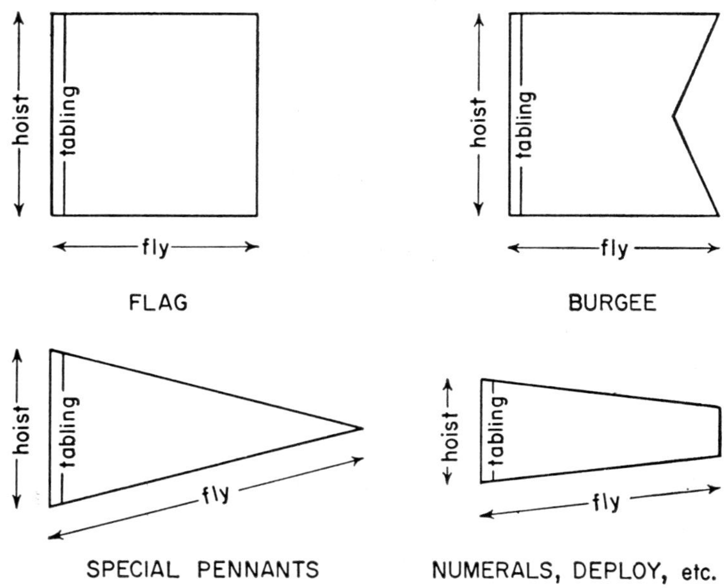

FIG. 7.5 PARTS OF A FLAG

An infrared transmitting set is presently installed aboard many ships of the Navy. The purpose of the equipment is to transmit signals by means of infrared light. The transmitters may be operated as a steady source for "point of train" purposes, or they may be flashed in Morse Code for signaling or recognition. The beacons may be operated separately or simultaneously by either of two manual keys.

The beacons of the infrared transmitting set are designed to show a beam through 360 degrees in bearing and from 20 degrees below the horizontal to zenith in elevation. The radiated signals cannot be detected by the unaided eye at distances greater than 400 yards. The beacons do not show any white light; however, they are visible as dim red lights at short distances.

A nancy receiver is required for all types of nancy signaling equipment. All image-forming receivers perform the same basic task. They gather up invisible rays and convert them into light that can be seen.

There are two basic designs—the *phosphor receiver* and the *electronic re-*

ceiver. The most widely used phosphor receiver during World War II was a small, hand-held receiver less than 7 inches long and weighing less than a pound. Its sensitive element is a phosphor disk which, when properly charged, transforms nancy waves into visible light. The disk is charged by a small charging plate within the instrument.

7.9. Code Flag Facilities. Figure 7.5 shows the various flag types and their parts.

The *fly* is the length of the flag as measured from the staff to the outside edge.

The *hoist* is the vertical width of the flag when flying free.

FIG. 7.6 MAKING UP THE HOIST. *Official U.S. Navy Photograph*

The *tabling* is the double thickness of bunting—taped, bound, and stitched—which is at the hoist of a flag.

The *tail line* is a short length of halyard attached to the lower part of the tabling and carrying the snap hook. It serves as a spacer, separating the flags of a hoist for clearness in reading signals. (The tail line is not illustrated in Fig. 7.5.)

The ring is attached to the top of the tabling and snaps into the tail line of the preceding flag or hook of the halyard.

The tack line is a 6-foot length of braided signal halyard with a ring at one end and a snap at the other.

FIG. 7.7 HOISTING THE SIGNAL. *Official U.S. Navy Photograph*

The flags of a hoist are always read from the top down. When a signal is too long to fit on one halyard—when, in other words, more flags are required than can be made into a single hoist—the signal must be continued on another halyard. When a signal is broken into two or more hoists, it must be divided at points where there can be a natural space without affecting the meaning of the signal.

Flags are kept on flag boards (commonly referred to as "flag bags") as shown in Fig. 7.6. Such storage permits rapid makeup of the hoist. Figure 7.7 shows how the signal is hoisted as soon as each flag is attached.

Flags used on board ship are shown in color in the signal books. There are some differences between the flags used on merchant ships and those on men-of-war; some foreign port signals are made with flags not contained in international signal books. Certain flags, such as the "man overboard" and "breakdown," should be made up and hoisted ready for breaking instantly.

7.10. Other Signal Light Facilities. Men-of-war carry a number of lights which may be used to signal not only the status of the vessel according to the requirements of the Rules of the Road but for other purposes as well.

The *boom lights, portable towing lights,* and minesweeper *polarity signal lights* (red and green) are energized from various local lighting circuits.

The masthead light, range light, side lights, white stern light, and permanent towing lights are usually controlled from the running light switch box on the bridge.

The signal and anchor light switch box on the bridge controls the following lights (when installed):

- Aircraft warning lights
- Forward and after anchor lights
- Blinker lights
- Breakdown and man overboard lights
- Minesweeping lights
- Underwater task lights
- Speed lights
- Station keeping lights
- Steering light
- Wake light

In wartime, red and green convoy lights are also installed on merchant ships. Those would be energized from a special switch box.

Due to the wide variety and uses of ships' lights, a brief description of the most important ones is included here:

Aircraft warning lights. One red light is installed at the truck of each mast and extends more than 25 feet above the highest point in the superstructure. Where it is impossible to locate this light so that it is visible from any location throughout 360 degrees of azimuth, two lights are installed.

Blinker lights. These lights are installed only on the signal yardarm, outboard, one port and one starboard. On ships with more than one mast they are located only on the forward yardarm. Screens are fitted at the base of these lights to prevent glare or reflections from interfering with navigation of the ship. Lights are operable from signal keys located on the bridge.

Boom lights. Boom lights are permanently installed on the outboard end of each boat boom and are energized from receptacles on the ship's service lighting system.

Breakdown and man-overboard (not under command) lights. These are mounted on brackets extending abaft, and offset from the mast or structure to permit all-around visibility insofar as practicable. To facilitate pulsating these lights as a "man-overboard" signal, the rotary snap switch in the signal light switch box which controls them is fitted with a crank handle.

Minesweeping lights. These lights are installed on all minesweepers or ships fitted to sweep mines to warn other ships to keep clear of the sweep gear. The signal consists of a triangular display of three 32-point green lights, one on each forward yardarm and on the foremast. Lights on the yardarm are at least 3 feet outboard of the mast and the light on the mast is at least 3 feet above the lights on the yardarms. The light on the mast is installed on a bracket extending forward of the mast and offset to ensure visibility ahead and astern.

Polarity signal lights. Two lights, one red and one green, are installed close together to each side of magnetic minesweepers to indicate polarity or direction of magnetic field. These lights are mounted to be visible from 20 degrees forward to 20 degrees abaft the beam on each side of the ship. They are pulsed by contacts in the minesweeping control panel.

Speed light. This light is installed in ships of destroyer escort size and larger to indicate the ships speed. It is located at the truck of the mainmast, except where the height of the foremast is such as to interfere with the visibility of the light. In this case it is located at the truck of the foremast. If it is impracticable to locate this light so that it is visible throughout 360 degrees of azimuth, two lights are provided.

Stationkeeping lights. Two lights are installed on all minesweepers required to give sweep information at night. They are located in a vertical plane perpendicular to the keel so that accurate observations may be taken to aid bearings. These lights are mounted to be visible from 20 degrees forward of or 20 degrees abaft the beam on each side of the ship. If it is impossible to locate the lower light so that it can be seen on both sides of the ship, two lights are used, one on each side.

Steering light. This light is installed on specified ships and on ships where the pilothouse is more than 100 feet abaft the bow, unless structural interferences make use of this light impracticable. The light is located on the jackstaff or other spar or structure and is visible to the helmsman in the pilothouse. The light is installed on the centerline.

Stern light, blue. This light is installed near the stern of ships likely to be engaged in convoy operations. It is mounted to show throughout a total azimuth of 12 points—from astern to 6 points on each side of the ship.

Towing lights. Minesweepers, tugs, and other ships normally engaged in towing operations have permanently installed towing lights. Other ships have two portable towing lights, each of which is equipped with sufficient cable and a plug connector to permit energizing these lights from the nearest lighting receptacle connector.

Underwater task lights. These lights are installed on all ships engaged in underwater operations, such as minesweepers. They consist of three lights in a vertical line one over the other not less than 6 feet apart. The highest and lowest of these lights is to be red, and the middle light is white. The red lights are the same fixtures as those used for breakdown and man-overboard lights, whenever practicable, and the switching is arranged accordingly.

Wake light. This light is installed on the flagstaff or after part of the ship to illuminate the wake, and is so mounted that no part of the ship is illuminated by it.

A type of signal light that does not fit into the foregoing categories is the *life preserver light*. Some ships also carry floating lanterns for marking waters at night when necessary.

In addition, aircraft carriers have a complex arrangement of night-flight operations lights installed. These consist of such lights as deck edge lights, homing lights, parking lights, take-off lights, etc.

7.11. Messages and Signals. In naval communications, a message is a formal type of communication; like a postal letter, it is addressed to a certain destination(s), has a *text*, and an *ending*. The address is known as the *heading*. The signal's form is abbreviated or modified. The signal (or sign) conveys a meaning nevertheless. Signals may be used in the process of transmitting messages; examples of this are message handling signals and call signs.

On Navy ships, the captain must authorize all communications sent from the ship—signals and messages. Some signals are made as a part of the ship's routine without this specific authorization in each case: an example of this is the running lights which are always turned on at sunset when under way. Messages, on the other hand, are authorized and released by the captain (or by other officers delegated this authority). Messages which have been released are delivered to either the ship's radio or visual station for transmittal. At either place call signs, internal handling instructions (signals) are added; the recipient is then "called" and the message is transmitted and receipted for. The ship maintains complete files of communications transmitted and received.

7.12. What the Deck Officer Should Know About Communications. The precise procedure for sending and receipting for messages must always be followed by the deck officer in order to avoid confusion and dangerous communications delays. These procedures are described in many communications publications, some of which the deck officer must be familiar with and use. For the most part, however, he must rely on the understanding of procedures possessed by the communications officer, radio officer, or signal officer.

The deck officer of a man-of-war should be able to use expertly the following signal publications:

1. The Allied Naval Signal Book
2. The International Code of Signals
3. Wartime Instructions for Merchant Ships

The deck officer of the merchant vessel uses:

1. The International Code of Signals
2. Wartime Instructions for Merchant Ships

Vessels may communicate in a number of ways using signals from these books. The signals are designed mainly for the various visual means, sound, or radiotelephone; radiotelegraph may also be used.

The International Code of Signals is published in several foreign language editions. With this signal book, language is no barrier to communications. For instance, a French merchant ship communicating with a United States man-of-war would make up her message from the code of the French language edition of the International Code of Signals. The United States ship would look up the code meanings in her English language edition.

The Allied Naval Signal Book is used by United States men-of-war and NATO men-of-war. The International Code of Signals is used among merchant vessels of all nationalities and among men-of-war. The Wartime Instructions for Merchant Ships is used by ships in allied naval convoys. Each of the books contains types of signals not found in the others; messages or signals are sent, receipted for, and executed differently depending upon which system is in use. Signals from one book may, on occasion, be employed in conjunction with those of another book. The signal flags and pennants in each of the above books are basically those contained in the International Code of Signals with some additional flags and pennants in the other two books.

The deck officer should be familiar with the location and operation of all signal and communication equipment in the vicinity of the bridge. This includes the various signal (navigational) lights as well as the radiotelephone units, whistles, pyrotechnics, flags, pennants, and megaphones. It is necessary for the deck officers to reach this equipment instantly, day or night, and operate it correctly as the occasion demands. Larger ships have personnel to operate much of this equipment when ordered by the deck officer, but on smaller ships and on some merchant ships it is the deck officer himself who operates this gear.

8

Ground Tackle, Anchoring, and Mooring

Windlasses

A ship's windlass is designed primarily for handling anchor chain. The chain is guided over a fairlead and then around the capstan where it rides on a collar called a *wildcat*. This wildcat is essentially a sprocket designed to engage the links of the chain. The sprocket can be keyed to the capstan shaft and thus to the windlass motor, or it can be allowed to run free. A brake is also incorporated to control the wildcat when it is running free. Windlasses are arranged in many ways and combinations on various vessels, and are often provided with warping heads for handling line. There are four general classes, depending on the kind of drive: steam, electric, electric-hydraulic, and hand.

8.1. Steam Windlasses. Steam windlasses are usually of the horizontal shaft type. Reversing and speed control are obtained by use of a hand-operated piston valve which reverses and regulates steam flow to the valve chambers. Two wildcats and two warping heads are usually provided. The warping heads are keyed to the shaft, but wildcats are provided with locking mechanisms which permit free rotation of the wildcat when paying out chain. Chain payout is controlled by hand-operated brakes with drums attached to the wildcat.

8.2. Electric Windlasses. There are two sub-types of windlasses with straight electric drive. The first sub-type has a horizontal shaft arrangement like that of the steam windlass and has the horizontal shaft, gearing, and motor all self-contained on a single bed plate.

The second sub-type has a vertical shaft arrangement with the shaft leading from the wildcat to the motor and reduction gear in the windlass room below deck. When a warping head is provided, it is mounted in a vertical position above the wildcat, the wildcat and warping head shafts being concentric.

8.3. Electric-Hydraulic Windlasses. Electric-hydraulic windlasses are particularly adapted for anchor handling with its varying load due to the wide range of speed and torque. This hydraulic drive was developed to overcome all the operating and installation objections inherent with either steam or direct electric-driven windlasses. The electric-hydraulic windlass is similar to the electric type except, instead of having the motor coupled direct to the reduction gearing, the power is transmitted from the motor through a hydraulic transmission with a variable stroke which results in a wide range of output shaft speed.

Electric control is required only for light starting duty as the motor is started under no-load condition. The motor is direct-coupled to the pump unit of the hydraulic transmission, also known as the A end. Fluid under pressure is delivered through piping from the A end to the hydraulic motor unit B end. This B end is coupled to a suitable reduction gear which drives the windlass shaft.

Windlass speed is determined by varying the stroke of the pump A end. This is done by control handwheels located on the weather deck and at the pump. These handwheels also control the direction of rotation of the windlass and are suitably marked. The stroke at which the A end is set determines the quantity of hydraulic fluid delivered to the B end which, in turn, determines the speed at which the B end rotates.

A locking head located above the gear casing permits disconnecting the wildcat shaft from the drive for dropping anchor. On each wildcat shaft, usually just below the main deck, is a lined brake controlled by brake handwheels on the main deck and in the windlass room. The brake is capable of controlling and stopping a free fall of anchor and chain at 15-fathom intervals *up to 60 fathoms*. When disengaging the locking head, the brake should first be engaged.

Vertical warping or capstan heads are provided; they are located port and starboard on the main deck. These are driven through vertical shafting from the main reduction gearing. Locking heads are not provided on the capstan shafts; hence they rotate during the windlass operation. When using the warping head, sometimes called the capstan, the wildcat shaft locking head should be disengaged and the wildcat held by the brake.

8.4. Hand Windlasses. The use of hand-operated windlasses is limited to small vessels where the weight of the anchor gear is such that it may be handled in a reasonable time and without excessive effort on the part of operating personnel.

8.5. General Operating Instructions and Precautions for Power-Operated Windlasses. Although used intermittently and then only for a short period, the windlass must be capable of handling all required loads under most severe conditions.

Windlass brakes are designed to hold the weight of anchor and chain up to 60 fathoms and, when paying out, to stop and hold the load. In order to perform these services, the brakes must be kept in good condition and properly adjusted. Loss of anchor and chain in a majority of reported cases is ascribed to failure of the brake to hold due to maladjustment.

Anchors

When an old-fashioned anchor (Fig. 8.1) is let go in fairly deep water, it strikes the bottom crown first and immediately falls over until it rests on the

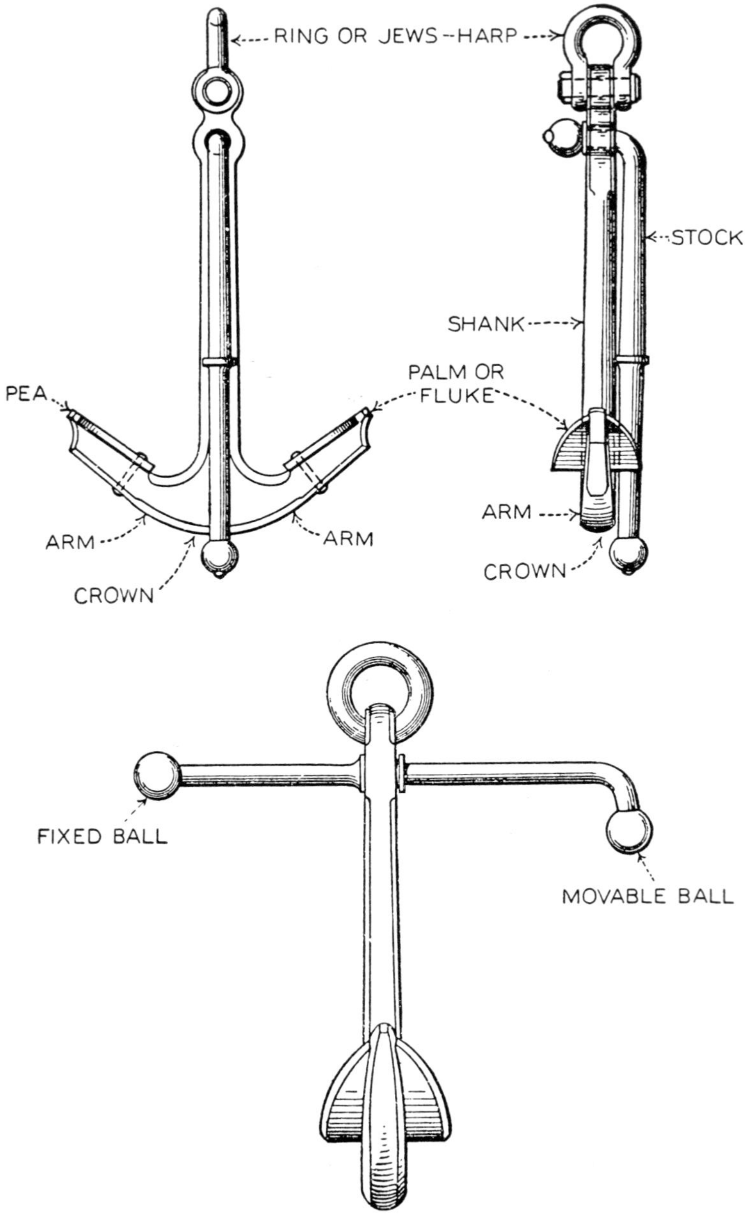

FIG. 8.1 OLD-FASHIONED ANCHOR

end of the stock. From this position any drag on the chain to one side "cants" or capsizes it, pulling the stock down horizontally upon the bottom and pointing the flukes fair for biting. As the drag continues, the fluke is forced into the ground. If the anchor is well designed and enough scope of chain is used so that the pull is approximately parallel to the bottom, the heavier the pull, the deeper the fluke digs in. To obtain this effect a long enough scope of chain must

FIG. 8.2 LARGE NAVY STOCKLESS, 40,000-LB BOWER ANCHOR ON *U.S.S. MIDWAY* (CVA 41) CLASS

be used so that the pull of the anchor will be approximately along the bottom. Use of too short a scope will lift the shank of the anchor and break out the fluke in a series of jumps.

Stockless anchors present an enormous advantage in ease of handling and stowing. On account of the absence of a stock, they can be hoisted directly into the hawse-pipe and stowed there ready for letting go quickly. Some features common to stockless anchors are:

1. The arms are pivoted upon the shank and can swing from 30 to 45 degrees on either side.
2. The palms are in the plane of the arms instead of at right angles to it.
3. As a result of this construction, both flukes should bite if either one does.

4. The arms carry a shoulder with a sharp edge at the crown which takes on the bottom and throws the arms downward to ensure the flukes biting.

The stockless anchors most commonly encountered are the Baldt, the Dunn, or the Norfolk (Fig. 8.2) which is a Navy-manufactured anchor of the Dunn

FIG. 8.3 LIGHT-WEIGHT TYPE (LWT) NAVY ANCHOR. LST STERN ANCHOR, 3,000-LB SIZE. *Official U.S. Navy Photograph*

type. The Navy stockless is used on most men-of-war on account of its convenience in handling. Stockless anchors do not have as much holding power as an old-fashioned anchor of the same weight or of the same fluke area, which is a better measure of holding power than weight. With too short a scope of chain or even under a steady pull, a stockless anchor has a tendency to disengage its flukes by gradually turning over and rolling them out. It also has a tendency to clog or ball with mud in a muddy bottom. If this occurs and the anchor breaks out, the arms may pivot to an angle where it is impossible for the flukes to bite again. It can offer no resistance to dragging except its weight.

Many experiments have been made to overcome the foregoing disadvantages of the stockless anchor while preserving its handiness. Some of the best results obtained have been found in the design of the lightweight type of anchor (Fig. 8.3) which is rapidly coming into use for many types and sizes of ships. This anchor has extra long and sharp flukes and an anti-rolling rod called *the stock* through the crown. Under a steady pull it shows a tendency to bury itself deeper in the bottom like the old-fashioned anchor and a disinclination to roll. In fact, difficulty in breaking it out before getting under way may be a disadvantage. It is claimed that this anchor has about three times the holding power of an old-fashioned anchor of the same weight and about ten times the holding power of similar stockless anchors of the conventional types. The stock through the crown does not interfere with stowing it in the hawse pipe.

8.6. Anchor Gear. *Chain Cable.* Chain is made to Navy and commercial link standard dimensions which are the same, 6 wire diameters long by 3.6 diameters wide. Anchor chains furnished in time of war are of all types: die-lock, forged and welded, stud-link, and cast steel. Die-lock and high-strength welded steel types of chain are considered standard and will replace other types. (See Fig. 8.4.)

FIG. 8.4 SWIVEL-SHOT, LARGE-CHAIN, 3½" HEAVY-DUTY DIELOCK. *Official U.S. Navy Photograph*

Connecting Shackles. Detachable links are used for connecting the shots of anchor chain. They have replaced the old Navy standard U-shaped shackles and also the Kenter connecting shackle. Detachable links of commercial design are also furnished for use with cast-steel chain. (See Fig. 8.5.)

Bending Shackles. Bending shackles for use with cast steel chain (conforming to American Bureau of Shipping requirements) and with die-lock chain are used for attaching the anchor to the chain cable. (See Fig. 8.6.)

Mooring Swivels. Forged steel swivels with two detachable links attached at each end are for use in mooring. (See Fig. 8.7.)

Housing Chain Stoppers. Navy standard housing chain stoppers are used for holding the anchor taut in the hawse pipe or for riding to an anchor or holding the anchors when the anchor chain is disconnected for any reason. The large chain stopper wrenches are used for equalizing the strain on the stoppers when riding to an anchor with more than one stopper in use and for securing the anchor in a hawse pipe. (See Fig. 8.8.)

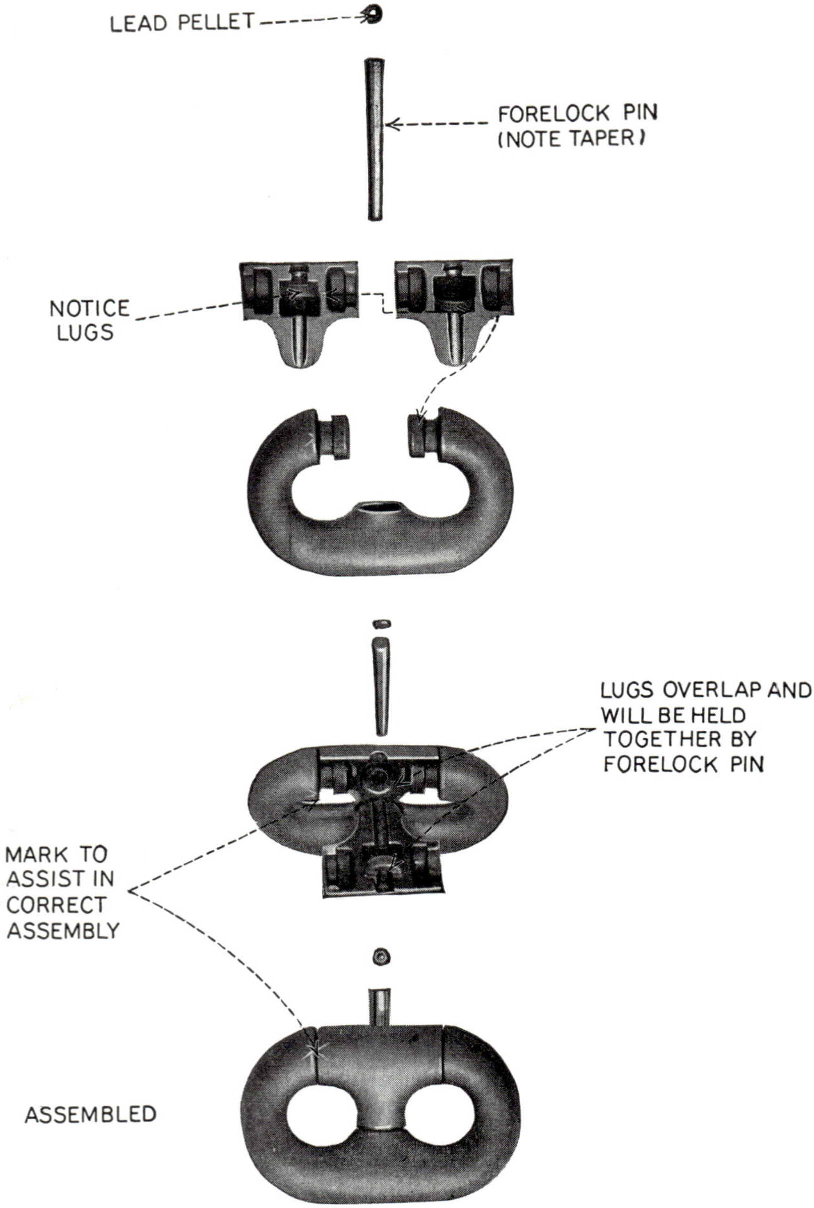

FIG. 8.5 DETACHABLE LINK

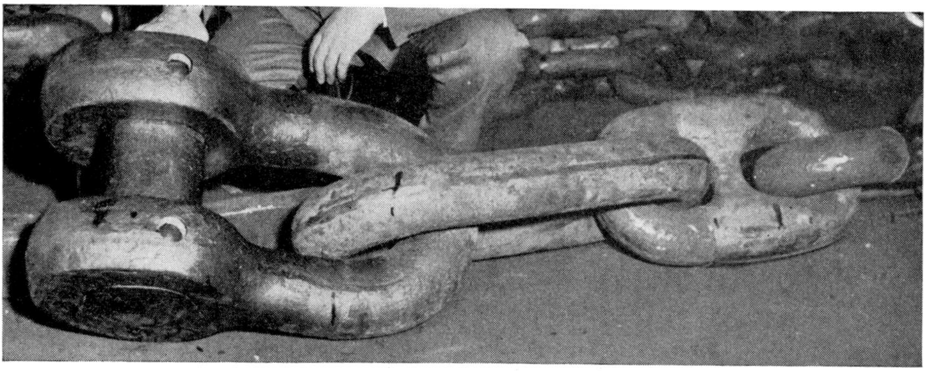

FIG. 8.6 LARGE-SIZED BENDING SHACKLE FOR 3½" HEAVY-DUTY DIELOCK CHAIN. *Official U.S. Navy Photograph*

Mooring Shackles. Forged-steel shackles are used for attaching the anchor chain to mooring buoys. An additional special lightweight mooring shackle, not possessing the full strength of the anchor cable, is used for some ships.

FIG. 8.7 MOORING SWIVEL, LARGE SIZE. *Official U.S. Navy Photograph*

Shackle Tool Sets. Tool sets including spare taper pins and locking plugs are provided for use in assembling and disassembling detachable links.

Clear Hawse Pendants. A wire rope pendant, 5 to 15 fathoms long, fitted with a thimble at one end and a thimble length of open link chain and a pelican hook at the other end, is used in clearing a hawse which has been fouled by the anchor cables.

Dip Ropes. A fiber rope pendant fitted at one end with a thimble and a dip shackle large enough to engage a link of the anchor chain is provided for use in mooring or clearing a hawse.

Outboard Swivel Shots. Standard outboard swivel shots consisting of detachable links, regular chain links, a swivel, end link, and bending shackle are fitted on most vessels to attach the anchor cable to the anchor. These vary in length up to approximately 5 fathoms and are also termed bending shots. The taper pin in the detachable link in the outboard swivel shot is additionally secured with a wire-locking clip.

Chain Cable Jacks. A cable jack consisting of a lever mounted on an axle and two wheels is used for handling anchor chain of 2¾-inch size and larger. An anchor bar of the pinch-point crowbar type is used for smaller sized chain.

Mooring Hooks. These hooks are used on destroyers and smaller ships for facilitating mooring to a buoy.

FIG. 8.8 CHAIN STOPPER. *Official U.S. Navy Photograph*

8.7. Die Lock and High-Strength Welded Steel Stud-Link Chain. Die-lock and high-strength welded steel chain are standard. These types of chain are now made to the same physical specifications. They have been tested in service and have proved their superiority. Cast-steel, die-lock, and high-strength welded steel chain are capable of withstanding great shock, have uniform dimensions, and the elastic limit of the links is high.

Under the usual service conditions, the links of these types of chain do not stretch or become deformed, and the chains will operate smoothly over the wildcat during the period of their entire useful life. Cast-steel chain can be distinguished by the fact that the studs are solid and an integral part of the links, and each common link in the shot is identical. High-strength welded steel chain in some types is made up with alternate solid-forged links having integral studs. Every other link has the stud welded in place. In one type of high-strength, welded steel chain which is constructed of alternate solid-forged and welded links each welded link is reforged after welding, and the entire chain has the appearance of being made of solid-forged (or cast-steel) links with integral studs throughout. The studs of the die-lock chain are also an integral part of the link; they are, however, split through the middle. The fact that the studs of cast-steel, die-lock, and high-strength welded steel chain cannot fall out is a great advantage because this eliminates the danger of the chain kinking and the pounding of links on adjacent links.

Die-lock links are made of two-forged pieces, both roughly U-shaped. Two stems of one piece contain a series of paralleled indentations, giving them the appearance of screws. The socket piece has holes at each end of the U. In joining the two pieces to form a link, the pierced socket section is heated, then the stems of the other section are thrust into the holes. The socket section is then pounded with a drop hammer, forcing its material around the indentations in the stems in die blocks.

8.8. Chain Identification Marks. Each shot of the chain usually bears a serial number that is stamped, cut, or cast on the inner side of the end links of each shot at the time of manufacture. In the case of cast-steel chain this number is preceded by the letters C.S. If an end link is lost or removed from a shot, this identification number should be cut or stamped on the side of the new end link of the altered shot. Cast-steel and some types of high-strength welded steel chain have these markings on the studs of alternate links only.

Each shot of die-lock chain has a serial number and date of manufacture stamped on the inner side of the end links. The studs of such chains are marked "U.S.N." on one side of the stud and the wire diameter of the chain on the other side.

8.9. Anchor Identification Marks. Every anchor, except small boat anchors and lightweight type (LWT) anchors 100 pounds and less in weight, when purchased for Navy use and before delivery to respective naval shipyards, has cast or cut in its crown a serial number. This serial number should not be confused with the weight number which also appears on the anchor. It is the practice on stock or old-fashioned anchors for one side of the crown to be marked with the initials of the inspector, the name of the manufacturer or commercial name of the anchor, and the serial number of the anchor. In the case of lightweight type anchors, this legend may appear on the shank.

On the opposite side of the crown the weight of the anchor in pounds, the year of fabrication, and "U.S. Navy" are cast, stamped, or cut by the manufacturer. The same practice is adhered to in regard to the stockless anchors, except that the markings appear on each side of the flat of the crown.

Ground Tackle

Ground tackle consists of all of the anchors and anchor chain in a vessel as well as the necessary handling gear. The sizes, weights, and amount supplied to vessels in service are determined after consideration of the characteristics and operating conditions of the vessels and of past experience.

8.10. Navy Type Detachable Link (Fig. 8.5). The Navy detachable link has been adopted as standard for use as a connecting link for joining shots of anchor chain on naval vessels. It has replaced the old U-shaped connecting shackle and also the Kenter connecting shackle which consisted of two interlocking half sections held in place by a stud and a taper locking pin.

The Navy Type Detachable Link consists of a C-shaped link with two coupling plates which form one side and the stud of the link and a taper pin which holds the parts together and is locked in place at the large end by a lead plug. The chief advantages of the detachable link are:

1. They are approximately the size and shape of the common links of stud link chain with which they are designed to be used, and they will, therefore, ride as smoothly over the wildcat in both the flat and vertical position as the common links.
2. The use of long shots (40 or 45 fathoms) immediately inboard of the outboard swivel shot is not necessary with the detachable link, as there is no disadvantage in having it on the wildcat when the chain is up and down and the windlass is under the strain of breaking out the anchor.
3. End links enlarged sufficiently to take the eye of the old U-shaped shackles have hitherto been necessary. The detachable link can be threaded through a common link which permits the omission of these enlarged links in conjunction with the old U-shaped shackle. In the past this has been the main cause of cables slipping on the wildcats. The omission of the enlarged end links and the old-type shackles in cable not only eliminates this jumping but permits the pockets of the wildcats to be designed to closer tolerances.
4. Because both ends of the detachable link are closed, they do not catch on the hawse pipe or deck fittings. This feature has reduced the number of losses of anchors and chain.

When assembling detachable links, care should be taken to ensure that the parts are correctly matched. Detachable link parts are not interchangeable, and matching numbers are stamped on the C-link and on each coupling plate to enable proper identification for correct assembly. The matching surfaces should be slushed with a mixture of 40 percent white lead and 60 percent tallow by volume prior to final assembly. Vessels outfitted with cast-steel anchor chain are furnished detachable links of Navy and also commercial manufacture, such as E-Z joining links (Naco), Esco connecting links, and riveted (Naco) connection links.

8.11. Riding and Housing Chain Stoppers (Fig. 8.8). Riding and housing chain stoppers consists of a turnbuckle inserted in a short section of chain, a slip or pelican hook attached to one end of the chain, and a shackle at the other end. The stopper is secured by the shackle to a permanent pad on the vessel's deck. When in use, it is attached to the anchor chain of the ship by straddling a link with the tongue and strong back of the pelican hook. A certain number of standard chain stoppers are supplied to naval vessels as a part of their equipment. It is the latest practice in the case of destroyers to fit one stopper in the way of each chain cable. On auxiliary vessels and vessels of merchant type, two in the way of each chain cable are fitted if it is practicable to do so. On destroyers, the stopper is made of the same strength as the anchor chain. Because the windlass on these vessels has only one wildcat, it is intended that one chain shall be held by the windlass and the other by its chain stopper. In the case of other vessels, chain stoppers are furnished of such size that the strength of each is equal to about 40 percent of the chain cable with which it is used. The stoppers should not be solely relied upon for holding the anchor. Upon anchoring, the wildcat brake band should first be set up tight, and the stoppers then used to back up the brake band. The wildcat should then be disconnected from the engine.

8.12. Purposes of Chain Stoppers. The purposes for which chain stoppers may be used are as follows:

1. To ride to when at anchor, in addition to the use of the brake band on the windlass.
2. For letting go the anchor more quickly than can be done by the brake band when vessels are anchoring in formation.
3. As an emergency fitting in case the brake band of the windlass should get out of order.
4. To hold a chain from running out, while it is being taken off the wildcat, to permit another chain to be put on for heaving in. On destroyers, to hold a chain being bitted or unbitted or brought to the windlass.
5. To hold the anchors taut in the hawse pipes when housed, except in cases where the end of the anchor shank or the anchor shackle projects above the deck line (nonstandard).
6. To hold an anchor chain when disconnected for the purpose of attaching the mooring swivel. Lashing can be resorted to in addition.

Anchoring

Letting go a single anchor is perhaps the simplest method of securing a ship to the bottom, and if the holding ground is good she should ride easily in bad weather provided ample scope of chain is used. The disadvantages are that in a strong current or in a gale she may sheer considerably, and also when a ship is anchored, it swings to the combined effects of the wind and current. Therefore it is necessary to have an unobstructed area equal to a circle whose radius is the length of the ship plus the scope of chain used. If, for some reason, the anchorage does not afford such an area, the ship must be moored.

8.13. Letting Go. In modern ships with heavy ground tackle, the anchors are commonly housed in the hawse pipe and secured by chain stoppers which engage the chain by a slip or "pelican" hook.

To ensure that an anchor will let go *immediately* after the housing chain stopper is released, prepare as follows: connect up windlass wildcat, slack and release the outboard housing chain stopper and then engage the pelican hook of this housing chain stopper to the first horizontal chain link abaft the link previously engaged, release friction brake, heave in until windlass wildcat has the strain, cast off the after chain stopper(s), walk out anchor until the outboard housing chain stopper has the strain and there is one link of slack chain abaft stopper, set up friction brake lightly to prevent the slack chain from going into the chain locker, and disconnect windlass wildcat.

If the drift between the hawse pipe and the chain locker is considerable, it is well to rouse up a few links of chain and lighten the slack forward to a point just abaft the stopper. Care must be taken that all is clear below decks and in the chain locker.

To let go, the bale shackle pin is pulled out, and then the bale shackle of the chain stopper is knocked off the pelican hook with a sledge.

Always bear in mind that the anchors may be unexpectedly required when on soundings, in narrow channels, restricted waters, or working around docks, etc. If they are ready for *instant* use, they may save worry and trouble. The anchor should always be let go with the ship moving slowly either ahead or astern, to avoid paying the chain down on top of the anchor.

8.14. Anchoring in Deep Water. Where it becomes necessary to anchor in very deep water, it is absolutely essential that the ship should be going dead slow. As the anchorage is approached at very slow speed, the usual practice is to walk out the anchor to within 5 to 20 fathoms from the bottom at the proposed anchorage, fasten the stopper to the chain, and disengage the windlass, making the anchor ready for "letting go," and then let go. Maintain only enough headway to avoid paying the chain down on top of the anchor. The details of handling the windlass for anchoring in this way will vary with the type of windlass used, but it will be found that even where the ship is dead in the water and where the anchor is let go with only a few fathoms of drop, the weight of the chain alone will cause it to run out violently. In extreme cases, where the depths run to 40 and 50 fathoms, it may be advisable not to "let go" but to "walk out" the chain by the windlass engine until the anchor is on the bottom and the necessary scope of chain out.

8.15. Anchoring at High Speed. If obliged to let go at a higher speed, or if for any reason it does not seem safe to check the ship with a short scope, the chain should be allowed to run until the ship loses her way sufficiently to make it safe to snub her. There is no great harm in running out 75 or even 90 fathoms of chain and afterward heaving in to a shorter scope; and it should be remembered that, in cases where the headway has to be checked by bringing up on the chain, the danger is less with a long scope than with a short one.

The danger connected with letting go while under considerable headway is often overlooked because the damage resulting does not necessarily show itself at once. The excessive strain may distort and weaken the links of the chain without actually parting them. The result is that the chain may give way at some time under a comparatively moderate stress. The practice of reducing the ship's headway by means of her ground tackle may introduce strains sufficient to cause fracture and will in any event be very apt to strain the chain beyond its maximum safe load equal to the proof load.

8.16. Scope of Chain for Maximum Holding. The scopes given in the following table are the "optimum" scopes for maximum holding. If longer scopes are used, the chain may be stressed beyond its safe service working load; if shorter scopes are used, the anchor will tend to drag before developing the full safe load on the chain. These figures apply substantially, regardless of the size of the ship, provided the ship is furnished with a properly balanced outfit of ground tackle and is given a safety factor of 4 on the ultimate strength of the chain. The scopes shown for the greatest depths could be obtained only by bending additional shots to the standard lengths of chain cable.

Scope Table

Chains	Depth in fathoms (outboard lip of hawse pipe to bottom)									
	5	7½	10	15	20	25	30	35	40	45
Cast-steel chain (fathoms).....	64	78	91	110	127	142	155	166	178	188
Die-lock N.E. steel chain or 1.25 manganese steel chain (fathoms).................	74	90	104	127	146	164	178	192	204	216
Die-lock nickel steel chain (fathoms).................	78	95	109	133	154	174	188	202	216	228

It is a common rule under ordinary circumstances to use a length of chain equal to five to seven times the depth of the water. This is satisfactory in depths of water not exceeding 18 fathoms. This amount of chain is perhaps enough for a ship riding steadily and without any great tension on her cable. On the other hand, if conditions necessitate, the chain should be veered when anchored in shallow depths to the maximum indicated in the scope table.

If greater holding power than that given by one anchor with the scope of chain shown in the table is necessary, it is better practice to drop a second anchor even with moderate scope of its chain than to rely upon the one anchor with a longer scope. Of course, in the case of extreme necessity when the greatest holding power is necessary, all anchors should be dropped and the chain veered the greatest possible scope. If there is ample sea room, it would be better to reduce the scope to the amounts shown in the table and accept the possibility of dragging anchor rather than risk breaking the chain.

8.17. Anchor Chain Marking. *1. Painting.* Anchor chain shall be painted as follows to serve to identify the length of cable paid out:

One link on each side of the 15-fathom detachable link shall be painted white.

Two links on each side of the 30-fathom detachable link shall be painted white.

Three links on each side of the 45-fathom detachable link shall be painted white, etc.

Detachable links shall be painted as follows: 15-fathom D link, red; 30-fathom D link, white; 45-fathom D link, blue; 60-fathom D link, red, etc.

The exception to the foregoing is that all of the links in the last 15-fathom shot inboard shall be painted red, and all of the links in the next adjoining 15-fathom shot shall be painted yellow.

On auxiliary vessels where the distance between hawse pipe and wildcat is short, and consequently, only a short time is available for painting the chain while it is being heaved in, it may be desirable to limit the number of painted links on each side of the detachable link to one.

2. Marking. Anchor chain may be marked by turns of wire on the studs of certain links. The number of links counting away from the detachable link is used as a marker for that shot.

The first link at each side of the 15-fathom detachable link has one turn of wire around the stud.

The second link at each side of the 30-fathom detachable link has two turns of wire around the stud.

The third link at each side of the 45-fathom detachable link has three turns of wire around the stud, etc.

8.18. Care to Prevent Bending Chain. Anchor chain when being used should not be subjected to short bends. Care should be taken to ensure whenever possible that anchor chain is not subjected to bending, such as may occur when the cable is lying across a vessel's stem, when a vessel is riding to a single anchor in a strong wind or current and the vessel is *"horsing"* or tacking back and forth, or when the cable is rove through a buoy ring or passed over a bolster of small radius. Chain is not as strong when subjected to such transverse bending. This is especially true in the case of die-lock detachable links, and therefore extra precaution should be taken to prevent subjecting these links to transverse bending.

8.19. Care of Ground Tackle by Ship's Force. Anchors, chains, and appendages should be kept in good condition by the ship's force. The chain cables should be overhauled whenever necessary and precautions taken to see that the various shots are properly marked and in good order. As the chain comes in when getting under way, each link should be examined for cracks and for other defects.

8.20. Periodical Inspection and Painting of Ground Tackle in Service by Ship's Force. Once each quarter and more often if necessary, all anchor cables in sizes up to and including 1½ inches should be ranged on deck and examined throughout their entire length. If necessary, they should be scaled and cleaned of rust and other foreign matter. Detachable links should be disassembled and examined for excess wear or corrosion, and, where conditions warrant, the links should be replaced by new ones. Before reassembly, the links should be white-leaded. The detachable link located in the outboard swivel shot is fitted with a corrosion-resisting steel locking wire which serves to hold the taper pin in position. Disassembly of this link requires the removal and probable destruction of the locking wire, and the availability of replacement wire of the same type should be established prior to removal for inspection. Shackle bolts, locking pins, and swivels should be carefully examined and put in order, and such parts as require it should be coated with the special black chain paint furnished vessels for this purpose. In cold weather it is desirable to apply some heat to counteract the natural thickening of this paint. This may be done by an immersion electric heater or a steam coil. Experience has also shown that, when left standing for a considerable period, the turpentine substitute may evaporate to a considerable extent, with the resultant thickening of the paint. Vessels receiving anchor chain coated with green paint from stores should leave this coating intact and cover it with black chain paint.

Chain of sizes in excess of 1½-inch wire diameter should be overhauled, wirebrushed, and placed in a good state of preservation as often as is necessary. At least once each 18 months all anchor chain cable, regardless of size, including shackles and shackle pins, and detachable links, should be examined, overhauled, and placed in a good state of preservation. Shackles and shackle pins and detachable links should be refitted and greased or white-leaded, and identification marks should be restored if necessary. To distribute the wear uniformly throughout the entire length of the cable the shots should be shifted to a new position as necessary. In the case of vessels having cable the shots of which are connected with detachable links, 40- or 45-fathom shots may be shifted to any position in the cable which, in the Commanding Officer's opinion, will tend to distribute the wear evenly throughout the cable. In the case of vessels the shots in the cables of which are connected with U-shaped shackles, the 40-fathom shot should remain the first shot inboard of the outboard swivel shot regardless of wear. If serious defects are discovered during this overhaul of the anchor cables, the defective shots should be shifted to the bitter end of the cable until replacement can be accomplished. See the *Bureau of Ships Manual* for complete instructions for maintaining anchor equipment.

8.21. Weighing Anchor. In heaving in, the windlass and chain can be relieved of considerable strain by a judicious use of the engines and rudder. To do this, the forecastle detail must keep the bridge fully informed as to how the chain "tends"; whether the chain is "taut" or "slack"; when the anchor is at "short stay"; when the chain is "up and down"; and when the anchor is "aweigh." As the anchor is hove in, the report is made to the bridge, "Anchor in sight, sir," "Clear or foul anchor," as the case may be; "Anchor clear of the water, sir," "Anchor is up, sir." The Captain will then direct, "Secure the anchor" or "Get the anchor ready for letting go."

In case the chain tends across the bow, it may be cleared by stopping the windlass and going astern.

8.22. Foul Anchor. While modern double-fluke anchors are much less likely to foul than the old type, they occasionally give trouble in this way. A ship whose anchors house in the hawse pipes may be greatly embarrassed by the lack of facilities for lifting the anchor to a point where it can be hung securely and where the chain can be handled conveniently. As a rule in such cases, put a chain stopper on the chain, disengage the wildcat and let anchor go, engage wildcat, and heave in.

Under conditions such that the anchor may be expected to foul, it is a good rule to "sight" it frequently; and indeed this is advisable under any conditions when a ship remains at single anchor for a long time. It is especially important if bad weather is found to be approaching after lying for some time under circumstances which make it probable that the anchor may be foul. For sighting, the anchor may be weighed and another anchor let go when the chain of the first one is "up and down."

An anchor sometimes becomes so well dug in or so fouled in rock or coral that it cannot be raised in a normal manner. Here the use of a wire strap around the crown may be needed. Anchorage charts often reveal locations where anchors have been lost, such as St. George's Channel, Bermuda. If necessary to anchor there it might be wise to consider fitting a crown strap and a buoyed work wire to the anchor before letting go. An anchor that has been lost with some chain attached can usually be recovered by the use of grapnels or an *anchor hawk*, as described in NAVSHIPS 250-880-5.

8.23. Anchoring by the Stern. It is often convenient and sometimes necessary for a vessel to anchor by the stern. For this, an anchor of moderate size is usually sufficient. It is an excellent plan to carry such an anchor at the stern ready for letting go. A shackle should be kept at hand, and a thimble turned into the end of the best wire hawser. This hawser can be shackled up to the anchor, and all made ready for letting to in a very short time. When a ship comes into a narrow or crowded anchorage with a fair tide where perhaps unanticipated difficulties are found, it might be of the greatest value.

8.24. Anchoring by the Stern, Using Bower Anchor. If it should become necessary to use a bower anchor over the stern, the simplest way to deal with the situation is to walk the anchor out until it hangs outside the hawse pipe with its ring just clear. At the same time make preparation for unshackling at the outboard swivel shot and hold the chain forward of its connecting shackle by a chain stopper. Pass the end of the best wire line out through the stern chock, outboard of everything, and take it forward and shackle securely to the free link of the outboard swivel shot. In this as in all other cases where the anchor is to be used without its own chain, it is very important to use a "weighing line" and buoy-rope on the crown of the anchor. Let go when ready by knocking off the bale shackle of the chain stopper.

If it is desired to ride by the bower anchor chain rather than by the wire line, pass the end of the best wire line forward from the stern-chock as before and stand by to shackle it to the chain when ready. Run in at slow speed to the point where the anchor is to be placed, keeping the ship under control; let go, and veer to 60 or 90 fathoms as desired, bringing to with the anchor chain shackle well inside. Pass the end of the wire line inside the hawse pipe. Stopper the chain well and bend on an easing-out line just forward of the chain shackle. Take the easing-out line to the warping head and take the strain, disconnect the chain, and shackle the end of the wire line to the end link. Take the hauling end of the wire line to the after winch or capstan. Ease out forward, heave in aft. Let the easing-out line go with the chain, standing clear of the end. Range ship ahead and let the other bower anchor go when in position. Heave the wire line and chain aboard aft and secure on deck as practicable. Ship is now anchored fore and aft.

In many harbors, a swell sets in on the beam of vessels riding to the wind which causes them to roll incessantly. In such cases, a stream anchor planted off the bow with a line from the quarter admits of springing around, head to the swell. In the tropics, a similar plan adds much to the coolness of the ship by bringing the wind abeam.

8.25. Riding to a Single Anchor. A vessel at single anchor in a strong tideway is likely to sheer considerably. This movement brings the current first on one side and then on the other and drives the vessel across the stream until brought up by her chain often with a violent shock. This may be prevented in a great measure by holding her with a steady sheer away from her anchor, by putting the rudder over as far as may be necessary, and keeping it there. The stern is driven over to one side and she is canted across the current and held there.

A ship is never in greater danger of dragging her anchor or parting her cable than when driving down with a slack chain, broadside on or partially so to wind or tide. Such a situation may of necessity arise in anchoring or may come about in sheering, as described above. It frequently happens in squally weather, where a ship swings in one direction during a lull just in time to be caught by a strong squall on the beam and be driven bodily off. It may be brought up with the chain taut across the stem.

In lying at an anchorage where such situations may arise, the greatest watchfulness should be exercised. Steam must be kept on the steering engine, a man must be at the wheel, an ample scope of chain should be veered, and a second anchor should always be ready for letting go at a moment's notice, even though there seems no chance of its being needed.

When the conditions are such that there is a possibility of dragging the anchor, a lookout should be posted to ensure instant notice if she begins to drag. The drift lead is useful, though not always to be trusted. This is a heavy lead kept on the bottom with its line made fast to some place well forward convenient for observation and left hanging with considerable slack. If the ship drags, the line tautens and tends ahead.

So long as a ship is fairly steady, a drift lead will usually give notice in case of dragging, but if she sheers about considerably, it cannot be relied upon. The farther forward it is used, the better, because the bow moves much less than the stern in sheering.

Good bearings of objects on shore are more reliable than the drift lead, and a range is best of all. Both of these are less trustworthy when the ship is sheering about than when she is steady because a range will open out when the ship swings and may seem to indicate that she is dragging. Its indications may be checked by watching the heading. Radar ranges to fixed objects ashore may also be used to detect dragging.

There are times when unexpected and unusual swells, seas, and currents set in toward an anchorage. In such cases the only thing to do is to get up steam promptly and shift to a safe anchorage or stand out to sea. The sailing directions should always be carefully read and every effort made to obtain the latest weather reports. There are also times when nearby ships will swing in opposite directions when there is little current or wind. In such cases each ship may heave in sufficient chain to avoid fouling; at other times one or more boats placed at the stern will exert sufficient power to push the ships clear or hasten their swing.

It is always advisable to keep a detachable link on deck where it can be reached conveniently for slipping suddenly if an emergency arises and to be sure that the pins can be driven out without difficulty. Tools for unshackling should be kept in a convenient place and never removed. A buoy and a buoy-rope at hand complete the preparations for slipping at short notice. In an exposed anchorage subject to sudden gales, these precautions are, of course, especially important.

If a vessel or other danger is seen drifting down upon you when lying at anchor in a tideway, by giving the ship a cant with the rudder thus bringing the current on the bow, and veering the anchor chain roundly, you may sheer well over across the tide and probably be clear of danger.

If an anchor is known to have dragged in a clay bottom, it should be picked up as quickly as possible; for it is certain to be "shod" (balled in mud) and to have lost much of its proper holding power. In letting go where the bottom is of this kind, it is important to give a good scope in the very beginning to prevent even the little dragging that is commonly to be expected as the anchor digs down to get its hold.

The plan of bending two cables together to obtain a "long scope" is not recommended if such a scope would result in exceeding the safe limit in the scope table. The fact must not be overlooked, however, that a defective link or shackle may result in disaster where a single cable is in use, and this may make it wise to let go a second anchor in cases where no chances can be taken. The vessel would then be moored.

Mooring

A ship is moored when she has two anchors down at a considerable distance apart and with such scope of chain on each that she is held with her bow approximately midway between them. A ship moored requires an unobstructed area reduced to a circle with a radius only slightly larger than the length of the ship.

8.26. Reasons for Mooring. There are two basic reasons for mooring: (1) To reduce the radius of the circle of the unobstructed area in which the ship will swing, by using an ordinary moor. Since it is desirable to make a taut moor under these conditions, the anchors should be so placed with respect to any current that a straight line connecting the two anchors would be parallel to the direction of current flow. (2) To snub the bow of a ship and prevent it from sheering in a current, gale, or hurricane, by using a bridle or hammerlock moor. Under these conditions, it is necessary to use a slack moor so the angle between the chains is about 90 degrees, and to place the anchors so that a straight line joining them would be perpendicular to the direction of current flow or expected wind.

8.27. The Ordinary Moor. In the ordinary moor the ship stands against the current (wind) to the proper position and lets go the first anchor which must always be the riding (upstream) anchor at that time. She veers on the riding chain, carefully laying out the chain so as to keep it taut and tending ahead as she drops down with the current (wind) to the position for letting go the lee or downstream anchor. When that position is reached, the second anchor is let go. She now veers on the lee chain and heaves in on the riding chain, taking care to lay the lee chain out properly until she is riding midway between both anchors with the desired scope of chain to each.

A mooring swivel is frequently used to prevent the anchor chains from fouling one another when a moored ship swings to the wind or current. When the mooring swivel is used, it is impossible to veer both chains.

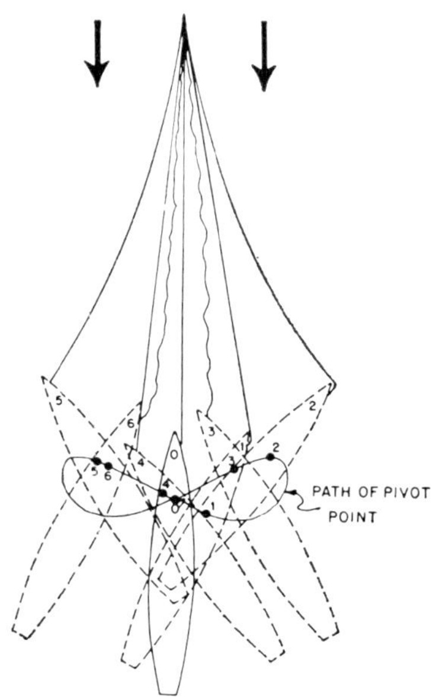

FIG. 8.9 HORSING MOVEMENT OF A SHIP RIDING TO A SINGLE ANCHOR. This diagram shows the violent figure-eight horsing motion of a ship riding with a long scope of chain out to a single anchor during a high wind.

8.28. The Bridle or Hammerlock Moor. In riding out a gale or hurricane, a vessel will often sheer violently back and forth across the wind (Fig. 8.9) and here the rudder has little effect in holding her with a steady sheer. This tacking back and forth is often called "horsing." Violent horsing can be cut in half by dropping a second anchor under foot with minimum scope out to it to act as a snubber, but even such reduced horsing may become excessive in a violent storm. In such cases, it is well to pick up the second anchor and redrop it at one extreme reach of a sheer, still using only a short scope, and then ride to this bridle (Fig. 8.10). The two chains will now work together to snub

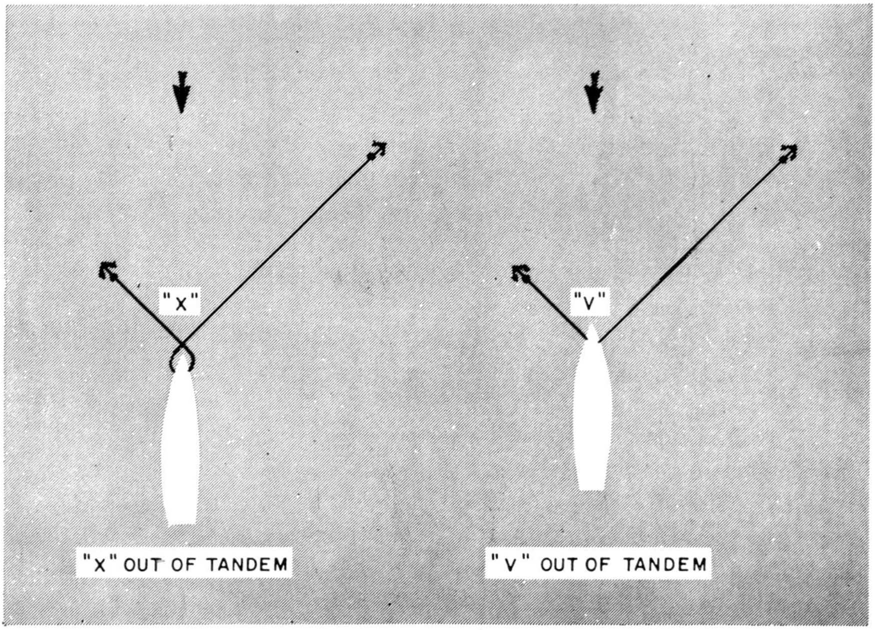

FIG. 8.10 ANCHORS ACTING AS A BRIDLE

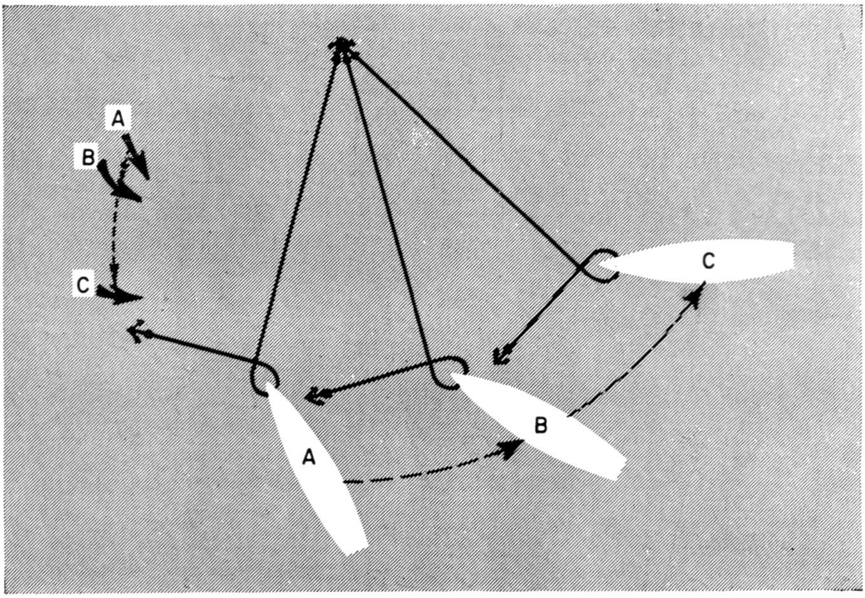

FIG. 8.11 EFFECT DESIRED AS SHORT-SCOPE ANCHOR DRAGS PROPERLY

the bow and hold it steady. The "horsing" will be almost completely eliminated and the main engines can be used with precision to offset the greater portion of the wind. The wind velocity—ship speed curves (Fig. 8.12) are of value here. Should the wind be from a hurricane or typhoon, and should the bottom be sand or mud, it is sometimes feasible to permit the short-scope anchor to drag

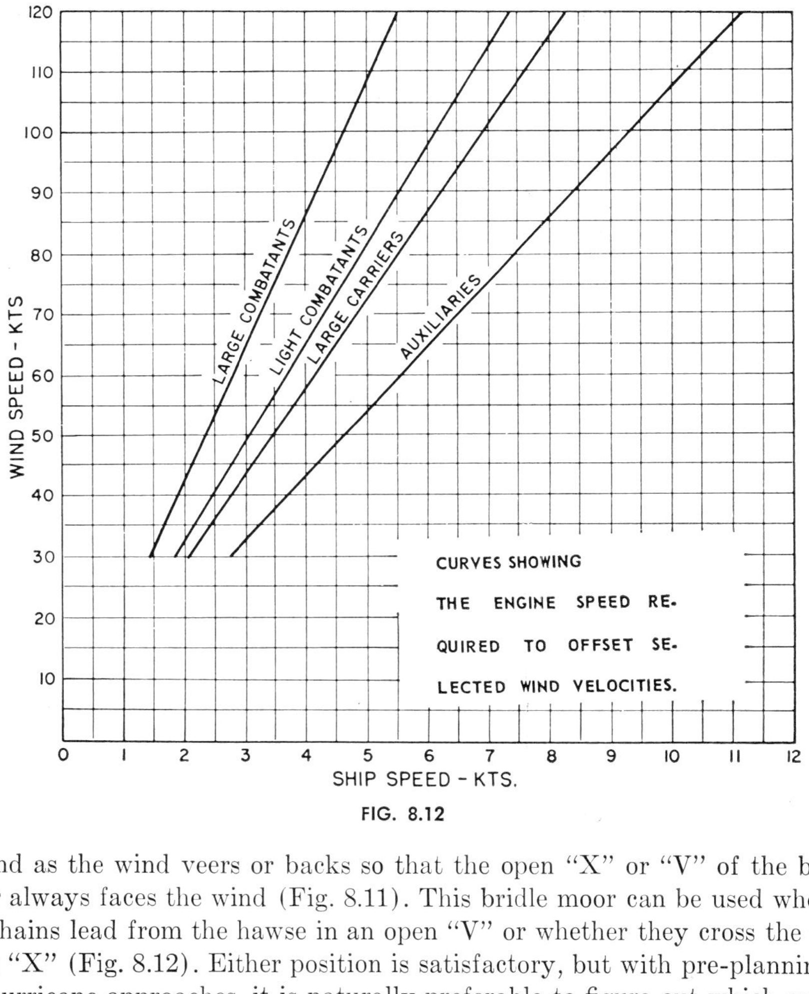

FIG. 8.12

around as the wind veers or backs so that the open "X" or "V" of the bridle moor always faces the wind (Fig. 8.11). This bridle moor can be used whether the chains lead from the hawse in an open "V" or whether they cross the stem in an "X" (Fig. 8.12). Either position is satisfactory, but with pre-planning as the hurricane approaches, it is naturally preferable to figure out which anchor should be used as the riding (long-scope) anchor in order to ride out the entire wind shift with an open "V" hawse as the short-scope anchor drags around.

8.29. Length of Scope When Moored to a Buoy. The need for veering chain when moored to a buoy during adverse weather is most important. If the ship is snubbed up close, the pull is generally upward, tending to break the buoy loose from its moorings. In addition, the lack of substantial weight of chain permits the transfer of the ship movements to the buoy to be horizontal with the weight of chain providing a cushion for the surges. Scopes as great as 45 fathoms have been used to good advantage, and where circumstances permit, the chain to the mooring buoy should be veered to produce a catenary such that the chain is never straightened out by the ship's movements. The Hammerlock Moor can also be used at a buoy to control "horsing," but care must be exercised to veer adequate chain before dropping the second anchor to ensure that it is dropped well outside the anchor clump to which the buoy is secured.

8.30. Wind Velocity—Ship's Speed. The number of knots of engine speed required by a vessel to offset any given headwind velocity should be valuable information to seamen and navigators. This will be particularly true during storms at anchor where such information could assist a master in his efforts to use his engines to prevent dragging anchor. It could also be of value when secured to a buoy. A navigator of a ship under way could use the information to assist in computing the retardation of his ship being caused by a steady headwind. It could also be used by a task force commander to compute the loss of speed by the guide of his formation, and by a master while fueling or transferring stores alongside a station ship, while underway, to compute the difference in loss of speed between his vessel and the station ship.

Four representative types of ships were selected and the curves for them (Fig. 8.12), as developed by the Bureau of Ships, are shown. It is believed that a person having general knowledge of the superstructure and hull design of ships could make an acceptable interpolation from these curves for any vessel.

The curves in Fig. 8.12 were developed using the basic formula:

$$\text{Wind load} = R_w A V^2$$

where R_w is equal to wind friction factor. (R_w is generally taken as 0.004.)
V is equal to wind velocity in knots.
A is equal to projected area above the waterline in square feet.

Three sets of calculations with varying R_w were used. The plotted values illustrated represent the mean of these values.

Since the curves represent the engine speed required to offset completely any given wind speed and hold the ship in equilibrium against the wind without any assistance from the anchors, an engine speed of perhaps 1 to 3 knots less than that selected from the curves should be used by a ship at anchor; otherwise, the ship would be in equilibrium with the force of the wind, and the anchor chains might cease to hold her head into the wind. Also there could be danger of overriding the anchors and fouling the ground tackle.

9

Cargo Handling and Replenishment

This chapter will describe various methods of handling cargo, most of which are common to the Merchant Service as well as the Navy. During time of national emergency, a knowledge of cargo handling on all types of vessels may become necessary.

The Navy is engaged in research to make cargo handling faster, safer, easier, and cheaper. Consequently, many improvements and modifications will be found on newer types of Naval ships that may not be on other cargo vessels. The Navy has many ships designed for specific tasks that have rigs peculiar to those types. Where rigging of a particular ship differs from the standard gear, it will be described.

9.1. Topping and Spotting Booms. Standard booms for cargo handling are 5 ton; for replenishment, 10 ton. Before cargo can be worked, the ship's boom must be topped, spotted, and secured in position. Operating personnel should be familiar with nomenclature and rigging procedures. (See Fig. 9.1.)

The rigging procedures are as follows:

1. Request power for winches and test them (steam winches should be thoroughly drained before use).
2. Assign men to winches, guys, whips, topping lift, and gypsy heads.
3. Lay out guys and preventers to proper fittings.
4. Lay out topping lift wire or bull line along the deck. Take five or six turns with topping lift wire, or bull line, around gypsy head in the opposite way from the whip (if the whip goes over the top of the drum, run the topping lift wire underneath the gypsy head). A man should be positioned as back-up man on the topping lift.

Tending the gypsy is the key job in topping or lowering booms. Five or six round turns about the gypsy are recommended when working with wire rope. Fewer turns are likely to slip, and more turns are likely to form slack in the wire. These slack turns may fall over the edge of the gypsy and cause the boom to drop. It is a good practice to assign one or two men to back up the man on the gypsy. These men can keep the wire from kinking, keep it clear of the winch, and aid the man on the gypsy in case of trouble.

5. Assign one man to overhaul the whip as the boom is topped.
6. Raise boom to the desired height.
7. Secure topping lift as follows:

a. *Single Topping Lift.* Shackle bull chain to pad eye as shown at (31) in Fig. 9.1, and slack off on the bull line until bull chain takes the strain. Throw bull line off the gypsy head and secure it to the topping lift cleat with a minimum of three round turns and three figure eights.

b. *Multiple Topping Lift.* Apply stopper chain to topping lift wire, using stopper (rolling) hitch and two half hitches. Take turns around the wire with the remainder of the chain and hold it. Surge the topping lift

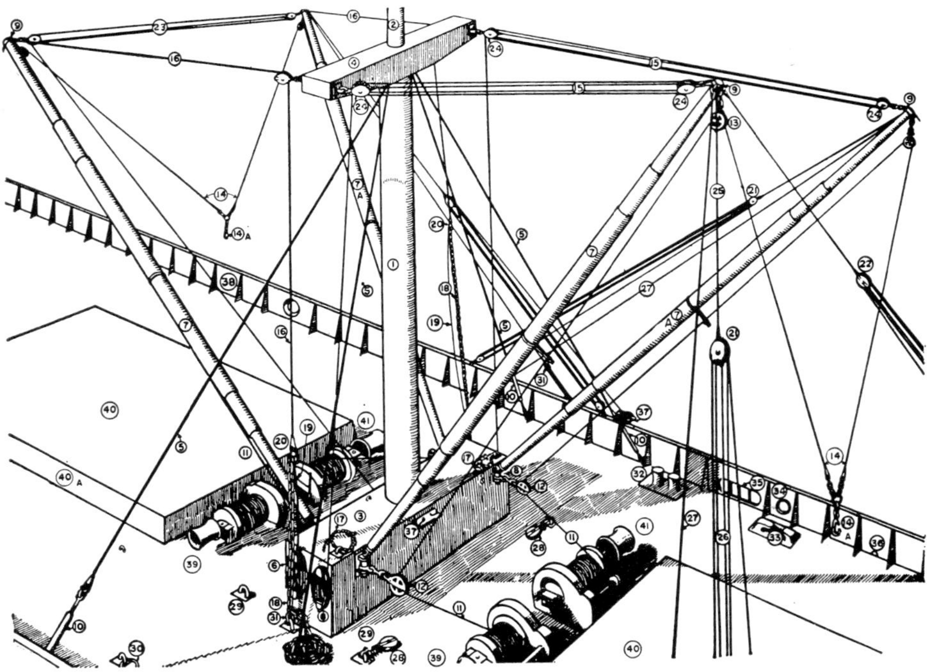

FIG. 9.1 CARGO HANDLING GEAR

NOMENCLATURE FOR FIGURE 9.1

1. Mast
2. Topmast
3. Mast table
4. Crosstree
5. Shroud
6. Topping lift cleat
7. Hatch boom

7A. Yard boom

8. Gooseneck
9. Linkband
10. Turnbuckle
11. Cargo whip
12. Heel block
13. Head block
14. Cargo whips

14A. Cargo hook

15. Topping lift (multiple)
16. Topping lift (single)
17. Stopper chain
18. Bull chain
19. Bull line
20. Bale
21. Outboard guy
22. Inboard guy
23. Midship guy
24. Topping lift block
25. Guy pendant
26. Guy tackle
27. Preventer
28. Snatch block
29. Pad eye
30. Pad eye and ring bolt
31. Shackle
32. Bitts
33. Open chock
34. Closed chock
35. Freeing port
36. Scupper
37. Cleat
38. Bulwark
39. Hatch winch
40. Cargo hatch

40A. Hatch coaming

41. Yard winch

wire until the stopper takes the strain and belay it as described for the single topping lift. Remove the stopper.

8. Spot booms in a working position by hauling on the guys. The yard boom is positioned over the pier, clear of the ship's side. The hatch boom is spotted slightly past the centerline over the hatch.

9. Set up on outboard guys and preventers. Guys should be slightly more taut than preventers. Set the inboard or midships guys as taut as possible by hand. Shackle the cargo whips to the cargo hook and pick up a load. Raise the load until the angle formed by the whips is about 120 degrees. Now equalize the outboard guys and preventers by easing off the guy tackles. As outboard guys and preventers are being equalized, take in all slack in the inboard or midship guys. It is a good practice, when originally spotting the booms, to swing them slightly wider than desired. When guys and preventers are equalized, the booms will move inboard into position.

There are several methods of raising and lowering booms. The standard practice is to apply the topping lift wire directly to the drum of the winch. This is the safest method, but the time required may be prohibitive.

Cargo ships being constructed, and some already in service, have special topping lift winches installed on the masts and king posts. These winches offer greater speed in raising and lowering booms. To top booms you merely run the winch until the boom is at the desired angle.

Another method of topping and lowering single topping lift booms is by means of the cargo whip which is led from the head block through a fairlead block at the base of the mast, then shackled to one of the top links of the bull chain. By taking in on the whip, the boom is raised; by slacking off, the boom is lowered. This is the least desirable method and should not be considered unless the other methods cannot be used.

9.2. Yard-and-Stay Method. In the yard-and-stay method of cargo handling, two booms are used. One of these booms is called the hatch boom and it plumbs the hatch. The other is called the yard boom and it is rigged out over the side so that it plumbs the dock or pier. (See Fig. 9.2.)

The cargo whips coming from the hatch and yard winches are rove through their respective heel and head blocks and are shackled to the same cargo hook.

If the whip has a thimble spliced in the end in the usual manner, it may be impossible to reeve the whip through the block, making it necessary to remove the whip from the winch drum so that the winch end may be rove through. The Navy consequently has adopted the following method: A large eye is spliced in the whip and the thimble held in place by a wire rope clip. It is then an easy matter to remove the thimble, reeve the whip through the block, and replace the thimble.

The winch controls are usually located in such a position that one man can operate both winches and have an unrestricted view of the hold.

A load is moved from hold to pier in the following manner: The yard whip is kept slack as the hatch whip hoists the load from the hold and clear of the coaming. Then, by heaving around on the yard whip and paying out on the hatch whip, the load is racked (swung) across the deck and over the side. When the load is plumbed under the yard boom, the hatch whip is slacked off and the yard whip lowers the load to the pier.

Nearly all methods of rigging yard-and-stay cargo-handling gear for heavy lifts require that the cargo whip be doubled-up and a runner block used.

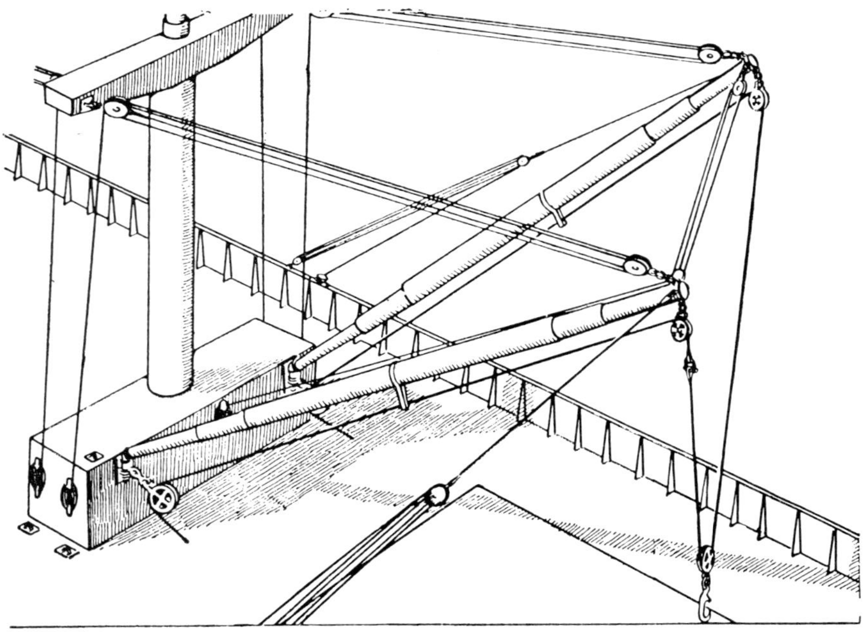

FIG. 9.2 YARD-AND-STAY WITH MIDSHIP GUY

Doubling-up the whip accomplishes two things: it doubles the load that may be lifted by the whip, and it reduces the load on the winch by half.

Most yard-and-stay rigs use ⅝-inch wire; therefore a block with at least a 12-inch sheave must be used for a runner block. Larger whips, of course, will require larger runner blocks (¾-inch wire requires a 14-inch block).

The end of the whip may be secured in several ways. The best method is to shackle the eye of the whip to the becket of the head block.

9.3. Yard-and-Stay Double Purchase. The chief advantage of the yard-and-stay double purchase is that lifts as heavy as the safe working load of the cargo booms can be handled at nearly the same rate as ordinary 1- or 1½-ton drafts. Light filler cargo encountered during the operation can be handled with scarcely any loss of time.

The only difference between this rig and the ordinary yard-and-stay is that both cargo whips are doubled up and the runner blocks shackled to the cargo hook.

9.4. Single Swinging Boom with Double Purchase. The single swinging boom with double purchase is considered one of the best methods of rigging for handling loads beyond the capacity of a *single whip* up to the capacity of a single boom. It is quickly and easily rigged and has the added advantage of flexibility. Load may be placed at any point in the square of the hatch or on the deck.

The yard boom will be the one to be rigged, so the hatch boom is topped up and secured out of the way. (See Fig. 9.3.)

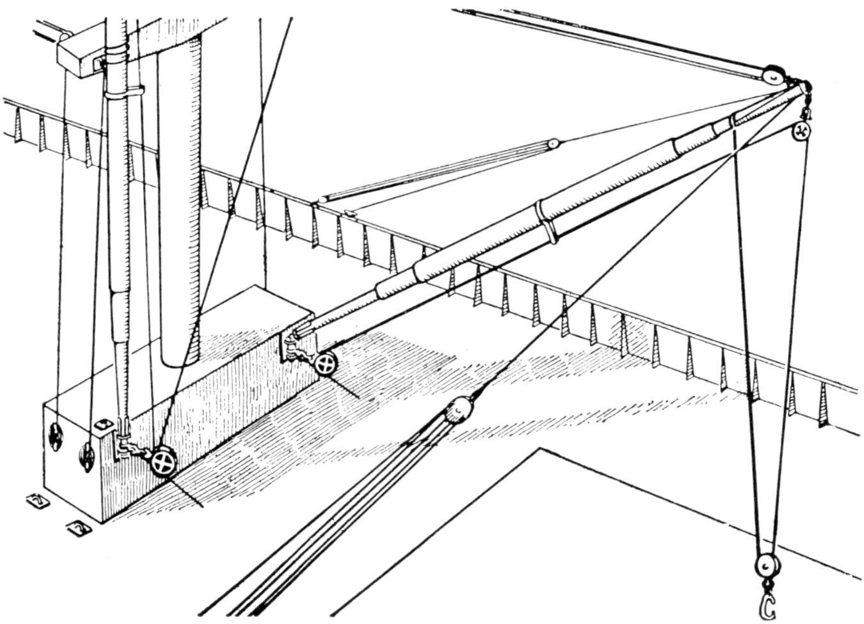

FIG. 9.3 SINGLE SWINGING BOOM WITH DOUBLE PURCHASE

The procedure for rigging the boom is as follows:

1. Strip the hatch whip from its drum and replace it by the yard boom's topping lift wire. Make sure the topping lift wire has a fairlead. This can only be done with a boom which has a miltiple topping lift.

2. See that the yard whip is long enough to permit doubling-up (250 to 300 feet).

3. Double-up whip.

4. Remove preventers from yard boom, and lead guys to proper fittings.

5. Top up the boom and swing into position by hauling on guy tackles. The hauling part of the guys may be fairled to winches at adjacent hatches, or men may be assigned to haul on the guys when swinging a load.

9.5. Two Swinging Booms. A load greater than the capacity of a single boom may be handled by using two booms working together as a single swinging boom. In this case, the whip of the two booms should be fastened to oppo-

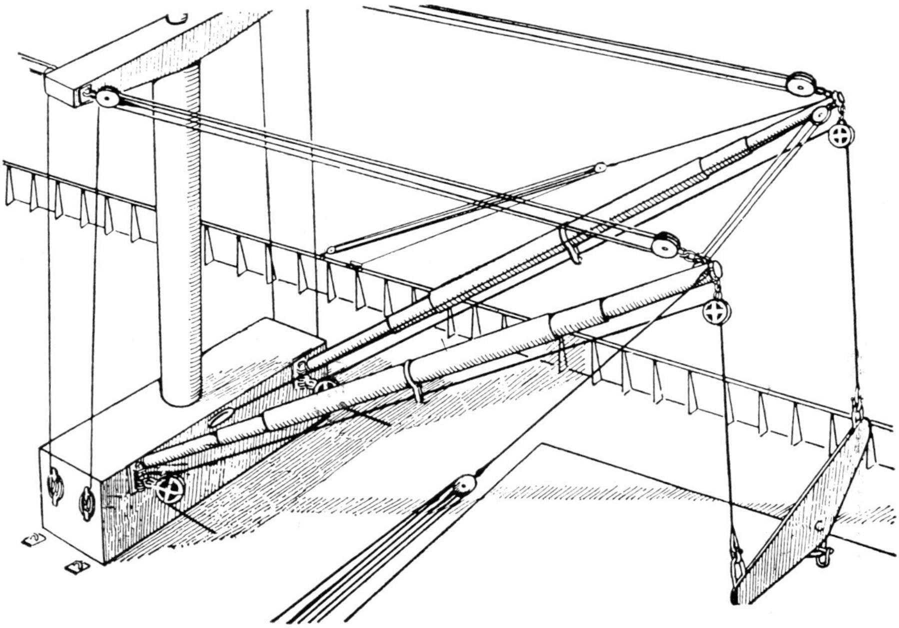

FIG. 9.4 TWO SWINGING BOOMS

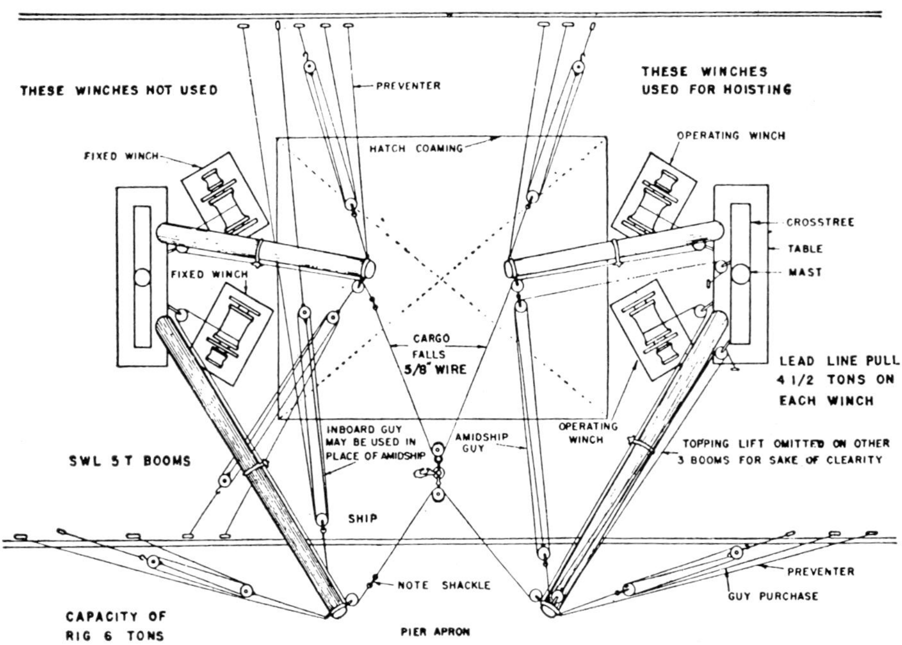

FIG. 9.5 RIGGING A DOUBLE-GANGED HATCH

site ends of a lifting bar or strong back, as illustrated in Fig. 9.4. The lifting bar serves to equalize any difference in winch operation.

To move a load from the hold to the pier, it first is hoisted clear of the coaming. Then, by using the guys, both booms are swung in unison until the load is over the pier. The load is then lowered to the pier. Swinging the load is a difficult operation, and it may be necessary to set the load on deck to change the position of the booms. Because this rig is cumbersome and difficult to handle, it should be used with great caution.

9.6. Block-in-Bight Method of Rigging a Double-Ganged Hatch. Many ships have double-ganged hatches, i.e., they are equipped with two pairs of ordinary cargo booms. Handling heavy lifts at a hatch in this manner is facilitated by rigging all four booms as illustrated in Fig. 9.5.

The rigging procedure is as follows:

1. Reeve the forward hatch whip through a runner block, and shackle the eye to the eye of the after hatch whip. Reeve the forward yard whip through a runner block, and shackle it to the after yard whip.
2. Run the shackles joining the two sets of whips to within a few feet of the head blocks of the after booms.
3. Shackle the two runners blocks to the cargo hook.
4. Heavy lifts slightly less than the sum of the safe working load of two parts of the cargo whips may now be loaded or discharged by the usual yard-and-stay method.
5. This rig has the advantage of being quickly rigged without the necessity of lowering the booms, and only two winches are required for its operation. In addition, the gear may be readily singled up for ordinary light loads.

9.7. Unrigging and Securing for Sea.

1. Assign men to winches, guys, whip, topping lift, and gypsies.
2. Cast off preventers.
3. Remove topping lift wire from cleat as described below.

   a. *Single Topping Lift.* Remove bull line from cleat, place it in a snatch block, fairleading it to the gypsy. Take five or six turns around the gypsy in the same direction as the whip (over the top), and top up boom until the bull chain is slack. Unshackle the bull chain and lower boom to its cradle.

   b. *Multiple Topping Lift.* Pass the stopper chain on the topping lift wire and remove the figure eights from the cleat. Surge the topping lift wire until the stopper takes the strain, then shift the wire to the gypsy. Heave around on the wire until the stopper is slack. Remove the stopper and lower the boom. If the cleat is large enough and conditions warrant, the boom may be lowered to the cradle by surging the wire around the cleat instead of transferring it to the gypsy. However, only experienced men should attempt this.

Regardless of the type of topping lift, men on the guy tackles must keep all the slack out of the guys to prevent the boom from swinging while it is being lowered and cradled.

While the booms are being lowered, cargo whips should be tended to prevent turns from piling up on the drum of the winch.

When both booms are cradled, all gear should be secured. Whips are rewound smoothly on the drum of the winch and the cargo hook is secured to a ring or cleat with a slight strain. Guys are secured to the heel block, or fittings on the mast table, then set taut. The hauling parts of the guys are coiled over the guy tackles and tied off. Topping lift wires or bull lines are secured to cleats with the remainder of the wire coiled and hung on a cleat. Bull chains are shackled to pad eyes on deck. If the ship is being made ready for sea, all running rigging and cargo-handling gear is secured.

9.8. Safety Precautions. Because topping and lowering booms are dangerous operations, safety must be emphasized. Men must be cautioned to stay from under booms while raising or lowering operations are in progress. The deck should be kept as clear of loose gear and lines as possible. A clean and orderly deck is safest.

The stopper chain of the topping lift must be properly secured; otherwise serious accidents may result. A rolling hitch and two half-hitches with several round turns are recommended. They are applied quickly, hold securely, and may be removed easily.

Topping lift wire should be secured about the topping lift cleats with a minimum of three round turns followed by three figure eights. To prevent the last few turns from slipping off the cleat, mouse the last two figure eights. Never half-hitch a topping lift wire around a cleat. It may tighten and become virtually impossible to remove. In lowering a boom by surging, the mousing and the three figure eights are removed, after which the three or four turns are gradually surged.

When shackling a bull chain to a pad eye, the shackle should be inserted beneath the first slack link in the chain. Otherwise, this loose link and the shackle may crowd the bottom of the chain, thus causing dangerous distortion and strain.

9.9. Jumbo Booms. Tanks, landing craft, harbor boats, crash boats, locomotives, and other extremely heavy cargo required by forces in the field present difficult problems in stevedoring operations at advanced bases. Loading a heavy lift is a simple matter at ports in this country; however, the problem does not end at these ports. At overseas bases these heavy lifts must be offloaded, and shoreside equipment or floating cranes are not always available. Often the ship's gear must be used for this purpose.

Most modern ships are fitted with jumbo booms of 30- or 50-ton capacity, normally located at one of the largest hatches of the vessel. Some Navy ships are equipped with 60-ton booms to handle the largest size LCM's. Installations on "Mariner" type cargo ships include a 50-ton boom at the No. 3 hatch and a

50-ton boom at the No. 6 hatch. See Fig. 9.6. Other modern ships are similarly equipped. Many ships used in task-force operations are provided with heavy lift gear at practically all hatches for the expeditious discharge of such heavy equipment as landing craft, tanks, and bulldozers. Practically all cargo-handling personnel operating in the field will have occasion to work heavy lift rigs, and for this reason must understand rigging and operating procedures of jumbo booms. See Fig. 9.7.

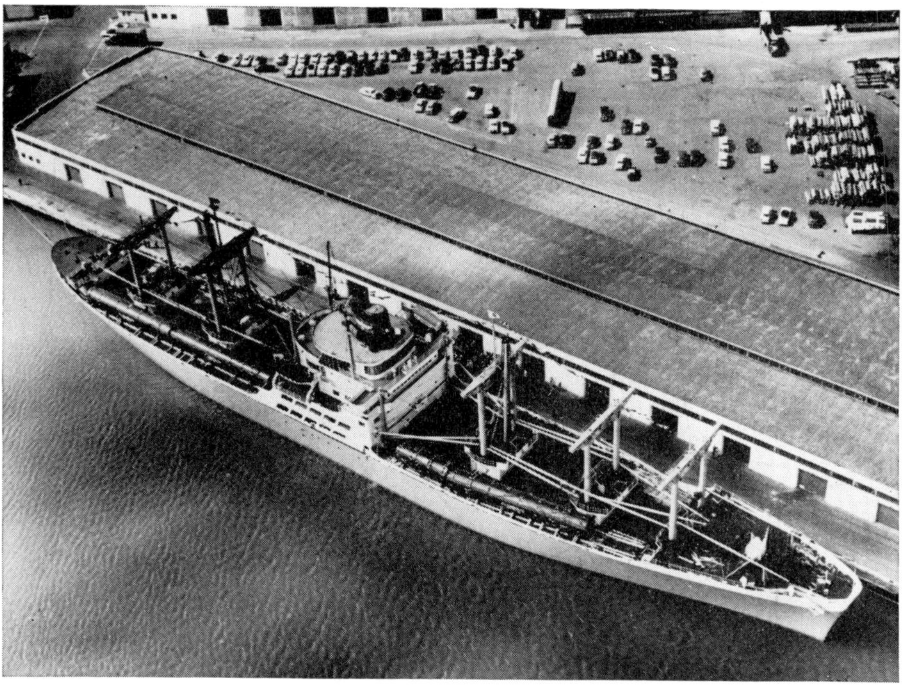

FIG. 9.6 BOOM PLACEMENT ON MARINER TYPE SHIP. *Official Photograph, Maritime Administration, United States Department of Commerce*

Most jumbo booms are carried in an upright position, collared to the mast or king post and fully rigged, with topping lift, load purchase, and guy tackles already secured.

This first step in rigging a jumbo boom is to lead all purchases to power. Four sources of power are required. The load purchase and the topping lift wire are led through heel blocks to the winches at the hatch to be worked. The guy tackles are led out to proper fittings, and the hauling parts of the guys are led to adjacent sources of power. The winches at adjacent hatches may be utilized, depending on the location of the boom.

To free the boom for use, it is necessary to send men aloft to release the collar that secures the boom to the mast.

Before making a hoist with a jumbo boom, all gear should be thoroughly checked to make sure blocks are running free and that none of the lines are chafing. Turns of wire on the drums of winches should lay tight and evenly around the drum. Guy tackles should be free of twists, and hauling parts of guys should fairlead to sources of power. Hasps and hooks of snatch blocks should be moused securely with seizing wire. Stays, shrouds, and preventers must be checked and tightened if necessary. This is extremely important, for it is possible to bring down a mast in attempting to handle a heavy lift.

Before a jumbo boom is operated, swing the ordinary cargo booms at the hatch clear of the working area. Generally, it is sufficient to swing these booms

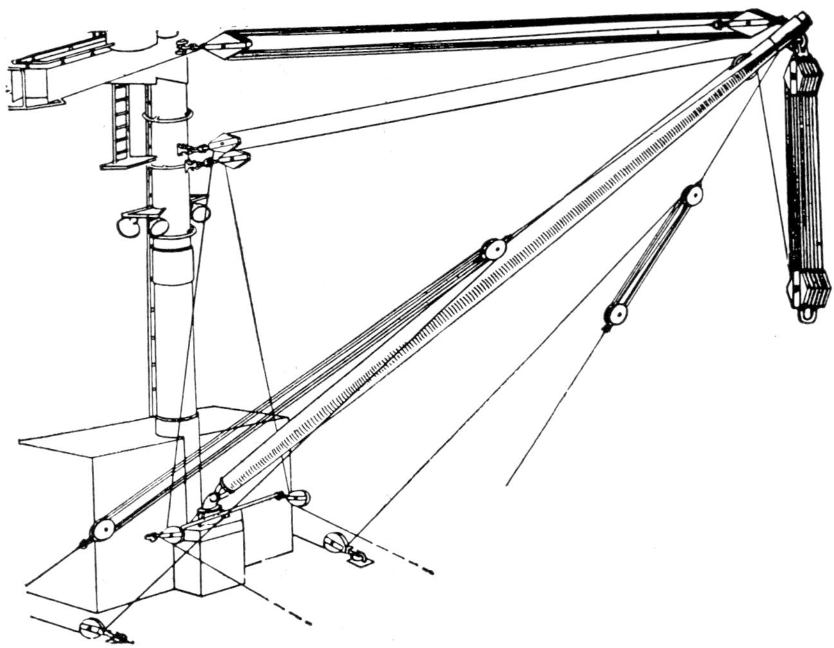

FIG. 9.7 JUMBO BOOM

outboard against the shrouds and secure them with the guys. In working deck cargo, however, it may be necessary to top the booms very high in order to clear the deck space.

The head of the boom must be plumbed directly over the load and the slings carefully slung and shackled to the lower purchase block. Then the load is hoisted a few inches off the deck and all gear carefully checked for any indication of undue strain. Hoist the load carefully until it is clear of the hatch coaming. By heaving around on the guy tackles, the boom is now swung over the ship's side, and the load may then be set on pier.

One of the greatest difficulties in working a jumbo or heavy lift boom is handling the guys. Every change in position of the boom must be accompanied by a change in adjustment of the guys. When a boom is topped, the guys must be slacked off; when it is lowered, the guys must be taken in. To swing a boom, one guy must be heaved in on and the other slacked off, and this requires co-ordination between the men handling the guys.

When a boom is swung outboard or inboard, one guy may be considered as a "hauling" guy; the other, as the "following" guy. The latter is generally the troublemaker. Green hands often fail to ease off on this guy smartly enough and it parts with disastrous results. It is good practice to allow a small amount of slack in a following guy, but never enough to permit the boom to slap about.

A heavy lift suspended outboard from the head of a jumbo boom may cause a ship to develop a considerable list. This places a great deal of added strain on the guys. The boom has a natural tendency to swing outboard in the direction of the list, and if this is not properly controlled, a guy tackle may easily carry away.

9.10. Precautions. Rigging and operation of cargo booms used for heavy lifts require skill, care, and common sense. There are many precautions to be observed, and to neglect any one is to invite trouble.

1. Don't overload. Make certain that the rig will make the lift safely; rig carefully and check each piece of gear as it is rigged. Check stays and shrouds.
2. Plumb the load directly under the boom head. Sling carefully and use dunnage or other suitable chafing gear at points where there may be chafing.
3. Check every part of the rig before picking up the load. Hoist the load a few inches off the deck, and check the rig for indications of undue strain.
4. Hoist, swing, and lower the load slowly and smoothly. Jerking causes terrific strain in the rig and can easily part something. Hoist loads only high enough to clear the coaming and bulwark. A particularly heavy load raised too high will affect the stability of the ship and may cause considerable list. Listing increases the strain on the guys and preventers and therefore the danger of parting. If something does part when a load is raised high, the effect will be worse than if the load were lower.
5. WATCH while a load is being moved, and keep every part of the rig under constant observation. LISTEN for any change in sound. A wire or fiber rope will normally hum under strain, but when it starts to squeak or squeal, LOOK OUT. A faulty block may give warning by squeaking or groaning.
6. Keep unnecessary personnel out of the area; those concerned with the operation must keep alert.
7. LOOK ALIVE AND STAY ALIVE.

All safety precautions should be strictly observed by all hands at all times. The following list contains common sense precautions that all cargo handlers must observe.

1. Wear safe clothing and shoes. Do not wear trousers that are too long, and do not wear rings while at work.
2. Use the accommodation ladder or brow for boarding and leaving the ship.
3. Climb ladders in the hold only when hoist is not in motion.
4. Use the walkway on ship's side away from the side on which the hoist is operating.

5. Secure hatch rollers properly.

6. Lower blocks, crowbars, chain slings, bridles, etc., into the hold by cargo falls or other lines.

7. Pile hatch covers in an orderly manner.

8. Lay strongbacks flat so they will not tip over on personnel or be dragged into hatches or overboard by slingloads.

9. Stand in the clear when strongbacks and hatch covers are being handled on the deck above.

10. Stand in the clear away from suspended loads.

11. When steadying loads, do not stand between load and any fixed object. Always face the load and keep feet and hands in the clear.

12. Stand clear of slings being pulled from under loads by cargo falls.

13. When using a dragline to move cargo, stand out of the bight and clear of the throw of the block and hook.

14. Be especially attentive when handling objects with sharp or rough edges.

15. Learn to lift properly to prevent strain.

16. Always use a light when entering dark places.

17. Never walk backwards while working with or around cargo on board ship.

18. Step down from elevations—never jump down.

19. Bend over projecting nails in dunnage to prevent puncture wounds.

20. Report to your supervisor any defect in tools, materials, appliances, and gear.

21. When short pieces of dunnage are required, use only the proper cutting tools.

22. Report all injuries (even scratches, cuts, and splinters) to your supervisor and get immediate first-aid or medical attention.

23. Know the location of fire-alarm boxes and fire-fighting equipment.

24. Do not engage in horseplay, practical jokes, or arguments.

Replenishment at Sea

Replenishment at sea is the term applied to the transfer of fuel, munitions, supplies, and men from one vessel to another while ships are under way. The first significant replenishment operation at sea in the U.S. Navy was in 1899, when the collier *U.S.S. Marcellus,* while being towed, transferred coal to a warship, the *U.S.S. Massachusetts.* Since that time, many methods have been tried and abandoned. Those described in this chapter have been adopted as the most feasible and are currently used in the fleet.

The cargo of a replenishment ship is intended for delivery to a base, replenishment group, or to the fleet at sea. It is the fleet-issue-loaded vessel with its varied cargo, stowed for quick and easy handling at sea, that will be discussed in this chapter.

9.11. General Discussion of Replenishment. The cargo of the replenishment ship is determined by one of the following considerations: (a) Requisitions prior to loading, (b) anticipation of fleet requirements, or (c) need for issuing provisions and stores in standard units. A package or kit which contains a specified grouping of items in fixed quantities, usually pallet loaded, is considered to be a standard unit.

There are four general principles for the loading of ships replenishing the fleet at sea to ensure maximum efficiency in unloading.

1. Lots of homogeneous cargo should be stowed, if possible, in several holds, so that they may be off-loaded at as many transfer stations as possible.
2. Provision must be made for adequate passageways and working areas in and around the cargo to permit quick segregation of lots, checking, and separate handling of heterogeneous types of supplies. Loading must be planned so that the remaining cargo can be readily reshored at the completion of replenishment to reduce the danger to personnel from shifting cargo.
3. Bulky and heavy items must be placed near loading areas and in holds that can accommodate their transfer most readily. The hatch opening, the height of the hold, and the fact that certain types of receiving ships can receive bulky items only at certain stations must all be considered.
4. Replenishment must be accomplished at the highest possible tonnage rate per hour and in the shortest practicable time consistent with safety.

Normally the receiving ship maneuvers to take station on the delivering ship and adjusts course and speed as necessary to maintain station during the operation. However, when large aircraft carriers are to be replenished, the replenishment ship may complete the final phase of the approach.

The delivering ship may use line-throwing guns with all types of receiving ships except carriers. The danger of firing gun lines into the hanger deck areas where supplies are being received makes it necessary for carriers to fire lines to the delivering ships.

Except for gear actually rigged on the receiving ship, such as fairlead blocks, riding lines, etc., the delivering ship furnishes all equipment during replenishment.

The zero end of the distance line is secured at or near the rail of the delivering ship, and the other end is tended on the receiving ship.

Each station on both ships has a telephone talker, and a signalman who is familiar with prescribed signals.

There are several methods which can be used to transfer cargo at sea. The tabulation below provides the load capacities of these methods under normal operating conditions. These figures must be reduced when transferring in rough or heavy seas.

Method	*Maximum Capacity per Load (pounds)*
Burton	3500
Housefall	2500
Double housefall	2500
Modified housefall	2500
Wire highline	800
Manila highline	600

9.12. Burton Method. The procedure for rigging a burton station on the delivering ship, using the port boom for transfer (burtoning), and using the starboard boom as the hatch boom (hoisting cargo out of the hold) (Fig. 9.8) is as follows:

1. Secure the inboard end of the *bridle messenger* to the bridle and fake it down on deck. Lead the eye-spliced end outboard in the vicinity of the transfer stations.

2. Fake down the *burton whip messenger* on deck and attach its snap hook to the bridle.

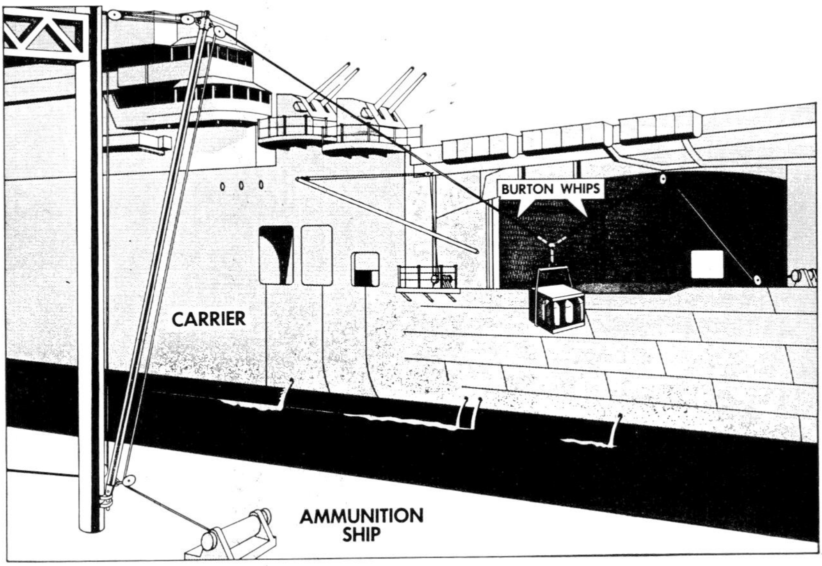

FIG. 9.8 BURTON RIG

3. Fake down the telephone and distance lines and attach to the bridle.

4. Secure the thimbled eye of the burton whip, which is a 6 x 37 high-grade plow steel wire rope, $\frac{5}{8}$ or $\frac{3}{4}$ inch in diameter and 350 to 600 feet long, to the triple swivel and hook of the engaged boom. Reeve the bitter end of this wire through the head and heel blocks. Secure the bitter end of the whip to the winch drum and spool the whip onto the drum.

5. Secure a *preventer* of at least 1-inch high-grade plow steel wire to the inboard side of the boom head and secure the pins of all shackles with seizing wire. Top up the boom so that the cargo whip plumbs the desired transfer point on the port side of the main deck. Set up guys and belay. Lead the preventer to the starboard side of the ship as close to a 90 degree angle from the boom as possible, or a provided fixed position. Apply strain and belay the preventer.

9.13. Burtoning Procedure. When the receiving ship completes her approach on the delivering ship, a heaving line (or line-throwing gun line) is sent over by the delivering ship. This line is attached to the bridle messenger, which is then hauled aboard the receiving ship. Attached to the messenger is the bridle and the lines attached to it. As soon as the bridle is received, the telephone and distance lines are secured and walked clear of the receiving station. The receiving ship takes the burton whip messenger from the bridle and then secures her burton wire to the messenger. The burton wire is hauled back to the delivering ship and secured to the triple swivel hook. The ships are then ready to transfer cargo.

Burtoning stations on the receiving ship are rigged in accordance with plans which designate the burton point for each type and class of ship. The burton whip block is secured to the burton fitting. The bitter end of the whip is rove through the whip block, and then led to the designated winch drum or gypsy head. The thimbled eye of the burton whip should be on deck ready for passing when the burton whip messenger from the delivering ship with attached snap hook is received on board.

The beckets, or sling, of the load are placed on the cargo hook of the burton whip. The delivering ship then hoists the load clear of the deck and rail. The receiving ship takes in on her burton whip as the delivering ship slacks away and the load is worked across. When the load is suspended over the receiving ship's deck, her whip is slacked and the cargo lowered to the deck.

Successful burtoning necessitates teamwork between the winchmen of both ships. Steady tension should be maintained on each whip regardless of movement due to rolling and yawing.

Stress in rigging may be reduced to a minimum by keeping the load as low as possible (consistent with sea conditions) and hoisting it just high enough to clear the rails of the two ships. As the load crosses between ships, it should be lowered to trace a catenary yet be maintained at a sufficient height to prevent immersion.

On completion of the replenishment operation, the burton whip is returned by means of the burton whip messenger to the delivering ship. The messenger, along with the telephone and distance line, is attached to the bridle and payed out with the bridle messenger. All messengers are returned in the same manner as received, only in reverse order. Thus, the time required for rerigging on the delivering ship will be reduced, and preparations for transferring cargo to the next receiving ship will be expedited.

9.14. Housefall Method. Rigging for housefall transfers can be done in several ways, making use of one or two booms. The housefall boom and whip can be plumbed over the center of the hold, thus serving the twofold purpose of lifting cargo to the deck and then transferring it to the receiving ship. This method, however, reduces the rate of transfer because of the longer distance the hook travels. Rigging procedure described requires two booms, as shown in Fig. 9.9.

Rigging the normal housefall as described below requires the use of two booms of the delivering ship, one boom located at the active hold and one boom at the hold forward of the active hold:

1. Secure the thimbled eye of the cargo whip to the triple swivel and hook, reeve the bitter end of this wire through the head and heel blocks of the boom at No. 2 hold and spool it onto the winch drum. The position of the boom, the guys, and the preventer are the same as for burtoning.

2. Secure temporarily the housefall block (a runner block of 12 or 14 inches) to the bulwark with a short piece of 2-inch manila line outboard, opposite the center of the hold.

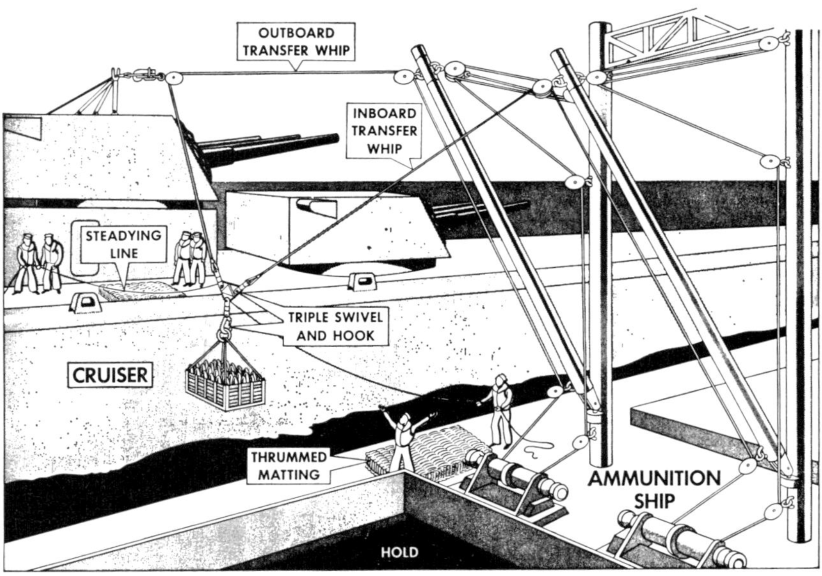

FIG. 9.9 HOUSEFALL RIG

3. Secure the thimbled eye of the transfer whip to the triple swivel and hook. Reeve the bitter end of the whip through the housefall block (outboard of all projections and rigging), walk it forward, reeve it through the head and heel blocks of the boom at No. 1 hold, and then spool it onto the winch drum.

4. Secure galvanized preventer wire (at least 1-inch high-grade plow steel) to the inboard side of each boom head. Top up the booms so that its whip clears all standing rigging or projections when making transfers to the receiving ship.

5. The remaining steps in rigging the delivery ship are the same as for burtoning.

9.15. Housefall Procedures. The receiving ship secures a gin block to the suspension point (same as for burton). A wire pendant (¾ inch 6 x 37 high-grade plow steel) with a thimbled eye, is run through this block. The thimbled eye remains on deck for attachment to the pelican hook of the housefall block.

When received, the hook is secured to the thimble of the housefall block eye.

A strain is applied to the wire pendant until the housefall block is two-blocked and the hauling part is then secured.

The housefall block messenger is detached from the bridle and used to haul the housefall block over as the delivering ship pays out on the housefall transfer whip. The block is then secured to the wire pendant and made ready for cargo transfer.

On the delivering ship the load is hoisted clear of the rail with the housefall cargo whip. The strain is taken on the housefall transfer whip, and the load is worked over to the receiving ship.

On completion of the transfer operation, the lines are passed back to the delivering ship in the usual manner. This type of rig proves advantageous when the receiving ship cannot keep good fore-and-aft position.

When loads must be kept higher above the water than is normally possible in housefalling, the housefall rig can be modified by the addition of a trolley block on the transfer whip.

9.16. Double Housefall. The double housefall speeds transfer of cargo to ships that do not have sufficient suspension points to handle more than one rig. However, double housefalling is somewhat slower than housefalling to two separate receiving stations.

In a double housefall operation the delivering ship uses two adjacent housefall rigs. Both housefall blocks are passed over the receiving ship simultaneously. In operation, one housefall rig alternately passes a loaded net to the receiving ship while the other returns an empty net to the delivering ship. The separation of the two housefall rigs on the delivering ship must be a minimum of 25 feet to prevent the outboard whips from fouling.

9.17. Wire Highline. The wire highline is usually used in transfers to destroyers and other small units.

To use the wire highline, the receiving ship must have a high attachment point. This is usually a pad eye welded to the ship's house structure. There is also an additional pad eye of 1-inch diameter located below about 12 to 18 inches from the first pad eye. The block used to fairlead the inhaul line is attached to this second pad eye. Sufficient deck space must be provided in the vicinity of the pad eyes to handle cargo being received.

The highline passes from a winch on the delivering ship through a block on a boom head and then across to a pad eye on the receiving ship (Fig. 9.10). A trolley rides the highline and is moved toward the receiving ship by an inhaul line (manually handled) and is brought back to the delivering ship by an outhaul line (winch operated).

A boom is normally used to provide a satisfactory lead for the wire highline. However, any other point of suspension on the ship's structure will serve if it is sufficiently high and strong. The highline is normally 350 feet of $\frac{5}{8}$- or $\frac{3}{4}$-inch wire (high-grade plow steel) with a thimbled eye on its outboard end. The following description is of a boom rigged highline:

1. Reeve the inboard end of the wire through a trolley block, through the head and heel blocks of the boom, and then spool it onto a winch drum.

2. Attach a pelican hook to the thimbled eye with a ⅞-inch shackle.

3. Shackle the manila outhaul line (2½ inch) to the inboard end of the trolley block and then run it through an 8-inch wooden block and a swivel attached to a becket on the underside of the head block. (See Fig. 9.10.) The outhaul line is finally taken through a fairlead to the gypsy head of a winch.

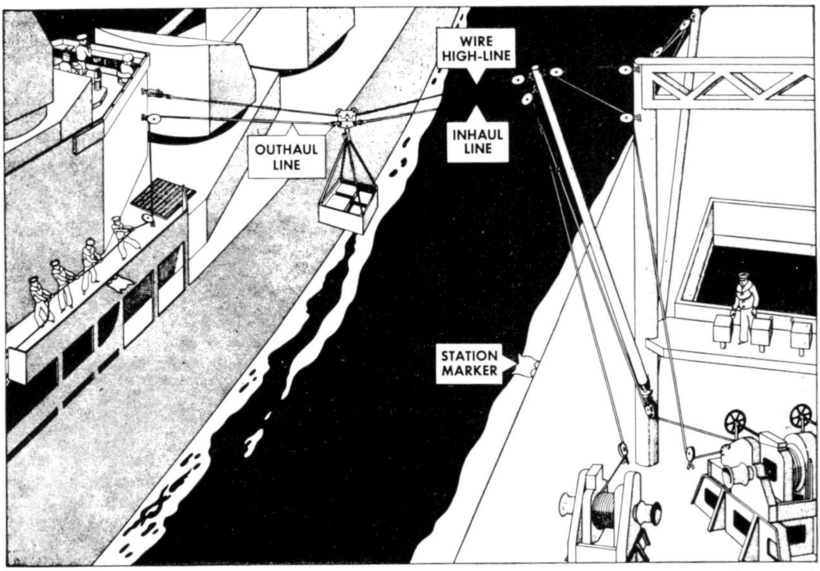

FIG. 9.10 WIRE HIGHLINE RIG

4. Equip the manila inhaul line (2½ inch) with a snap hook attached to the bridle. Fake down the center section of the line on deck clear for running and shackle the standing end to the outboard side of the trolley block.

Little preparation is required on board the receiving ship for the highline method. Below the highline pad eye, an 8-inch snatch block is secured to take the manila inhaul line. Additional snatch blocks are rigged to fairlead the inhaul line clear of the landing area.

9.18. Highline Procedure. When the inhaul line comes on board the receiving ship, it is detached from the bridle and led into the blocks provided for it. The pelican hook is secured to the highline pad eye, establishing the highline connection.

The load is now hooked to the trolley block and a strain is taken on the highline, thus lifting the load clear of the deck and rail. The load is hauled across to the receiving ship by slacking the outhaul line and taking up the inhaul line. When the load is suspended over the landing area, the delivering ship slacks off on both the highline and outhaul line, setting the load on the receiving ship's deck.

It is important always to keep a good catenary in the highline to avoid unnecessary strain when a load is suspended.

9.19. Manila Highline. The manila highline can be used in transferring provisions, ammunitions, personnel, and light freight. Preparation for rigging is essentially the same as for a wire highline, and a 12-inch snatch block attached to a pad eye at the delivering station is sufficient. It is kept taut during transfers either by manpower or by a capstan. The entire rig is relatively easy to set up and is the safest method now available for transferring personnel. Considering the manpower involved, however, helicopter transfer is always preferable when practicable.

To transfer individuals singly or in pairs, the only safe rig is the manila highline: all lines can be tended by hand. Heaving in lines by hand, with a sufficient number of men standing by for emergency, is the best method to ensure against the highline parting from sudden strains caused by rolling ships.

The burton method is a means of rapidly transferring 4 or 5 men in a skip box. It is used only when the situation demands quick transfer of a relatively large number of personnel and time does not permit individual transfer by the standard manila highline method.

9.20. General Safety Precautions for Transfer and Refueling

1. Personnel assigned to transfer stations must be adequately trained in all phases of safety procedures and precautions. They should wear helmet liners.

2. Because transfer stations on receiving and delivering ships are in exposed locations, personnel working close to ship's sides where solid bulwarks are not installed must wear kapok-type life jackets. If it is necessary to use inflatable-type life jackets, they must be inflated.

3. During heavy weather, personnel working on weather decks should wear life jackets.

4. Personnel must be cautioned to keep clear of suspended loads whenever possible.

5. Ample provision must be made to prevent the shifting of cargo, with its risk to both personnel and material.

6. Wire highline may not be used to transfer personnel. When manila highlines are used to transfer personnel, a capstan may not be employed to tend the line.

7. In handling ammunition, it must be remembered that carelessness and haste, in addition to causing accidents, often results in rendering ammunition unserviceable even when in containers. Ammunition must always be handled with greater regard for safety than general cargo.

8. In the transfer of personnel where water temperatures are low, "immersion suits" should be worn.

9. Whenever practicable, a rescue ship should be stationed astern of ships replenishing at sea for the purpose of rescuing personnel lost overboard.

10. During night replenishment flashlights (life-jacket type) should be

pinned to the left breast of each life jacket in use. They are not to be lighted unless the order is given to do so.

11. Plastic police whistles should be issued to each man wearing a life jacket during night replenishments. They are worn on a lanyard around the neck, with the whistle tucked inside the life jacket to prevent fouling in lines or gear.

12. A lifebuoy watch should be stationed in the after part of the ship with a 24-inch buoy fitted with a float light.

9.21. Fueling at Sea can be conducted by using either the close-in or span wire rigs. The choice is governed by the kind of ship delivering the fuel and the conditions under which replenishment is made. The close-in and the span wire (formerly Elwood) rigs differ primarily in the method of extending the hose to the receiving ship.

Only fleet oilers are provided with the installations necessary for transferring fuel by the span wire method. However, it is contemplated that several classes of ships undergoing conversion and modernization in the future will be equipped to fuel other vessels using this method. For the present, transfer of fuel between vessels other than fleet oilers requires the employment of the close-in rig.

9.22. Close-in Rig. The hose in the close-in method of rigging is supported by boom whips and bight lines which lead from saddles on the hose to booms or other high projections on one or both ships. (See Fig. 9.11.)

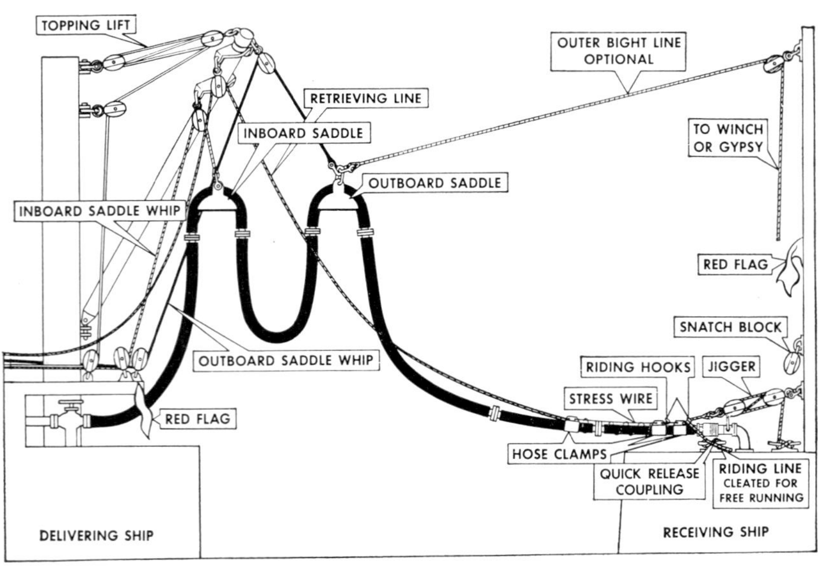

FIG. 9.11 CLOSE-IN RIG FOR FUELING

9.23. Rigging the Oiler for Close-in Refueling. The following is a check-off list that may be used as a guide in making up the hose and lines for close-in fueling.

1. The *hose* (6-inch) consists of the following: Oiler's onboard end, one 50-foot length, collapsible. Inboard saddle section, one 20-foot length, wire-stiffened. Section between saddles, one 50-foot length, collapsible. Outboard saddle section, one 20-foot length, wire-stiffened. Extension, one 50-foot length, collapsible. Outboard section, two 20-foot length, wire-stiffened. (Two hose clamps and hooks are attached to this section approximately 9 and 13 feet from the outboard end.) Total length of hose, 230 feet.

2. The *inboard saddle whip* is of 5-inch manila, 40 fathoms long, fitted at one end with a thimble eye and a 10-inch shackle. This whip is shackled to the inboard saddle and rove through a 14-inch snatch block on the outboard side of the boom head. It is then led to the inboard gypsy head.

3. The *outboard saddle whip* consists of 300 feet of ¾-inch wire rope (6 x 37 high-grade plow steel) or an equal length of 5-inch manila. It is fitted at one end with a thimble eye and a ⅝-inch shackle. The whip is shackled to the outboard saddle and rove through the block shackled to the pad eye below the head block. The whip is then led forward of the king post and to the drum (wire) or gypsy head (manila) of a winch.

4. An *outer bight line* is used only when fueling vessels larger than destroyer types. It is 5-inch manila 50 fathoms long, with a thimble eye and 1-inch shackle at one end. To the other end is taper-spliced a 15-fathom length of 2½-inch manila, then 15 fathoms of 21-thread manila, and to that 30 fathoms of 6- or 9-thread manila. The thimbled eye of the outer bight line is shackled to the outboard saddle, outboard of the outboard saddle whip.

5. The *retrieving line* consists of 50 fathoms of 3½-inch manila, with a thimble eye and a ¾-inch shackle at one end. The thimble eye is shackled to the hose clamp inboard of the hose messenger and led through a 12- or 14-inch snatch block on the forward side of the boom head, and through a fairlead to the inboard gypsy head. The inboard gypsy head is used alternately to serve both the retrieving line and inboard saddle whip.

6. The *hose messenger* is a 3½-inch manila line 40 fathoms long, with a thimble eye and a ¾-inch shackle at one end. To the other end is taper-spliced a 15-fathom length of 21-thread and to that 30 fathoms of 6- or 9-thread manila. The thimble eye of the messenger is shackled to the hose at 3-foot intervals. The seizing at the end of the hose should be two turns of 21-thread.

7. The fueling boom should be swung out 90 degrees and topped up so that the head of the boom is just clear of the ship's rail.

8. The hose should be topped up as follows: Two-block the inboard saddle. Top up the outboard saddle to a point just below the inboard saddle. Top up the end of the hose with the retrieving line to a point just below the outboard saddle. Lead the hose messenger to the superstructure deck and fake it down athwartships. The bights of the messenger should not be more than 10 feet long, and each bight should be stopped to the life rail.

9.24. Close-in Fueling Procedures. A 12-inch snatch block, through which the hose line messenger is led, is provided at each fueling station of the receiving ship. This block is placed inboard of the ship's side and about 6 feet above the receiving deck. On receiving ships larger than destroyers, a 14-inch snatch block is secured at the highest convenient point above the point where the hose will be taken aboard. This is used to fairlead the outer bight line, which helps support the outboard hose saddle. A riding line about 3½ fathoms long, made of 4- or 5-inch manila, is provided at each fueling station of the receiving ship. One end of this line is eye-spliced and secured to the hook or shackle of a jigger tackle. The other end is left free but, when the hose is aboard, will be secured to a cleat. When fueling a destroyer, if the Stewart elbow is not used, the end of the hose is shoved into the trunk and lashed securely.

As the receiving ship steadies alongside, heaving lines or line-throwing gun lines are sent over from each station on the delivering ship to corresponding stations on the receiving ship. By means of these first lines the telephone cables, distance lines, hose line messengers, and outer bight lines are started over. If the oiler has difficulty getting her lines across, the receiving ship may use her own line-throwing guns, when requested to do so by the oiler. As soon as the oiler's telephone jackboxes reach the deck of the receiving ship, connections are made and communications established.

The oiler pays out the hose messenger by hand as the receiving vessel draws it on board. On the receiving ship the messenger is led to the snatch block provided for it and, finally, to a winch; or it is led on deck for heaving in by hand. The oiler pays out the retrieving line and saddle whips, allowing the hose to come across assisted by the outer bight line from the receiving ship (if such a line is being used). As the end of the hose comes on board, the receiving ship cuts the stops securing it to the messenger, one by one, until the riding line hook is within easy reach. The bight of the riding line is then slipped over the riding line hook and the riding line is set taut. The hose end now is ready to be coupled to the receiving ship's hose or to be lashed in the fueling trunk. After this has been done, the messenger is restopped to the hose, removed from the snatch block, and the bitter end is returned to the oiler. The oiler tends the messenger as the ships open or close distance.

If an outer bight line is used the receiving ship takes it to the 14-inch snatch block provided and tends it. As the ships roll, the hose bight may dip in and out of the water unless the outer bight line is used to raise and lower the outboard hose saddle. When the ships roll in opposite directions, the hose will rise up suddenly, and the bight line (as well as the oiler's saddle whip, which is also helping to support the saddle) will stretch out horizontally. If these lines are not slackened immediately, they will break under the strain.

The outer bight line (tended by the receiving ship) and the outboard saddle whip (tended by the oiler) need constant handling by alert and well-trained personnel. Fast-moving winches will have to be used. Winchmen on both ships, working together, attentive to the outboard saddle, should try to keep the two lines in the form of an upright V.

When the outer bight line is not used, the outer hose bight is controlled by the oiler's outboard saddle whip alone.

When fueling is completed, the engineering force on the receiving ship gives the STOP PUMPING signal, disconnects the hose (after a back suction has been taken or the hose blown clear), and lashes closed the necessary valves or replaces the end flanges or hose caps. The hose is eased out on the bight of the riding line. As the outer bight line or hose messenger is being eased out, the oiler heaves in and two-blocks the inboard and outboard saddles. The oiler stops off the inboard saddle whip, removes it from the winch drum, and belays it to a cleat. The retrieving line is placed on the same drum, and with it the hose is hauled aboard. Finally, the receiving ship returns the outer bight line, the telephone and distance lines, and the messengers.

9.25. Span Wire Rigs. In the span wire rig, the hose is extended by use of a single-span wire stretching between the two ships. The hose hangs from trolley blocks which ride along the wire. This permits ships to open out to between 140 and 180 feet. The greater separation is safer, better for station keeping, and conducive to easier maneuvering. These factors not only allow commanders a wider latitude in choosing a fueling course, but they also facilitate the use of antiaircraft batteries should the need for them arise. The span wire method, with its higher suspension, affords protection for the hose in rough weather. With this method, the hose may be rigged in two different ways —the manila rig or the all-wire rig.

9.26. Manila-Rigged Span Wire. The manila rig is simpler than the all-wire rig, but requires more personnel at each station during the fueling operation.

The following is a guide to be used in making up the hoses and lines for fueling by the manila-rigged span wire method. (See Fig. 9.12.)

1. The *hose* (6 inch) consists of the same number of sections and is joined in the same way as the hose in the close-in rig. The riding line hook is attached to the hose 9 feet from the outboard end. For destroyer fueling, manila straps are used to help support the outboard end of the hose at the riding hook and at the bitter end.

2. The *inboard* and *outboard saddle whips* are made up and rigged as in the close-in rig.

3. A *retrieving line* of 4- or 4½-inch manila, 61 fathoms long, fitted at one end with a thimble eye and a ¾- or ⅞-inch shackle, is fairled through a 12- or 14-inch snatch block on the forward side of the boom head. From there it goes through a fairlead to the inboard winch gypsy head.

4. The hose line *messenger* is 60 fathoms of 4- or 4½-inch manila fitted at one end with a thimble eye and a ⅞-inch shackle. To the other end is taper-spliced a 15-fathom length on 2½-inch manila, then 15 fathoms of 21-thread. Added to that is 30 fathoms of 6- or 9-thread manila. The messenger is shackled to a fitting on the riding hook hose clamp and is stopped to the hose with 21-thread at intervals of 3 feet. A wire pendant connects the riding hook hose

clamp with the outboard hose clamp to prevent the hose from taking any strain from the riding line.

After the riding line is secured, the messenger is unshackled and returned to the delivering ship by means of a hose messenger retrieving line. By retrieving the hose messenger during the fueling operation, the time required for the breakaway is reduced, and preparation for fueling the next ship is expedited. The retrieving line is 50 fathoms of 21-thread.

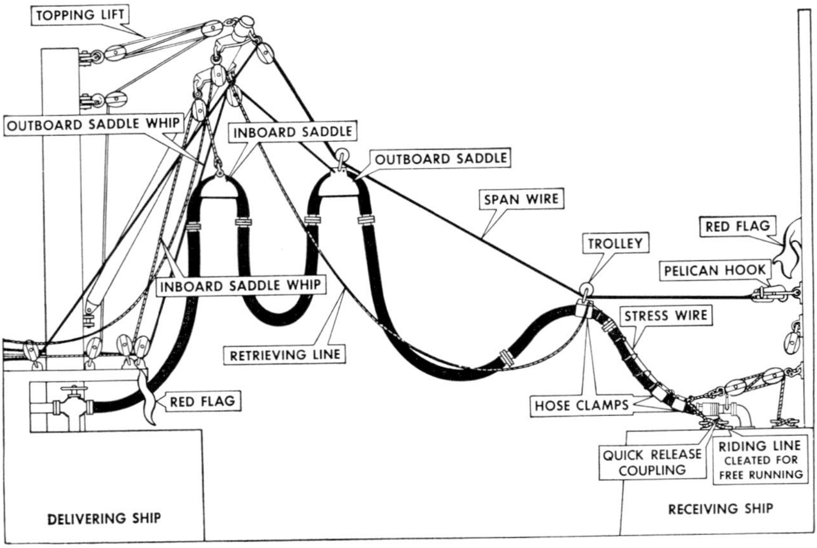

FIG. 9.12 SPAN WIRE RIG FOR FUELING

5. The *span wire* is 6 x 37 high-grade plow steel, ¾ inch in diameter and 450 to 600 feet long.

The wire boom falls are single whipped and the span wire is reeved through two trolley blocks and (if used) the free trolley. To the undercarriage of each trolley block is attached a swivel and a shackle.

6. One *trolley block* is connected to the outboard saddle slings and the other to the hose clamp. When a free trolley block is used, it is connected to a manila strap and ring about 10 feet from the outboard end of the hose.

7. The fueling *boom* is swung out 90 degrees and topped up so the head of the boom just clears the ship's rail.

8. The *retrieving line* is shackled to the hose clamp so it is outboard of the span wire trolley block and inboard of the wire pendant.

9. The hose is topped up as in the close-in rig.

10. A ⅞-inch diameter pelican hook is fitted to the span wire eye. The span wire is led to the superstructure deck level and the span wire is stopped securely to the hose messenger 200 feet from the inboard end of the hose.

There should be seizings 5 and 8 feet from the pelican hook. One method recommended is to form bights of 1/4-inch wire rope at these points by tucking the ends of the 1/4-inch wire into the span wire several times. The bights should be about 3 inches. The messenger is stopped to these bights with flat seizings of about four turns of small stuff. The stop securing the pelican hook to the messenger should be run through the long link of the pelican hook, and the hook should be allowed to hang a few inches below the messenger to facilitate cutting the stop.

9.27. Rigging the Receiving Ship for Span Wire Rig. At each fueling station on the receiving ship there must be the following:

1. A 1-inch tie-in pad eye for the span wire pelican hook, located above and approximately in line with the point at which the hose is to come aboard.
2. Another pad eye or wire strap (at least 6 inches below the tie-in pad eye) on which a 12-inch snatch block is placed to fairlead the hose messenger.
3. A third pad eye below the second, for shackling the end of the hose riding line.

9.28. Span Wire Procedures. The lines are passed in the same manner as in the close-in method. On the receiving ship, the hose messenger is led through the 12-inch snatch block, and as much of the messenger as possible is run in by hand. Then it may be taken to a winch. Aboard the oiler the messenger is payed out by hand, and the stops holding the messenger are cut in sequence as rapidly as necessary. The span wire is payed out from the drum as the messenger hauls it across.

When the span wire comes aboard the receiving ship, the pelican hook is made fast to the tie-in eye, and the stops securing it to the messenger are cut. The oiler now begins to tend the span wire, making sure that a good catenary is maintained at all times. The span wire should never be slacked off enough for the hose to touch the water.

When the hose messenger is cut free from the span wire, the receiving ship resumes heaving it in while the delivering ship pays out on the retrieving line and saddle whips. The saddles should be positioned so the span wire will carry the weight of the hose, but the saddle whips can be used to control the height of the hose.

When the hose end comes within easy reach of the men on the receiving ship's deck, the stops seizing it to the messenger are cut, one by one, until the bight of the riding line can be slipped over the riding line hook. Then the end of the hose is tended and controlled by the receiving ship, and the messenger is returned to the delivering ship where it is made up for the next receiving ship.

When all is secure, the receiving ship opens the hose valve and notifies the oiler to begin pumping.

On the receiving ship upon completion of fueling the valve is closed and the hose uncoupled. Then, by using the riding line, the end of the hose is eased out clear of the ship's side. The oiler two-blocks the inboard and the outboard saddles, stops off the inboard saddle whip, removes the whip from the winch gypsy head, and belays it to a cleat. The retrieving line is then placed on the winch gypsy head, and the hose is hauled aboard the oiler. When the hose has been retrieved, the span wire is slackened by the oiler, and the receiving ship trips the span wire pelican hook and eases out the end of the span wire on the bight of an easing out line rove through the long link of the pelican hook. The span wire pelican hook is not to be tripped until the span wire has been slacked.

The oiler then hauls in the span wire and the telephone and distance lines to complete the operation.

9.29. Safety Precautions While Fueling. The span wire is secured to the winch drum by one wire clamp. In case of a casualty or other emergency causing the ships to veer apart, the wire must be payed out from the winch drum to the securing clamp. As the strain increases, the wire will slip free of the clamp and drop over the side with minimum possibility of casualties to gear and personnel.

A $5/8$-inch shackle must be used for attaching the span wire to the pelican hook and for securing the stress wire between the hose clamps. This shackle acts as a safety link which parts before the full breaking load of the $3/4$-inch wire can be imposed on the boom, thereby minimizing the possibility of failure of the boom.

9.30. All-Wire Span Wire. The all-wire span wire rig for fueling at sea is used with a wire saddle line and a wire retrieving line. This can be used only if enough winches are installed at the stations to be rigged. A minimum of three winch drums is required.

The all-wire rig can be used to replenish destroyers or larger. When fueling a DD, however, a 10-foot section of 4-inch hose is attached to the outboard section of 6-inch hose, whereas a quick release coupling is used for larger ships.

Part II

SHIPHANDLING

10

General Principles of Ship Control

Shiphandling is the highest form of seamanship. It is an art because the forces involved are so many, so variable, and so different from ship to ship, under all conditions of wind and sea. But it is, nevertheless, a very necessary, pleasant, interesting, and satisfying art.

Mastery of the art requires not only alertness, coolness, and foresight but also constant study and much experience, including study of the forces involved, of the general shape and position of the under and above water parts of the ship, and experience in handling the ship under many different conditions.

Many naval and merchant ships are twin or multiple screw and some have twin rudders. The ability to handle these ships is founded on a thorough understanding of the forces set up in a single-screw ship when the propeller revolves, the rudder is "put over," and the ship moves through the water. We shall discuss, first, the propeller and rudder forces in a single-screw vessel and then go on to consider these forces in a twin-screw vessel.

10.1. Forces that Affect Maneuvering in a Single-Screw Ship.[1] The action of a propeller in a single-screw ship brings into play many unsymmetrical forces. In order to understand these, it is necessary to have some idea of the wake behind the ship and of the manner in which a propeller generates forces. As a vessel moves through the water, she experiences skin friction due to the viscosity of the water and tends to drag some of the water with her. If we measure the velocity of this water relative to that of the ship at increasing distances from the hull, we find that close to the hull the relative velocity is small because the water clings to the ship. The relative velocity increases as the distance from the hull increases until a point is reached where the water has no motion with respect to the surrounding sea. Its velocity relative to the ship equals the velocity of the ship. The boundary layer includes the water from the hull to the point where the relative velocity equals that of the ship. The width of this layer varies in some cases from zero at the bow to several feet near the stern. The net effect is that the boundary layer, a body of water, is given a forward motion by the passage of the ship.

Owing to this frictional drag upon the surrounding water, there is found aft, in the vicinity of the ship, a following current or wake called *the frictional*

[1] Based, in part, on *Propeller Action in a Single Screw Ship*. Courtesy of the Director, David Taylor Model Basin.

wake. The frictional wake is, in most cases, greatest at the surface in the vertical plane through the keel and abaft the ship. It decreases downward and outward on each side, as shown in Fig. 10.1. Streamline and wave patterns affect the velocity of the wake, but their effect is small except in destroyers and motor torpedo boats.

The propeller revolves in this wake. Since the wake water is moving forward relative to the sea, the propeller, in effect, is advancing into a moving body of water. Its speed is less than that of the ship. Thus, if a ship is moving at

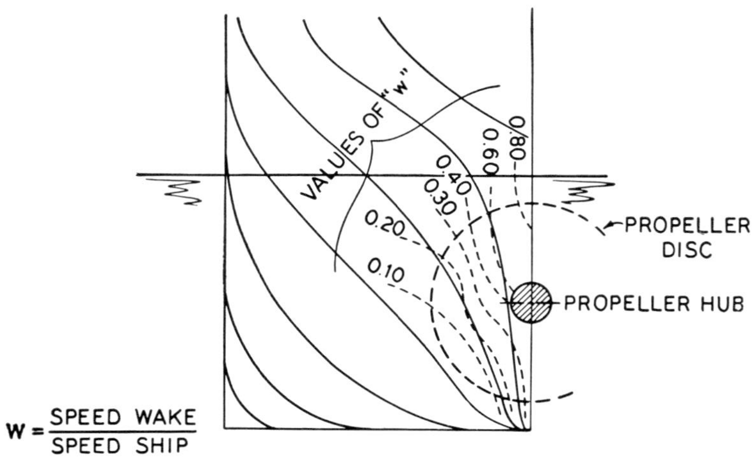

FIG. 10.1 WAKE DISTRIBUTION FOR SINGLE-SCREW SHIP

15 knots and dragging a wake with her at 3 knots, the propeller is only advancing at 12 knots *relative to the wake.* The ratio of the wake speed to the ship's speed is called the *wake fraction:* in this case $3/15 = 0.20 = W$.

The speed of the advance of the propeller through the wake is given by

$$V_A = V(1 - W), \text{ where } V = \text{ship's speed in knots}$$

In the preceding case,

$$V_A = 15(1 - 0.20) = 15 \times 0.8 = 12 \text{ knots}$$

The wake speed of 3 knots is only an average speed and actually varies from place to place. Behind shaft struts, skeg, or rudder, the wake speed may equal that of the ship. It is this variation in wake pattern which causes the unsymmetrical propeller forces.

The wake pattern has been measured on many models. A typical wake distribution appears in Fig. 10.1. The curves on the right side are similar to those on the left. The figure shows by contours of W values the distribution of the wake velocity (the fore-and-aft components) over the propeller disc. Along the line labeled $W = 0.60$, the speed of advance of the propeller through the wake is only 40 percent of the ship's speed. See example above.

In addition to the fore-and-aft motion, the water moving aft alongside the ship has an upward and inward flow under the counter due to the general rise of the water as the stern moves forward.

Analysis of Propeller Action. The maximum thrust is developed at about 0.7 of the radius from the centerline of the shaft. We shall discuss the forces generated at this point. The velocity of the blade section relative to the water is the resultant of two component velocities:

1. A forward motion through the water at velocity V_A or ship speed minus wake velocity.
2. A rotational motion of the propeller which is given by $2\pi RN$. R is the radius under consideration (0.7), and N is revolutions in a unit of time.

The resultant velocity of the water V_0 is a combination of the forward and rotational motions. (See Fig. 10.2.) The effect of V_0 striking at the angle of

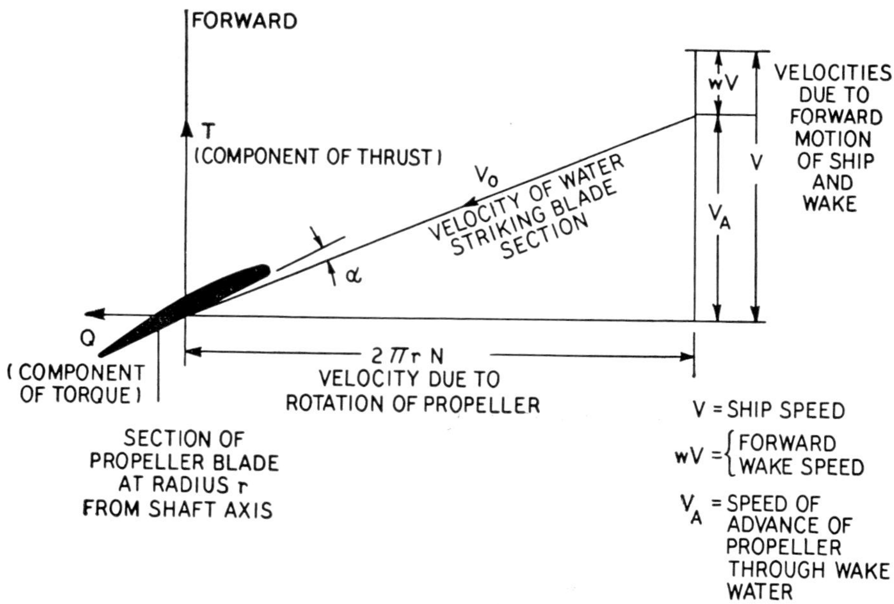

FIG. 10.2 SIMPLIFIED VELOCITY DIAGRAM FOR PROPELLER BLADE SECTION

attack α is to develop lift and drag just like the forces on an airplane wing. The direction of V_0 to the face of the blade section at the angle of attack produces forces which can be resolved into two components, a fore-and-aft force, the thrust T, and a torque Q. The former, T, propels the ship and the latter, Q, *generates a reaction or transverse force through the shafting which tends to force the stern to port or starboard.* If we look at Fig. 10.2 in another way, we could consider V_0 as one of the forces, all acting in the direction of the arrow and on the after surface of the blade. In fact, the forces act on both sides of the blade. If the angle of attack α between V_0 and the surface of the blade is small, the component forces T and Q will be small. If the angle is large, i.e., when WV, the forward wake speed, is great, the force V_0 will strike the surface of the blade at a more effective angle. T and Q will be large. WV, the wake speed, varies as shown in Fig. 10.1, which explains the unsymmetrical forces acting on the hull. The amount of work done by each blade of the propeller will vary with its position in the disc.

There are four regions where the maximum change in force occurs:

1. As blade A (Fig. 10.3) approaches the vertical point, Fig. 10.1 shows it will pass through a region of relatively high wake speed and therefore low values of V_A. (Fig. 10.2 shows that α will increase as V_A drops in

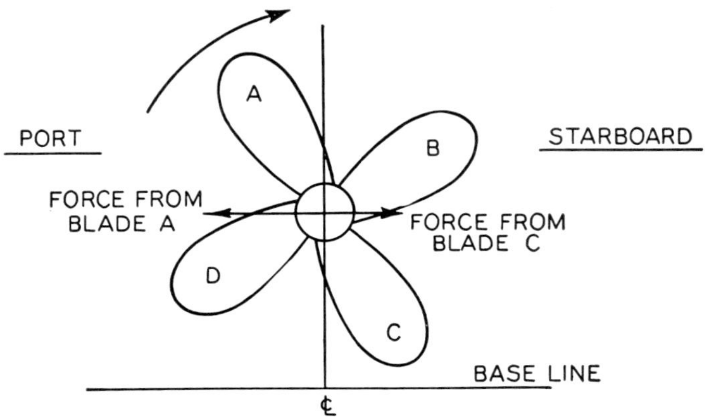

FIG. 10.3 VIEW FROM ASTERN SHOWING FORCE ON SHAFT AXIS

value. V_0 will act on the blade at a larger, more effective angle of attack. T and Q will increase.) The increase of the torque Q *reacts* through the shafting on the stern of the ship, forcing it to port with a right-hand propeller. If we revert to our explanation of V_0 above, it is clear the V_0 is acting from starboard to port against the rear side of the blade A. Because a right-handed propeller is being considered, V_0 will have two components: T, the thrust propelling the ship through the water; and Q, which generates a transverse or athwartship *force* now directing the stern to port.
2. Blade C will pass through an area opposite to that of blade A. A transverse force to starboard will be exerted. (But as the wake speed in the lower part of the disc is much less than in the upper, the angle of attack is smaller than on blade A, and the force to starboard will not be as great as that to port. The resultant force of these two forces is to port.)
3. Blade B will move downward against the upward flow of water under the counter. This flow is equivalent to an increase in N in the formula $2\pi RN$. The angle of attack α, the velocity, thrust, and torque increase.
4. Blade D moves up with the flow and experiences a decrease in the above factors.

It is clear that B overbalances D, and the ship's head tends to fall off to port. There are two more factors that affect the steering of the ship:

1. The propeller imparts a helical motion to the slip stream which impinges on the rudder even when amidships. That part of the helical slip stream above the axis of the propeller tends to move the stern to starboard, and the part below the axis tries to move the stern to port. The resultant force depends on the area of the rudder above and below the centerline of the shaft and the uniformity of the slip stream.
2. The next factor is the submergence of the propeller. If the vessel is in ballast or at light displacement, the propeller may break surface, causing a decrease in the transverse effect of blade A. When the ship has little or no speed, blade A which is near the surface may draw air and again decrease the transverse force.

The ship is therefore subject to several opposing, variable forces. Her actual behavior will depend on the magnitude of these forces. General experience shows that many single-screw vessels with a right-hand propeller turning ahead tend to fall off to port. There are exceptions, and no hard and fast rule can be laid down. Observe your own ship.

10.2. Getting Underway. With the ship stationary or just starting to move, the wake does not exist or is negligible. The top blade A may break the surface and thus lose some of its usual transverse force to port. If it does not break the surface, air may be drawn down with the same effect. The lower blade still acts to force the stern to starboard. The rudder, even when amidships, receives the helical slip stream at an effective angle high up on the port side. If the rudder has a larger area above the axis of the propeller than below, that force tends to move the stern to starboard. The result of these forces may be that the stern will move to starboard.

When backing, the forces due to blades A and C are reversed. Blade A may break the surface, but, regardless, C, acting to port, will predominate. There is no helical slip stream thrown against the rudder. Most of the water which passes through the propeller disc comes from the free surface, and the rudder exerts no steering force until the ship gains sternway. The upper part of the discharge flow from the propeller strikes the starboard underwater body of the ship at a good angle. The lower part strikes the keel on the port side at a poor angle. It is probable that the force due to the upper part predominates, and the result of these forces tends to push the stern to port.

10.3. Handling Ships with Controllable Pitch Propellers. An increasing number of small ships, such as the new LST's, tugs and the Navy's latest non-magnetic ocean minesweepers (MSO), have controllable pitch propellers. The blades of the propellers are rotated by a hydraulic mechanism in a plane parallel to that of the propeller shaft as described in Chapter 4. Thus the blades can be adjusted to take more or less bite, or can be reversed in pitch. It is this latter feature particularly that adds maneuverability; the new minesweepers, for example, can be stopped in less than two ships lengths when going ahead full power.

The forces acting on the controllable pitch propeller are the same as those described above for conventional propellers. The shiphandler uses his rudder and engines in the conventional way except, instead of speeding up, slowing, or reversing his engines, he adjusts or reverses the pitch of his blades by a control mechanism on the bridge. Since the response to change in propeller blade pitch is instantaneous the shiphandler must become accustomed to this disappearance of dead time when backing. Another novelty to the seaman trained to handle conventional vessels will be his ability with controllable pitch propellers to move the ship quickly with high power. This is done by keeping the shaft revolutions high and the propeller pitch low. An increase in propeller blade pitch then applies power to the ship very suddenly.

10.4. Steering a Twin-Screw Vessel, Single Rudder. A twin-screw vessel has two propellers, one on either side of the centerline. Generally, they are out-turning, i.e., the starboard one is right-handed and the port one left-handed. They turn in opposite directions to balance the propeller forces and enable the ship to steer a straight course with no rudder.

A multiple-screw vessel normally has four propellers, two on a side out-turning and so controlled that those on a side go ahead or astern as a unit. As the action of a multiple-screw vessel is similar to that of a twin-screw vessel, only the latter will be discussed.

The steering of a twin-screw vessel is considerably simpler than that of the single-screw ship. It will be found that the strong tendency of the single-screw ship to back stern to port does not hold with the twin-screw vessel, and that the latter backs with equal facility in either direction, barring the effect of wind, waves, and currents.

The various forces affecting the action of the single-screw ship are still present to a degree in the case of a twin-screw vessel. In many cases they are considerably less because the forces from one screw are balanced by similar but opposite forces emanating from the other screw. In addition, there is a new force due to the movement of the screws around the centerline. It will readily be seen that with one screw going ahead and the other astern, there results a turning moment that tends to throw the *bow* to the side of the backing screw.

One powerful force should not be overlooked. It is the momentum of the ship, ahead or astern, acting through the center of gravity. If a twin-screw ship is going ahead and one screw is backed, two opposing forces are set in motion, namely, the force of the backing screw acting in one direction at a certain distance from the centerline and the weight of the ship acting in the opposite direction. These are in addition to the forces due to the action of the wake on the rudder if it is "put over."

The steering of the twin-screw ship will be considered under the following headings, no wind or sea:

1. Ship and Screws Going Ahead.
2. Ship Going Ahead, Screws Backing.
3. Ship Going Astern, Screws Backing.
4. Ship Going Astern, Screws Going Ahead.
5. One Screw Going Ahead, Other Screw Backing.

10.5. Ship and Both Screws Going Ahead, Single Rudder. In this case, with the rudder amidships the ship will steer a steady course. The transverse forces of the two propellers are equal and opposite in direction. As the shafts are offset equally, no turning moment is felt.

When the rudder is put over, it will receive some of the discharge flow from the propeller on that side but not as much as in a single-screw ship. The principal force which turns the ship is that set up by the wake against the forward side of the rudder.

If one screw is stopped with the rudder amidships, the turning moment of the revolving screw will take charge, and the ship will turn toward the side of the stopped screw. The discharge flow of the revolving screw does not strike the rudder.

10.6. Ship Going Ahead, Both Screws Backing, Single Rudder. The steering effect of the rudder is the only force turning the ship from a straight course. All other forces are equalized. The effect of the rudder is reduced as the headway is lost until there is no steering control when the ship is stationary.

If one screw only is backing and the other stopped with headway on, the turning moment of the backing screw added to the momentum of the ship going ahead will swing the stern away from the backing screw. If there is deadwood forward of the screw, the discharge flow will strike the underwater body and increase the swing. Some carriers have such deadwood, but destroyers do not.

10.7. Ship Going Astern Both Screws Backing, Single Rudder. If the rudder is amidships, the various forces are equalized, and a straight course can be steered. If the rudder is put over, the pressure of the water that you are backing into against the back side of the rudder will enable a course to be steered. However, most of the water which passes through the screws comes from the free surface and thus has little effect on the rudder.

If one screw is stopped, the turning moment of the backing screw is added to the effect of the rudder when it is put over away from the revolving screw. The swing may be slowed or stopped if the rudder is put over toward the screw. The effect of the rudder is to counteract the effect of the screw, and how effective it is will be dependent upon the size of the rudder and speed of the engine.

10.8. Ship Going Astern, Both Screws Going Ahead, Single Rudder. The ship will respond to the rudder, i.e., a left rudder will throw the stern to port unless excessive sternway is on. The transverse forces of the screws will be equalized. The steering effect of the rudder when going astern will be reduced gradually as the ship loses headway, until all steering control is lost before the ship has lost sternway. This is because the discharge flow from the propellers will interfere with the flow of water against the back of the rudder.

10.9. Ship Stationary, One Screw Going Ahead, Other Screw Backing, Single Rudder. The rudder will have little effect until head or sternway has been gained. The turning moments of the two screws will be additive, but they may not be great when the shafts are close together. If the ship has no deadwood, she may turn easily. In narrow waters the two screws should be operated at such speeds that the ship does not gain head or sternway when going ahead and backing at one- or two-thirds speed. This balancing of forces will enable the captain to move the ship ahead or astern as desired by varying the speed of the backing or ahead engine. As a general rule with one engine ahead the same amount as the one astern, the ship will slowly make headway. The rudder may be used to increase the swing when some steerageway has been gained. If the ship has some deadwood, the turn can be made more easily by "seasawing," i.e., moving ahead and then moving astern, using the rudder to increase the swing.

10.10. Twin Rudders on Twin- and Multiple-Screw Vessels. Twin rudders are not a new invention but rather a recent improvement of an old development. They are now installed on many naval vessels, large and small, and vary in position, shape, and size. On battleships and carriers the rudders are not directly behind a pair of propellers but extend nearly to the keel level. There may be deadwood forward of the propellers. On destroyers they are in line with the propeller shafts but the bottom edge of the rudder only extends to the level of the propeller hubs.

The rudders on destroyers receive most of the upper half of the discharge flow of the propellers when going ahead. The lower half passes under the rudder. The installation of twin rudders has improved the maneuverability of naval ships. The general rule when handling twin-screw destroyers with one rudder is to order the proper rudder after the ship has gained head or sternway. The installation of twin rudders has changed this rule, and the shiphandler should now order right or left rudder before the engines are moved. The rudder should be put over to take advantage of the discharge flow from the ahead propeller. This flow acts against the forward side of its rudder and thus creates a powerful moment of force to turn the ship.

The improved turning characteristics of destroyers are appreciated when turning in narrow channels and going alongside a nest or tender where large angles of approach must be used. The maneuvers to shove off from a nest, tender, or pier under awkward conditions of wind and tide are facilitated too.

The key to all these ordinarily difficult maneuvers is the decisive effect of the discharge flow from the ahead propeller on the rudder astern of it. If the other propeller must be operated astern, it may do so without affecting the turn adversely because the water passing through its disc comes from the free surface and does not impinge on the rudder to any great extent. Hence, the vessel can be turned to port from dead in the water, for instance, by ordering the "left full rudder, ahead two-thirds speed" on the starboard engine and "back, two-thirds speed" on the port engine. The speed of the port engine can be varied thereafter to allow the ship to gain steerageway as and if desired. The rudder and the starboard engine need not be changed until the turn is completed.

10.11. Turning Characteristics. The standard method of finding any ship's turning characteristics is to turn her in a number of complete circles under

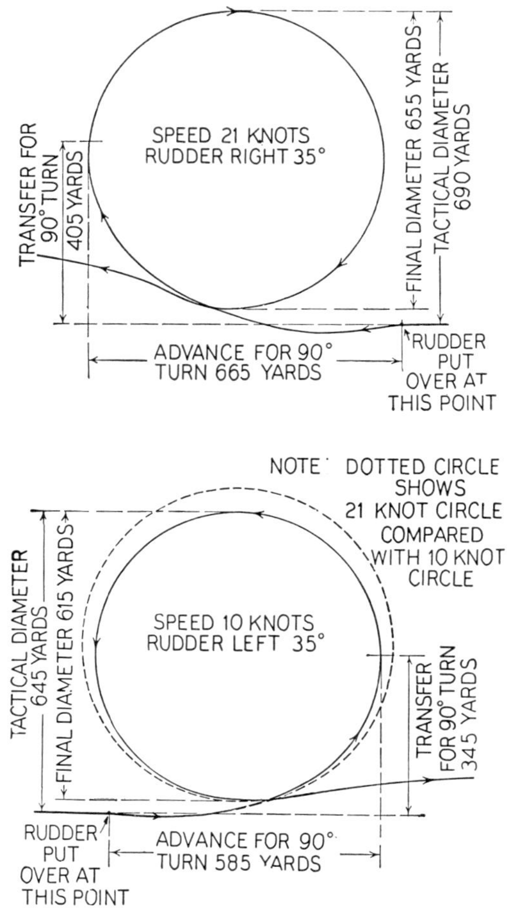

FIG. 10.4 TURNING CIRCLES

varying conditions and to record the results for each turn. The variables used are: right or left rudder of various degrees; steady speeds of different value; and differences in draft and trim. When taking turning data, the effects of wind and sea are noted and allowed for. Most changes, of course, are not as much as 360 degrees, but by studying the complete turning circle the ship's

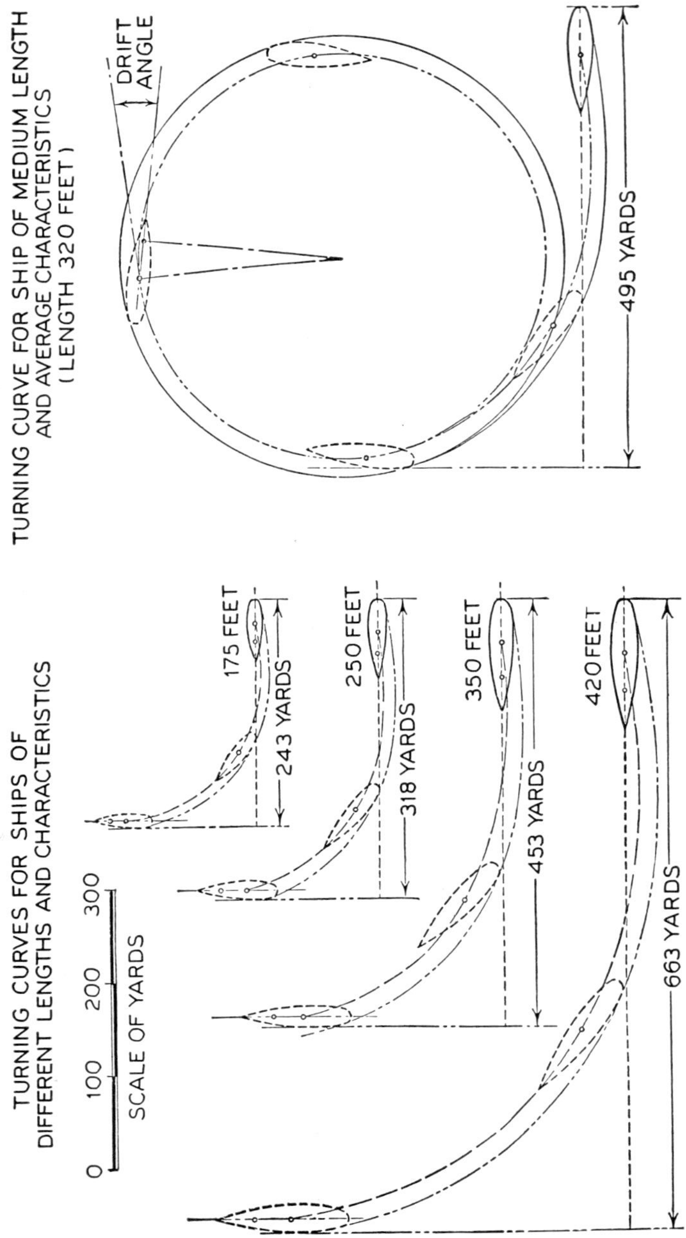

FIG. 10.5 TURNING CURVES

behavior for turns of any extent can be determined. In considering the track actually followed by a vessel during a turn, certain terms must be defined. These terms may be understood more easily by a simultaneous study of Figs. 10.4 and 10.5. Figure 10.4 shows actual turning circles of the *U.S.S. New Mexico*. One figure is the circle made at 21 knots with 35 degrees right rudder; the other, the circle made at 10 knots with 10 degrees left rudder. Figure 10.5 illustrates some differences in the turning curves made by ships of different lengths and characteristics.

10.12. Definitions. *a. Turning Circle.* The path followed by the *pivoting* point of a ship in making a turn of 360 degrees or more. For the ordinary ship the bow will be inside and the stern outside this circle.

b. Advance. The distance gained in the direction of the original course. The advance will be a maximum when the ship has turned through 90 degrees.

c. Transfer. The distance gained at right angles to the original course when the ship has turned through 90 degrees.

d. Tactical Diameter. The distance gained to the right or left of the original course when a turn of 180 degrees has been completed.

e. Final Diameter. The distance perpendicular to the original course between tangents drawn at the points where 180 and 360 degrees of the turn have been completed. Should the ship continue turning indefinitely with the same speed and rudder angle, she will keep on turning in a circle of this diameter. It will always be less than the tactical diameter.

f. Kick. The distance the ship moves sidewise from the original course away from the direction of the turn after the rudder is first put over. The name is also applied to the swirl of water toward the inside of the turn when the rudder is put over to begin the turn.

g. Drift Angle. The angle at any point of the turning circle between the tangent to the turning circle at that point and the keel line of the vessel.

h. Pivoting Point. That point about which the ship turns when the rudder is put over.

The turning circle is the path followed by the pivoting point during the turn. The pivoting point is in the horizontal centerline of the ship, and its position on that line depends on the shape of the underwater hull and especially on how much the after deadwood is cut away. The pivoting point moves forward if the ship is trimmed down by the head and aft if it is down by the stern. This characteristic is illustrated by the standard motor launch. When light it pivots well aft on account of the weight of the engine, and when heavily laden it pivots well forward. The pivoting point may also move aft along the keel line to some extent if the ship is deep in the water and forward if she is light. It is normally in the forward one-third length of the ship. When its position is once determined, it does not vary enough in the ordinary ship under different conditions of load and trim to cause any difficulty in ship handling.

10.13. Other Forces Affecting Turning. Every seaman knows that the wind affects turning. The freeboard and superstructure act as a sail area whose effect must be considered, especially at low speeds. In most ships there is more superstructure forward than aft, and even in cases where it is equally distributed, there is usually a higher freeboard at the bow than at the stern. The general effect of this construction is that the forward part of a ship acts as a headsail. This has some effect on a vessel going ahead but is very important in backing. A vessel when backing will back into the wind almost invariably. The propeller acts as a pivot, and the bow and superstructure under the pressure of the wind will fall off. The stronger the wind, the stronger the tendency to back into the wind. This tendency can be used to facilitate a turn when maneuvering in shallow waters.

The condition and relative direction of the sea affect both the progress and steering of the ship by their effect on the underwater body. Any sea forward of the beam will retard the motion of the ship over the ground to a greater or less extent, while any sea from abaft the beam will accelerate it. The general effect of the sea on steering is to cause a ship to seek the trough. If the sea is on the bow or quarter, it may be necessary to carry a definite amount of either right or left rudder in order to maintain the course. This will cause some loss of speed on account of the rudder effect. In a gale it is usually necessary to slow down when "bucking" a heavy sea. The slow engine speed opposed to the force of the waves causes the ship to lose steerage way. Under these conditions the tendency of the ship to "fall off" into the trough is very pronounced. Full ahead on one engine and one- or two-thirds astern on the other will usually straighten her up. There may be anxious moments, however. With a following sea the tendency of the vessel is to yaw, and an excessive amount of rudder may be needed to keep her steady. This excessive rudder will slow her down and may counteract the effect of the following sea in increasing her speed over the ground. A ship with no deadwood aft yaws more than one whose underwater body is continuous to the rudder. All the effects of the sea just described are more noticeable at low speeds than at high speeds and may vary if the wind and sea are from different directions.

Current affects the underwater body of the ship. It is especially important because its existence may not always be realized. Known ocean currents may be shifted, accelerated, diminished, or even reversed by winds blowing steadily in one direction over a long period of time. Currents in harbors, straits, and bays are caused by the action of the tides. The currents at the entrances to certain harbors (the Golden Gate) are strong at times and run at an angle with the entrance course. The current may be reduced or reversed by the tide. The direction and probable force of currents in ports and along the coasts may be determined approximately by the study of tide tables and current charts, but every effort should be made to verify the data found in these publications because the effects of wind and weather may make them inaccurate. Observation of the shape of the shore line and of the direction in which buoys and other anchored navigational aids are leaning will give a good check on the force and direction of the current running at any given time.

The general effect of a current on the underwater body of a ship is to move her bodily in the same direction in which the current is running. When turning in a current, the ship, at the completion of the turn, may be well down in the direction of the current from her position when the turn was started. When held at any point, as by an anchor, the ship usually assumes the position where the current has the least underwater area on which to act. For this reason, an anchored ship heads into the current unless the wind or sea is strong enough to overcome its effect. For the same reason, a ship at anchor will swing with the change of the tidal current. By means of spring lines, current can be used to cant a ship or to move her toward a dock. Steering is always easier when heading into a current than when going with it, except in narrow channels.

Eddies and whirlpools will be found where two currents meet and may become a source of danger, especially to low-powered ships.

Shallow water will modify the normal action of screws and rudder in steering or turning a ship. She may be sluggish in answering her rudder or she may take a sudden sheer to one side. High speeds can be made in shallow water by the use of excessive power, but large waves are formed by the turbulence which causes destruction to shipping and water-front facilities. The best seamanship in harbors and rivers is constant watchfulness, foresight, slow but steady speed, having an anchor ready for letting go, and some consideration for other craft.

10.14. Restricted Waters. A thorough knowledge of the maneuvering characteristics of the ship is necessary when handling her in restricted waters. The master of a single-screw vessel can turn, dock, or place her alongside another vessel with apparent ease if he has this knowledge and takes advantage of the wind or current. The master of a twin-screw vessel with high power should be able to do better than he who commands a single-screw ship. Destroyers are frequently asked to maneuver in locations and under conditions in restricted waters that no other larger vessel in the Fleet would attempt. These maneuvers could be facilitated by using an anchor at short scope.

10.15. Casting in a Narrow Channel. The expression to *cast* means to turn a ship in her own water. The ships in a fleet turn in this manner when getting under way together in a crowded anchorage and when the fleet is headed in the wrong direction. Single vessels in restricted anchorages often have to turn in their own water too because of nearby anchored vessels or a restricted maneuvering space.

The problem of turning twin-screw vessels with electric drive or reciprocating engines and a single rudder is not a difficult one. Go ahead on one engine and back on the other, using the rudder when head or sternway has been gained. If the vessel is fitted with twin rudders which are directly behind the propellers, order *"hard over"* rudder before going ahead on one engine. Back the other engine at such speed as necessary to prevent headway or sternway being gained. Turbine ships are harder to *cast* than ships with reciprocating engines because they are sluggish. It is necessary to *seesaw*, and it is more difficult to turn in one's own water.

Single-screw vessels can be turned quite easily in light winds in restricted waters. Take advantage of the tendency in right-handed vessels to back to port. The first move is to go ahead full with hard right rudder but reverse the engines before much headway is made. Shift the rudder after headway has been lost and back down a short distance and then go ahead *full*. The rudder should be ordered *right full* before the engines ahead begin to turn over. The writer once saw a small collier turn in a space of two or three times her length in this manner. In stronger winds it is advisable to turn so that the tendency to back into the wind can be used to increase the turn.

Most seamen know that an anchor can be used to facilitate and expedite a turn in a restricted space, but few seamen use an anchor for this purpose. In these days of steam and electric windlasses, anchors can be hove in without any delay or effort. High-powered vessels generally use their twin screws and powerful engines to turn in places where a single-screw, low-powered steamer uses an anchor.

The anchor is dropped underfoot at short scope. If low powers are used, the anchor will drag somewhat, but the strain on the chain will not be injurious. The engines can be operated ahead and astern as before, but only slow speeds should be used and little steerway gained. The turn should be made to starboard by pivoting on the anchor when going ahead and by the tendency of the stern to swing to port when backing. A careful check of the chart should be made to ensure that a dragging anchor does not foul a submarine cable.

10.16. Navigating in Narrow Channels.[1] A vessel will be set off the nearer bank when proceeding along a straight, narrow channel, especially if the draft of the vessel is nearly equal to the depth of the water. This effect is particularly noticeable in narrow reaches with steep banks such as certain sections of the Panama Canal and is called *bank cushion*. As the ship moves ahead, the wedge of water between the bow and the nearer bank builds up higher than that on the other side, and the bow is forced out sharply. The suction of the screw, especially with a twin-screw ship, and the unbalanced pressure of water on the quarter lower the level of the water between the quarter and the near bank and force the stern toward the bank. This is called *bank suction*. The combined effect of bank cushion and bank suction may cause the ship to take a sudden and decided sheer toward the opposite bank. If a single-screw steamer traveling at very low speed with her starboard side near the right bank takes such a sheer, she may be brought under control by going ahead full with right full rudder. The added steering effect may overcome the bank suction. A twin-screw vessel under similar conditions has a fair chance to recover from such a sheer by going ahead full on the port engine, stopping or backing the starboard screw, and putting the rudder full right. Should the

[1] Based on *Ship Handling in Narrow Channels* by Lt. Comdr. Carlyle J. Plummer (T) U.S.C.G.R. Courtesy Cornell Maritime Press, New York.

sheer carry the vessel across mid-channel, the starboard anchor should be dropped and snubbed if necessary. All engines should be reversed as the first anchor is dropped.

10.17. Turning in a Bend. There are several factors which affect a ship trying to turn in a sharp bend in a narrow channel. Two of these have been described, *bank suction* and *bank cushion.* Both are strong when the bank of the channel is steep; they are weakest when the edge of the channel shoals gradually and extends into a large shallow area. The tendency of the ship to continue along her original course when the rudder is put over will be felt in the shoaling case. If the bank of the channel is abrupt and the ship deeply laden, a bank cushion will act against the tendency to continue on her course. The river or canal currents are strongest *in the bend,* and there may be eddies or counter currents on the lee side of the point. Turning in a bend requires a knowledge of how these forces act. It is seaman-like to use the forces which are favorable and to avoid those which are opposed. The currents may change direction while moving through along canal or channel and slack water may be encountered.

A head current is the safest because a ship can be stopped very quickly, but a fair current enables the ship to proceed at good speed with very little speed on the engines. Bank suction increases with engine speed. Bank cushion increases with the ship's speed. The force of the current against the quarter can be used to turn the ship; therefore it is advantageous to proceed with the current.

If the current is ahead, the best position to start the turn is from the middle of the channel. The eddy under the point and the increased current in the bend are both avoided. Proceed at a very slow speed over the ground so that the ship can be stopped quickly by the engines and the current, and maybe an anchor or two.

There are three choices in *making* a sharp bend with a fair current:

1. *Hug* the point.
2. Stay *in the bend.*
3. Proceed on the bend side of the middle of the channel.

If the ship *hugs* the point (Fig. 10.6) A, the helmsman will require a small amount of rudder toward the bank to steer a straight course. Less rudder will be necessary as the channel begins to bend and the ship moves away from the bank. This signal "less rudder" is a great help in determining when to begin the turn in clear as well as in foggy weather. However, slack water or eddies may be encountered around the turn. These forces may make it very difficult to prevent a sheer toward the near bank, particularly in shallow water when laden. The stern may feel the current under the quarter and thus increase the sheer.

If the master decides to make the turn *in the bend,* i.e., away from the point, B (Fig. 10.6), the question arises when to turn. If it is started too late, the

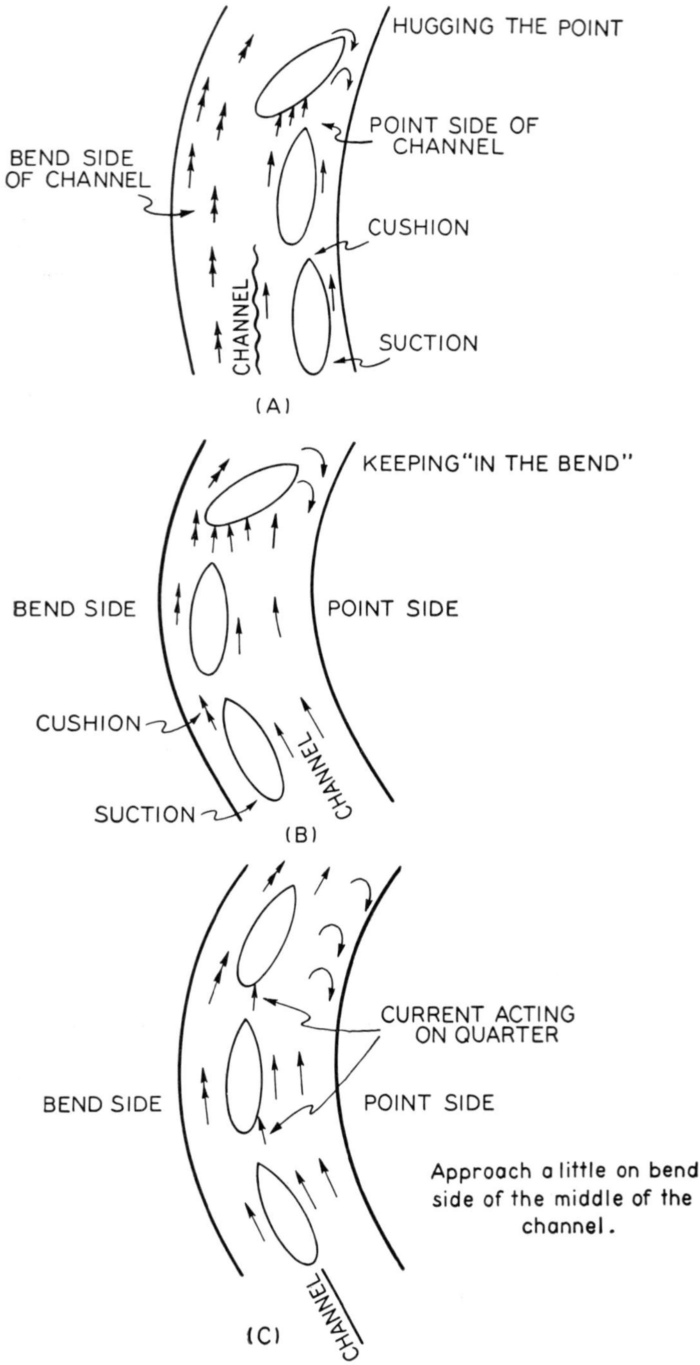
HUGGING THE POINT
POINT SIDE OF CHANNEL
BEND SIDE OF CHANNEL
CUSHION
CHANNEL
SUCTION
(A)
KEEPING "IN THE BEND"
BEND SIDE
POINT SIDE
CUSHION
SUCTION
CHANNEL
(B)
CURRENT ACTING ON QUARTER
BEND SIDE
POINT SIDE
Approach a little on bend side of the middle of the channel.
CHANNEL
(C)

FIG. 10.6

ship may ground on the bank *in the bend*. If he starts too early, there is a grave danger that the bank suction on one quarter added to the force of the current on the other may give the ship a rank sheer. The bank cushion under the bow will increase the sheer. If the bow should enter the eddies under the point, the ship may pivot and eventually ground on both sides of the channel at the same time.

Perhaps the safest way to turn with a fair current is to approach the turn on a course a little to the bend side of the middle of the channel (Fig. 10.6), C. The eddies under the point and the increased current *in the bend* can be avoided, and the force of the current against the quarter can be used to assist the turn.

Two ships should not attempt to pass in a narrow channel in a bend. The ship which has a head current should stop and wait for the other to clear the bend.

10.18. Orders to the Wheel (Under All Conditions). Orders to the wheel and to the engine room telegraph must be given firmly and distinctly and repeated by the steersman or engine order telegraph operator in the exact words given as a check to show that they are understood and are being obeyed. A standard phraseology should be used to ensure a uniform result for changes in course and speed. In giving many of the commands to the steersman, the first word gives the direction so that the wheel can be started immediately, and the second gives the amount of rudder to be used.

"Right (left) standard rudder!" Standard rudder is the amount used to turn with a certain tactical diameter. It varies with the ship and is determined beforehand for each vessel so that all ships in a formation will turn together.

"Right (left) full rudder!" This is used when it is desired to make as short a turn as possible. The number of degrees to use for full rudder is also determined beforehand and is always greater than that used for standard rudder. As the full throw of the rudder is about 35 degrees to each side, full rudder is set a few degrees less in order to ensure that the rudder will not jam hard over against the stops.

"Right (left) 5 (10, etc.) degrees rudder!" This command is used when a more gradual turn is desired than would be the case if either standard or full rudder were used.

"Right (left) rudder!" This order starts the wheel over in the desired direction immediately and must be followed by other orders as an obscure situation develops. It should seldom be necessary to use this order.

With an experienced steersman all the foregoing orders may be followed by an order to steady on a certain compass course. The steersman will carry this out without further orders and report when steadied on the new course. With an inexperienced steersman or when the new course cannot be determined beforehand, the following orders are used:

"Rudder amidships!" This further slows the swing and is a warning that the new course is being approached.

"Meet her!" This order requires opposite rudder of about 10 degrees to stop the swing.

"Steady!" or "Steady as you go!" These are given the steersman when it is desired to keep the ship on the heading she has at that instant.

The object of these orders is to steady the ship on the new course without letting her swing past it with consequent loss of position and unnecessary use of rudder. The exact times at which the various orders should be given for each ship must be determined by trial and experience. One valuable point to note is that a ship with heavy weights, such as turrets near the bow and stern, requires more rudder to start and is harder to stop due to the momentum acquired during the turn from the flywheel effect of these weights.

Certain other orders to the wheel are used on occasion.

"Shift the rudder!" This is an order to change from right to left rudder or vice versa. It is often used while going ahead and backing in restricted waters to assist in a turn.

"Right (left) handsomely!" This order is used for small rudder angles to make slight changes of course. It is rarely used today and probably is unfamiliar to most steersmen.

"Nothing to the right (left)!" This is given when the course to be made good is a shade of the course set, and all small variations in steering must be kept to the right (left) of the compass course. It is frequently used to avoid obstructions, buoys, or passing ships.

All orders to the steersman in regard to the course must refer to the compass by which he is steering at the time and must be in the form "Course, Zero, Zero, Five." The steersman is not concerned whether the course is true or magnetic and must never be required to apply compass corrections of any kind. The officer of the deck should check the compass course upon assuming a new course and frequently thereafter.

The steersman must repeat all orders exactly as given and must report when they have been carried out. When he is relieved, he must report the fact to the officer of the deck and report the course being steered. "Very well" is the proper acknowledgment to any report made by the steersman. The expressions "Aye, Aye," "O.K." or "All right" are incorrect and should not be used.

10.19. Orders to the Engine Order Telegraphs (Annunciators). Orders to the engine order telegraphs are in three parts;

1. The first part designates the engine, as "Starboard (port) engine" or "All engines." This puts the annunciator man on the alert.
2. The second part of the command gives the direction in which the engine order telegraph is to be moved, as "Ahead" or "Back."
3. The third part of the command gives the speed at which the engine is to be moved.

Thus: "All engines ahead full"; "Port engine back two thirds"; "Starboard engine ahead standard!"

Every order to the engine order telegraphs must be repeated word for word by the operator. When the engine room has acknowledged the order by the repeat back system on the telegraph, and the shaft revolution indicators show the engines are in the process of carrying out the order, the operator should then report what the engine is doing. Thus, upon the order from the officer of the deck, "Starboard engine, back one third," the operator repeats, "Starboard engine, back one third, sir." At the same time he rings up "back one third" on the starboard engine order telegraph. When this is repeated back from the engine room on the engine order telegraph, he reports, "Starboard engine answers back one third, sir."

10.20. Man Overboard. If the experiment is tried of throwing over from the bow a light buoyant object, it will be found that by the time this object reaches the stern it will be clear of the ship by a considerable distance. The surface wash will throw it off from the side. A man falling overboard may feel this wash to a certain extent, but he sinks in the beginning far below its influence into the suction of the screw. Moreover, his first instinct is to swim back toward the ship.

One or more, preferably more, life buoys should be thrown over at once. If a little presence of mind is exercised here, it is often possible to throw one of these very close to the man and if possible between the man and the ship. At the first alarm a number of men (previously instructed through drills, etc.) go aloft to try to keep the man in sight; and as quickly as possible a quartermaster follows them with a pair of good binoculars and a set of semaphore flags.

The ordinary life buoy is so small that often the man in the water cannot see it, and it is of little or no assistance to the lookouts who are trying to keep him in sight. This is a serious and often fatal defect. It is well to keep a number of these small life buoys about the deck to be thrown overboard on the instant by anyone who may be near them; but in addition there should be available packets of sea dye marker for daytime use and battery-operated water lights for nighttime. These latter are necessary to serve as markers not only for the man but to keep the spot in sight from the ship. At least 50 percent of all life buoys kept about the deck should be equipped with the battery-operated lights mentioned above. The sea dye markers should be kept handy to each life buoy.

One method of maneuvering to recover a man overboard is to go full speed astern as soon as the man is clear of the screw and to lower a boat as soon as speed has been reduced sufficiently. The boat pulls back in search of the man and is guided by signals from the lookouts aloft who have semaphore flags, provided they have succeeded in keeping the man in sight. Failing this the boat cannot go far wrong if it pulls back on a course opposite the original heading of the ship, for, although the steamer in backing will probably throw her head to one side, she will not usually gain a great amount of ground in that direction before coming to rest.

In most conditions of the sea, a boat may be lowered with reasonable safety from a ship at a speed of 4 knots. If the weather is moderate or the sea calm or from such a direction that there is no occasion for maneuvering to lower the boat, all this is simple enough. If conditions are such that turning wholly or partially is necessary, many officers put the rudder hard over, keep the engines turning ahead at the same or even greater speed than before, and describe a circle, thus coming back, with the ship to a point near that at which the man went over.

Observations upon turning circles of a large number of steamers show that a steamer turning with hard over rudder will pass within a short distance, rarely so much as the ship's length, from the point where the rudder was put over. No doubt the symmetry of the curve may be considerably modified by wind and sea, but not sufficiently to prevent a return to the area of the starting point. The time required for the full turn will vary with the length, the speed, the weather, and the maneuvering powers of the vessel, especially the turning circle, the time required to describe it, and how close the ship will come to a marker thrown over just before putting over the rudder. It will, of course, be understood that in turning the speed must be regulated according to conditions of the weather. It would not do, for example, to come up into a heavy sea at full speed.

If the conditions are such—due to lack of a proper marker, or to fog, or to any other cause—that difficulty is to be anticipated in finding the man, it is probably better to stop and send the boat back along the course opposite the original heading. This emphasizes the need for having a compass in the boat. In case of fog, the vessel should avoid changing her position while the boat is away. The compass is thus a guide for finding the way back—assisted, of course, by sounding the whistle, firing guns, etc.

There can be no question that in weather too heavy to admit of lowering a boat the one method that can give hope of saving a man is to turn and attempt to pick him up with the ship. In using this method the use of cargo nets over the side, attended by strong swimmers wearing immersion suits and with safety lines attached, is of immeasurable value.

Motor whaleboats are kept ready at sea for instant hoisting out and use as lifeboats.

Special rules are laid down for cases of man overboard in formation. All officers concerned must be familiar with these. Generally speaking, they provide for the necessary maneuvers to keep clear of other ships while picking up the man and for signals notifying other ships of the situation.

There are several different methods which ships may use to pick up a man directly without lowering a boat:

Method No. 1. This involves the use of the Williamson Turn.

Put rudder over with full rudder toward the side from which man fell overboard (if known). When ship's head has changed 60 to 90 degrees (varies for type of ship), shift the rudder and turn to the reciprocal of the original course.

Speed may be increased as appropriate. This method is relatively slow, and in poor visibility there is always the chance of losing sight of the man. However, it is an excellent maneuver to ensure getting back on the original track if the exact time the man went overboard is unknown or doubtful. It is also recommended for use in low visibility, particularly at night.

Method No. 2. A single turn toward the side from which the man fell overboard is another method. Hold original course for one minute; then, using full rudder and maximum speed available, turn until ship is headed for man. This method is fast and should be used if man is in sight.

Method No. 3. A two-turn maneuver is somewhat simpler for an inexperienced shiphandler in that it ensures the ship coming back to the spot where the man should be. A 180-degree turn with full rudder is made toward the man. Ship is steadied on reciprocal of original course until man is about 30 degrees abaft the beam. Another 180-degree turn with full rudder is then made again toward the man. The ship is steadied on the original course and should then be headed for the man. This method is not so fast as Method No. 2.

It is recommended that ships fitted with a Dead Reckoning Tracer mark the trace at the time the man goes over. A course to steer back to this point on the chart can then be obtained.

A method becoming increasingly popular in the Navy for use at night is for the ship to turn immediately toward the side from which the man fell overboard and train a searchlight on the life ring that was thrown over. When the turn is completed and an estimate is made of where the man went over, a life raft is released and the light shifted to the raft. The ship then circles slowly, keeping the illuminated raft as a reference point for the man and searching for the man at the same time on both sides.

An additional recovery method used extensively by submarines and by some surface ships is the "Y" method. It is called this because the ship's track is that of a Y. Rudder is put over hard to the side from which the man fell overboard and engines are backed FULL. As sternway is commenced rudder is shifted. When bow is headed near bearing of man, engines are stopped, rudder shifted and engines ordered ahead, maneuvering to pick man up.

Handling Steamers in Heavy Weather

In the days of sail and in the early steam vessels, most of which also had after sails, the conventional way to handle a ship when the weather was too heavy for her to proceed on her course was to bring her head up until she had the sea on the bow and to hold her there with the rudder and sail, with little or no resultant headway. If she fell off—as from time to time she did—and started to gather way, the hard down helm and after sail would bring her promptly back to meet the sea. Thus she came up and fell off making some little way through the water, but none of it against the sea, and, in the main, drifting steadily to leeward. For such bluff-bowed ships, this was and is the ideal way of riding out a gale. But a modern steamer, whether a man-of-war, liner, or tramp, carries no sail and is commonly long and sharp. The propeller acts as a drag, tending to hold her stern-up to the sea, and this tendency is assisted by the excess of draft which such steamers usually have aft. To hold such a steamer bows on to the sea, she must be forced into it—not at great speed, perhaps, but sufficiently great to maintain steerageway. This can strain the ship severely and causes grave doubt as to the wisdom of such a method of lying to. The opinion of late years is that a steamer should run slowly before a sea or lie to with the sea astern or on the quarter; and this view is supported both by theoretical considerations and by a convincing amount of practical experience.

10.21. The Approach of a Tropical Storm. When a master is forewarned of the approach of a tropical storm, his first thought must be the location of the center and the estimated track of the storm. The geographical position of his vessel with respect to the proximity of land or shoal water, and whether his vessel will be in the dangerous or navigable semicircle of the storm, must be determined at once.

He should make an early decision and use all necessary speed to gain the safest possible geographical location before the storm is upon him. Once the center of the storm is near him, he should then be free to reduce speed, to avoid damage to his vessel. It may then be desirable to proceed at dead slow engine speed, barely maintaining steerageway, or even lie to for several hours until the storm passes. It might be desirable and safe to maintain a fair speed downwind.

If a master is unable to gain a satisfactory position with respect to shoal water, with the winds he can expect in a tropical storm, he may be forced to oppose the winds with appreciable engine speed, accepting risk of damages, to avoid being pushed into shoal waters.

On the other hand, if arrival in the navigable semicircle of a tropical storm before the storm is upon him is his only concern, he should not force his vessel in any continued effort to do so, but should ride out the storm as best he can.

10.22. Controlling a Ship in Very Heavy Weather. The easiest position for a ship in a very heavy sea would be that which she would herself take if left at rest and free from the constraint of engines, helm, and sails. A ship, if left to herself in a seaway will usually fall off until she has the sea abaft the beam, the propeller acting as a drag and holding her stern-up. In this position she will roll deeply, but easily, and will drift to leeward, leaving a comparatively smooth wake on the weather beam and quarter, rolling deeply, but in most cases easily, and taking little or no water on board. If oil is used along the weather side and astern, the wake can be converted into an "oil-slick" and danger of seas breaking on board effectually prevented.

If a ship rolls dangerously, she may be kept away more from the wind and sea by using a drag over the stern, or by turning over the engines just fast enough to give her steerageway; for it seems to be established, as the result of experience, that a steamer may safely run with the sea aft or quartering, *provided she runs very slowly.* Clearly this is not "running" in the old sense of that term, according to which a vessel going before the sea was forced to her utmost speed with the idea of keeping ahead of the waves, which were expected to "poop" her if they overtook her. It is evident from the statements of a large number of shipmasters who have tried the experiment of slowing down or stopping when running before a heavy sea, that this maneuver, so far from resulting in the disaster which many seamen would expect from it, had an extraordinary effect in easing the ship and keeping her dry.

The explanation of this seems to be that a ship running at high speed through the water draws a wave after her which follows under her counter and rolls along toward the waist on either side, tending continually to curl over and break on board. This wave is reduced to insignificant proportions in running dead slow.

10.23. Roll and Pitch. Another point which enters into the behavior of a vessel going before the sea is that as she rolls and pitches she buries first one bow and then the other, increasing the pressure on the bow so buried. If she is being driven through the water, her head will be forced off, first to one side and then to the other, causing her to yaw badly with a continual tendency to broach to. This cannot be met by the rudder, because, at the very time the bow is buried, the stern is lifted more or less out of the water, and the rudder loses, for the moment, its steering power. As the stern is lifted, there also comes a racing of the propeller which is in itself a serious danger at high speed. There seems no question that the dangers connected with running, so far from being increased, are greatly reduced if not altogether removed, by slowing or stopping.

It will of course be understood that in this matter, as in all others connected with seamanship, due regard must be had for the peculiarities of the individual ship and that the maneuver which is safest for a majority of ships may be dangerous for certain ones. Thus a ship whose cargo may shift should not be allowed to roll excessively; nor should a warship whose heavy guns or missiles are carried high above the center of gravity. On whatever course the vessel may be kept, this rule may be regarded as of universal application; that, other things being equal, *the lower the speed at which she is run, the easier she will ride.*

10.24. Relation Between Ship and Waves. Attention is invited here to an important relation, not always recognized, between a ship and the waves in which she floats. For every ship (in a given condition as to trim, etc.) there is a perfectly definite "rolling period"; a period, that is to say, in which she will make a complete roll, *without regard to whether she is rolling 10 degrees or 40.* So, also, in the case of a seaway, there is usually a fairly regular interval of time between wave crests passing a given point. If the point is a ship in motion, her motion may increase or decrease the interval between the waves so far as she herself is concerned; but this will not change the *regularity* of the interval. If it happens that this interval coincides with that required for the ship to complete a roll, each wave as it passes her will add its rolling impulse to the accumulated effect of those which have preceded it, and the ship will roll more and more deeply until she reaches *the maximum roll of which she is capable.* She will not capsize (if properly designed and is undamaged) because there are forces at work to resist the rolling, and these increase as the depth of roll increases, until the rolling forces and the resisting forces balance. But she will continue to roll to the maximum limit until something is done to break up the synchronism between her period and that of the sea. This can be accomplished, *provided the ship has headway,* by changing the course or the speed, thus changing, not the real, but the apparent, period of the waves. By running more nearly into the sea—meeting the waves—the apparent period is shortened; by running more nearly before it, the period is lengthened; but in either case it is *changed* and will no longer agree with the rolling period of the ship. The same effect is produced by a change of speed. If, therefore, it is judged from the violence of the rolling on a given course that the period of the waves is coinciding with that of the ship, the course or speed or both should be changed.

The length of the ship, as compared with that of the waves, is also a very important factor in the behavior of the ship, especially when she is running more or less with the waves or meeting them. It often happens that a small ship, in a long sea, will be perfectly comfortable where a larger and longer one makes very bad weather. The small craft climbs up and slides down the waves, accommodating herself to their slopes, and pitching only as the slope changes; but the longer craft, partially spanning the crests and the hollows of the waves alternately, one end being poised on the crest of one wave while the other end is buried in the adjoining one, may be making very heavy weather. A few years ago a large aircraft carrier in the Philippines was badly battered by a typhoon, but a destroyer escort, which passed through the same gale at very nearly the same place, was perfectly comfortable. A Cuban revenue cutter, less than 100 feet long, was caught in one of the heaviest cyclones of recent years some 20 miles south of Cienfuegos. She rode it out not only without discomfort, but without damage to a light dinghy which she carried rigged out at davits overhanging the side and only 8 feet above the water. There are many other similar cases on record. It is not unusual to hear that a vessel had foundered in a gale and that her boats have ridden out the same gale in safety. The great difficulty here is to off-load the boats and get them clear of the ship. Once clear, they are often much safer than the ship.

10.25. Bringing a Ship Bows-On. If, when a steamer is before the sea or in the trough, it is decided to bring her up to it, bows-on, she should first be slowed until she has barely steerage way, and should then be brought up as gradually as possible. To put the wheel over with considerable speed on and bring her up with a rush—slapping the sea in the face, as it were—would result in serious damage, if not in foundering. After getting her up to the sea bows-on, the greatest watchfulness is required, first, to avoid falling off into the trough of the sea, as she will try to do the moment she loses way; and second, to avoid driving into the heavy, breaking seas, which will threaten her now and again. There is reason to believe that many of the phenomenal "tidal waves" reported as having suddenly overwhelmed steamers in mid-ocean have been simply the exceptionally heavy waves which build up from time to time in any long-continued gale; and that their destructive power was due to the fact that the vessels were driven into them instead of being allowed to drift before them and ride over them unresistingly. An officer should always be kept at the engine-room telegraphs, in lying to bows-on, and an engineer standing by below, to obey his signals instantly. So long as she heads up to it, the more slowly she turns over, the better. If a heavy sea is seen bearing down upon her, she should be stopped altogether. If she falls off, it will be necessary to increase the speed a little to bring her up, but she must be slowed again as soon as possible.

10.26. Using a Sea Anchor. The use of a *sea anchor* has been advocated by many writers on seamanship, and it was commonly assumed that by its use any vessel may be held head to sea and enabled to ride out a gale. No doubt this can be done if the sea anchor is large enough and strong enough. If used with a modern steamer to keep her head to sea, it acts against the drag of the screw, which, as has been seen, tends to keep her stern to it. To overcome this the anchor must be very large and very strongly built; otherwise, its effect will be to keep her in the trough of the sea.

With small ships, and especially with yachts, a sea anchor has been used with good results. Such a ship, riding to leeward of a sea anchor of fair size with an oil bag hauled out to a block on the hawser well clear of the stem, and drifting slowly astern, will ride out almost any gale with safety and comfort. Indeed, as has been said above, this is the ideal position, in very bad weather, for any vessel which can be made to take and keep it. But it is doubtful if a large steamer could be made to do this without the use of an anchor too unwieldly to be handled conveniently in a heavy gale.

10.27. Practical Sea Anchors. In cases where a light drag is needed and no sea anchor is available, a boat may be used, with a hawser made fast to a span from the bow and stern ring bolts, and to a belly band amidships. A long spar (or a number of spars lashed together) may be used, also slung by a span. If a heavy awning can be added to such an improvised anchor, it will help to break the sea.

There are cases recorded of vessels having been kept head to sea by paying out their chain cables, unbent from the anchors. Where the water is shallow enough for the chains to drag on the bottom, they are especially helpful. A good-sized manila hawser, paid out *on the bight*, both ends being kept on board, makes a very convenient drag—perhaps the most convenient that could be devised. With both ends leading in through the stern chocks, it would be extremely helpful for holding her stern-on, and with one end at the stern and the other at some point near the beam, she could be held with the sea on the quarter. A block on the hawser would admit of reeving a line for hauling oil bags out and in.

In twin-screw ships, the propellers have not as much drag as with single screws, and such ships can sometimes be held up the sea without being driven into it dangerously, by turning over the lee screw very slowly. This is often the best way to lay a twin-screw ship to, although there is nothing in the nature of the case to prevent such a ship from riding easily with the sea astern or quartering.

10.28. Shoal Water. Where the depth of water admits, a stream anchor or a good-sized kedge may be let go with a long scope of cable, either chain or rope, and allowed to drag on the bottom. In comparatively shoal water, a vessel may ride out almost any gale, *at anchor,* provided she has a sufficient scope of cable. Under such circumstances in very deep water, *one anchor with two cables on end* may be used. The engines may be used, but very cautiously, to relieve the strain on the cable if this should be thought to be too great. In more shallow water, the bridle or hammerlock moor could be used.

10.29. The Calming Effect of Oil. The effect of oil in calming a rough sea has been known from the earliest times. The action of the oil is not only to prevent the breaking of waves, but to a considerable extent also to prevent them from forming. Its effect, when used on an angry sea, is described by all who have tried it as magical. Even in a surf, while it cannot altogether prevent the waves from breaking as they are driven in upon the shoals, it greatly reduces their violence and will often enable a boat to land when otherwise it would be out of the question.

Almost any kind of oil will give good results, but some kinds are very much better than others. Animal and vegetable oils are best; for example, sperm, porpoise, linseed, olive, and cotton seed; and fairly thick and heavy oils are better than lighter ones. Oil of turpentine is probably the best of all. Mineral oils are much less effective; however, a very thick sticky oil, or one that tends to thicken or congeal in cold weather, may be improved by thinning with petroleum. Soapsuds has a remarkable effect in preventing the formation of waves, but it does not keep them from breaking when formed.

Any method will answer for dispersing the oil which produces a slow and steady flow. A convenient way is to fill the closet bowls with oakum and oil, or to place a slightly opened can where it will give a slow drip into the bowls and out through the waste-pipes. A still simpler way and one frequently used is to fill a canvas bag, 1 to 2 feet square, with oakum and oil, and punch a number of holes through the canvas with a sail needle. Such a bag may be hung over the side at any point where it is found to give the best result. If there is danger of its being thrown back on board by the sea, its lanyard may be led through an eyebolt or a shackle in the side. If for any reason, a very rapid flow is wanted, a hose may be led through a scupper or over the rail, and the oil poured into it through a funnel; or, in a sudden emergency, the oil may be thrown or pumped over the side. The quantity used need not exceed a few gallons—4 or 5 at most—even for a large ship riding out a prolonged gale.

It should be noted that the rate at which the oil spreads is slow in comparison with the speed even of a vessel drifting. Thus a vessel lying with engines stopped can make a "slick" to windward but not to leeward—except, perhaps, very close alongside—because she drifts faster than the oil can spread. So, in running, a vessel can leave a slick astern and to some extent on either hand, but can do nothing to calm the waves ahead of her. She can, therefore, avail herself of the benefits of oil if she is running more or less before the sea, but not at all if steaming into it. (See also Section 12.10.)

10.30. Summary. We may sum up the various methods of handling a ship in heavy weather with the statement that the ship will usually be safest and most comfortable when stern-to, or nearly stern-to, to the sea, and *drifting before it.*

If, by the use of sails, a drag, or any other means, she can be held bows-on, *while being still allowed to drift,* this is probably the best way to lay her to; but if she cannot be held up without being forced into the sea, it will be because of the natural drag of the stern and propeller, and in this case advantage should be taken of this drag to hold her more or less directly stern-on, letting her drift in this way.

Even if the position she takes up in drifting is nearly in the trough of the sea, it will usually be found that she is easier in this position than in any other; however, the use of oil, as described above, is especially important in such cases.

If the position which she takes in drifting proves to be one in which she rolls dangerously, then she may run just fast enough to steer, *but no faster,* and so keep the course which is found most comfortable.

One final word of caution. The effect of "free surface" water in bilges and compartments is particularly important during heavy weather when a ship's stability is severely tested. Pumps must be kept operable to remove water; electrical switchboards must be kept dry. This points up the need for making watertight closures well in advance of the onset of the worst part of the gale. To minimize free surface effect (see Chapter 3) it is advisable, particularly when bad weather is expected, to keep all tanks containing liquid either full or empty.

11

Docking and Mooring—Handling Alongside

The expressions *docking* and *mooring* are sometimes used synonymously. A ship *docks* when it goes alongside a wharf or pier or when it enters a drydock. A ship *moors,* according to naval usage, when it secures to a mooring buoy or lays out two anchors with or without a mooring swivel. A ship also *goes alongside* a wharf, pier, or another vessel and is described in the navy deck log as being *moored* at a certain pier or alongside another ship.

Mooring Lines

The lines used to secure the ship to a wharf, pier, or another ship are called *mooring lines.* Mooring lines have a dual purpose. They must be as light as possible for easy handling and, at the same time, strong enough to take considerable strain while coming alongside and to hold a ship in place when secured. Five-inch manila or smaller nylon is used for mooring lines in destroyers or smaller vessels. Larger ships may use 8- or even 10-inch lines. The manila lines may be reinforced or replaced by heavier lines or wire hawsers when the ship is finally securing alongside. Nylon lines are being introduced in the Fleet for all ships.

The mooring line which runs through the bull nose or chock near the eyes of the ship is called the *bowline.* The corresponding line aft is the *stern line.* These lines should lead well up the dock to reduce the fore-and-aft motion of the ship. Other mooring lines are either *breast lines* or *spring lines.* They are called bow, waist, or quarter breasts and springs, depending on the part of the ship from which they are run. Breast lines are run at right angles to the keel and prevent a ship from moving away from the pier.

Spring lines leading forward away from the ship at an angle with the keel are *forward (bow, waist* or *quarter) springs.* Springs leading forward or aft prevent a vessel from moving aft or forward, respectively. Two spring lines, leading in opposite directions in the same vicinity, act as a breast line from the pier to the ship.

If a ship moves ahead or astern with lines out, a breast may become a spring, and spring lines may change their leads. To prevent confusion and to add to the efficiency of line handling, lines are numbered from forward aft, according to the position where they are secured aboard ship. A ship may use fewer or more lines as necessary in which case the numbers are changed accordingly. Because numbers are shorter and more precise than names, they are generally used while the lines are being handled. The names are used to a greater extent when the ship is secured and the use and lead of each line becomes definite. Figure 11.1 shows the names and numbers for seven mooring lines.

Lines can be of the greatest assistance in making or clearing a pier. Prior to a ship coming alongside, the required lines with eye splices or bowlines in the ends should be led through the chocks up and over the lifelines. Heaving lines

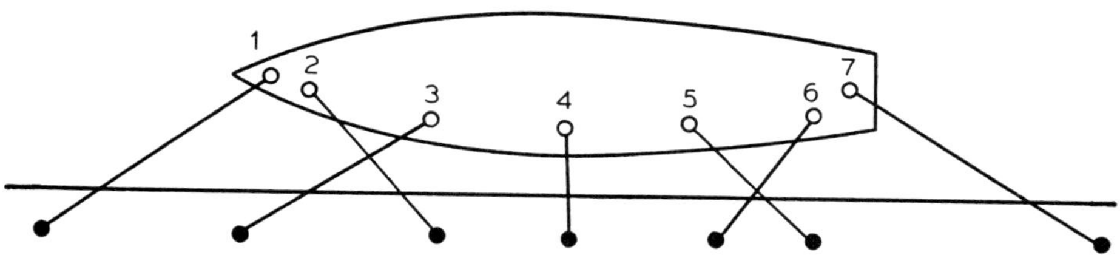

FIG. 11.1 (1) BOWLINE (2) AFTER BOW SPRING (3) FORWARD BOW SPRING (4) WAIST BREAST (5) AFTER QUARTER SPRING (6) FORWARD QUARTER SPRING (7) STERN LINE

that have been successfully passed should be made fast near the splice and not at the end of the bight where they will become jammed when the eye is placed over the bollard. Heaving lines should be passed as soon as possible; then the heavy lines, the bights of which are necessarily hard to handle, may be run later when the vessel is farther up the pier and nearer her berth.

As a large ship works her way up the pier or into the slip, the lines should be *fleeted* up the pier in short steps, thus keeping them in position for use.

If two bights or eye splices are to be placed over the same bollard, the second one must be led up and through the eye of the first and then placed over the bollard. This makes it possible for either to be cast off independently of the other. This is called *dipping the eye.*

The ship in (a), Fig. 11.2, is lying off a pier with a bow breast line secured to a bollard. If the line is led to a winch and a strain put on it, the bow will swing toward the pier, and the stern will move out. It should be noted, however, that the stern does not go out as much as the bow comes in. Because the ship is not held rigidly at the pivoting point, the mass as a whole will respond to the force acting on the bow, and the resultant motion will be like that shown in the figure.

If, in the above case, the stern is held by a line to the pier as shown in (b), the pivot is transferred to the stern, and the whole ship moves toward the dock. This requires much greater effort than to turn the ship near her natural pivoting point as in (a).

If the bow and quarter breasts are hove in at the same time, the ship will be breasted in bodily but at greater expenditure of work than in the preceding cases.

If the ship has way on, either ahead or astern, her momentum enters into the problem of her behavior. In Fig. 11.2(c), the steamer is moving forward

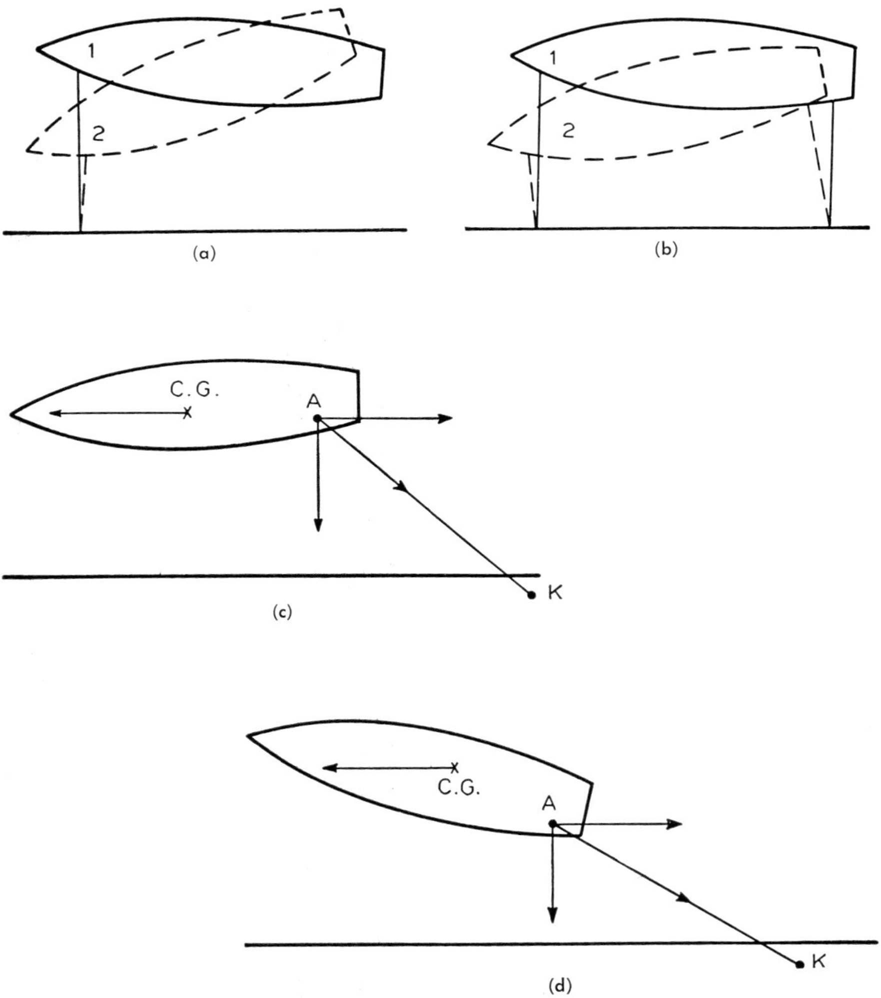

FIG. 11.2 HANDLING STEAMERS AROUND A PIER

parallel to the face of the pier (with engines stopped and rudder amidships). The after quarter spring "A K" is taut. The motion of the ship will be that resulting from her momentum along the original course and the tension along A K. The tension on A K may be resolved into two components. One retards the ship along the line of her original course and thus directly opposes the momentum, and the other moves her toward the pier. The stern will swing in and the bow out. It is important to note, however, that the momentum which is concentrated at the center of gravity forward of the pivot (A) opposes the turning and tends to keep the ship parallel. Thus, as a matter of fact, the ship does not turn much but comes in nearly parallel to the pier as shown in Fig. 11.2(d).

If the vessel in Fig. 11.2(c) is moving ahead with an after spring from the bow instead of from the quarter, the forces acting are similar to those in Fig. 11.2(d) but with an important difference. The momentum increases the turning effect of the spring, instead of opposing it. This will be clear if (a) and (d) in Fig. 11.2 are compared. The result is that the bow of a ship moving ahead on an after bow spring turns sharply in toward the pier, and the stern will swing out.

If the ship is moving astern instead of ahead, the conditions are reversed in that a forward quarter spring will turn the stern sharply, whereas a forward bow spring has little turning effect. The reason is because its point of application is near the ship's pivot point.

The vessel in Figs. 11.2(c) and (d) could go ahead on her engine(s) and put her rudder left to throw her head in. Since the steering effect of the rudder is due to the discharge current against the rudder and since the stern cannot move to starboard because it is held by the spring, left rudder can have comparatively little turning effect. Right rudder, on the other hand, will help to throw the stern in.

If in Fig. 11.2(a) the rudder is put left, it will throw the stern out and increase the rapidity with which the bow turns in. If put right, it will oppose the turning but not enough to overcome it.

In any case, if the line is made fast at the ship's natural pivoting point, and the engine(s) are turned over ahead or astern, the ship will spring in bodily. Her heading can be controlled by putting the rudder over, which throws the stern to either side, as desired. The ship swings under the influence of the rudder while coming bodily in on the spring, and it is often possible to come alongside quickly and smartly by using this line only.

Orders to the Men at the Lines

COMMAND	MEANING
Slack (Slack Off) the Bowline (Number One):	Pay out the line specified, allowing it to form an easy bight.
Take a Strain on One (or Number One):	Put number one line under tension.
Take in the Slack on Three (or Number Three):	Heave in on number three line but do not take a strain.
Ease Three:	Pay out number three enough to remove most of the tension.

COMMAND	MEANING
Check Three:	Hold number three line but not to the breaking point, letting the line slip as necessary.
Hold Two:	Take enough turns so that the line will not give.
Double Up and Secure:	Run additional lines or bights of lines as needed to make the mooring secure.
Single Up:	Take in all lines but a single standing part to each station, preparatory to getting underway.
Stand by Your Lines:	Man the lines, ready to cast off or let go.
Let Go:	A command to slack off smartly to permit those tending the mooring lines on the pier or other ship to cast off.
Take in One (or Number One):	Retrieve line number one after it has been cast off. When used by the conning officer it means to slack one, cast it off, and then pull it back aboard. When used by the officers in charge on the forecastle it is preceded by the commands "slack one" and "cast off one" and means merely to retrieve line number one and bring it back on deck.
Cast Off:	A command *to those tending the mooring lines on the pier or on another ship* to disengage or throw off the lines from over the bollards or cleats.

In securing alongside a dock, wharf, or pier, special attention must be paid to the state and range of the tide. When securing at high water, enough slack must be left in the lines to ensure that at low tide they will not part, carry away bollards, or in extreme cases list the ship to a dangerous degree or even capsize small vessels.

11.1. Making Landings. The names *dock, pier,* and *wharf* are used almost interchangeably. These are all structures connected to the shore with enough water up to them for vessels to come in or alongside. A pier is built at right angles to the shore; a wharf is parallel. Both are sometimes called docks, although strictly speaking a dock is a structure used for drydocking a vessel. The space between neighboring piers is called a *slip*.

Wharves and piers may be built on piles which allow a fairly free flow of water under them and in the slips between them. Their underwater construction may be solid, in which case there will be no current inside the slips, but eddies of various sorts may be found. Warehouses or other buildings may be built on them which may result in varying the effect of the wind on the upperworks of a vessel.

Wind and current at right angles to a pier are always more dangerous than when they are blowing or running respectively along its face. In coming alongside, as in all shiphandling, the conditions of wind and current existing should be observed carefully, and they should be used to assist rather than to hinder when possible. Several cases of going alongside under different conditions of wind and current will be discussed.

11.2. Going Alongside, No Set On or Off the Pier (Single-Screw Vessel). A single-screw (right-handed) vessel can make a landing to port with little difficulty under these conditions. The ship should be headed for a point a short distance outboard of the place where the bridge will be when the ship is secured. The course of the approach should be at an angle of 10 or 15 degrees with the face of the pier. Slow speed should be used, and the engines should be stopped when there is sufficient headway to reach the berth. Enough headway to steer should be retained when the ship is almost abreast of her berth; the engine can be backed to stop the ship and to swing her stern to port and then parallel the pier. The ship can then be breasted in by the mooring lines and winches. If a single-screw ship must go alongside to starboard, the angle of approach should be about 10 degrees. The speed should be less than that used for a port landing but still enough to keep the ability to steer. The point for which the ship should be headed during the approach should be the final position of the bridge when secured.

The port anchor may be dropped at short stay during the early part of the approach. The anchor is then dragged over the ground. The anchor will add to the steering ability, reduce the speed of the ship, and give the master better control of the bow and the stern when the final landing is being made. The difficulty at that time is that reversing the engine will force the stern away from the pier and the bow toward the pier when a *close* landing is contemplated.

The engines should be stopped later than would be done with a port landing. The dragging anchor will enable the ship to be stopped without backing with much power or for a long time. The anchor will probably enable the master to lay the ship alongside her berth without backing at all or perhaps with just a touch astern.

Every effort should be made to get a stern line out as soon as possible so that this line can be held when the engine is backed and the stern prevented from swinging away from the pier.

11.3. Going Alongside Port Side to Being Set On to the Pier. The point for which the ship should be headed at the start of the approach should be farther away from the pier than the one used with no set. The angle of approach should be 20 to 30 degrees, and the speed of approach should be higher than above without attempting a high-speed landing. The amount of set ought to be watched constantly. If it is apparent that the set is more than anticipated and that, if the ship continues, she will strike the pier, there are four choices:

1. Head farther away from the pier.
2. Back clear and try again.
3. Stop with the ship parallel to the pier while there is still some open water between the ship and the pier.
4. Proceed at greater speed in order to reduce the time during which the ship is subject to the wind and current.

The conditions near the berth, the proximity of the berth, and the course will determine what action is advisable. If there are ships alongside of the pier ahead and astern of the allotted berth, it is probably wise to head up or back clear and try again. If there is a clear space astern of the berth, it may be possible to make a landing short of the correct berth and, once alongside, move up to the berth. There is always the fourth choice in which case the discharge current from the backing screw may cushion the impact somewhat.

11.4. Going Alongside, Starboard Side To—Set On to the Pier. If a single-screw ship attempts a starboard landing with a set on to the pier and finds that she is too close, the four choices of action are still available but number two should be changed to: stop with the ship heading slightly away from the pier. In this position she can back, and, in so doing, the stern will swing away from the pier. The ship will parallel the pier, lose her headway, and the discharge current will cushion the impact. The first choice, to head up, is not as satisfactory as before because the original approach course should make a smaller angle of about 10 degrees with the face of the pier. There is not as great an angle through which to head up, as was available in a port landing.

11.5. Going Alongside a Pier—Being Set Off. Should it be seen that the vessel is being set off the pier, the ship's head can be pointed well up into the pier and the approach made with more speed. The bow lines are gotten out, and, when close to the pier, the rudder is put away from the pier. For a single-screw vessel, port side to, the landing is easily made by backing down on the screw. For a single-screw vessel making a pier to starboard, the approach should be made by snubbing the port anchor as previously described. In addition to snubbing the port anchor, a starboard quarter breast can be used to hold in the stern as the ship forges slowly ahead on it. In a twin-screw vessel, the outboard engine is backed full.

11.6. Going Alongside—Current from Ahead. When a current is running parallel to a wharf, the ship should be headed into the current and then brought alongside. Slack water is the most favorable condition; yet the current, if not too great, can be used and the berthing accomplished without trouble, except in the case of the largest vessels. With current from ahead, a vessel can use more speed through the water without increasing its speed relative to the wharf; therefore, the ship will have better rudder control with the same degree of safety.

The ship, making little headway along the face of the wharf, is brought in fairly close and parallel. The vessel must not be canted in because she might come in too fast and cause damage. A forward bow spring is sent well ahead and up the wharf.

When in position relative to the pier, all forward motion relative thereto is stopped, and the vessel is slowly dropped back on the spring. The amount of tension on this spring determines the rapidity with which the ship drifts in to the wharf. The rudder may be used to swing the stern.

Should the current be very strong, the ship should go a little way above the wharf and drop the outboard anchor. A forward bow spring should be run ashore. By veering chain and holding the spring, the ship will swing slowly in toward the wharf. By using the rudder and adjusting the strain on the anchor chain, perfect control can be maintained. When alongside, a bow breast and forward quarter spring are got out as soon as possible.

11.7. Going Alongside—Current from Astern. Making a pier or wharf with a fair current is difficult and should be avoided when possible. If there is swinging room and if other reasons do not forbid, much time and fuel will be saved and danger avoided by dropping an anchor, swinging with the tide, and making the pier as previously described with the current from ahead. If this is not practicable, tugs should be used if available. If a tug is not available, the approach should be made as slowly as possible and as near to the pier as is safe. When about in position, the bow is canted out a few degrees from the pier. An after quarter spring is got out as soon as possible and the engines backed to keep from parting the line. Backing on the inboard engine of a twin-screw ship forces more water between the pier and the ship, thus cushioning her as she comes in. Care must be taken to prevent the stern from swinging away from the pier. The use of after spring lines and stern lines with the outboard engine backing will prevent this movement.

11.8. Clearing a Pier—No Wind and No Current. Clearing a pier is less difficult than making a pier. The first step is to slack all lines carefully and observe the effect. If the ship does not drift out, it will be necessary to force the stern away from the pier.

In a single-screw ship with the starboard side to the pier, the engine is backed. This swings the stern rapidly to port. If the bow is forced into the pier, right rudder is used to clear it as the ship goes astern. When the stern is about 50 feet out, the bow will be pointed in toward the pier. A quarter breast which becomes a spring as the ship continues to go slowly astern is now held. This action will bring the bow out. When pointed fair, the ship casts off and goes out ahead.

If the port side is to the pier, an after bow spring is used as the ship goes ahead slowly on it. Left rudder throws the stern well out. The ship is cast off and backed down slowly with right full rudder until clear. As the stern will gradually turn in toward the pier, it will be necessary to stop when parallel to the pier and several beams' width out from it. The vessel now goes ahead with right rudder, and the bow falls off as required.

A twin-screw vessel can shove off from a pier easiest by holding the after bow spring and slacking off all other lines. The outboard engine is turned over, slow ahead, until the inboard propeller is clear of the pier. Fenders can be used as necessary on the bow. Once the inboard propeller is free, let go all lines, and back both engines slow. The discharge current from the inboard propeller will breast the vessel out, particularly if the pier is a solid one. The officer conning should glance aft to note any tendency of his ship to start swinging either way. He should use the engines for steering until sufficient sternway is reached when the rudder can be used. The distance between the pier and the bow should be noted and the rate of turn regulated to prevent touching. The inboard screw discharge current when it reaches the bow tends to keep the bow off.

11.9. Clearing a Pier—Being Set On. This is a difficult situation. To go out ahead without the use of tugs or an outlying anchor is risky. The ship must be taken out astern. A single-screw ship with her starboard side to the pier should go ahead on an after bow spring, and when the stern is well out, cast off and go astern full. If she is port side to, it is advisable to wait for a change in conditions.

A twin-screw vessel should first go ahead on the outboard engine, holding the after bow spring. This will throw the stern out against the set. Then the ship should cast off the spring and back immediately on both engines. The wash of the inboard engine will tend to keep her off, and the speed of the inboard engine can be varied to keep the bow clear.

11.10. Clearing a Pier—Being Set Off. If the ship tends to drift out when all lines are slacked, the clearing of the pier is very simple. Continued slacking away on the lines will ease the ship off the pier. All lines are then cast off and the ship can proceed.

11.11. Clearing the Wharf—Current from Ahead. Ease off on all lines except the forward quarter spring. The stern will come in and the bow will go out where it will catch more of the current. While going ahead very slowly to keep the stern and the propellers away from the wharf, the bow will continue to swing. When far enough out, the ship may cast off all lines and proceed.

Should the stern not be clear to go ahead on the engines, a bow breast should be kept under a light strain, and, when the ship is headed correctly, the breast should be checked while slacking the quarter spring. The stern will swing out and clear, and all lines can then be cast off and the engines put ahead. To use the bow breast in this manner, the bow must be farther out than the stern; otherwise the ship will come back alongside.

Should the wind prevent the bow from falling off, it will be necessary to go out astern as previously described for a ship being set on to the wharf.

11.12. Clearing a Wharf—Current from Astern. Holding an after bow spring and easing out on a quarter breast will let the stern swing outward. When pointed correctly all lines are cast off, and the ship goes out astern.

11.13. Working into a Slip. The best time for entering a slip is when the water is slack. The procedure is the same as in making a pier. In entering, an anchor under the forefoot for snubbing is of great assistance, since it gives greater steering control.

Should there be a good current running, the docking should be done with tugs; but if they are not available, the ship can make a landing on the end of the pier, heading into the current.

The ship is then warped into the slip. The springs for this purpose should be as short as possible to reduce the radius of the swing and should be as nearly perpendicular to the keel lines as practicable to produce the best springing effect. To protect the ship's side and the pier, a camel or suitable fenders are placed at the knuckle of the pier where the pressure is localized. The camel will distribute this pressure over many of the ship's frames, thereby preventing the crushing of them. The ship goes slowly ahead until the knuckle of the pier is amidships. Then the after bow spring is held and the rudder is thrown hard over toward the pier, aiding in the turn. After entering the slip, if being set off, the lines should be got out and walked up the pier in short steps and used as necessary to hold the ship in. If being set on, lines are run to the opposite side of the slip.

When backing into the slip the same principles of using the springs apply. The lines, especially the bow breasts, must be very strong, and good seamanship dictates that more than one be run. The additional ones are preventers and are used as alternates in the shifting of the lines for better leads. The vessel is dropped back until the knuckle of the pier is amidships, and with the quarter spring held the bow breast is eased. When the ship is pointed correctly, the engine is backed slowly. Deck winches are a help and almost a necessity in the case of a single-screw ship backing to starboard.

The most difficult problem is making a landing on the upstream side of a slip, with the wind and current from the same direction and at right angles to the pier, its underwater structure solid, and a large warehouse built upon it. Under these conditions entering the slip at slow speed will invariably result in the ship's stern being swept across to the other side of the slip before it can be controlled, with almost certain damage to the ship, the pier, or other shipping. Such a landing is sometimes possible with a high-powered vessel by entering the slip at a high enough speed so that the stern will be out of the effect of wind and current before it can be swept across the slip and then backing full to kill the headway. Such a maneuver requires expert judgment as to the instant of backing and perfect coordination with the engine room in order to avoid ramming the head of the slip. This is a fascinating maneuver for young shiphandlers, especially in destroyers, and many have come to grief trying it. It is always risky and should be attempted only in case of necessity.

11.14. Mooring to a Buoy. The ship should approach slowly with the current from ahead, and the buoy should be kept on a constant bearing. The buoy should be picked up on the lee bow if there is any wind so that the bow will drift toward it rather than away. It may be wiser to pick it up on the starboard bow in the case of a single-screw ship because of the tendency of her bow to drift to starboard when the engine is backed.

Most large ships moor by shackling the end of the chain, unbent from an anchor, to the ring of the buoy. Small ships may pass the end of the chain through the ring, haul it in, secure it on deck, and ride to the bight, but this is not recommended because the chain can be damaged where it is passed through the ring. In case the mooring is made habitually, a heavy wire mooring pendant with an eye and shackle on the end may be made up. In any case a hook rope is first passed to hold the bow in position while the moor is being made, and a man is placed on the buoy to handle lines and shackle up. Under favorable conditions, the man can be lowered over the bow and the hook rope passed down to him, but it is better to send him to the buoy in a boat. The same boat can carry the hook rope and a messenger fastened to the end of the chain. The hook rope is secured first. Care should be taken not to put too much strain on it or the buoy may be capsized and the man working on it thrown into the water. Next the messenger is run from underneath through the ring and brought back on deck to the capstan. The messenger is used to haul the end of the chain down to or through the ring. The shackle and pin, if used, can be taken in the boat or preferably lowered from the bow with lines, thus making it easier for the man on the buoy to handle them.

Some destroyers send out the bight of a heaving line in a boat. The bight has been cut and a ring and snaphook inserted. The man in the boat reeves one end of the heaving line through the ring on the buoy and snaps the ends together. The mooring party on the forecastle hauls the heaving line through the ring rapidly. A heavier messenger follows as the ship approaches the buoy. The anchor has been secured and enough chain roused up to allow the end to be reeved through the bull nose and to hang down to the water. As the ship gets closer to the buoy, the messenger which is made fast to the end of the chain is hove in. The chain reeves up through the ring on the buoy and is hove up on deck and secured. Such a mooring can be made in 1 to 2 minutes. It is important in all moorings to bring the ship up to the buoy and not attempt to pull the buoy to the ship.

Large ships, such as cruisers, with heavy chains usually moor close to the buoy with a wire. The anchor chain is then shackled to the wire and slides down the wire where it is made fast to the ring on the buoy.

A buoy can be picked up with a fair tide under some circumstances. It will require skillful shiphandling and speedy work by the man on the buoy unless the ship has twin rudders. When the vessel swings, it may put a severe strain on the moorings. In some cases, time will be saved if it is possible to drop an anchor, swing, and pick up the buoy with the current from ahead.

11.15. Slipping a Mooring. A strong manila line or flexible wire is run through the buoy ring and back on deck to use as a slip rope. A strain is taken on it, and the chain is unshackled. Should the ship be riding to a bight of the chain, an easing-out line is used to ease the chain through the ring while the chain is hauled in. The ship is now riding to the slip rope, and unmooring is completed by letting the end of the slip rope go and reeving it through the buoy ring.

11.16. Mooring to Buoys, Bow and Stern. It is sometimes necessary to moor bow and stern to two mooring buoys in order to avoid any swing in a restricted space. When possible, the approach should be made against the current and on the side from which any wind or current present will tend to set the ship down on the line of buoys. The bow is moored to the upstream buoy in the usual manner. Meanwhile another boat should be used to carry a line to the second buoy to hold the ship's stern from swinging off. This line should be no heavier than necessary so that it can be handled easily, and care must be taken always to keep it away from the screws. The end of a wire of sufficient strength may then be carried out and shackled to the buoy ring. The first line may be taken in or kept as a preventer and for use in unmooring. The final moor should be taut in order to prevent the vessel from ranging ahead or astern. This may be ensured by heaving in on the stern line with an after winch or by veering on the buoy chain, taking in the required amount of stern line, and then heaving taut on the chain.

In unmooring it is the usual practice to take in the stern line first, although this moor may allow the ship to swing in either direction before taking in on the final lines.

11.17. Winding Ship. It is seldom necessary to wind ship at a pier. When it must be done, the most satisfactory method is by the use of tugs, especially in the case of large ships. Tugs give better control and assist the engines to turn the ship. With a small ship or when no tugs are available, it is quite possible to wind ship by the use of the engines or by taking advantage of the current.

It is safer to pivot on the bow and thus avoid possible damage to the rudder and screws against the pier. An after bow spring should always be held on the stern lines slacked or let go. With a current from astern the stern will usually start out by itself. With a single-screw ship, starboard side to, backing down slowly will usually start the stern out. If port side to, going ahead dead slow on the after bow spring will have the same effect. With a twin-screw ship, backing the inboard screw or going ahead on the outboard one or a combination of the two should start the stern out. The swing is made on the after bow spring, but a forward bow spring on the other bow should be led from well aft to assist in controlling the swing and to take the strain after the swing is past 90 degrees, as shown in (a) and (b) of Fig. 11.3. The bow is kept clear of the pier by backing a little as needed. Should a strong current be running, a long after breast may be used to slow down the first part of the swing. During the latter part of the swing the engines may be used to slow the swing and to prevent the ship from slamming into the pier. After the winding is completed, the ship can be spotted in position with the engines or by hauling her ahead of the vessel's original position.

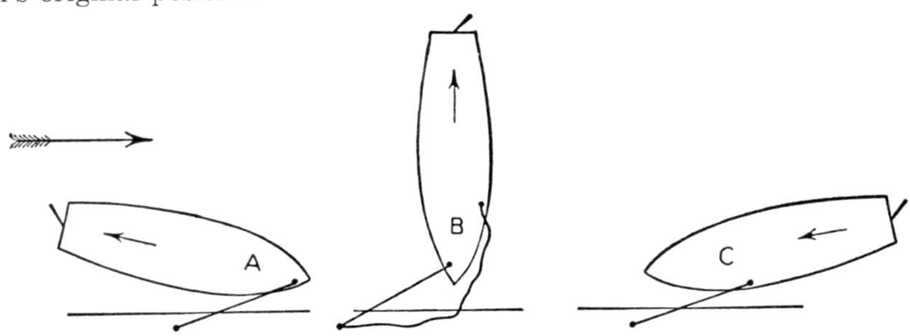

FIG. 11.3 WINDING SHIP

11.18. Going Alongside Another Vessel. In going alongside another ship at anchor or at a mooring buoy, the same general rules apply as for going alongside a pier. The anchored vessel may be riding to the current. Her stern should be watched carefully during the approach, as it may yaw, and allowance must be made for the effect of the wind. On the windward side, the ship coming alongside may expect to drift down fast on the anchored vessel. On the leeward side, a close approach must be made or the landing may be missed.

There is a double possibility of damage when going alongside another ship. Projections, such as bridge wings and sponsons, must be watched carefully to make sure they do not foul. There should be no headway or sternboard on the vessel coming alongside when she finally touches. Fenders or puddings should be used by both ships and, if possible, should extend far enough horizontally to cover several frames.

In clearing from alongside another vessel, the same precautions should be observed. If the wind is favorable it may be possible to drift away from the anchored ship and go out ahead or astern as desired. If this cannot be done, it is usually handier to spring the stern out and back away, but when going out astern care must be taken not to rake the anchored vessel's side with the bow.

11.19. Destroyers. More naval officers learn practical shiphandling in destroyers than in any other type of vessel. This is especially true of coming alongside, as destroyers are often required to dock, go alongside tenders, or secure in nests at a mooring buoy. For this reason, it is well to note carefully some of the peculiarities of the type.

Several kinds of destroyers will be found in the Navy ranging from the early 1,600-tonners to the latest 4,000-ton destroyer leader with the power and armament of a small cruiser. All types have certain common characteristics which affect their handling. The 2,200-tonners will be discussed here.

They are high-powered in comparison to their displacement and respond quickly to the engines. They have plenty of backing power. With one engine going ahead at one-third or two-thirds speed and the other backing, at the same speed, they will normally make headway if the backing revolutions have not been standardized. Destroyers have a narrow beam compared to their length; consequently the propeller shafts are close together. The fact that most of the after deadwood has been cut away gives these ships a small turning circle.

All destroyers have much more surface exposed to the wind forward than aft. This constructoin acts as a permanent head sail and gives the bow a tendency to fall off before the wind and the stern a tendency to back into it. The light construction of destroyers makes it imperative to avoid heavy pressures and sharp blows on their hulls when coming alongside. The twin rudders directly behind the propellers of the 2,200-ton class greatly improve their maneuvering characteristics. Even the 1,600-ton class maneuver better than the old "four-pipers." There is an understandable fascnation in handling destroyers which often leads to the taking of unnecessary chances. A good destroyer officer is one who handles the power at his disposal daringly when he needs to do so, but does not invite disaster by rashness. It happens occasionally that an engine will not follow the signal through fault of personnel, and engine order telegraphs have been known to break down while on full speed ahead or astern. Should such an accident happen when the commanding officer is charging into a landing at high speed and trusting to his backing power to stop in time, merely to show himself to be a smart shiphandler, the result may be a smashed bow or some worse accident caused not by an effort to perform some important service but merely by bravado. Situations are sure to occur when the extra power available must be depended upon. It is the height of folly to invite more of them than necessary. Handle a destroyer with caution normally, and with boldness when necessary.

It is customary to set a destroyer's standard speed at 15 knots for maneuvering. Experience has shown that this gives ample power both ahead and astern for working at close quarters. It has the added advantage of making shiphandlers familiar with the results to be expected when any particular speed is ordered.

The destroyer officer gets constant practice in going alongside under various conditions of wind and tide. The general principles for handling steamers apply to destroyers as well as to any other vessel, but the peculiarities of the type must always be kept in mind.

The light fittings, large power, and small lines make it especially important to exercise care in warping or springing around piers. It is often necessary to work the engines in opposite directions with great power, but great strain must not be brought on the lines by imparting motion to the vessel.

The speed to be used in approaching a pier should be in keeping with the space available ahead or astern of the landing and with weather and harbor conditions. It often happens that the landing must be made when other vessels are moored ahead and astern of the assigned berth. Caution is obviously essential.

The pier should be approached at a slight angle (10 to 20 degrees) slowly but steadily, and the engines stopped with sufficient headway to overrun the landing slightly. When it is certain that the ship's momentum is sufficient to carry the bow into heaving line distance, the engines are stopped. Heaving lines are passed as soon as possible, and the mooring lines hauled ashore. The engines are worked to bring the ship in to a securing position, care being taken that springs and breast are properly led and that neither bow nor stern is brought in too sharply, because any localized pressure may result in damage to the frail shell plating.

When the current is running strong and parallel or nearly parallel to the pier, the wind may be considered of secondary importance, for the destroyer is long and narrow and when placed at an angle to the axis of the current will rapidly be carried toward or away from the pier, depending upon that angle.

If the current is forcing the vessel on to the pier, as will often happen if piers are athwart the stream, it is important that engines be maneuvered to keep squared up with the pier as the vessel drifts in with the current. Under such circumstances sufficient fenders must be ready in place to avoid serious local damage that may be caused to the shell plating.

If the current is running away from the pier, speed must be used, and lines must be got out smartly. Once the lines are out, the stern must be brought in by springing upon the bowline and by skillful use of engines and rudders.

To make a landing with a fair tide is very difficult, as the slightest cant toward the pier will bring the current under the inboard quarter. This force will require two-thirds or full speed of engines and lines to bring the stern in, especially if strain be brought on a forward line which is an error commonly made (every destroyer officer learns that it is useless to attempt to spring in on a bow spring with fair tide). The ship should be brought as nearly parallel to the pier as possible, preferably with a slight cant outward, and the after lines got out smartly and held. If a line can be secured aft, it will act as a spring, and the current will bring the ship in readily. The bow will take care of itself. It will ordinarily save time with this condition of current to turn and stem the current or to drop an anchor and swing before approaching the pier.

Going Alongside a Vessel at Anchor. This is very similar to going alongside a pier except that conditions of wind and current to be met are usually more favorable. The destroyer must keep clear of an overhanging stern or of any projections from the side of the vessel approached and if possible select a part of the ship's side where there are no projections. A whaleboat or a triced-up accommodation ladder may inflict serious damage to the destroyer's upper works. The greatest danger is the yawing of the vessel at anchor. Destroyers yaw very freely when anchored and riding to wind or current.

There is rarely any difficulty in going alongside another vessel at anchor. The fact that she is at anchor makes it reasonable to suppose that there will be ample maneuvering room astern, except when she rides with her stern close to the beach. Caution and not high speed is all that is required. The approach should be made fairly well clear with a slight cant inward, and the stern should be brought in and the bow carried out by backing the outboard screw and by use of the rudders as forward lines are passed. Here a spring leading forward from the after forecastle chock of the destroyer can be held and the ship breasted in easily by the current. Care must be exercised not to get the current on the inboard bow. Unlike the situation in going alongside a pier, the forward lines are the more important, for the stern will be taken care of by the wind or current to which the anchored vessel is riding. The engines and rudder can be used to assist in paralleling the two vessels as they draw together.

If a destroyer must approach a nest or tender at a wide angle, the bow should be placed about the width of the destroyer from the bow of the outboard destroyer or tender. The destroyer should then be stopped. A bowline should be sent across and held as a pivot, but not hove in until the ships are parallel. The approaching destroyer now goes ahead one third or two thirds on the inboard engine after the rudder has been put over, hard, away from the nest. The powerful turning force of the discharge flow from the ahead propeller against its rudder is to be utilized. The outboard engine is backed at the same speed and changed as necessary to prevent gaining head or sternway. After the ship is parallel to the nest, the bowline can be hove in assisted by a stern line, and a gentle landing made.

11.20. Tugs and Pilots. Normally a ship's master or commandng officer will have a pilot aboard when tugs are used to assist in shiphandling. There may be times, however, when the commanding officer must direct the tugs himself. For this reason Fig. 11.4 is shown. These tug signals are those officially adopted as standard by the Navy, although some local variations will undoubtedly persist. It is always advisable to confer, when practicable with the tug skipper, before the operation, agreeing on the signals to be used, methods of making fast the tug, etc.

11.21. The Mediterranean Moor. The Mediterranean Moor (Med Moor) is essentially a method of mooring a ship perpendicular to a mole or pier using lines to secure the stern and two anchors to hold the bow in place. As the name implies, the Med Moor is used almost exclusively in many Mediterranean ports where pier space is at a premium.

TUG BOAT SIGNALS

HAND WHISTLE (Police Type)

FROM STOP TO HALF SPEED AHEAD	1 BLAST
FROM HALF SPEED AHEAD TO STOP	1 BLAST
FROM HALF SPEED AHEAD TO FULL SPEED AHEAD	4 SHORT BLASTS
FROM FULL SPEED AHEAD TO HALF SPEED AHEAD	1 BLAST
FROM STOP TO HALF SPEED ASTERN	2 BLASTS
FROM HALF SPEED ASTERN TO FULL SPEED ASTERN	4 SHORT BLASTS
FROM HALF OR FULL SPEED ASTERN TO STOP	1 BLAST
CAST OFF, STAND CLEAR	1 PROLONGED 2 SHORT

Notes:

1. A blast is 2 to 3 seconds in duration.
   A prolonged blast is 4 to 5 seconds in duration.
   A short blast is about 1 second in duration.
2. In using whistle signals to direct more than one tug, care must be exercised to ensure that the signal is directed to and received by the desired tug. Whistles of a different distinct tone have been used successfully to handle more than one tug.
3. These signals may be transmitted to the tug by flashing light. However, flashing light signals should be restricted to use only when hand whistle or hand signals cannot be used.
4. Normally these whistle signals will be augmented by the hand signals given below.

HAND SIGNALS

HALF SPEED AHEAD OR ASTERN—Arm pointed in direction desired

FULL SPEED (Either) —First describing arc (as in "bouncing" an engine telegraph)

DEAD SLOW (Either)— Undulating movement of open hand (palm down)

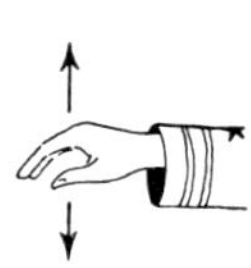

STOP (Either)—Open palm held aloft facing tug

TUG TO USE RIGHT RUDDER—Hand describing circle as if turning wheel to right (clockwise) facing in the same direction as tug

TUG TO USE LEFT RUDDER—Hand describing circle as if turning wheel to left (counterclockwise) facing in same direction as tug

TUG TO RUDDER AMIDSHIP—Arm at side of body with hand extended, swung back and forth

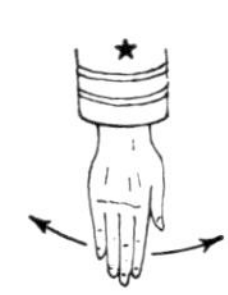

CAST OFF, STAND CLEAR—Closed fist with thumb extended, swung up and down

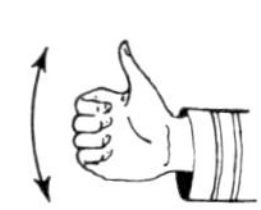

Note: Tug shall acknowledge all of the above signals with one short toot (one second or less) from its whistle, with the exception of the backing signal which shall be acknowledged with two short toots and the cast-off signal which shall be acknowledged by one prolonged and two short toots.

FIG. 11.4 TUG SIGNALS

This type of moor has several advantages. First, it saves space in a harbor or port where pier space is small. Second, it provides a strong moor for high winds and rough weather. Third, it eliminates many of the problems associated with mooring alongside in nests. Fourth, each ship has its own brow to the pier. The Med Moor has two major disadvantages. There is a strong possibility of anchors becoming fouled with other ships when trying to get under way. For this reason it is often advisable to get under way in inverse order of entering port. The second disadvantage stems from the first in that it is difficult to get under way quickly in a crowded harbor.

This discussion will be limited to the destroyer. Cruisers and merchantmen also utilize the Med Moor. Carriers, however, usually anchor outside the harbor due to limited maneuvering space.

Since the Med Moor is made perpendicular to the pier, it is important that the conning officer determine how far from the pier the anchors will be dropped. The first consideration is the length of the ship. The destroyer is about 390 feet long—130 yards. Next, it is necessary to determine the scope of chain to be used. The shortest chain on a destroyer is 105 fathoms. In making the Med Moor, it is necessary to choose a scope of chain that will allow enough room from the pier for the ship to maneuver freely but allow a margin for error while using the anchor. In general, 75 fathoms is chosen for a destroyer and this allows a reserve of 30 fathoms. Thus we find that the drop distance from the pier is equal to the length of the ship plus the scope of chain (75 fathoms equals 150 yards)—280 yards.

With this distance determined, the conning officer takes a course approximately parallel to the pier. Speed is reduced so that when the ship is approximately 50 yards short of the position abreast of the berth, it has only bare steerageway. At this point the outboard anchor or the one opposite the pier is dropped from the wildcat. If the starboard anchor is dropped, RIGHT full rudder and a twist on the engines is applied so as to keep the anchor chain clear of the ship. The starboard anchor chain is veered until the ship reaches a point 50 yards on the opposite side of the berth. At this time, the other anchor is dropped. Since a destroyer has only one wildcat, the first anchor is dropped from it so that the scope of chain may be shortened at any time and the other is dropped from the compressor. The compressor is a movable constriction in the chain pipe which serves to check the anchor chain. Then the second chain is veered and the first taken in while the ship twists and backs into the berth. To obtain its final position the ship backs gently against the catenary of the anchor chain which is on the wildcat. When the moor is completed, there should be an equal amount of chain to each anchor and in an optimum situation, an angle of about 60 degrees between the anchor chains. When the ship is close to the pier, the conn is generally shifted to the fantail where the distance to the pier can be obtained more accurately. The stern is secured to the pier with a stern line and quarter lines which are crossed under the stern. A stronger moor may be obtained by using the towing line as the stern line and reinforcing the quarter lines with wire. Once the stern line is secure, the moor is tautened by heaving in and equalizing the anchor chains. There should be moderate strain on the anchor chains and they should be standing well out of the water so that a wind from the bow will not damage the ship.

When getting under way, the ship will use its engines to keep its stern perpendicular to the pier and will heave up on the second anchor dropped. This

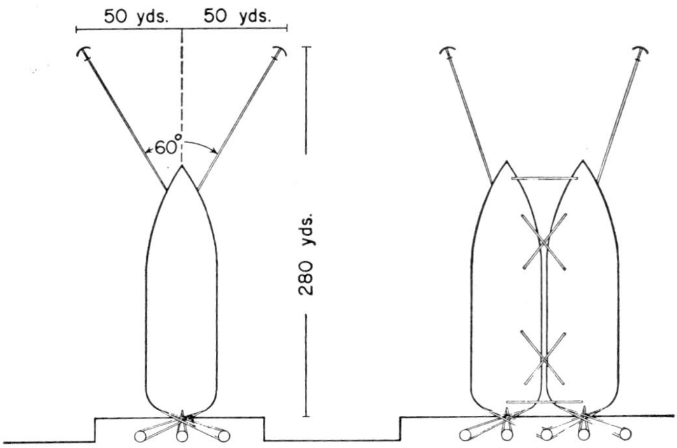

FIG. 11.5 MEDITERRANEAN MOOR

anchor is "heaved right on in." Then the remaining anchor is heaved to short stay and then "heaved on in." The ship is then free to proceed out of port.

Destroyers often moor in a Med Moor test. This is accomplished by having the first ship drop her inshore anchor only and then back into a stern first position against the pier. Remaining DD's moor alongside with standard mooring lines to the other ships. The last ship in drops her offshore anchor only and twists into the stern first position. The final result is a nest which is moored in a Med Moor.

In some harbors, such as Barcelona, Spain, a gale which sends seas into the harbor entrance will cause a rapid rise and fall of the sea level within the port. A firmly moored ship under these circumstances may part her lines to the wharf, and the resultant fore-and-aft surging may carry away the brow. It is recommended that, if circumstances warrant, the brow be removed and all lines to the wharf be eased or taken in. A line or two tended on a winch will usually keep the stern under control as the ship surges. Nylon lines with their greater elasticity should be of help.

12

Towing and Salvage

The rescue of ships in distress and the refloating of those who have grounded is a highly specialized calling to which men devote their lives, but every seaman should have a basic knowledge and understanding of towing and salvage. This chapter provides all that the average seafarer need know in order to tow or be towed in an emergency or to meet the emergency of a sudden grounding.

Towing

12.1. The Towline. Generally speaking, the longer and heavier the towline, the easier the towing will be. A decided dip or catenary gives the same advantage here as in the case of a vessel at anchor riding with a good scope of chain. The weight of the catenary acts as a spring, preventing variations in the tension from being thrown upon the towline in sudden jerks.

Wire rope has proved very satisfactory for heavy sea towing. The advantages of using wire rope are that it is convenient for casting off, takes up comparatively small space when stowed, and does not deteriorate if properly dried and oiled before stowing.

Manila is satisfactory for light or moderate towing. It is heavy enough to give a good dip when used in sufficient length, but it is not too heavy for convenient handling. Its buoyancy is a distinct advantage, particularly where lines are to be hauled over considerable distances. Nylon is becoming increasingly popular for relatively light towing because of its resiliency, ease of handling, and long life.

For towing a vessel as small as a 2,200-ton destroyer in rough weather—and it must not be overlooked that rough weather may be encountered in almost any towing operation—the full length of an 8-inch manila hawser or a 1-inch diameter wire rope will be none too much.

Where the tow is a vessel whose displacement is comparable with that of a cruiser or aircraft carrier, the towline should be made up of 2½- or 2¼-inch diameter wire rope connected to a good length of her own anchor cable. The length that is needed will vary with circumstances, but it is far better to have too much than too little.

A point of some importance in towing in a seaway is to keep the ships in step. In other words, use such a length that they shall meet the waves and ride over them together. If the length of the line is such that one vessel is in the trough of the sea as the other is on the crest, the line will slacken for a moment and then tauten with a sudden jerk; whereas if they meet the waves at the same time, the tension of the line will remain comparatively steady.

12.2. Securing on the Towing Ship. In securing the towline, consideration must be given to the possible necessity for letting go in an emergency, such as the sudden threat of collision.

For convenience in letting go, it is desirable to have a break in the line near the stern. It would be advisable to have a shackle connecting two parts of the line at or near this point, together with some arrangement like a pelican hook for slipping quickly. The possible whip of towlines and bridles when released at any point may be overcome by the use of preventers.

If the towing ship is comparatively large and has a chock at the stern, the line should be brought in through it. It is a good plan to use a short length of chain for the lead through the stern chock, shackling outside to an eye in the end of the towing hawser and inside to a towing bridle. The chain through the stern chock not only takes the chafe, but by its flexibility does away with the dangerous nip which might be thrown into wire if the tow chanced to take a rank sheer onto the quarter.

Where the chain is not used for taking the chafe in the stern chock, the towline must be fully protected by chafing gear which should be a long and bulky pudding. The stiffness of such a puddening reduces the sharpness of the nip which, without the pudding, would be thrown upon the towline from time to time by the sheering of the ships. A towline leads down, not up. Worm, parcel, and serve manila. Use canvas, hides, burlap, and old rope on wire towlines.

Where the strain is not too heavy to be taken by one pair of bitts, the towing line may be secured as shown by Fig. 12.1. Figure 12.1(A) shows how a towrope should not be secured. Here the greater strain comes on the forward bitt, which might result in the bitts lifting from forward and being torn out.

In Fig. 12.1(B) the greater strain is taken by the after bitt and, though the forward bitt has some strain, they should hold under ordinary conditions.

When one pair of bitts is not strong enough, the line can be taken to as many as three sets of bitts if available. To divide the strain it is advisable to take one turn around the first bitts, two around the second, and three around the third, thus leaving the line free to render slightly and so equalize the strain.

Where pelican hooks are used for letting go, as shown by Fig. 12.2, the strain is taken momentarily on the hook, relieving the shackle so that it can be disconnected and towing hawser slipped when ordered. This arrangement entails practically no delay. An arrangement with a pelican hook taking the steady strain of towing offers the quickest emergency release.

Where pelican hooks are not used, a strap may be used on the wire or chain outside the shackle, and a heavy purchase hooked to the strap and taken to a winch. For letting go, the strain is taken by the winch long enough to disconnect at the shackle, after which the strap is cut as ordered (Fig. 12.2(C)). A preventer must be used to prevent a dangerous whip upon letting go.

(A)

(B)

FIG. 12.1 LEAD TO TOW. (A) The wrong way to secure a towrope. (B) The right way to secure a towrope. (Direction of tow →.) *Official U.S. Navy Photographs*

There are some conditions under which it is convenient to use a span on the towing ship. The two parts are brought in through the quarter chocks. Generally speaking, this makes it rather easier for the towing ship to steer, and the advantage gained in this respect may become important in cases where a small

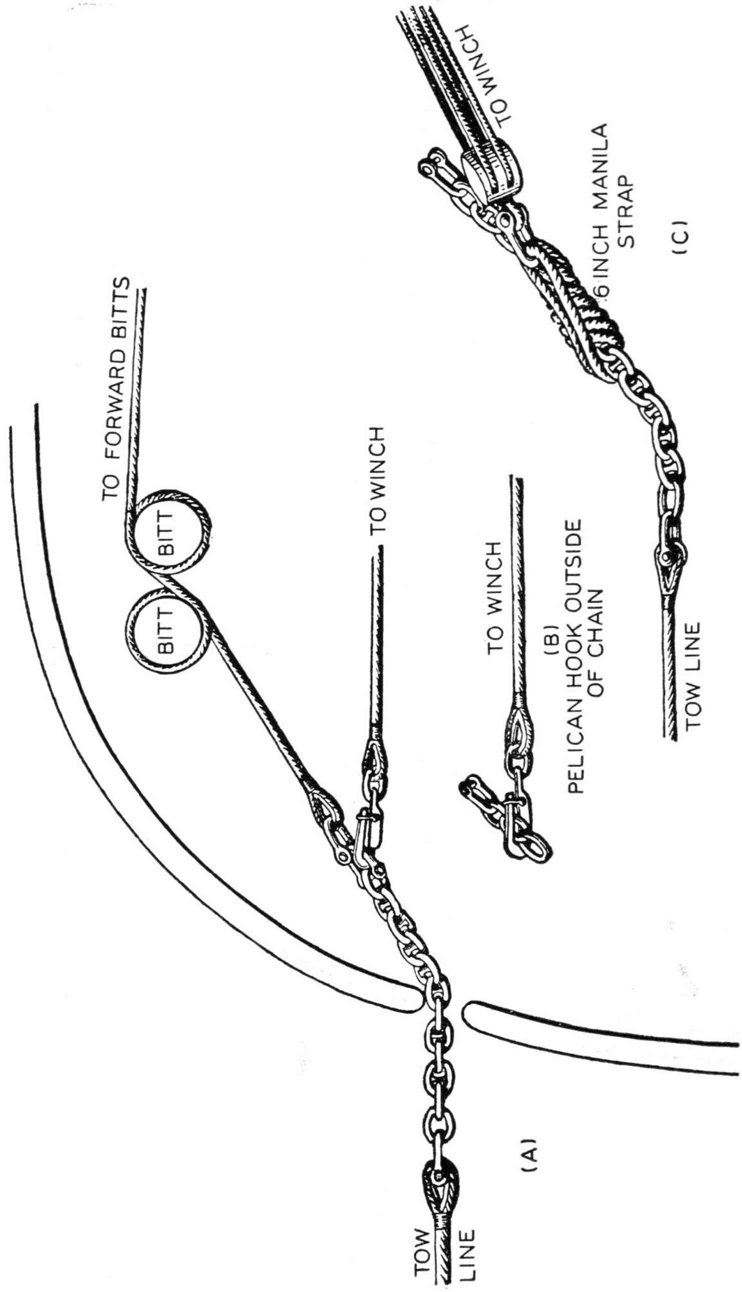

FIG. 12.2 TOWING BY BITTS

ship is dealing with a heavy tow. Where the line leads from a chock directly over the rudder, it binds the stern so that it can only swing in obedience to the rudder by dragging the tow with it. A large ship can take care of this situation by the power of her steering gear, assisted if necessary by the propellers; but a small ship with a heavy tow and with the line leading through the stern chock will steer very sluggishly if she steers at all. Tugs which are specially fitted for towing have their bitts well forward of the rudder to allow the stern to swing; the fittings abaft the bitts allow the line to sweep freely across from one quarter to the other.

Where a span is used, it may be of chain, wire, or manila. In this as in other cases, arrangements must be made for letting go quickly in emergencies.

A convenient plan is to bring the towline in through a quarter chock and bend a hawser from the other quarter to it at such a point outside that the two parts shall form a span of convenient length. The lines may be secured around bitts as previously described. This plan has the advantage that by letting go the second line we get rid of the span at once and have to deal only with the towline itself.

12.3. Towing Engines. Vessels designed especially for towing, such as seagoing tugs, are fitted with towing engines which carry the towline on a drum and pay it out and haul it in automatically to keep the towing tension constant. For a steam towing engine this tension is controlled by a differential valve which is set to remain centered with any desired strain on the towline. If the tension rises momentarily, the valve moves in one direction. The drum revolves and pays out line until it has the required strain, and then the valve centers itself and the drum stops revolving. If the tension decreases, valve and drum move in the opposite direction, reeling in the line until the proper tension is restored. With electric towing engines an automatic controller accomplishes the same result. In this way the line is paid out or reeled in just enough to meet the condition prevailing at any given moment, and the average length of towline remains virtually constant. The drums of towing engines must be very rugged and are fitted to heavy foundations built into the frame of the ship. They are placed at a distance from the extreme stern, the pivoting point, in order to lessen the interference with the steering of the towing vessel. Guiding chocks and bollards and a long quadrantal chock for the towline extending along both quarters of the ship at deck level are installed for the same reason.

The standard towline for vessels fitted with a towing engine is 300 to 350 fathoms of wire rope $1\frac{1}{2}$ inches in diameter. It is stowed on the drum of the towing engine when not in use. When towing is finished, the towline is cast off by the towed vessel and automatically reeled in by the towing engine. The eye of the towline is generally fitted with a shackle and open link for securing aboard the tow.

12.4. Securing the Tow on Board. On board the tow the hawser is usually secured to the anchor cable, although there may be many conditions under which some other arrangement will be provided. If the anchor cable is not used, it is desirable to use at least a short length of chain to take the chafe in the chock in the same manner as already described for securing on the towing ship.

Where the anchor cable is used, the hawser is secured or shackled to it and the cable veered away to the desired length, after which the windlass brakes are set up and springs or chain stoppers are used to take the real strain of towing, as in Fig. 12.3. It is well to have a shackle between the windlass and the point to which the springs or chain stoppers are secured and to keep tools at hand for unshackling if it becomes necessary to let go in an emergency. Generally speaking, the tow should not let go in this way except in case of extreme

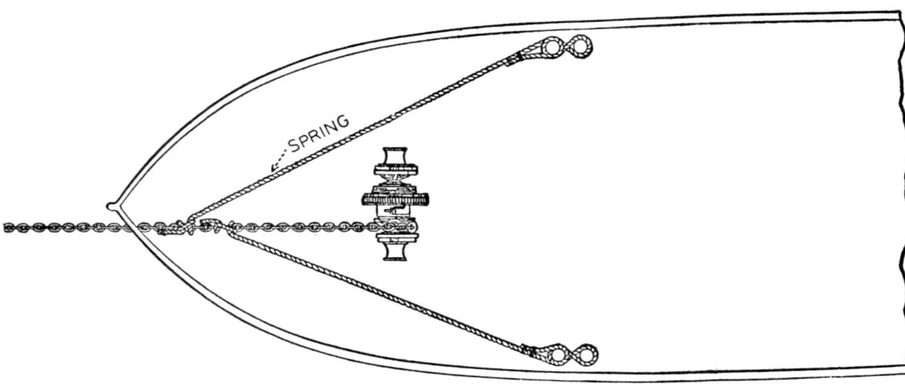

FIG. 12.3 BOW OF SHIP TOWED

emergency, as the line weighted with a considerable length of heavy anchor cable would sink immediately, hanging as a dead weight from the stern of the towing vessel. In this position it would be extremely difficult to handle and would be in danger of fouling the propellers. This applies only to cases where the tow is a vessel of some size and where she is towing by her anchor cables. It is evident that, where a large ship is towing a small one, the natural way of casting off is for the tow to let go, which leaves the line to be handled by the large ship.

12.5. Taking a Disabled Vessel in Tow at Sea. In good weather, this maneuver presents no special difficulty and calls for no extended discussion. The lines are run and secured as already described. The towing vessel starts ahead slowly on the course upon which the disabled vessel happens to be heading and uses every precaution to prevent a jerk on the line. It waits before changing course until both ships have gathered way and are moving steadily with a good tension on the towline.

In bad weather, towing should not be attempted unless exceptional circumstances make it necessary. The running of lines in a heavy sea is attended by considerable difficulty, especially if the vessel to be towed is unable to assist by placing herself in a favorable position. Moreover, in really heavy weather, it would be necessary to proceed so slowly that little or no time would be lost by waiting for the weather to moderate.

Disabled vessels lie in different positions relative to the wind, depending on the size and position of the superstructure, the trim and, perhaps, drag due to damage. If there is more superstructure forward than aft, the vessel will lie with the wind from abaft the beam to astern. Vessels with much superstructure amidships will lie with the wind abeam and those with superstructure aft, such as tankers, lie head into the wind. All such vessels make leeway of 1 to 3 knots and, if lying at angle to the wind, headway or sternway. Vessels down by the head tend to head into the wind and vice versa.

If a tug with a deckhouse forward and a flat stern is the towing vessel, she should approach down wind and just clear of the bow, except when the vessel lies head into the wind. In this case, the approach of the tug is still down wind but just ahead of the vessel, using her engines to keep position and clear. The sketches in Fig. 12.4 show that the tug is able to work close to the disabled vessel and still keep clear by a kick ahead at intervals.

It will be considered in the discussion which follows that one vessel is going to tow another and that the weather is rough enough to call for the use of all reasonable precautions but not rough enough to make towing impracticable. It may be assumed that the disabled vessel will be lying with wind and sea a little abaft the beam, this being the position which a steamer usually takes when lying in a seaway with engines stopped. The other vessel places herself on a parallel heading either to windward or to leeward. In considering which of these positions is to be preferred, we must remember that considerable time will be required to run the lines; and that during this time both vessels will be drifting. A vessel which is light will drift faster than one which is loaded, and the drift of a vessel in ballast-trim often amounts to several knots. If the lighter vessel is to leeward, she will drift away from the other and make it very difficult to run the lines. It may be said that, as a general rule, if there is any important difference in the rate of drift of the two vessels, the lighter one should be to windward when the work of running the lines is begun. The towing vessel then places herself to windward if she is drifting faster than the other vessel and to leeward if she is drifting more slowly and on the same heading as the disabled vessel.

Where the difference in the rate of drifting is considerable, the time available for running the lines after the work is once begun will be short at best. Every precaution should be taken to prevent delay, with a clear understanding being established between the ships and all preparations made before the towing ship takes her position. In communicating between two ships, megaphones are of the greatest value. Under any except the most unfavorable conditions they should make it possible to perfect a thorough understanding of what is to be done and how. To a great extent they also may take the place of signals between the two ships after the towing begins, although a code should by all means be adopted and will be useful under many conditions. It is an excellent plan when feasible to send an officer on board the tow to remain there. First acquaint him with the plan to be carried out and provide him with a list

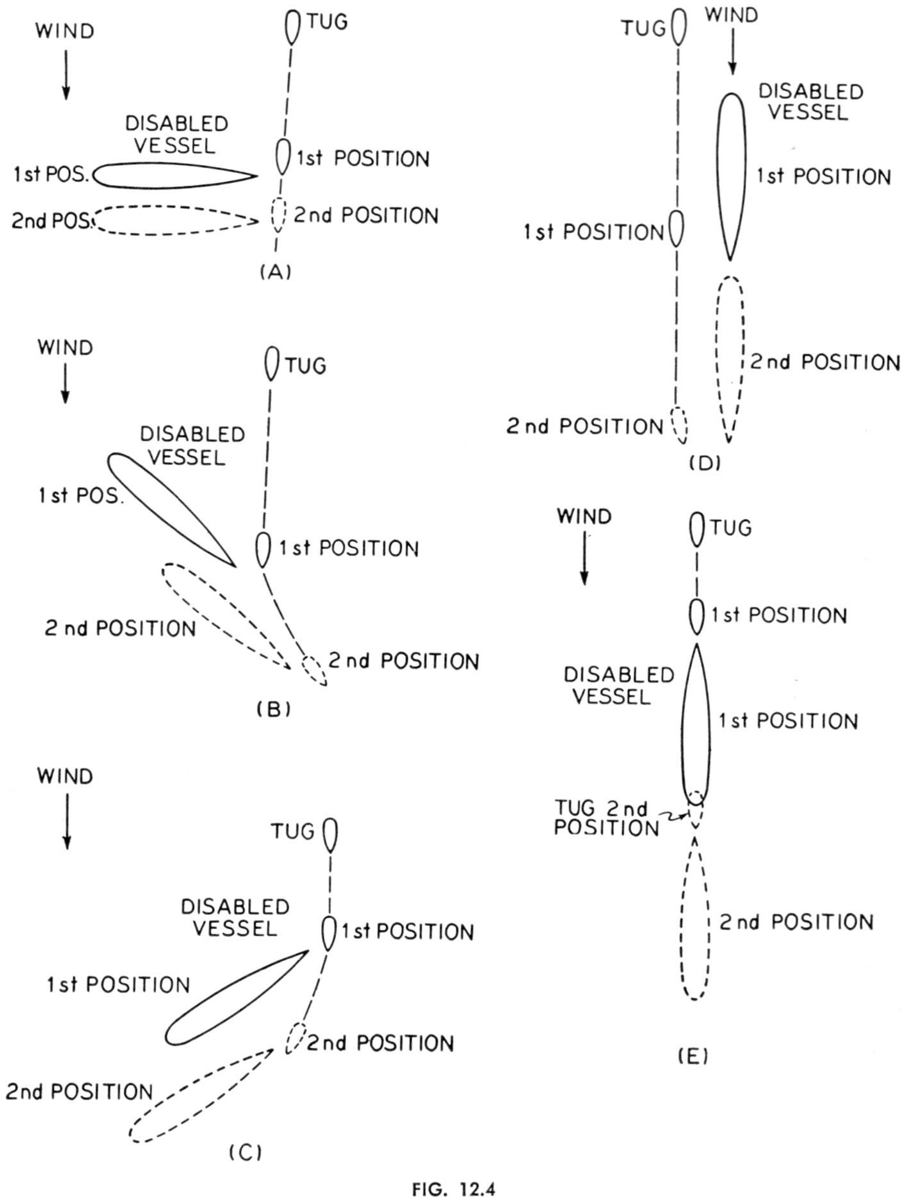

FIG. 12.4

of whistle and sight-signals for handling the lines and the ships. If no boats are to be used, a paper may be floated across to the tow giving full instructions and a list of the signals. This may be sealed up in a bottle and attached to the rope or the float. Whistle signals are preferable to flags because they can be used at night or in a fog.

The following is suggested:

Code of Sound Signals for Towing

A short blast must not exceed 2 seconds in length.
A long blast must not be less than 6 seconds in length.

I am putting my rudder right	1 short blast
I am putting my rudder left	2 short blasts
Go ahead	2 long
Stop	1 long, 2 short
All fast	2 long, 1 short
Haul away	2 short, 1 long
Let go	2 long, 5 short
Pay out more line	1 short, 2 long
Avast hauling	3 short
I am letting go (emergency)	5 short, 5 short, 5 short

12.6. Handling Lines. The first line to be run will be a light one by means of which the heavier ones can be hauled across. A 3-inch manila is a convenient size to begin with and, if new, so much the better because it will float freely. If a boat is to be used, it should be lowered with the crew and the greater part of the line in it and made clear as quickly as possible. The line should be paid out as the boat pulls away for the other ship.

The line may be floated alongside the disabled ship without much difficulty. The best way to do this will depend upon circumstances, but a common way is to float a good length of the line by life belts, casks, or any other means and to steam slowly around the disabled vessel, dragging this astern and causing it to foul the disabled vessel. If proposing to take up a position on her weather bow, it is a good plan to steam along to leeward fairly close aboard, cross the stern, and come around parallel to her heading. This will cause the line to foul her stern, which entails a little trouble in shifting it forward, but it leaves the towing ship in position without further maneuvering. Similarly, if proposing to take a position on the lee bow, pass along to windward, cross the stern, and come around to leeward. The line should be picked up without difficulty.

There may be special circumstances which will make it desirable for the disabled vessel to run lines, but under ordinary circumstances it is more convenient for the towing vessel to run them. Having got the first line across, the heavier lines are run and made fast to the anchor cable of the vessel to be towed where possible. A good length of cable is paid out—20 to 45 fathoms is none too much for heavy work—and the line made secure on both ships as has been described. Chafing gear is used liberally wherever it is needed. In the meantime, full instructions about starting are given to the chief engineer, and, when all is ready, the engines are started ahead as slowly as possible and stopped the moment the line begins to tauten out. Then a few more turns are made and so on until the inertia of the tow is overcome and both ships are moving slowly with a steady tension on the line. The revolutions are then increased little by little, and the course changed gradually as may be necessary.

When the tow is finally straightened out and moving steadily, the speed is worked up to that at which it is thought wise to continue.

In all changes of course the tow puts her rudder at first to the side opposite that of the leader and so steers around into the leader's wake.

If the sea makes it dangerous to tow to windward, it is worth while to consider whether a port cannot be made on a course which will present fewer difficulties, even if the distance is much greater.

After settling down to a steady rate of towing, the lines should be examined, the strain divided as evenly as possible, chafing gear renewed wherever necessary, etc. Hands should be stationed night and day to watch the lines on both ships, with axes and unshackling tools ready for slipping hurriedly if necessary. It is well to have a light messenger line between the ships for hauling messages across and for use in running a new line in case of necessity. This line should be left slack and should have ample length to allow for the fact that if the towline parts the leading ship will forge ahead considerably before she can be stopped.

SALVAGE

12.7. Groundings. Salvage in its broadest sense includes the salvage (recovery) of cargo, the removal of wrecks and the refloating of grounded ships. Only the matter of grounded ships will be discussed herein since the salvage of cargo and the removal of wrecks involve highly specialized techniques not of major interest to seafaring men generally.

The first thing the captain or salvage officer of a grounded ship should do is to notice whether the ship is lively, i.e., is affected by the swells. If so, it may be possible to refloat her at once by sallying ship, backing full speed, and pulling by any large tugs or other vessels which may be available. Should these measures be unsuccessful, the next step is to send out an anchor astern with a wire to hold the ship in her original grounding position. Put and keep a heavy strain on the wire. Swells tend to force ships farther up the beach and to turn them broadside to the sea. Such movements will make salvage more difficult.

The next step is to sound all around the ship to determine what part of the ship is grounded and the loss in draft. The average loss in draft times the tons per inch immersion as taken from the ship's curve will give the total loss in buoyancy in tons. This weight is the problem. The loss in buoyancy must be reduced to the point where the pull of beach gear and tugs will be sufficient to refloat the ship.

Salvage men have calculated that a pull of about 30 percent of the remaining lost buoyancy is required to refloat a ship aground on a sandy bottom with a gentle slope; 50 percent when the bottom is hard or gravelly; 60 to 80 percent for coral; and 80 to 150 percent for rocky bottom. A large amount of weight will probably have to be removed from the ship to reduce the lost buoyancy to a manageable amount.

Personnel, fuel, and water can be removed more easily and quickly than stores, cargo, ammunition, spare parts, and guns. The decision to remove any or all of these weights, as well as when to move them, will depend on a number of factors. In all probability the determining factor will be the weather for offshore groundings, although the state of the tide at grounding and the time of the next tide or perhaps the next spring tide may be equally important. In some parts of the world the range of the tide is so great that ships grounding at part tide are refloated at high tide. Where the rise and fall of the tide is not that great, the time and date of the next spring tide may be the deciding factor as far as time is concerned.

Some other physical conditions must be known and weighed also. Was the ship fully loaded? What compartments, bottoms, and tanks have been holed? What kind of a bottom is under the ship? Currents can scour the sand from under a ship in one place and pile it up in another. While all the physical conditions are being compiled and analyzed, it is wise to send out additional anchors to hold the ship in her grounding position.

12.8. Planning and Methods. The planning must be carefully done so that the weights to be moved, the equipment to be used, and the dredging to be done will be coordinated and completed at the end of the time available. Some fuel and water will be required for engines. All of the weight removed cannot come from the double bottoms and fuel tanks or the stability of the ship may be adversely affected. Many men are required to handle stores, spare parts, cargo, and ammunition. These men must have deck space to handle these items and space alongside for the barges to receive the stores, fuel, etc. The plan must coordinate the times for removing weights, for laying out the beach gear on deck, for dredging alongside, and for receiving barges.

The following measures may be available and used:

1. Remove fuel.
2. Remove water.
3. Remove cargo.
4. Remove stores.
5. Remove spare parts.
6. Remove ammunition.
7. Remove guns or missiles.
8. Transfer some men.
9. Dredge and scour alongside.
10. Tunnel under the ship.
11. Rig pontoons.
12. Rig beach gear.
13. Expel water from holed compartments.
14. Services of salvage vessels.
15. Services of tugs.
16. Twist the ship.

When large ships are stranded, a trench can be dredged along each side of the ship if the bottom is sandy or muddy. The trench should be made deep enough to receive a large part of the sand and mud upon which the ship is resting. If this sand and mud will crumble and move into the two trenches, the ship may be floated in her grounded position. This method has been successful occasionally.

European salvage men have had some success with scouring the bottom from under the ship. A small vessel with her propeller well immersed is secured to the grounded ship in such a position that the discharge current from her propeller scours the sand from under the ship. In some cases the engines of the ship itself have been used to remove sand and mud from under the ship. Care must be taken that a new shoal is not formed astern. The engines cannot be operated in the other direction, i.e., astern, for very long because sand may be washed under the ship and the problem of refloating made more difficult.

Divers with high-pressure hoses are used, after trenches have been dredged, to start the movement of sand and to wash away the remaining sandy supports. Divers are used also to tunnel under the ship so that the chains which hold the pontoons in place can be rigged.

The Navy stores and maintains large pontoons at certain yards on both coasts. These pontoons have a lifting capacity of some hundreds of tons. They have been used to reduce the amount of the lost buoyancy of grounded ships. The pontoons are rigged under the quarter and bows with regard for the propellers, shafting, rudder struts, and skeg which are easily damaged.

It is probable that certain compartments were flooded when the ship grounded. These compartments can be nearly freed of water by forcing air into the compartment. Care must be taken that the pressure of the air does not rupture the compartment. The pressure of the air should be a little higher than the pressure of the water due to the distance below the surface.

12.9. Beach Gear. One of the most useful pieces of equipment now available to a grounded ship is beach gear. It is in reality a development of the older methods of laying out anchors and cables for salvaging ships. The present-day beach gear consists of:

1. A large, special anchor with oversized palms fitted with a crown line and buoy for recovery.
2. 15 fathoms of chain attached to the anchor.
3. Two or three 100-fathom wire cables of galvanized plow steel.
4. A shot of chain.
5. Shackles.
6. Wire stoppers.
7. A four-sheave wire tackle.

The anchor, sometimes backed by a second one, is planted well out and in the direction the ship will have to move when it is refloated. The chain is used between the wire cables to keep the direction of pull as nearly horizontal as possible. The end of the second wire is led through a chock or an opening cut in the side of the grounded ship. The tackle is laid out on deck and aligned with the wire so that the pull when applied will be straight with no nip or bend in the wire. The hauling part of the tackle is now led to a ship's winch or to a salvage winch installed to supplement the ship's gear. In the salvage of the *Missouri,* nine sets of beach gear were used in addition to three sets laid out from each of two salvage vessels which were pulling too. Each set of beach gear can exert a pull of 50 to 60 tons. Such gear kept under heavy strain has been a large factor in salvaging many vessels.

The services of salvage vessels are valuable because they have a large amount of gear, including beach gear, which can be used. The personnel are trained for salvage work; their anchor gear and winches are oversized and therefore valuable for this work.

There is a tendency to use any tug which is near when a ship grounds, under the impression that prompt, energetic action is desirable. Often this is not true because small tugs are of little use for pulling on large vessels aground. Tests of the pulling power of tugs have been made, and it has been found that they exert a pull of about 1 ton per 100 horsepower. Large seagoing tugs have a pull of 10 to 15 tons and Naval Fleet tugs about 30 to 35 tons.

It is very probable that the lost buoyancy of a large ship is so great that she exerts a pressure of hundreds of tons on the bottom. The sand under her has been packed so tightly that it has a consistency of low-grade concrete. We may say that the ship exerts a powerful suction on the bottom. One of the ways of breaking this suction has already been mentioned, namely, the crumbling of the sand under the ship into dredged trenches alongside. After much of the weight of lost buoyancy has been removed, and all preparations completed to refloat the ship, large tugs are used occasionally to twist the ship and thus break the suction. Small tugs are useful now to hold large tugs and salvage vessels in position to exert their best pull. In these operations cross currents can be very annoying.

The salvage plan should include the measures necessary to ensure the safe voyage of the refloated ship to the nearest base or shipyard. Anchors and chains must be returned, the holed compartments should be patched and shored, and the stability and trim of the ship must be satisfactory. If her engines cannot be used, she will have to be towed. These measures may require the reloading of fuel, water, stores, and even ballast in order that a further disaster may not occur.

12.10. The Use of Oil (Fig. 12.5). Oil was probably used to calm troubled waters long before Pliny the Elder (A.D. 23–79) wrote in his Natural History that "all sea water is made smooth by oil." (See Section 10.29.)

The value of oil for this purpose was noticed again when whaling vessels in the eighteen forties were cutting up carcasses, alongside, in tempestuous weather. About 1885 the Hydrographic Office addressed a letter to all Masters of Merchant vessels, asking them for their views and experiences with oil at sea. A large number of letters were received and publishd in 1886. These letters recounted the destruction suffered when the crests and even seas pooped a vessel or crashed on board.

The experiences of these seamen show that oil prevents the crests from breaking and delays the development of the smaller waves which are superimposed on the larger ones. Only the underlying swell remains. The beneficial action of oil is due to the increase of surface tension.

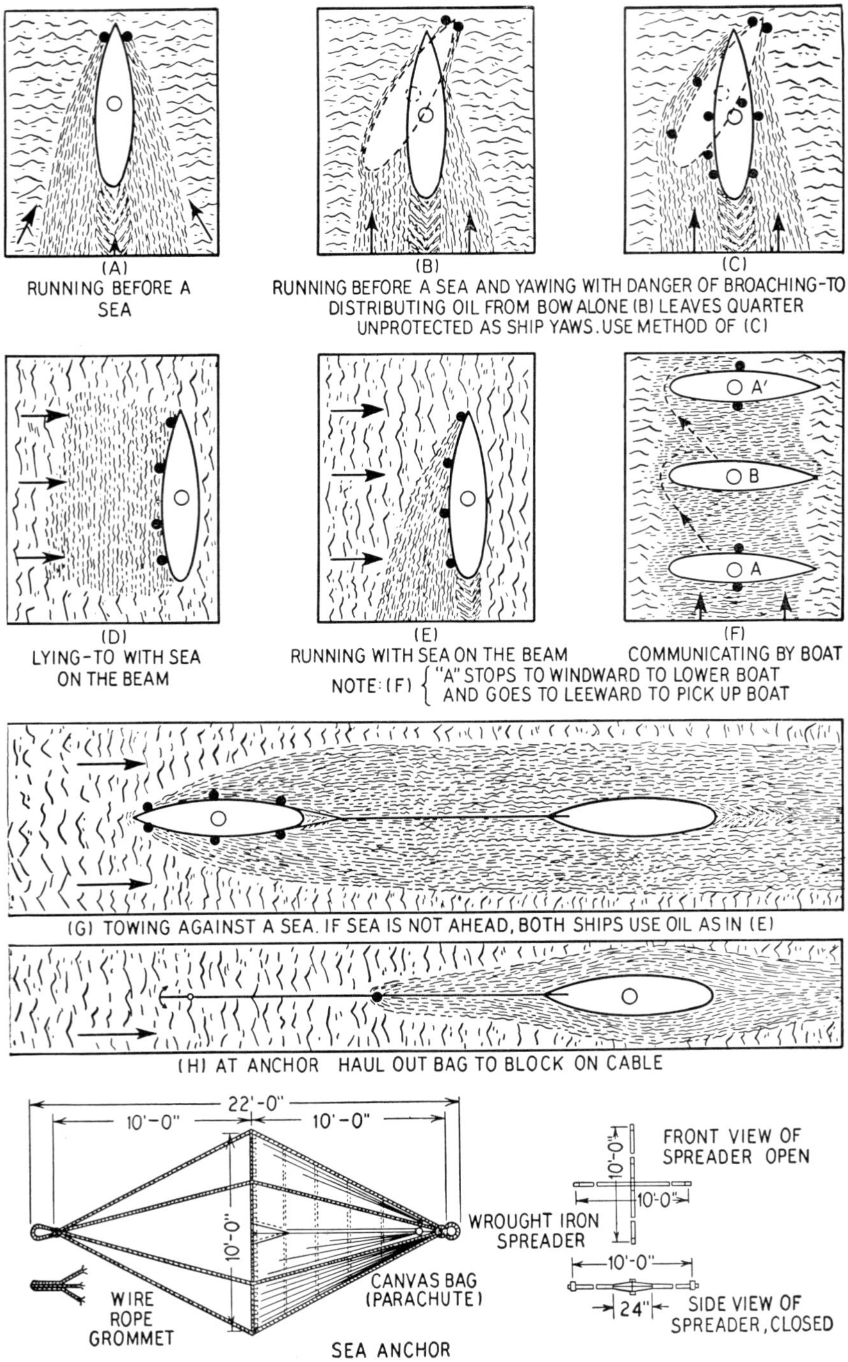

FIG. 12.5 THE USE OF OIL

12.11. Rescuing the Crew of a Wreck at Sea. There are no hard and fast rules which can be laid down for rescuing the crew of a wreck; only what is considered the best practice by experienced and capable seamen may be stated. So many elements control the application of the general rules, such as sea, wind, urgency of immediate assistance, maneuverability of the assisting ship, and the training and experience of the boat crews, that each case must be decided according to circumstances.

After having made contact and established communications, find out how urgent the case is and how much help may be expected from the crew of the wreck. If it is at night and weather conditions indicate an improvement or at least no worse weather, and the master of the disabled vessel feels he can hold on and you feel you can maintain contact, wait for daylight.

Under any circumstances, when the rescue begins, determine the comparative drift of the two vessels and whether or not there is any wreckage about the disabled vessel. If there is wreckage, discern how it will hamper your boat work. If your ship drifts faster than the disabled vessel, go to windward; but if the opposite is the case, go to leeward. Both vessels should distribute oil freely. If for any reason the wreck cannot use oil, the rescuing vessel should steam around her and run oil freely to create a slick into which the wreck will presently drift.

Before the rescue work begins, the boat should be equipped with two sharp hatchets in brackets, one at the bow and one at the stern; one ring life preserver with stout heaving lines made fast; two spare life jackets stopped with sail twine under each thwart; two spare oars; two small oil bags; and a small tin of storm oil.

If the weather is very rough, extreme precautions will be called for in lowering the boat and getting her clear. The ship should be held with the sea on the bow to give a lee for the boat and to reduce rolling as much as possible. The crew, with life belts on, is lowered in the boat. Frapping lines are used around the falls to steady the boat, and fenders are rigged to prevent the boat from being stove in if she swings in heavily.

Assuming that the boat gets off and makes the trip to the wreck in safety, the officer in charge must decide how he will establish communication and take off the passengers and crew. It is out of the question to go alongside to windward; and if he goes alongside to leeward, not only is there a risk of being stove by the wreckage which is likely to be found floating under the quarter, but there is the much more serious danger of being unable to get clear of the side again. A vessel lying in a seaway with engines stopped drifts to leeward at a rate which is always considerable and may amount to several knots. A boat alongside such a ship to leeward is in exactly the same position as if she were alongside a dock against the face of which a strong current is setting. As a rule the boat must never be brought actually alongside the wreck. She may either lie off to windward and keep well clear and hold up head to sea or to leeward and hold on with a line from her bow to the wreck. If obliged to go alongside, the stem may be allowed to touch, with all being ready to back off if the boat shows a disposition to get broadside-on. The people on board the wreck put on the life belts, go down a line one at a time, hand over hand, and are hauled into the boat. In most cases the most favorable point for working will be under the lee quarter or the lee bow. This depends upon the way the wreck is lying with reference to the sea. It is sometimes possible for people to lower themselves or be lowered to a boat from the head-booms or from an overhanging main-boom when they could not be rescued in any other way. So serious is the question of avoiding actual contact with the wreck that many officers consider it best for the rescuing ship to go to windward and drop the boat down with a line, putting only two or three men in the boat.

13

Boat Handling

As ship's boats are normally designed for a variety of purposes, their handling will be discussed under the various conditions in which they are used.

The power boats of a ship, including landing craft, do the greatest part of their work in port or off the beach, running from ship to shore. Under these circumstances it is normally a safe and simple matter to hoist and lower them with a crane or boom. When hoisting or lowering at sea or at anchor during rough seas, certain precautions must be observed in order to prevent the boat from being stove in, swamped, or the crew thrown overboard. If these precautions during operations in rough seas are understood and observed at all times, there should be no difficulty in hoisting and lowering boats under more favorable sea conditions.

With the exception of those boats which are provided with their own davits, large boats and heavier landing craft are hoisted and lowered by means of boat cranes or booms which hook on the slings rigged in the boat. The slings are attached to hoisting eyes which are built into the strongest sections of the boat. Davits which are used for many of the smaller boats can be considered as nothing more than two cranes which perform the same job in a slightly different manner.

Davits are used to swing the boat to the lowering position and then after it has been hoisted to swing it back on board. The actual raising and lowering could be done by manning the boat falls with sufficient men. However, the falls are generally taken to some source of power to accomplish raising the boat and then to a belaying point, such as a cleat or the gypsy head of a winch, for lowering.

The various types of davits used have been described in Chapter 6 and will not be discussed further here, with the exception of the radial davit, which is the most common aboard small combatant ships and is also the most difficult to use.

Prior to swinging the boat out, the boat plug is checked and reported to the man in charge as being in place. If the boat is resting in chocks, it is hoisted clear of the deck and all preparations made to swing it out. When it is clear of the deck, the boat is shifted aft so that the bow will clear the forward davit. This davit is rotated and the bow pushed out so as to clear the side. The rear davit is then rotated and the stern pushed over the side. The boat is then ready for lowering. This type of davit provides a rapid and simple method of swinging out small boats such as the motor whaleboat and the pulling whaleboat.

Care must be taken in hoisting the boat out of the chocks. In hoisting it clear, it is advisable to hoist the stern of the boat first to avoid any danger of striking the propeller or rudder against the deck. After the stern has been hoisted clear, the commands are "avast heaving" and next "pass the stopper." The hauling part is then stopped-off as shown in Fig. 13.1 by means of two half hitches and two or more turns, and by holding the bitter end of the stopper and the fall firmly together by hand. When the stopper has been passed and

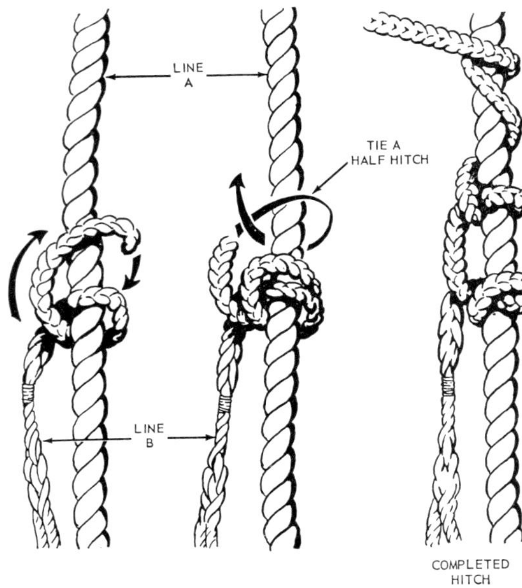

FIG. 13.1 STOPPING-OFF THE FALL

secured, the command "walk back" is given, and the strain is gradually released on the hauling part and taken up by the stopper. When all the strain has been transferred to the stopper, the order "up behind" is given. The men on the hauling part quickly run forward with the slack of the fall to the davit where the line is belayed on a cleat at the command "belay." Figure 13.2 shows the proper manner of doing this. Emphasis is placed upon passing a round turn first and a half hitch last. It is well to mention here that many seamen prefer to pass two round turns first, in lieu of one, for added protection. This procedure places the weight of the boat on the whole cleat rather than on just one of the horns which could conceivably shear off. When the after fall is properly belayed, the bow of the boat is raised and secured in like manner.

After the boat has been lifted clear of the chocks, the order "launch aft" is given and the boat is moved aft far enough to let the bow or stem of the boat clear the forward davit. When it is clear, the order "launch forward and bear out" is given. Then, as the bow is pushed out, and the boat is pushed forward, the bow passes over the side and between the davits. When it is far enough forward for the stern to clear the after davit, the command "bear out aft" is given, and the stern is pushed over the side of the ship. The davits are then placed at right angles to the ship, and the boat is ready for lowering.

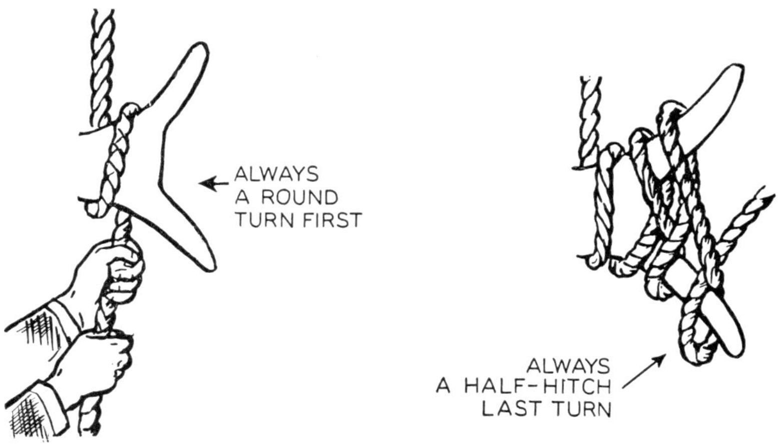

FIG. 13.2 SECURING THE FALL TO THE CLEAT

13.1. Lowering and Hoisting Boats by Radial Davit. *Lowering Away.* Because the placing of the boat in the water when anchored or moored is normally a simple operation of lowering it until it is waterborne, lowering away will be discussed only while under way, as this is the most dangerous and difficult. The proper use of the sea painter and the rudder is essential to keep the boat from being thrown against the side of the ship. The use of the sea painter will be described later in this chapter. In lowering a boat in heavy weather, steadying lines called *frapping lines* (Fig. 13.3) must be used. In using frapping lines, one end is secured to something solid on deck. The bight is passed around the falls and the end is brought back on deck and tended by a turn or two. The purpose in using the frapping lines is to keep the boat from swinging wide as the ship rolls. Another means of keeping the boat from swinging out as the ship rolls is the use of *traveling lizards* (Fig. 13.3). The traveling lizards are kept in hand in the boat, after a turn is taken around a thwart. Under no circumstances are the lizards to be secured in the boat.

When ready to lower, the man in charge of lowering takes a position between the davits. When lowering by means of radial davits, only the most experienced men should be used on the cleats when slacking the falls. Care must be taken to prevent the lines from jumping the cleats. At the command "lower away together" the men on the cleats remove all the turns but the round turn and then gradually pay out the falls. The boat should go into the water on an even keel or slightly by the stern. If the sea is rough, the boat should be held clear of the water until a trough appears in which to set it down. If the boat is set down on the crest of a wave, the hoisting gear will be subjected to a heavy strain when the sea drops out from beneath the boat. As soon as the boat is waterborne, the command "up behind" is given. At this order the men on the cleats remove the final turns and slack the lines so that the boat will ride immediately to the sea painter.

The coxswain of the boat should have his crew fend the boat away from the ship during the descent. Small boat fenders between the boat and the ship will also help. He will have his engineer start the engines during lowering to be

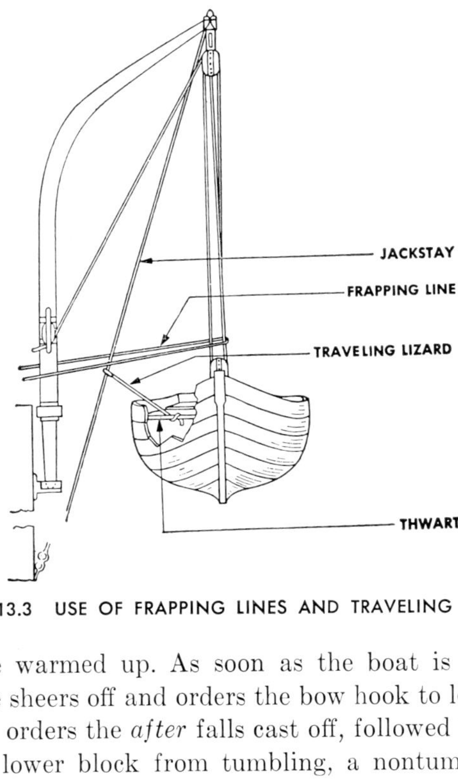

FIG. 13.3 USE OF FRAPPING LINES AND TRAVELING LIZARD

sure that they are warmed up. As soon as the boat is waterborne and the blocks are clear, he sheers off and orders the bow hook to let go the sea painter. The coxswain then orders the *after* falls cast off, followed by the *forward* falls.

To prevent the lower block from tumbling, a nontumbling block is used; i.e., one designed to keep from turning over as it is hoisted back on board and the weight of the boat is no longer on it. However, even with this type of block, care must be taken in bringing it back on board. If the hauling part is pulled, the parts of the falls will not pass through the sheaves, and the block will turn over. Another safety factor is the swivel hook attached to the lower block. This swivel permits the removal of twists in the falls. If an attempt is made to raise the boat without removing the turns, the grind of the turning block may shear the shank off the hook. If not, the friction on the parts of the falls will be so great as to increase materially the difficulty of raising the boat. The block is of the automatic releasing hook type described in Chapter 6.

One thing to be stressed is the necessity of using the lanyard in hooking and releasing so that the hands are kept clear of the block. In addition to the danger of fingers being caught between the hook and the ring, there is the danger of a hand being mauled between the heavy block and the boat as a swell raises the boat unexpectedly. In hooking on, always lead the lanyard through the hoisting ring and then use the lanyard to draw the ring on the hook. The hook should then be held closed by means of the lanyard until the boat is clear of the water and the weight of the boat is on the hook. To prevent the hook from accidentally tripping, the lanyard should be bent around the shank of the hook.

Hoisting In. Assuming that the boat is approaching the side of the ship to be hoisted in, all preparations should be made in advance for receiving it. The davits should be swung out over the side at right angles, and the blocks should be lowered near the water from the davits, crane, or boom. The sea painter should be dropped by means of a light line. The most important point is to be sure that the painter is secured to the boat at the proper time.

When in position the boat coxswain should order his men to hook on the blocks. When this has been accomplished, he reports this to the man in charge of the "hoisting in detail." The man in charge must not commence hoisting until it is reported by the coxswain that the blocks are hooked.

When all is in readiness for hoisting, the man in charge gives the command "set taut." The men on each winch then take the slack out of the falls. Because it is often difficult for the winch men to tell when the slack is out of the falls, the order "heave around together" and then "avast heaving" is given when the slack is out and the falls are taut. When the slack is out, the proceedings are stopped, and the man in charge checks to see that all is in readiness for hoisting, i.e., there are no dips or turns in the falls. When everything is ready, he gives the order "heave around together." If one end of the boat is hoisted faster than the other, the command is given, "avast heaving forward (aft)"; then "heave around together" when the boat is again level.

If possible the boat is stopped at deck level to disembark the personnel. When the boat has been hoisted high enough to clear the rail, the order is given "avast heaving," then "pass the stopper." When stoppers are passed and secure, the tension on the hauling part of the falls is eased off by the command "walk back together," at which the men on the winches slack the falls by rotating the turns around the gypsy heads. When the stoppers have taken the strain, the command "up behind and belay" is given. On this command the men at the winches throw off the turns and quickly take the slack to the cleats where it is belayed. Speed is essential, for at this time the weight of the boat rests entirely on the stoppers. When the falls are belayed, the stoppers are removed and the boat swung in.

13.2. Lowering and Hoisting Boats by Crane or Boom. A lee is first made and it is preferable in this case to have no way on the ship. Steadying lines are secured to the bow and stern of the boat and are tended by the boat handling detail on deck. The safety runner is also rigged. The boat is then lowered until just clear of the water, and at the proper moment lowered quickly into the water. As soon as the boat is waterborne, the ring of the slings is run clear of the hook by a pull on the safety runner. Figure 13.4 illustrates the use of this safety runner. When the ship is rolling or pitching, steadying lines

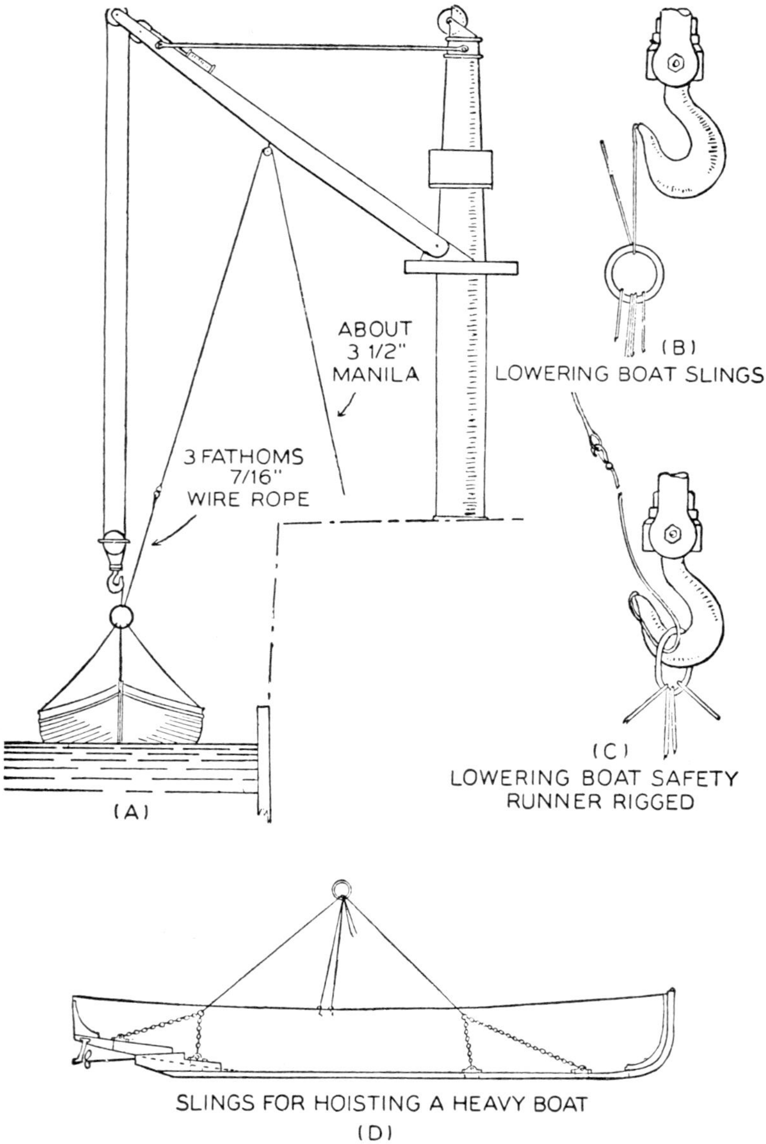

FIG. 13.4 SAFETY RUNNER IN USE WITH BOAT SLING

should be used on the crane blocks. The boat's engine should be running before the boat takes the water. Round fenders should be hung over the bow and quarter of the boat.

When all preparations have been made, for hoisting, including handling the slings from the bight of the safety runner, the boat is worked up under the crane. The bowline, stern line, and steadying lines are passed, and the crane block lowered. The slings hang slack in the bight of the runner while the legs are shackled to the hoisting chain bridle of the boat.

FIG. 13.5 MOTORBOAT HOISTED ABOARD USING A GRAVITY DAVIT. *Official U.S. Navy Photograph*

When hoisting in a seaway using a crane or boom, there are three principal difficulties to be overcome.

First, after hooking on and hoisting has commenced, the boat may retouch the water, as the ship rolls, with a violent jerk which may prove destructive to hoisting gear. To avoid this, advantage is taken of a quiet moment. When the ship has begun to roll toward the boat or the boat is on the crest of a wave, the block is quickly lowered, the ring of the slings is run onto the hook by the safety runner, and hoisting is commenced.

Second, the rolling of the ship may cause the boat to swing into the side of the ship as it is being hoisted. This danger is met by the use of fenders hung over the side of the boat. By properly tending the steadying lines, taking in and holding the slack as necessary, the swing will be somewhat reduced. The action of the steadying lines is similar to that of frapping lines. Where the boat is making headway through the water or the ship is pitching badly, a long bowline and stern line should be used to help reduce the surge fore and aft as the boat is being hoisted.

Third, after hoisting and swinging in, difficulty is sometimes encountered in plumbing the boat into the chocks. It is advantageous, especially where the boats stow close to the side of the ship, to have four steadying lines, two forward and two aft. These lines should lead from opposite sides of the boat at the bow and stern so as to cross each other, thus providing a better lead for steadying the boat into position.

Figure 13.5 shows a motor boat being hoisted aboard using a gravity davit. With this type of davit, the problem of plumbing the boat directly into its chocks is eliminated. Many other problems encountered when using either cranes or booms are also eliminated when using this type of davit.

13.3. Use of the Sea Painter. The sea painter is made fast close to the center to the forward thwart and leads out over the gunwale on the inboard side of the boat. It is rigged so to facilitate sheering the boat off from the side of the ship when it is necessary to get away. On many occasions it is necessary to hold the boat alongside the ship in order to embark additional personnel. Many seamen recommend that a lanyard be attached to a bow shackle or ring at the stem which can be passed around the painter and then hauled tight to facilitate holding the boat alongside during this operation.

FIG. 13.6 WHALEBOAT BEING LAUNCHED. *U.S. Coast Guard Official Photograph*

13.4. Handling a Power Boat. A Navy power boat crew usually consists of a *coxswain*, a *boat engineer*, a *bow hook*, and a *stern hook*. The coxswain is in command of the boat, subject to the supervision of any regularly assigned boat officer, or the senior line officer present in the boat. The engineer operates and cares for the engine. The bow hook handles forward lines and falls when making fast or letting go and acts as forward lookout when under way. The stern hook handles the stern lines and falls and keeps an eye out aft when under way.

In the motorboat, and in landing craft, the coxswain operates the engine himself. The clutch lever, the throttle, and the wheel are situated so that the coxswain can operate all of them. In the wooden motor whaleboat and in the motor launch, however, the engine is located amidships, while the coxswain is stationed aft at the tiller. In this type of craft the boat engineer operates the engine according to bell signals received from the coxswain.

Word of mouth engine orders are never given in a power boat. The noise from the engine would normally distort or drown out completely all voice orders. A system of bell signals has been devised as follows:

Number of strokes	*Meaning*
1	Ahead slow speed.
2	Engine running, propeller shaft disengaged. When engine is not running, two strokes mean, "Start the engine." When engine is running with shaft engaged, two strokes mean, "Throw out the clutch to disengage the shaft."
3	Back slow speed.
4	Full speed in direction propeller is turning at the time the signal is given.

13.5. Handling a Power Boat Underway. Steering a power boat is much the same as handling a single-screw ship, although the reactions of the boat to the engines and rudder are more pronounced. When under way in choppy seas, speed should be reduced somewhat, not only to avoid shipping seas, but also to reduce the strain on the hull and on the machinery due to the racing of the screw when the stern rides clear of the water. Navy boats are more than adequately powered and they may be swamped by running them too fast against the seas. When heading into the sea, it is possible to make fair speed by careful nursing, i.e., watching the seas and slowing, or even stopping for a moment as heavier seas bear down upon the boat. If the boat engineer has sufficient experience to regulate the speed in this way (assuming he can see), it is best to leave speed changes up to him. As in ships, the boat sometimes may be made to ride much easier when, instead of plunging head-on into the sea or running directly before it, a course is made with the sea on the bow or quarter. If running more or less across the sea, it is well to head up momentarily to meet heavy waves.

A large motor launch or landing craft has a high bow, and turning against the wind and sea is difficult. A large turning circle therefore may be expected and should be allowed for in confined waters.

Attention should be paid to weight distribution, especially in a head sea. Too much weight forward may cause the bow of the boat to plunge into the waves and possibly swamp; too much aft will cause the boat to fall off. When running before a heavy sea, weights aft will reduce yawing, but too much weight aft will cause the bow to ride too high. Wind and current should be observed and allowed for when leaving the ship or landing, as well as the compass course and time to destination, in order that the proper course may be steered in reduced visibility. In approaching any object in the water which may be damaged or injured by contact, such as a seaplane or target drone, always maneuver for a position that, when stopped, the boat and the object will be separating. A coxswain should be careful to slow in passing small open craft, men working on floats, or divers and swimmers, so as not to give them his wash. Neglect to do this can often prove dangerous to the other personnel near or in the water. Never pass a pier head, a bow, or stern of an anchored ship too closely.

13.6. Making a Landing. In making a landing, whether at a pier or a ship's accommodation ladder, it is a common mistake to keep too much way on the

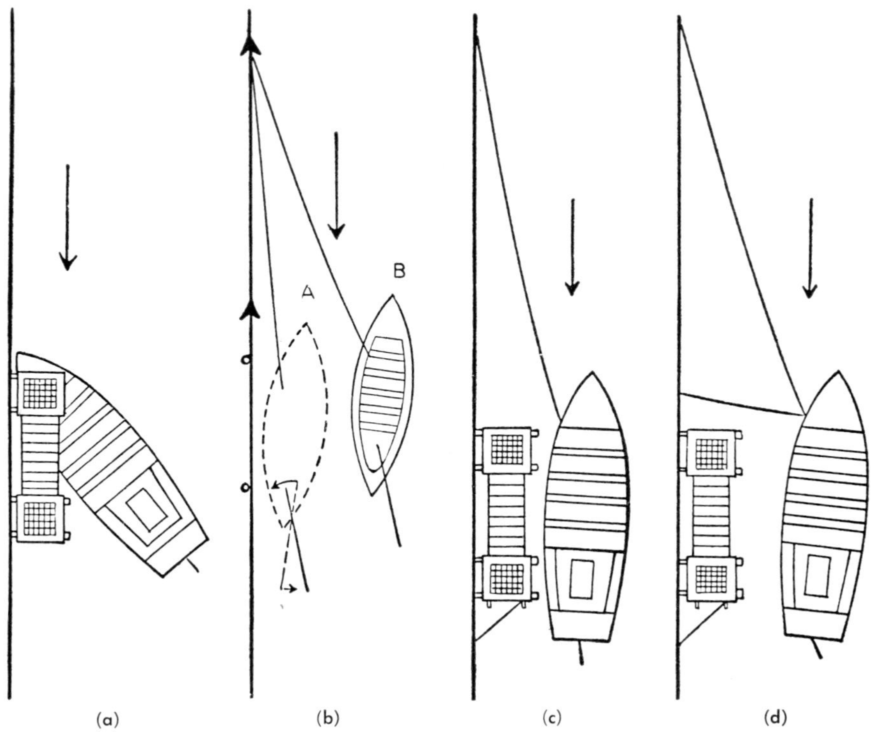

FIG. 13.7 LANDINGS. ACTION OF SEA PAINTER

boat (conditions of load and trim materially affect momentum). The landing should be approached at such an angle and at such a speed that, should the engine fail to back, control of the boat is still maintained and it can be sheered away by rudder action alone without damage. The engines may, and often do, fail to respond promptly. The backing throws the stern off to port (in a right-handed screw) which should be taken into consideration when determining the angle of approach. In coming alongside a ship's accommodation ladder in a current or in a heavy sea, care must be taken not to catch the tide or sea on the outboard bow, as this will sweep the bow in forward and perhaps underneath the lower platform of the accommodation ladder. Under these circumstances the boat may be swamped or damaged (Fig. 13.7a). The landing should be made by the aid of a boat line from forward, the boat being kept off a little from the side until the line is fast and then eased in by the rudder.

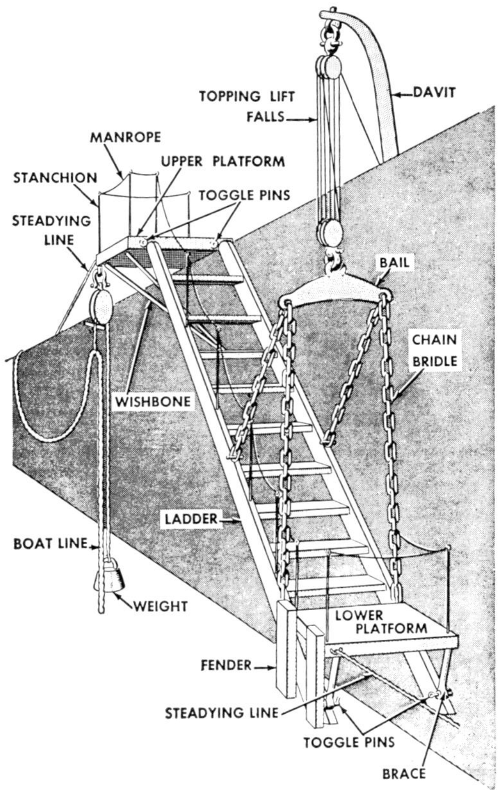

FIG. 13.8 ACCOMMODATION LADDER

A power boat coming alongside in a rough sea or in a strong tideway should always be required to take a boat line. Crews of power boats frequently make their landings at an accommodation ladder by the aid of boat hooks alone. This is done by taking hold of anything that is within reach, and holding on, often with great difficulty and with the ever-present danger of a man falling overboard between the ship and the boat. A boat, lying at the accommodation ladder in a tideway and secured by a boat line made fast to a cleat on the inboard bow of the boat, can be controlled by a touch of the rudder, which sheers the stern out or in and thus catches the current on one bow or the other (Fig. 13.7b). Where a ship is rolling in an open roadstead or riding with the wind ahead with waves surging aft alongside the ship, the boat rises and falls dangerously at the lower platform and contact with it may damage or capsize the boat. The need of a long boat line, in effect a sea painter, is here emphasized. By using the boat line in combination with the breast line and by judicious use of rudder and engines, the boat may lie alongside the accommodation ladder without coming into contact with it (Figs. 13.7c and d). Fenders should always be carried and used freely. See Fig. 13.8 for a rigged accommodation ladder.

13.7. Embarking and Disembarking. It is sometimes impossible, because of heavy seas, to make a landing at the accommodation ladder. In this case the passengers may come in over the boat boom. The rudder of the ship is put over to one side or the other, preferably to the side opposite that of the anchor. The ship will yaw back and forth but will usually yaw more on the side away from the rudder, thus creating a partial lee under her quarter. During each weather yaw of the ship, the boat pulls up under the quarter by means of a long bowline previously rigged and boat passengers can climb up cargo nets hung for this purpose from the boat boom or from the ship's side. As the ship yaws back, the boat drops back on the painter and awaits the next lee. Oil may be spread easily from the bow of the ship.

Another method of embarking is by a cargo net rigged under a crane or boom. As the boat comes under the crane or boom, the net is lowered and seized by the passenger, who is then swung aboard. When the ship is so equipped, the aeroplane whip of a crane with its fast hoisting speed should be used. This method is applicable under way in a heavy sea. It should be used only as a last resort, however, because timing must be precise and great danger is always present to the disembarking passengers.

The cargo net is also used in the embarkation of a large number of persons or troops at sea from the ship to a number of boats or landing craft. Cargo nets of sufficient length are hung over the side to reach into the boats. The boat comes alongside and is held there, while the foot of the cargo net is hauled into the boat and kept there while troops or personnel embark. The men who are already hanging on the net then drop into the boat.

Personnel may in like manner be transferred from the boat to the ship, but care must be taken in a heavy sea that the foot of the cargo net is taken into the boat to prevent personnel from falling between the boat and the ship and being crushed.

In the event of an emergency, such as recovering a number of people struggling in the water, cargo nets hung over the side of the ship may expedite their recovery. Nets used to handle personnel over the side should be fitted with 4 x 4 inch timbers lashed horizontally to hold the nets away from the skin of the ship, thus providing adequate foot and hand holds.

13.8. Securing Boats. Boats are usually secured to boat booms (Fig. 13.9), bows to guess warps, and sterns to a boat securing line which leads from the end of the boom well aft to the ship's side. This assists in holding the boats apart and in keeping them parallel to the ship. Sufficient slack should be allowed for roll and pitch when securing the boats.

In a heavy sea or storm, the boat boom may become unstable due to the roll and pitch of the ship, making it impractical to secure boats to it. The boats

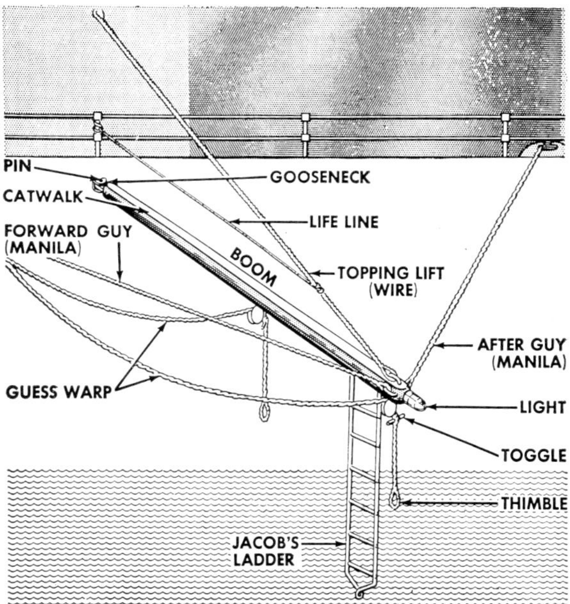

FIG. 13.9 BOAT BOOM

may, in this situation, be secured in tandem from astern for the duration of the blow. In either event, always use fenders and take precautions against lines chafing.

13.9. Handling a Boat Under Oars. A boat under oars, if properly handled, possesses much the same maneuverability as a power boat and, in the case of the whaleboat, considerably more seaworthiness. The following commands apply when handling a boat under oars:

Command	*Meaning*
Stand by the oars........	Lift oars off the thwarts, place blades flat on the forward gunwales, push oars forward until handle is over respective thwart.
Up oars.................	Lift oars to vertical position. Trim blades fore and aft with handle resting on footings.
Shove off the bow.......	Bowman lets go boat rope or sea painter or hauls in boat painter. Shoves off bow using boat hook.
Let fall.................	Let oars fall into rowlocks using crook of outboard arm to control the oars. Trim oars horizontally with blades trimmed fore and aft. Bowmen up oars before command of "let fall" or put out oars as soon thereafter as possible.

Command	*Meaning*
Give way	Move blades of oars forward and dip about half way into the water and start stroke. At end of stroke, blades are feathered fore and aft and pushed forward and another stroke is made.
Oars	Complete the stroke and level the oars horizontally with the blades trimmed fore and aft.
Back water	Row backwards.
Hold water	Complete the stroke, stop rowing, dip blade about half way into water and hold water to stop the way on the boat.
Stern all	When rowing in ahead motion, complete the stroke, then commence to back-water, gradually increasing the depth of immersion of the blades.
Way enough	When rowing in ahead motion, complete the stroke, raise oars with crook of elbow to about 30 degrees, swing blades forward and place oars in the boat.
Toss oars	Complete the stroke, come to "oars" raise the oars smartly to the vertical, rest handles on the footings and trim blades fore and aft.
In bows	The bowmen complete the stroke, swing their oars forward and boat the oars, then stand by with boat hooks or to receive the sea painter or boat rope.
Boat the oars	From "oars" or from "toss oars," place the oars in the boat with blades forward.
Out oars	Place oars in rowlocks directly from the boated position or from "stand by oars" position.
Stand by to give way	Term used in racing. The blades are pushed to forward position and slightly dipped ready for an instant start.
Give way port, back-water starboard (or vice versa)	The orders are followed to turn the boat without making way ahead or astern.
Give way port, hold water starboard (or vice versa)	This command will result in turning the boat with slight headway.
Trail oars	At this command, the blades of the oars are brought alongside the boat and left trailing in the water, in single-banked boats fitted with swivel rowlocks.

Large boats propelled by oars, such as the whaleboat, are normally steered by a sweep oar. A sweep oar is somewhat larger than an oar used to propel the boat. The coxswain using this oar can steer the boat with a great deal more maneuverability than the standard tiller, dependent on how much leverage is used or how deep the oar is set.

Boats under oars may also be steered by the use of a tiller (rudder); however, except for use in extremely heavy seas or during long periods of time, the tiller possesses no advantages over the sweep oar nor is it as efficient.

13.10. Handling a Boat Under Sail. It is far beyond the scope of this book to describe the handling of the numerous yachts, pleasure craft, and small commercial vessels which use sail as motive power. Ship's boats are work boats. They are not designed to sail and none carry sails except incidentally. Reasons of stowage forbid the use of false keels and special ballast. Because of this, ship's boats must depend upon their beam for stability and always have a tendency to make much leeway. An attempt will be made to give an elementary explanation of the principles of handling boats under sail with the hope that it will be sufficient for use in an emergency.

13.11. Terms Used. The direction of the wind is that from which it blows. To *windward* is into the wind; to *leeward* is down the wind. A *lee shore* is the shore to leeward. The *weather side* is that side exposed to the wind; the *lee side* is the opposite side. Naturally a boat heels away from the wind, so that the lee side is *down* and the weather side *up*. When the rudder is amidships the tiller or helm [1] is also. The coxswain may put his tiller up to windward and have *weather helm*, or if he puts it down he has *lee helm*. When a boat turns her head into the wind, as if she were going to tack, she is said to *luff*. The opposite of this is *bearing away*.

When sailing with the wind on one side, a puff of wind may strike the sail, causing the boat to heel and possibly capsize. To prevent this, the coxswain may luff by putting his helm down until she turns into the wind; the sails then cease to *draw*, and the boat comes back to an upright position.

It may be necessary in a heavy squall to let the sheets go; therefore, in a small boat *never belay the sheets*.

When a sheet is hauled in and the boom or foot of the sail is nearly fore and aft, the sheets are said to be *hauled aft*. In *setting the jib aback* or *backing the foresail*, the weather sheets are *flattened aft*. This may be required to give more turning effect in tacking and is also done in *heaving to*. To bring the sails more nearly parallel to the centerline of the boat they are *trimmed in;* the opposite is *easing off* or *starting the sheets*.

A boat cannot sail directly into the *eye of the wind*, but, depending on the boat and rig, sails at an angle of from four to six points from it. She must thus make a zigzag course upwind on *tacks*, or *boards*. This process is called *beating to windward*. She is said to be sailing *close-hauled* or *on the wind* on each of these tacks.

A sailboat is said to be on the *starboard tack* when the wind is coming over the starboard side, and the *port tack* when the wind is coming over the port side. When a boat is not sailing as close to the wind as possible with advantage, she is said to be *sailing free*. When the true wind is within two points on either quarter, she is said to be *running* before the wind. If, when sailing free, the wind is still forward of the beam, she is said to be on a *close reach;* if the wind is from abaft the beam, she is said to be on a *broad reach*. The *apparent wind* is the wind striking the sails which is generated by a combination of the ship's speed through the water and the *true wind*. In obeying the Rules of the Road the prevailing or true wind fixes the respective obligations of sailing vessels.

Tacking is bringing the boat on the opposite tack, head through the wind. *Wearing* consists in turning the boat from one tack to the other tack, stern through the wind. During this procedure, as the wind comes aft and the sails are trimmed flat, the boom is carefully and intentionally allowed to swing to the opposite side. The boom is then said to have been "*jibed*" over. Should the boat be sailing free and alter her course so as to bring the wind on the opposite side, carrying the stern through the wind, she has been "jibed," such as "jibing around a buoy."

[1] While the terms "helm" and "tiller" have been officially banned in connection with modern steamers, they are still applicable to sailboats and will be used here in accordance with original practice.

Trim. To do her best under sail, a boat must be trimmed in accordance with her build and rig. To effect this condition, the trim of the boat and sails must be altered, as necessary, to meet the varying conditions of sailing.

In sailing on the wind, a properly designed and trimmed boat should carry a slight weather helm; that is to say, she should have a slight tendency to come into the wind. If too much weight is carried forward, the boat trims by the head, the stern rises, offering less lateral resistance aft to the water. The bow being deeper, offers greater lateral resistance to the water, and in addition has the increased pressure of the bow wave on the bow. These forces form a couple tending to cause the boat to luff, which, to counteract, necessitates an excessive weather helm. Too much weight aft causes a corresponding tendency to fall off.

If the sails are too flat forward, lee helm is necessary to counteract the tendency to fall off; if too flat aft, weather helm is necessary.

In addition, as the boat heels, the forward component of the force of the wind on the sail acting on the center of effort is displaced to leeward of the keel line, which produces a leverage which tends to make the boat luff. This tendency is especially noticeable in a tall sloop rig, which endeavors to *"work out from under"* when struck by a sudden gust as she is making good headway close-hauled.

When running before the wind, weights should be carried aft to decrease yawing, but this may decrease the speed, if overdone.

After the boat has been underway for some time, the sail or halyards may stretch, or in wet weather, may shrink. This calls for appropriate setting up or slacking off on the halyards to correct the set of the sail.

13.12. Close-hauled—on the Wind. On the wind, a boat should carry a little weather helm. The sails should be kept well full, sheets not too flat, but everything drawing and the boat alive. It is a common mistake to get the sheets so flat that the boat, while pointing high, actually makes a course to leeward of that which she would make if kept away a little with sheets eased accordingly; and it is of course clear that, if kept away, her speed will be greater than when jammed up into the wind in the hope of stealing a fraction of a point. A boat of good draft with a deep keel and centerboard, and yachts designed for racing, with fin-keels 10 feet below their normal water line, will lie amazingly close to the wind with little leeway. Ship's boats, however, are not constructed on yachting lines and cannot be held up in the same way. The shape of the sail when close-hauled is very important; the leech should be almost flat, and some boats, to accomplish this, have battens which fit in pockets in the leech of the sail. A little curve or belly should be allowed in the luff. The cut of the sails, the way they are laced to the yards and booms, and the tautness of the halyards all affect the shape of the sails when drawing.

The sails being properly set, the luff of the sails is kept just short of trembling, with weather helm enough to let the helmsman "feel" that she wants to come into the wind. As the wind will vary more or less (in apparent, if not real, direction), it is necessary to be watchful and bring her up or keep her away from time to time in order that she may be always at her best. The sails should be kept fuller in rough than smooth water, as it is more important that the boat should be kept *going* so as to be always under command of the rudder. If a heavy breaking sea is seen bearing down upon her, she should be luffed to meet it and kept away again as soon as it has passed. If she loses way she becomes helpless at once. It is dangerous to be caught by a heavy sea on the beam; and, if the course to be made in rough water would bring the boat into the trough, it is the best plan to run off for a time with the sea on the quarter, then bring her up with it on the bow, and so make good the course desired without actually steering it at any time.

For a moderate squall, the boat should be luffed sufficiently to shake the sails without spilling them, thus keeping enough headway to retain control. If the wind becomes stronger, she must be luffed more decidedly and the sheets eased off. The sheets may, of course, be let go, and in an emergency this must be done at once, in addition to putting down the helm. For this reason it is a universal rule in boat sailing that the sheets should never be belayed or left untended in any weather.

13.13. Sailing Free. A boat sails her fastest on this point of sailing. The tendency to luff is strong, especially if the wind is fresh and the boat or sails are improperly trimmed. In a squall the situation is quite different from that in sailing close-hauled. Here the wind cannot be spilled by a touch of the tiller and the only prudent thing to do is to slack the sheets while luffing. In this procedure care must be taken not to jam the helm down hard for it causes the boat to heel dangerously to leeward, and as it turns into the wind the lee quarter and rail may go under, the end of the boom trip in the water, and the boat capsize. The same thing may happen in jibing if the boat is allowed to fill away too quickly on the new tack. The force of the wind would be much reduced by running off, but the trouble with this is that, if it comes too strong, there is no recourse but to lower the sail, and the chances are that it will bind against the shrouds and refuse to come down. Moreover, there is always danger that the wind will shift in the squall, and the mainsail may jibe with dangerous force.

The gaff-headed rig has the advantage over the tall triangular rig on this point of sailing, but the gaff-headed mainsail must be tended more carefully, as the efficient angle to the apparent wind exists within narrower limits than that of the jib-headed rig. In general, in sailing free the gaff-headed rig must be trimmed closer than a jib-headed rig in order to maintain its most efficient angle. This requires close watching of the apparent wind at all times. Underwater resistance may be somewhat reduced by partially raising the centerboard, if the boat be so equipped.

The tall jib-headed rig gains power as the wind hauls forward and the low gaff-headed rig gains power as the wind draws aft.

13.14. Running before the Wind. In a fresh breeze, this is the most dangerous point of sailing, because of the chance of an unintentional jibe. The danger increases if the boat yaws. From this follows the rule to keep the weight fairly well aft, though never at the extreme after end when running before the wind. Very careful steering is required; and, if the sea is heavy, the boom may jibe in spite of all the care that can be taken unless lashed to the lee rail or shroud by a "lazy guy."

Squalls are not as dangerous before the wind as when close-hauled or reaching, unless they are accompanied by a shift of wind. To reduce sail quickly in a gaff-headed boat, to meet this emergency, the peak of the mainsail may be dropped.

In running before the wind, the foresail is sometimes set on the side opposite the mainsail, a temporary boom being rigged by using a boat hook or an oar. A boat sailing in this way is sailing "wing and wing."

If the sea is rough, it is well to avoid running with the wind dead aft. To make a course directly to leeward, the wind may be brought first on one quarter and then on the other, the mainsail being clewed up or the peak dropped each time the course is changed, if the breeze is strong enough to make jibing dangerous.

A serious danger in running before a heavy sea is that of "broaching to." The boat will yaw considerably, the rudder will be often out of water, and the sails will be becalmed in the trough of the sea. The situation here is much like that of a boat running in a surf; and, as in that case, the yawing will be reduced by keeping the weights aft and by steering with an oar. The jib should always be set with the sheet flat aft. It helps to meet and pay her off if she flies to, against the helm. A drag towed over the stern is also helpful.

Another danger in running is that the boom may dip as she rolls and thus capsize the boat.

13.15. Tacking. In tacking, the same principles apply to a boat as to a ship. An after sail tends to bring her head into the wind and a headsail to keep her off; but all sails, so long as they draw, give her headway and so add to the steering power of the rudder.

It is clear that a short full boat will turn to windward better than a long and narrow one and will require a much shorter distance for coming around. Thus a short boat is preferable to a long one for working up a narrow channel.

When about to tack, the coxswain should let her fall off a little to fill the sails and gain good headway, and he should watch for smooth water and avoid luffing into a breaking wave. The rudder should not be suddenly put hard over, but should be put over enough to have a good effect at first and then more and more as the boat swings; by the time the boat is swinging rapidly the rudder should be over about 30 or 35 degrees and held there.

Under ideal conditions, a boat, close-hauled but with good way on, shoots into the wind as the tiller is eased down, making a good reach to windward and filling away on the new tack, without a moment losing headway. The main boom is hauled amidships in a two-masted boat and nearly amidships in a single-masted boat, and, as the jib and the foresail lift, their sheets are let go. The boat comes head to wind and as she pays off on the new tack the sheets are hauled aft and she is steadied on her course. Under less favorable conditions, such as a heavy head sea or a very light breeze, tacking is not so simple.

If the boat gets in "irons," the jib sheet must be held out on the old lee bow to pay her head around. Care must be taken not to make a "back sail" of the mainsail. If she gathers sternboard, the rudder is shifted, and, if necessary, an oar is gotten out to help her around. The statement is sometimes made that it is lubberly to use an oar in a boat under sail. The lubberly part is the getting into a position where an oar is needed.

Carrying the weights forward is favorable for tacking, but when a boat has sternboard she may be helped around by putting a few of the crew on the (new) lee quarter, where, by increasing the immersion of the full lines of the counter, they may add to the resistance and cause the bow to fall off.

Attention may again be called to the fact that in squally weather a boat is in a dangerous position whenever she is without headway, because she can neither be luffed nor kept away in the event of being struck by a heavy gust. If, through ignorance or carelessness, the sheets are belayed at such a time, the danger is greatly increased.

13.16. Wearing. In beating to windward, boats ordinarily go about by tacking, because in tacking they turn into the wind and gain ground to windward. In wearing around they turn away from the wind, losing more or less distance to leeward according to circumstances; still it is often possible to wear in winds so strong or water so rough that tacking ship's boats is impossible. It is often necessary to resort to wearing when maneuvering in close quarters, such as clearing a dock or avoiding a collision.

In wearing, the helm is put up and the mainsheet eased off in order to help in bearing away and to get the maximum effect of the mainsail in increasing headway. When the wind comes nearly aft the sheets are rounded in smartly in such a manner that both the sail and the stern pass through the wind at the same time. As the sails jibe over, the sheets are eased off slowly and gradually. Care should be taken at this point, especially with a sloop rig, that the boat be not allowed to come up on the new tack too quickly, as this may bring about a dangerous heel to leeward.

The details of the maneuver may vary considerably, according to the conditions of wind and sea and peculiarities of the boat as to rig and trim. In boats of more than one mast, it is best to sail dead before the wind, trim in the sails, jibe them, and ease them out on the new tack in the order of jib, foresail and mainsail.

In a fresh breeze, as jibing is dangerous, the mainsail should be doused, brailed up or the peak dropped before the wind comes aft, and set again in time to bring her to the wind on the new tack.

13.17. Remarks on Jibing. A sail is "jibed" when it is allowed to swing from one side to the other, the wind being aft or nearly so, and the sail full first on one side and then on the other. This may be done intentionally, as in wearing or in simply changing course, or it may come unexpectedly from a shift of wind or from the yawing of the boat. As it necessarily involves a violent swing of the sail, it puts a heavy strain upon the spars and rigging; it endangers everyone in its path and causes the boat to lurch more or less steeply to leeward. At this point the boat shows a strong tendency to luff on the new tack, and if not met with the rudder the boat may be knocked down and capsized.

It is important in jibing that the sails be trimmed flat before the stern of the boat is brought into the wind, and after the boom has jibed over, the sheets should be started slowly and gradually. The trimming of the sheets should be so timed with the swinging boat that the helmsman does not have to check his swing and wait for the sails to be trimmed in nor should the boat be allowed to run with the sheets flat aft. In either case the boat loses speed, and loss of speed is loss of control.

13.18. Reefing. When an open boat begins to ship spray and water over the lee rail, it is time to reef. A boat that is decked over may run with her lee rail awash; but when an open boat heels her gunwale close to the surface of the water, it must be remembered that a fresher puff may bear the gunwale lower without warning, and that the moment it dips, the boat will almost certainly fill and capsize.

The details of reefing will depend upon the rig, but a few general rules may be laid down. The men should be stationed before beginning, and should all be required to remain seated. The boat is then luffed, but not to the point where steerageway and control are lost. One hand lowers the halyards of each sail as much as necessary, another hauls down on the luff and shifts the tack. The sheet is hauled in a little to let the men get hold of and gather the foot. The clew earing, followed by the points, are then passed, and the halyards manned. The sail is then hoisted and the sheets trimmed as the boat fills away on her course.

If the boat has more than one sail, it is a good plan to reef them one at a time.

Handling Boats in a Surf

The proper handling of boats in a surf is an art in itself, calling for special judgment and skill that can be acquired only by practical experience. Some groups of seamen, such as fishermen along the New Jersey coast and the U.S. Coast Guard, have every opportunity to acquire this skill through experience and, in addition, are equipped with boats specially designed for the purpose. They are thus often able to take boats successfully through a surf so dangerous that it would be disastrous for the ordinary boat handler.

Amphibious operations require landings to be made regularly on beaches where more or less surf may be expected, and landing craft are designed especially for this work. In addition, their crews must be specifically trained for this type of work.

Due to the hazards of the sea any seaman is likely to find it necessary to take almost any kind of ship's boat or raft ashore through a surf. He may have a green or exhausted crew whose lives as well as his own depend on his performance; therefore it behooves each of us to learn all he can of the subject both by study and actual experience whenever opportunity offers.

A surf never looks as dangerous from seaward as it is, especially from a small boat. When there is any possibility of a surf, a beach should be approached with caution, and care should be taken to remain well outside the breakers until ready to make the attempt at running the surf. If there is any possibility of help from the shore, it should be awaited before running a heavy surf. If no help is available and it is necessary to run the surf unaided, two principles should be kept in mind. First, the boat *must* be kept end on to the surf to avoid broaching and capsizing. Second, the boat must be able to meet and resist the breakers to keep them from driving her toward the beach out of control or, in extreme cases, driving her under or throwing her end over end.

Methods of running the surf vary with the height of surf, type of beach, set of current, weather, type and trim of boat, gear available, and experience and condition of crew. The means and methods to be used must be decided after a consideration of these factors which by their nature preclude the statement of other than general principles. Although there are various methods for handling boats in a surf and accomplishing a landing successfully, only the one considered safest and simplest is discussed here.

13.19. Landing through a Surf. The most important consideration for the inexperienced coxswain is the necessity for remaining outside the breakers for a long enough time to study the surf carefully. Care must be exercised to ensure that the boat is kept far enough outside the outermost line of breakers to avoid being caught unexpectedly by a sea. When this is done, one will find that the large seas come in a more or less regular sequence, usually three or four in a series. Then follows a period of smaller seas during which there is a period of build-up. It is during this time that the entrance into the line of breakers must be made.

Having determined the period of the seas and decided on the run in, wait until the last sea of the large series breaks just inshore of the boat and then turn so as to present the bow seaward and back in. As each succeeding wave overtakes the boat, it may be necessary to pull ahead to meet it. With this method the oarsmen are normally faced so that the coxswain may best use them to control the speed and direction of the boat. Too much emphasis cannot be placed on the full utilization of the oars in steering the boat. It is possible for the boat to be kept headed directly into the seas by having first one side and then the other give way as may become necessary. As each sea passes, "stern all" and gain more distance toward the beach. With each overtaking sea the boat will be carried shoreward a considerable distance, even though the oarsmen are pulling against it. If even the smaller seas are of dangerous size, it will be necessary to impart a great deal of way to the boat in order to give it sufficient inertia to overcome the power of the sea and avoid broaching.

Broaching is most apt to occur when the seaward end of the boat is lifted by an onrushing wave, depressing the shoreward end in the relatively calm, motionless water which is immediately in front of the wave. We have, under these circumstances, one end of the boat deeper than the other and embedded in stationary water while the other end has a tremendous force acting on it. It is apparent that this force applied to one side or the other of the seaward end of the boat will create a powerful turning moment, one arm of which is equal to about the length of the boat. Thus, it is obvious that a great amount of power is necessary to overcome the forces which tend to cause broaching.

Considering the weight of the boat constant, since buoyancy is also a paramount feature, this power can be met only by rowing strongly against each oncoming wave. Weights should be located in the bow (seaward end) of the boat, but not in the extreme end. Oarsmen should use a short, fast, powerful stroke so that they may back-water as each sea passes with as little delay as possible.

It should now appear that the seaward end of the boat is the most important and the one on which adverse forces are apt to be most dangerous. Hence, it follows that, if there were some means of holding the bow steady while the overtaking seas pass, the problem would be less difficult. In practice there are two very handy devices for accomplishing this. These are the *drogue* and the *surf-line.*

A drogue is a conical-shaped bag about 2 feet wide across the mouth and 4½ feet long. It is towed mouth foremost by a 2½-inch line which is secured to the mouth by means of a bridle. A small line known as the tripping line is made fast to the apex, or pointed end. When towed mouth foremost, the drogue fills with water and offers considerable resistance; when towed by means of the tripping line, the resistance becomes negligible and the drogue passes through the water easily.

When a drogue is used in a boat landing through a surf, it must be carefully tended by men in the bow so that there is always a strain on the towing line when a sea overtakes the boat. When the sea passes it is desirable to "stern all," and the tripping line is hauled taut so that the drogue passes easily through the water. The coxswain and men tending the drogue must be alert to slack the tripping line well in advance of the arrival of the next wave in order to allow it to fill with water and exert the greatest resistance to keep the bow pointed seaward. A drogue is especially recommended when there is any current setting parallel to the beach.

A surf-line consists of a 2½- or 3-inch line made fast to an anchor just beyond the outermost line of breakers. This line should be about 150 fathoms long. The line which is coiled in the boat free for running is paid out by men in the bow so that there is always a strain on it when the boat is overtaken by a sea; as the sea passes, the line is again paid out.

A surf-line exerts a more positive force on the bow of a boat landing through a surf, but it is not recommended when there is any appreciable current setting parallel to the beach. The reason for this is, of course, that the farther the boat progresses (i.e., the longer the scope), the more it will be carried down by the current. Hence, there will actually come a time when the boat will be carried broadside to the waves or nearly so and be in grave danger of capsizing. Another disadvantage of the surf-line is that it actually stops the progress of the boat toward the beach each time the men in the bow hold it to permit a sea to pass under the boat.

13.20. Landing Craft (LCVP). The most interesting and important phase of LCVP operation is the run to the beach. In the surf, the coxswain and crew are really put to the test. There are a number of factors to be kept in mind when the student coxswain goes into the surf for the first time. These will be discussed, roughly, in the order in which they occur during a landing operation.

The coxswain should make certain that each crew member is in his place as he makes ready for the run. It is necessary that all men wear life jackets when the LCVP or LCM is launched. There may be no time to slip them on in an emergency. Even a champion swimmer can drown in a hurry if he happens to be knocked out for a moment in an accident, as when a boat turns over.

As the beach is approached, the rolling ground swell which begins to rise several hundred yards out from the shore determines the size of the surf. Once inside the breaker line, course should not be changed. Therefore, the boat should be lined up with the spot on the beach where it is to be landed. This should be done before the boat enters the surf.

The LCVP, handled by an expert, can cope with a 12-foot surf, but a 6- to 8-foot surf is high enough to cause plenty of trouble, especially for the beginner. Regulate the speed of the boat so that it rides in to the beach *just behind the crest of a comber.* If the boat is right on the crest it will be set down *hard* on the sand when the wave crashes and ebbs.

The boat should be kept *at right angles to the surf.* The LCVP is likely to broach if this rule is not observed. Usually the surf goes in parallel to the beach and if you hit the sand head-on the boat will ground safely.

The exact spot at which the coxswain aims his boat should be chosen with care. The LCVP was designed primarily to run aground on a sand beach. Any large stones or outcropping of rocks that might damage hull or ramp should be spotted in advance. Both coxswain and forward lookout will need to keep a sharp eye open for underwater obstacles.

If the boat should run aground on a sandbar some distance from the beach, the engine should be run slowly in forward speed until the hull is floated partly free by the next breaker. When the boat has this flotation, the engine speed should be increased. If the boat is not freed, the engine speed should be cut and an attempt made when the next incoming wave lifts the boat.

It is unwise to assume that the water is shallow all the way to shore if the boat grounds some yards out. Unless word is received from the beach party, the unloading of troops should not be attempted. The water may be 10 feet deep a few yards inshore from a sandbar that has stalled the boat.

After clearing all bars, the LCVP should be run on the beach at a good speed to ensure a good hold on the sand. When properly beached, the boat is at right angles to the surf and its keel is grounded along its entire length. In this position the boat is not likely to broach while loading or discharging cargo.

The engine should be kept turning over at about 1200 rpm's to hold the boat well up on the beach. Avoid letting the screw race wildly. Idle down when water recedes and the screw loses its bite between breakers.

Should the engine fail or some mishap occur when the boat is within the surf-line but not aground, the first thing to do is drop a stern anchor. This helps to hold the stern at right angles to the breakers if the line is payed out carefully. *Do not snub* the line, but let the boat surge toward the beach with each comber. Only when the boat touches the beach is the anchor line snubbed to prevent broaching. The flow of the tide is something to take into consideration if the boat will be beached for any length of time.

Several precautions may be taken to keep from broaching. *First,* be sure that the breaking seas are kept dead astern. Otherwise the stern will fall off to port or starboard as the water dashes against it. *Second,* drive well up on the beach so that the entire length of the keel is aground. *Third,* speed up engine in forward gear as incoming waves float the boat. *Fourth,* see that the antibroaching lines are thrown to the beach party at once. Figure 13.10 (a, b, c) show how these lines are used to prevent broaching.

No hard and fast rule can be laid down for the use of the antibroaching lines. The coxswain must think and act intelligently to allow for wind, different types of beaches, and other factors influencing broaching. Usually, however, it is wise to line up the bow with some object on the beach. Then it is immediately apparent if the bow or stern is moving. If this happens, the rudder should be put *in the direction in which the stern is swinging.* Then the engine should be speeded to drive higher on the beach to bring the stern around. Sometimes, if the surf is not high, it is possible to free the broached boat by engine power alone.

When retracting from the beach the coxswain tackles the most difficult part of the landing operation. It is during retraction that the beginner at boat handling is likely to broach or damage the rudder, screw, or skeg.

The coxswain will be most successful in getting away from the shore and safely beyond the breaker line if he observes the following:

1. The rudder should be set amidships before attempting to retract. In the LCVP this may be done by running the engine about half speed ahead. The discharge current or wash from the screw will force the rudder into an amidships position.

2. The bow should be lined up with an object on the beach. If this is done it will be easier to note any swing of the boat soon enough to correct the movement and hold her straight.

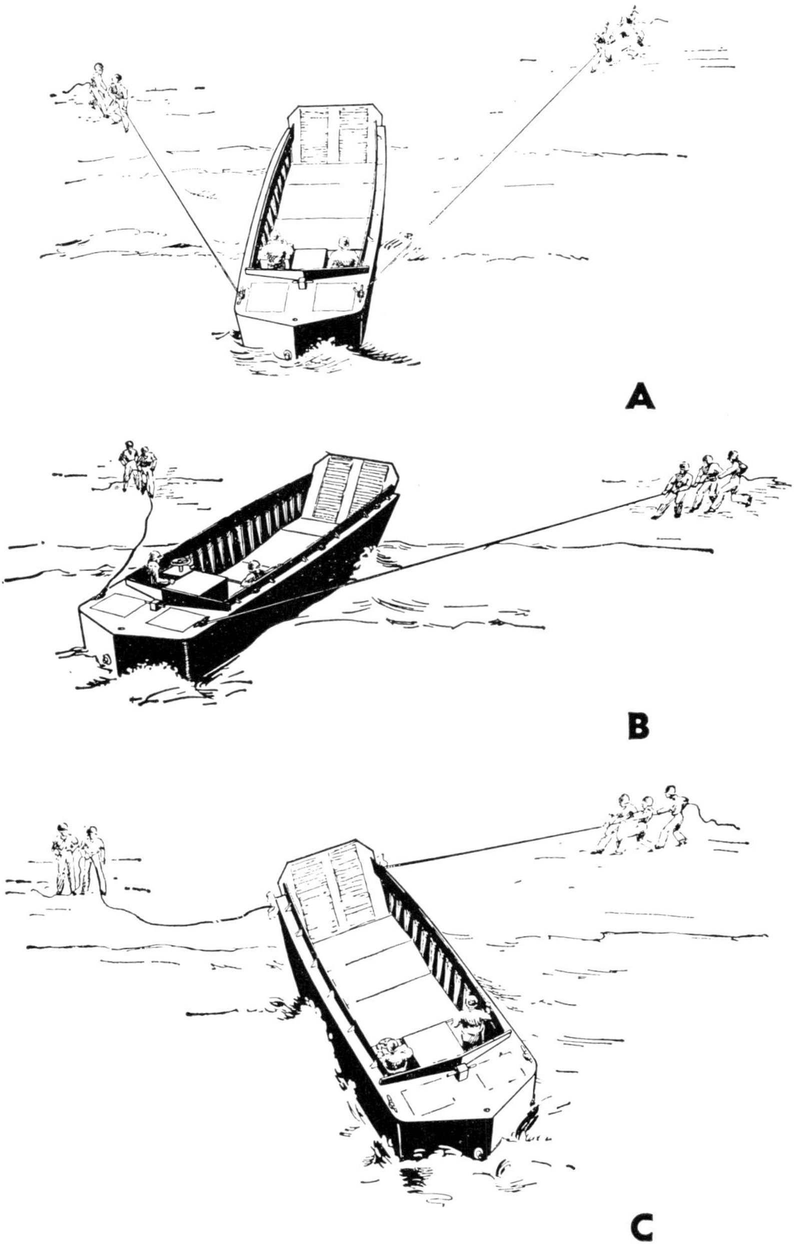

FIG. 13.10 BEACHING AN LCVP

3. Next, the engine should be shifted into reverse and a wave to float the hull should be awaited. When flotation is achieved the engine should be accelerated. Nearly always the boat will move backward a short distance.

4. When the wave recedes the engine should be prevented from racing needlessly as the screw loses its bite in the water; this will prevent the rudder and skeg from digging into the sand upon which they rest.

5. If the bow begins to swing, the steering wheel should be turned *in the direction of the swing.* This should bring the bow back. The wheel should then be turned back before this return swing is completed or the bow will move too far and require more maneuvering.

6. Once the LCVP is floating free and has passed any outer sandbars, it should be backed at right angles to the surf until outside of the breaker line.

7. Once through the breakers, and when the boat is on the crest of a wave, the rudder should be put over hard, the engine shifted into forward, and accelerated. This will cause the boat to pivot quickly and take the next sea on the bow.

13.21. Beaching and Retracting with the LCM. Most of the general rules laid down for running in to the beach in the LCVP are observed when piloting the larger tank lighter. The boat is kept at right angles to the surf and is driven ashore just behind the crest of a wave. It should be grounded well up on the beach along the entire length of the keel and the engines kept running with enough speed to hold the boat firmly beached while loading or unloading.

Once the coxswain is familiar with his boat, the tank lighter is easier to retract than the smaller, single-screw LCVP. This is due to the fact that the twin-screw design gives better control over the bow's tendency to fall off to port or starboard when backing through the surf. In retracting, the LCM's rudders are put amidships, both engines are reversed, and she is backed off slowly.

If the bow falls off to starboard there is no need to spin the wheel. The coxswain simply speeds up the port engine in reverse until the swing is corrected. Ease off on the throttles, however, as soon as the bow begins to come back to starboard. Otherwise it might continue its swing and fall off to port.

Like the LCVP, the tank lighter can broach in a few seconds. Because of her greater size, the LCM is apt to be more difficult to salvage. The same precautions to be followed with the LCVP help to keep the LCM at right angles to the surf.

13.22. The Motor Lifeboat. Prudent seamanship requires that a boat always be kept ready for immediate use as a lifeboat. Aboard all men-of-war and many passenger liners a boat of the motor whaleboat type is used as the ready lifeboat. The Coast Guard, however, to some extent uses the pulling whaleboat, but since the means of stowage and launching do not normally differ from the motored version, it will not be discussed separately.

Men-of-war, if so equipped, have two of their boats rigged as ready lifeboats, one on either side to expedite lowering. Smaller men-of-war and most auxiliaries have only one boat rigged as a lifeboat. Passenger liners like large men-of-war have two small boats rigged, normally situated on either side forward of the boat deck. The sea painter is kept rigged, and it becomes only a matter of releasing the gripes before the boat can be lowered. Most Navy lifeboats are rigged for lowering by radial davits as shown in Fig. 13.11, whereas most merchant ships use gravity davits. It must be pointed out that boats of the Merchant Marine are in themselves lifeboats and, as such, are rigged for immediate use. Thus, with but few exceptions all boats aboard merchant ships may be used as ready lifeboats with no special preparations.

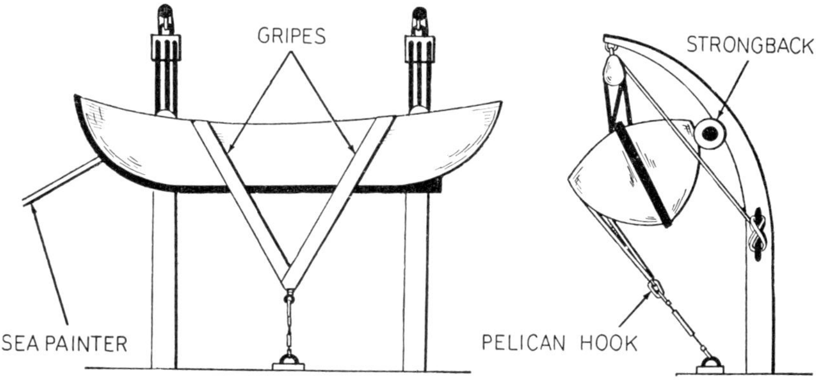

FIG. 13.11 READY LIFEBOAT

Since the gravity davit does not require that the boat be swung out to facilitate speed in launching, only the use of the radial davit will be discussed in any detail. As a general rule the davits are swung outboard and the boat is gripped against a spar spanned between the davit arms called a *strongback* or *pudding spar*. The boat is gripped up against two puddings (fenders) built up around the spar. The V-shaped grips (made of line or wire covered with canvas to prevent chafing) have their upper ends shackled to eyes in the strongback or pudding spar. The lower end of the V is attached to the deck, which is joined by means of a turnbuckle equipped with a pelican hook. The turnbuckle is used to take up any slack and keep the boat snug against the puddings. The pelican hook is used for quick release.

In order that the weight of the boat does not rest upon the falls, wire pendants called preventers are run from the davit heads to the hoisting eyes of the boat. These pendants are also equipped with pelican hooks to facilitate quick release. Aboard auxiliaries and merchant ships, the lifeboat generally rests upon chocks directly beneath the davit. To secure this boat for sea in its chocks, clamps equipped with a turnbuckle are used. The lower end is secured to the deck and the clamp is fitted over the gunwale. Several quick turns of the turnbuckle are all that is required to release the clamp and have the boat ready for instantaneous lowering.

14

Ice Seamanship

The increased commercial and military activity on and below the sea in high latitudes requires that all seamen today acquire a fundamental knowledge of ice seamanship. They should understand the basic principles of handling a ship in and near ice, the dangers involved, and how to avoid damage.

The Deep Freeze Operations of the Antarctic in connection with the Geophysical Year 1958 speak for themselves and have been highly publicized by the press. Not so highly publicized have been the replenishment of the Dew Line Operations of the Arctic.

In previous years, Arctic operations have followed a two-pronged approach with groups of ships departing in convoys from East and West Coast ports. They would steam around Alaska from the west, and Labrador from the east and into the central Canadian Arctic. In 1957 the pincers met. Three ships from the Pacific side joined an Atlantic-side icebreaker and charted the first practical deep-draft Northwest Passage. They then completed their circumnavigation of the North American continent.

The 1958 operation was conducted from Alaska's Aleutian Chain around Point Barrow eastward to the central Canadian Arctic. The threat of entrapment by the unpredictable polar ice pack closing fast against Point Barrow, the Pacific entrance to the Arctic, has been lessened considerably by the charting and development of the Northwest Passage, which provides an alternate route to the Atlantic.

The Atlantic group in 1958 consisted of five icebreakers, both Navy and Coast Guard, nine other Navy and Coast Guard ships, and twenty-two MSTS ships. Several of these are ice strengthened or designed primarily for polar operations. For the first time merchant ships were used in the Dew Line replenishment. A total of fifteen commercially operated ships took part in this operation.

On August 3, 1958, the nuclear-powered submarine *Nautilus* passed under the North pole en route from the Pacific Ocean to the Atlantic Ocean. The military implications of this voyage are clear, but the commercial applications are perhaps even more important. For example, a nuclear-powered commercial cargo or passenger submarine passing under the North Pole could reduce the normal distance between Tokyo and London from 11,200 nautical miles to 6,300 nautical miles. Similarly, the distance between all Pacific Coast and Atlantic Coast ports could be cut down by thousands of miles.

Ice seamanship has its own terminology. A few of the terms dealing with types of ice to be encountered are defined below.

Blink—A glare on the underside of extensive cloud areas created by light reflected from snow- or ice-covered surfaces; also observable in a clear sky. Blink caused by ice surfaces is usually yellowish-white in contrast to the whitish, brighter glare caused by snow surfaces. This distinction is sometimes difficult to perceive. In contrast to snowblink and iceblink, the sky is dark above bare land or open-water surfaces.

Bottom Ice—Ice formed on the bed of a river, lake, or very shallow sea irrespective of its nature or formation.

Calving—The breaking away of a mass of ice from its parent iceberg, glacier, or shelf ice formation.

Concentration (of Ice)—The ratio of the area of ice present to the total area of ice and water. Concentration is usually reported in tenths; for example, a water area may be five-tenths or six-tenths covered with ice. Descriptive terms most frequently used to describe the concentration are:

Open Water—less than one tenth ice cover.
Scattered Ice—one to five tenths ice cover.
Broken Ice—five to eight tenths ice cover.
Close Ice—eight to ten tenths ice cover.
Consolidated Ice—total ice cover revealing no sea surface.

Crack—A small, unnavigable, narrow break in sea ice that may reveal the sea-water surface. Cracks are usually caused by tides, temperature change, current, or wind.

Floe—Fragments of ice of no specific size. However, unlike the term *cake*, when floe is used with such qualifying terms as small, medium, or giant, a rather definite size is implied as follows:

Brash Ice—6 feet or the size of a pool table top.
Block Ice—6 to 30 feet or the size of a volley ball court.
Small Floe—30 to 600 feet or the size of a football field.
Medium Floe—600 to 3000 feet or the size of a long city block.
Giant Floe—3000 feet to 5 miles or the size of a small city.
Field—Ice ranging in area greater than 5 miles.

Floeberg—A mass of thick, heavily hummocked sea ice resembling an iceberg in appearance. Floebergs may be more than 50 feet in height. An iceberg in its last stages of disintegration may be mistaken for a floeberg.

Growler—A small fragment of ice that has been broken off an iceberg, generally greenish in color.

Hummock—A mound or hillock in pressure ice.

Fast Ice—All types of ice, either broken or unbroken, attached to the bottom of shoal areas, beached, stranded in shoal water, or attached to the shore. Fast ice may be classed as:

1. Ice foot.
2. Shore ice.
3. Stamuhka.
4. Bottom ice.

Heavy Ice—Any sea ice more than 10 feet thick.

Hummocked Ice—Ice piled haphazardly into mounds or hillocks. At the time of formation hummocked ice is similar to rafted ice, except that the former requires a greater degree of pressure and heaping than the latter. After hummocked ice and rafted ice have been repeatedly covered with snow and weathered, no distinction is then made between the terms, and hummocked ice is the term applied to both types.

FIG. 14.1 PACK ICE IN NORTH ATLANTIC. *U.S. Coast Guard Official Photograph*

Pack Ice (Fig. 14.1)—Any large area of floating ice driven closely together.

Pancake Ice—Pieces of newly formed ice usually between 1 and 6 feet in diameter. The raised rims and the circular appearance are a result of the almost constant rotation and collision of the cakes against one another. Small cakes up to about 18 inches in diameter are occasionally called *lily pad* ice.

Rafted Ice—A type of pressure ice formed by one cake overriding another, or rafting. Rafted ice has well-defined contours and may be regarded as recently formed.

Shore Ice—Ice that has been cast onto the shore or beached as a result of the action of wind, waves, current, tide, or the force of an adjacent ice area.

Young Ice—Newly formed ice between 2 and 8 inches thick in the transitional stage of development from crust to winter ice.

Iceberg (Fig. 14.2)—A large mass of land ice that has broken away from its parent formation on the coast and either floats in the sea or is stranded on a shoal.

Ice Crust—The next stage of development from slush. It is generally transparent, has some degree of hardness, and is frequently rubbery. New ice does not acquire hardness and strength until it has cooled to 16°F.

FIG. 14.2 ICEBERG IN THE NORTH ATLANTIC. *U.S. Coast Guard Official Photograph*

Ice Foot—Fast ice consisting of ice formed along and attached to the shore.

Lead—A long, narrow, but navigable water passage in pack ice. A lead may be covered by thin ice.

Puddle—A depression on sea ice filled with water. During the summer ice may absorb more heat than it radiates, with the result that the surface of the ice melts and forms small puddles. The water is frequently fresh enough for cooking and drinking purposes.

Slob—A dense form of sludge.

Slush—An accumulation of ice crystals which may or may not be slightly frozen together. The first sign of freezing is an oily or opaque appearance of the water caused by the formation of ice crystals.

14.1. Field Ice. Sea ice comprises nearly all forms of the ice found at sea except icebergs. Field ice has become for the seaman a general term embracing all types of sea ice except perhaps sludge and so-called local slob. More accurately, field ice is pack ice, compact or scattered. This ice has a tremendous surface area and a shallow draft and is greatly influenced by the wind and surface currents.

The ice along the edges of an ice field, particularly on the lee side, is misleading. The lee edge of the ice field is likely to be separated, whereas the windward edge is usually sharply defined and packed. The edge is often light and easily penetrated when the ice has been separated by wind, sea, and the sun. If a vessel enters an ice field, she may find that the ice will gradually become heavier and so closely packed that the speed of the ship must be reduced. Some ships return to open water before the situation becomes acute rather than run the risk of being beset or damaged. Fog is more prevalent over water partly covered with ice, which adds to the other risks of forcing a passage through an ice field. Grayish-green ice, according to whalers, is the hardest and should always be avoided. Field ice, then, should not be entered if another course is possible, especially when the opposite edge is not in sight. If it is necessary to pass through scattered ice, the utmost caution must be exercised because small pieces may be found, on contact, to be large and heavy. Sometimes in proceeding through scattered ice, a collision with the larger pieces cannot be avoided. The hull plating may withstand the shock, but there is always the danger of damaging the propeller or rudder.

There are two reliable signs of field ice. One is iceblink or ice sky. The presence of any appreciable area of field ice produces characteristic light effects in the sky above it which, once seen and recognized, can never be mistaken. The phenomenon is caused by the very high percentage of light reflected from the ice surface as compared with the surrounding sea. It is sometimes seen before the ice appears over the horizon. On clear days with sky mostly blue, iceblink is a luminous, yellow haze contrasting sharply with the rest of the horizon. It is brighter at the bottom and shades off upward. Its height depends on the proximity of the ice field. When the sky is overcast with low clouds, the yellow band is usually absent, and the iceblink appears as a whitish glare on the lower surface of the clouds above it. Both effects may occur at once under certain conditions of sun and sky. Either is a sure indication of field ice just over the horizon. The term *water sky* is just the reverse of ice sky. If the observer is completely surrounded by ice or nearly so, the bank of iceblink is sharply broken above the leads of open water. By contrast, water sky appears almost black.

The second, reliable sign of field ice is the abrupt smoothing of the sea and gradual lessening of the ordinary ocean swell. These are good indications of field ice to windward. As in the case of bergs, pieces of ice to leeward of the pack give warning of its proximity, but the absence of pieces does not mean that there is no ice near. Ships equipped with a helicopter or plane will find their scouting services invaluable.

Sea ice, other than fast ice in sheltered bays or along the coast, is continually in motion as a result of the wind, tide, and current. This motion may be the same for a time over a considerable area, but there are a number of factors tending to produce different motion of adjacent masses. Cakes, for example, vary in area and thickness so that the effect of wind and current differs on different masses of ice. Wind and current are subject to many variations: wind from the usual meteorological causes, and currents from tidal and other causes.

The swinging or turning of ice floes is due to the tendency of each cake to trim itself to the wind when the pack is sufficiently open to permit this freedom of movement. In close pack, movement may be caused by pressure from another floe. The wind may not be constant in direction because wind produces rotation as well as translation. This screwing or shearing effect results in excessive pressure at the jutting corners of floes and forms a hummock of loose ice blocks. Ice undergoing such movement is called screwing pack and is extremely dangerous to vessels.

Ice fields may open and shut like an accordion. There are always a certain number of lanes present; otherwise the ice could not move. Swell also tends to break up the ice. The vertical movement of the tide in narrow or shallow waters has the same fracturing effect. As a result of all these agencies, leads open and disappear, and the ice fields break up.

When moving floes are driven together or pressed against fast ice, bending, tenting, or rafting occurs. The amount of confusion depends on the pressure, its direction, and the composition of the ice. Definite ridges may thus be formed, the lines of which are at right angles to the force applied. Confused pressure areas of hummocky ice appear. The longer the pressure lasts, the greater the chaos. A release of pressure gives rise to lines of weakness in ice fields in the form of cracks or lanes. These are often parallel to pressure ridges, but owing to internal stresses an ice field does not necessarily crack in its thinnest part. Cracks are frequently found passing through ridges and hummocks of considerable height.

Any wind will tend to regroup ice that is more or less scattered over a considerable area. As the wind rises, the separate floes form lines in a direction at right angles to the wind direction. These chains break up when the wind changes and, after a time, realign themselves at right angles to the new wind direction. When the wind blows from the shore, a channel of open water usually forms between the coast and the ice or increases in width if already existing. On the other hand, a wind blowing on to a coast or on to fast ice tends to reduce the width of the channel previously existing. If the wind is strong enough, hummocks will be produced along a line approximately perpendicular to the wind direction.

The air temperature as well as the wind has an effect on the grouping of ice. If the wind which has regrouped ice is a cold one, the lowered temperature may cause further freezing so that the masses may freeze together in a new formation. Once frozen the ice will not be broken up so readily if the wind changes direction. On the other hand, if the weather is mild, cakes brought together by a change of wind will not freeze together.

Pack ice drifts with the wind or current or both. If the wind is strong, the ice will drift at an angle with the wind. The speed of drift may not depend entirely on the strength of the wind because the rate of drift may be affected by the presence or absence of open water in the direction of drift. Icebergs are affected by the current as well as by the wind. The current usually predominates but, if the wind is a force four or better, the berg may tack across the wind. Elkman's theory about drift of ice states that ice drifts about 45 degrees to the right of the direction toward which the wind is blowing in the northern hemisphere and the same amount to the left in the southern hemisphere. Observations reduce this angle to 30 degrees or very nearly parallel to the isobars on a weather map. The general direction of drift is with the current.

14.2. Cause of Freezing. A knowledge of the formation, growth, and decay of sea ice is desirable for an understanding of many of the problems in ice seamanship. The climatic factors bearing on the formation of ice naturally vary from place to place and from season to season.

In temperate and tropical latitudes, the ocean acts as a storehouse of heat from the sun. The visible and infrared wave lengths are largely absorbed in the surface layers, and the heat so stored is given off to the air at night and at other periods when the air is colder than the sea surface. In higher latitudes, however, as the nights begin to grow longer in the autumn, insufficient heat is stored in the short daylight period to compensate for the losses at night, and the temperature of the surface waters is therefore lowered. As the season progresses, the altitude of the sun becomes lower day by day; less radiation is received, and more is reflected from the sea surface owing to the low angle of incidence of the rays. Finally, the water reaches the freezing point and further loss of heat results in the formation of ice.

Conditions then become even less favorable for the retention of radiant heat from the sun, since ice reflects much more of the visible radiation than does water. Cooling of the air in contact with the ice is accelerated and, as this cold air spreads, more ice is formed.

14.3. Differences Between Arctic and Antarctic Ice. Differences in characteristics of the Arctic and the Antarctic are responsible for differences in the ice. An example is the low mean annual temperature of the Antarctic. The warmth of the Arctic summer has no parallel in the far South and, mainly because of this thermal difference, the ice sheets of the northern polar regions are unlike those of the southern. The margin of the Antarctic cap, overflowing its land support, is free to spread over the sea until fracture detaches huge strips, sometimes 10 to 20 miles long. In Greenland, by contrast, the edge of the inland ice ends on land, and icebergs irregular in shape are formed. The tabular or box-shaped berg is, therefore, in general, characteristic of the Antarctic; and the pinnacled, picturesque berg is typical of the North.

The Antarctic sea ice surrounds the continent, whereas the Arctic sea ice is a central mass surrounded by land. The ice moves around and outward from Antarctica and gathers in a belt formed by the meeting of southeasterly and northwesterly winds in the vicinity of the 60th parallel. There is a close similarity in the formation of this belt of ice to that formed in the Arctic in Davis Strait and eastward off Greenland. In the Antarctic it is unusual for sea ice to be more than 1 or 2 years old. The drift in both the Weddell and Ross Seas carries the pack out into the open oceans in a little more than a year.

In the Arctic, on the other hand, floes of great age are frequent. Ice formed off the Siberian coast takes from 3 to 5 years to drift across the polar basin

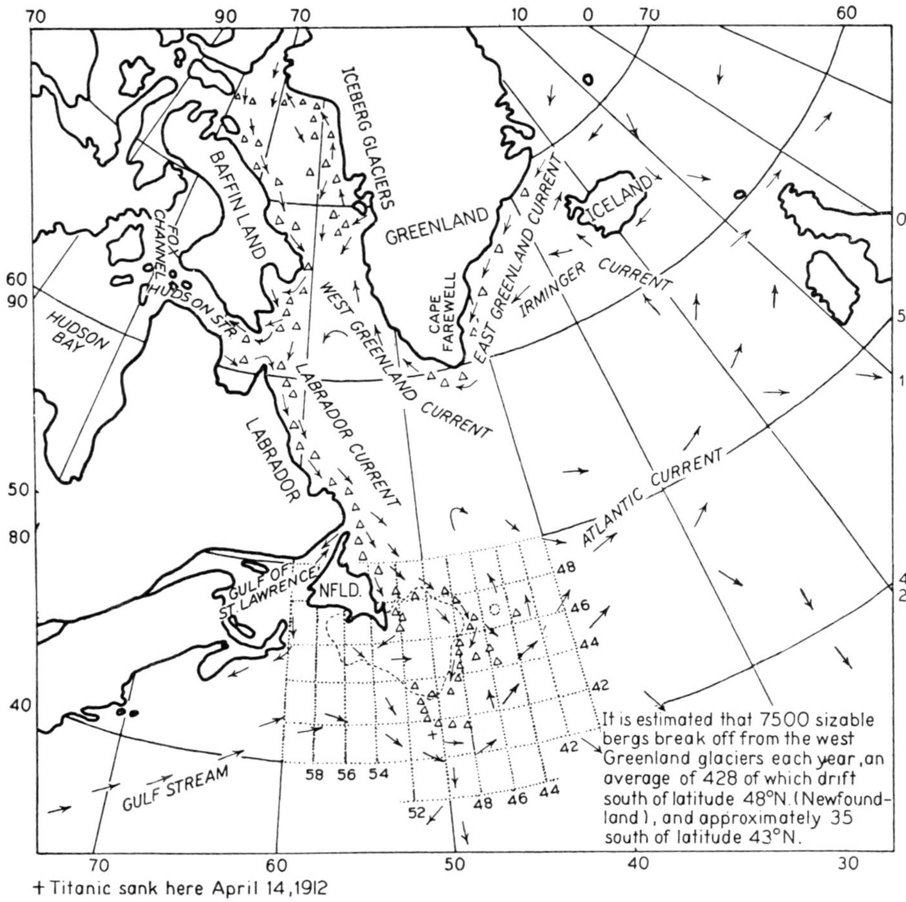

FIG. 14.3 ICEBERGS AND CURRENTS. *U.S. Coast Guard Official Photograph*

and down the eastern coast of Greenland. Ice of this age, therefore, becomes pressed and hummocked to a degree unknown in the Antarctic. The warmth of the Arctic summers also has its effect and the result is worn-down, more or less even, floes of great thickness known as "the polar cap ice." During the summer, melting on the surface is considerable, and pools of fresh water are formed on the floes. This is not a very marked feature off the east coast of Greenland, north of latitude 72° N, but in Baffin Bay the floes become covered with a maze of deep pools. In the Antarctic, surface pools on floes in the pack are almost unknown.

14.4. Passage in Ice. Modern steel merchant ships and most navy ships other than icebreakers are not suited for unassisted navigation in any but the most open of floating ice. The chief source of weakness is the bow plating, but others may be inferred from the following extract from the rules of the American Bureau of Shipping:

1. *General.* Vessels constructed with special strengthening which is at least as heavy as that described in this section will be distinguished in the text by the words "ice strengthening."

2. *Special strengthening,* for navigation in ice, should cover an area which extends from the stem to the midship three-fifths length (i.e., one fifth of the length from the stem) between lines which are, respectively, at least 3 feet above the load line and 3 feet below the light load line.

3. *Intermediate frames,* having a strength of at least 75 percent of the strength of the fore peak frames, should be fitted over this area. The intermediate frames should extend from the deck next above the strengthened area to a lower level than the top of the frame brackets of floor plate.

4. *Side plating* should be of moderate thickness forward to the strengthened area. The thickness of the shell plating over the strengthened area should not be less than 0.6 inch in vessels less than 250 feet long, and need not be more than 1 inch in vessels more than 500 feet long; the thickness for intermediate lengths may be obtained by interpolation.

5. *Rudder scantlings,* rudder stock, steering gear chains, etc., should be at least 10 percent above the ordinary requirements.

6. *Tailshafts* in single-screw vessels should have a diameter at least 5 percent and those in twin-screw vessels at least 10 percent greater than required by the rules.

7. *Propeller blades* made of cast iron or bronze should be replaced by ones made of steel.

8. *Sea connections* should be arranged so as to minimize the risk that attends their attachment to plating which is subject to ice damage. Main injections should be provided with steam connections for clearing the strainers.

Arctic-constructed ships or ships specially outfitted for extensive Arctic and Antarctic operations can normally navigate the majority of high latitude waters with icebreaker assistance. However, the proportion of icebreakers to the number of ships making such a transit is greatly increased when unstrengthened ships are used.

A few cargo ships, built with ice-breaking bows, form a class intermediate between the sea-going icebreakers and the strengthened cargo ships. Ships of the American Merchant Marine are not so equipped or constructed, since all major ports in the United States are warm water ports and heavy ice concentrations are not encountered.

14.5. Preparing Ship for Arctic, Antarctic Operations. The following check-off list to be attended to before leaving the home port for polar waters has been compiled from the combined experience of naval combatant types, naval auxiliaries, and merchant ships. Not every item, therefore, will be found applicable in individual cases, but careful consideration should be given to all.

1. *Screws.* Equip the ship with steel propellers. Provide a spare propeller for each shaft. If propellers have removable blades, see that all blades are interchangeable. Test the spare bosses in drydock to see that they fit the shaft. Provide wrenches for boss and propeller nuts. Since the propellers are especially vulnerable when operating through ice, consideration should be given to fitting the ship with some type of propeller guard. Have docking plans on board.

2. *Rudder.* Provide spare rudder assembly or temporary rudder and rudder stock. Fit steel-wire pendants from each quarter to after part of rudder blade. These will permit steering if the steering gear is damaged, and prevent losing the rudder if unshipped.

3. *Watertight integrity.* Test all bulkheads, peaks, and tanks by inspection or by air testing.

4. *Pumps and piping.* Test all sounding pipes and bilge and tank pumping pipes for leaks and fractures. Check the operating condition of main drainage pumps, electric submersible pumps and auxiliaries such as handy billies and P-500 pumps. Provide full allowance of spare parts. Renew any hose not in good condition. Clean all holds, scuppers, and bilges. After cleaning, take suction in each bilge well for an overall operating test. In loading cargo, see that no sand, sawdust, or coal dust is introduced into the holds.

5. *Fire mains.* Test all equipment, such as mains and cocks, and renew any defective ones.

6. *Cargo stowage.* For free passage of water to bilges and easy access to side plating in case of damage, stow cargo well away from the sides and tom in position. Load the ship so that she will be 3 or 4 feet down by the stern when in the ice. If in ballast, consideration should be given to the desirability of flooding the after hold most of the depth of the shaft tunnel in order to immerse the rudder and screws and minimize damage to them by ice.

7. *Underwater openings and projections.* Inspect all inlet and outlet fittings in drydock. Remove all projections, such as scupper guards and ringbolts, on the ship's side above and below the water line. These catch ice and slow down progress.

8. *Bracing in bow.* If feasible, install timber bracing in the forepeak, using horizontal "ice beams" extending from side to side at the load water line and bearing on fore-and-aft planks placed between the frames. Additional support to the fore peak bulkhead on the side toward No. 1 hold is desirable.

9. *Repair material.* Provide a supply of timbers, shores, quick-setting cement, sand, hull plates, angle irons, clamps, wedges, jacks, canvas, collision mats, etc., for the temporary repair of holes and leaks. Stow these near vital places most likely to be damaged. As outboard repairs below the water line may become necessary, consideration should be given to carrying shallow water diving equipment with accessories.

10. *Lookout stations.* Build a shelter in the eyes of the ship for forecastle ice lookouts. Rig a crow's-nest as high as possible on the mast and winterize with radiant heater and antiglare windows. Provide a protected conning station above the pilot house. Winterize the pilot house.

11. *De-icing gear.* Provide a number of hardwood or nylon-faced mallets at least 6 inches in diameter for removing ice. Scrapers can also be used, but they are more likely to remove paint with subsequent rusting.

12. *Sounding boat.* Install a portable echo sounder in a small boat for use in leading the ship into uncharted coastal waters.

13. *Mooring gear.* Provide the following:

(a) Deadmen made up of wooden planks (oak) of approximate dimension 3 x 10 inches x 6 feet. Deadmen are expended each time the ship is unmoored. It takes at least four at each mooring, and the ship may have to be shifted as often as once a day while unloading operations are being conducted.

(b) Straps made up of 6- or 8-inch manila or $\frac{7}{8}$-inch wire approximately 6 feet long with a large eye splice in each end. Straps are expendable with the deadmen and an equal number should be provided.

(c) Toggles of hardwood similar to a 4-inch mallet head with trailing lines. Each mooring line is secured to a manila strap with a toggle. Normally the toggle will be recovered, but a good surplus should be on hand to provide for losses when freezing makes it necessary to cut lines on unmooring.

(d) Mooring lines of size normally used. There should be no losses, but one has occasionally to cut off the eye if a toggle is frozen and cannot be withdrawn.

(e) Picks, shovels, and buckets with lanyards attached to be used by line handling party in burying the deadmen.

(f) A number of long wooden spars or telegraph poles for use as fenders and for construction of heavy temporary brows while alongside the ice.

The above gear is used for mooring to shelf ice or to bay ice in Antarctic operations as well as for general Arctic service. Also provide ice anchors, which are stockless single-fluked hooks, and ice axes, which have longer handles than conventional axes. Both the deadmen and the ice anchors have their place in mooring depending upon sea and wind conditions and the type of ice encountered.

14. *Main injections.* Install steam lines on intakes to prevent clogging with brash ice.

15. *Electrical equipment.* Add at least 25 percent to the allowance of 1.835 specific gravity storage battery acid normally carried. Thermally insulate below-water-line engine room bulkheads behind and above the main switchboards to eliminate condensation with subsequent water dripping on exposed elements of the board.

16. *Gas bottles.* Provide inside stowage for acetylene, oxygen, and other gas bottles in "stand-by," since if used directly from outside stowage in cold weather up to 75 cubic feet volume is lost.

17. *Small stores, ship's stores, or slop chest.* Provide ample supplies of warm clothing, footgear, smoked glasses, face lotion, and chap sticks. Allow 25 to 50 percent increase over normal consumption of food and candy. If tropics are to be crossed, arrange for cool stowage of candy bars.

18. *Recreational facilities.* Provide an adequate ship's library, recent motion pictures, and hobby shop equipment.

19. *Personnel.* Screen men before sailing if practicable and eliminate misfits. Carry a medical officer and dental officer if feasible. Require men wearing glasses to equip themselves with a spare pair before sailing.

20. *Miscellaneous supplies.* Provide poles and extra boat hooks for fending blocks of ice away from the ship's side and the screws in particular. Provide crowbars, which should be short, four sided, and wooden handled. Provide demolition charges, detonators, and fuses or cable and blasting machine. Allow for the possibility of being forced to winter in the ice by loading provisions for all hands for 15 months, and by providing sporting rifles, shot guns, ammunition, fishing gear, and vitamin supplies. Provide ice saws for cutting docks in ice floes, or freeing the ship if frozen in.

14.6. Preparations En Route to Polar Regions. The following check-off list covers items that should be completed before entering ice zones.

1. *Painting and lubrication.* Paint topsides and decks; grease the rigging with a light coating; put winter-grade lubricants in all machinery.

2. *Antifreeze.* Put ethylene glycol or alcohol in the cooling systems of motor boats and any other exposed internal-combustion engines.

3. *Batteries.* Fill all storage batteries in boats with 1.280 specific gravity electrolyte. Keep batteries as near full charge as is possible at all times.

4. *Water tanks.* See that no water tanks are more than 90 percent full. Owing to the risk of contamination with sea water from leaks caused by contact with ice, use potable water (if any) in the fore peak tank first. All water tanks adjacent to the outer skin of the ship should be equipped with heating coils.

5. *Towing gear.* Rig towing bridle forward for immediate use in the event of necessity of being towed by the icebreaker. Break out towing gear and keep it available on the fantail for possible use in towing another ship.

6. *Mooring lines.* Manila has a tendency to freeze or dry-rot in the center if exposed to cold for long periods. A line permitted to drag through snow and water becomes ice-coated immediately and is hard to handle, slippery in gloved hands and on winch drums and capstans. Therefore, keep mooring lines dry by stowing below decks or under canvas covers.

7. *Instructions.* Familiarize personnel with the operation of the main drainage system. Hold regular instruction periods on safety precautions in handling

cargo, life and personnel hygiene in frigid climates, survival, etc. If in the Arctic, indoctrinate personnel with regard to contacts with natives and compliance with game laws.

14.7. Shipboard Precautions. If the ship has been prepared for cold-weather operations, the chief dangers then are those resulting from low air temperatures and from water freezing on the topsides, either through sleeting and snowing or from taking aboard spray or green water in cold weather.

Keep decks swept clear of snow before it has an opportunity to form a crust or become trampled and hardened. This is important on the bridge and on the ladders. Scrapers should be used with care when removing ice close to electric cables and equipment. Salt-water hosing is a rapid means of melting snow and clearing decks but should be used only in nonfreezing weather after making sure that the overboard deck drains are not frozen. It is not advisable to use mixed steam and water for ship de-icing unless an abundance of steam is available. A better method is to use the condenser feed as a source of warm water and by a heat exchanger raise the temperature about 150°F. The water may then be used to cut into ice, making use of the weight of the ice to break it away.

Running rigging should be protected with canvas covers. Lowering a whaleboat with ice on the falls and cleats is dangerous. Canvas covers are a necessity for deck winches and similar appliances. They are also essential for open boats if the bilges are to be kept dry. Cover the deck space used for helicopter operations with a tarpaulin so that snow can be removed in a minimum of time.

Firemain cutout valves on firemain risers to weather decks should be secured and plugs drained at the lowest point between riser and plug. Fire hoses on weather decks should be drained and dried in a heated compartment before restowing in racks. Proportioners should be kept in a heated compartment adjacent to a hatch or door where access to weather decks will permit rapid connection to fire plugs. They should be drained after use, and dismantled, dried, oiled, and the chamber change-over valves reassembled.

In using fire hoses at freezing temperatures, satisfactory results can be obtained if good pressure is maintained; however, when the pressure is reduced or the hose is secured, the nozzle and plug become frozen.

Steam deck machinery should be carefully maintained to be operated daily, and condensed water drained out after use. If circumstances require that steam capstans and winches be ready for instant action in freezing weather, warm them up and leave them turning over at a slow speed.

14.8. Anchoring. It may be advantageous to lie at anchor when in brash, but as little of the cable as possible should be paid out. The anchor windlass should be kept ready for use in the approach of large masses of pack ice. When anchoring in rotten ice in shoal water, get into the ice as far as possible to avoid the swell. But if the water is deep and ice is present, anchoring should be avoided. It may be preferable to lie to and keep power available to move the ship as the shifting floes require. It is not advisable to anchor while in pack ice, as in most cases it is useless and will probably result in the loss of the anchor and cable.

Having decided to ride to an ice anchor, choose a strong floe which can shelter the vessel from surrounding ice. To ensure as nearly as possible obtaining the shelter of a natural dock, it would be well, in making fast to a floe, to take a position where a bight is formed by two strong projections. Such places can often be found. They offer at least moderate security in the event of other ice setting out toward the ship, the projecting angles of the floes receiving the first shock.

Lay the anchor from the side of the floe where a patch of open water is formed or where the surrounding ice is least packed. When riding to an anchor the movement of the ice must be continually observed. If there is a risk of the ice surrounding the ship, weigh anchor and move into a more open region off another floe. Keep the engines ready for use on the ship, ice can frequently be avoided and permitted to drift clear by judicious use of the engines while at anchor.

In selecting an anchorage in a bay or harbor which is open to drifting ice, the shallowest depths should be chosen provided other conditions are suitable. A vessel should not select an anchorage too close to a glacier cliff, since calving of the barrier may endanger the vessel or set up heavy swells making the position uncomfortable.

In bays or fiords where fast ice exists, the tidal currents may cause this ice to drift in and out of the harbor, rendering the anchorage unsafe. Fast ice in a harbor usually moves along a tidal crack and, under the force of onshore winds, may acquire violent motion. Vessels should quit moorings at the edge of fast ice whenever onshore winds blow.

14.9. Mooring and Unloading. Although ice conditions in the Antarctic are seldom the same from one year to the next, the general condition of the fast ice in the Ross Sea changes very little, particularly in regard to offering a clear "dock space" for mooring alongside.

The thickness of the fast ice in the Bay of Whales during the months of January and February was found to be approximately 12 to 15 feet, with a height of 3 to 4 feet and with sufficient strength to hold the weight of the equipment unloaded.

Breakups occur without warning, and ships moored to the ice edge must be prepared to get underway on short notice. Sometimes cracks will develop between the ship and the barrier, but the ice may not break up for several days. Prevailing winds, and currents coming from under the barrier, tend to cause the broken pack ice to drift to the westward. With this condition a starboard-side to mooring has been found to be the most desirable.

Prior to arrival alongside the ice, all gear should be put on deck in order and line handlers instructed as to how and where to bury deadmen and how to secure mooring lines. Secure manila strap and/or wire strap to each deadman, depending on the use of the hawsers and cables. Obviously if a deadman has both a manila strap and a wire strap, either wire or manila lines can be secured to it. At least four mooring lines should be ready to run with a toggle attached to the eye of each line.

FIG. 14.4 *U.S.S. NESPELEN* AOG-55 WITH BOW AGAINST ICE SHELF. *Official U.S. Navy Photograph*

It is normal practice to place the bow of the ship head-on against the ice and to hold this position by steaming ahead slowly (see Fig. 14.4). Line handling parties can then be disembarked onto the ice via Jacob's ladders. After passing over and securing the bowline and bow breast, the ship is warped around until she lays alongside the ice. Then the stern lines are put out and secured (Fig. 14.5). This procedure of placing the bow against the ice may not be considered advisable for large vessels. If the commanding officer prefers, the line handling party may be sent to the ice by boat and the ship held off until the deadmen are planted and all preparations made to receive the mooring lines. Then the ship can be brought alongside in the normal manner of making fast to a pier.

Plant deadmen (Fig. 14.6) well in on the ice shelf so that an almost horizontal pull will be made on the mooring lines when hauling the ship alongside. A trench for a deadman should be dug about 4 to 6 feet deep with sides at a slight angle as shown in Fig. 14.6 in order to give better holding power and to avoid the tendency to pull the deadman out before it is well frozen

FIG. 14.5 *U.S.S. NESPELEN* (AOG-55) MOORED ALONGSIDE ICE SHELF. *Official U.S. Navy Photograph*

into place. The deadman with the manila strap attached is buried in the hole and covered over with ice. A few buckets of water thrown on top of the fill will help freeze it in place in a few minutes. The mooring line is passed through the eye of the strap that protrudes up through its own part for quick release. Check toggles frequently to see that they are free for easy slipping. If wet snow or sleet falls, they may become frozen in place.

Four mooring lines distributed as shown in Fig. 14.6 are recommended. The number should be kept to a minimum to keep the ship safely secured and also to facilitate a hurried unmooring to clear the area during breakup. It has been found that many breakups occur at night when a limited number of men are available to slip the lines. Telegraph poles 12 to 16 feet long, hung vertically over the side of the ship, make the best fenders. There is usually some ground swell in the Bay of Whales which will cause a vessel to work up and down. Cane fenders have a tendency to ice up and may catch on the edges of the shelf ice.

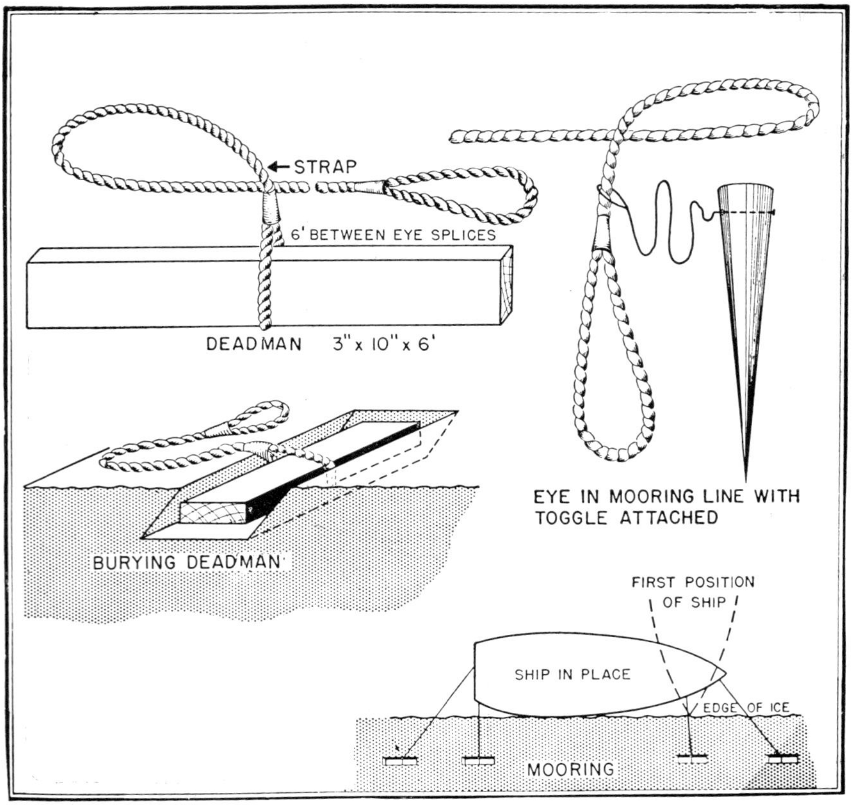

FIG. 14.6 MOORING TO ICE SHELF

FIG. 14.7 UNLOADING ON ICE BARRIER. *Official U.S. Navy Photograph*

The use of ice anchors in mooring alongside the Antarctic shelf is generally not recommended. The surface of the ice is too soft to provide adequate holding power. Mooring to timber with strap and toggle requires less manpower, makes weighing much easier and quicker, and eliminates the possibility of losing an expensive metal anchor. Figure 14.7 illustrates the height of barrier ice in comparison to the size of a ship.

14.10. Precautions. 1. To facilitate unloading, moor ship as close to the ice as possible.

2. In unloading heavy equipment, land it as far inboard on the ice as booms and cranes will permit so as to avoid having heavy weights on the edge of the ice near the ship.

3. Skidding of heavy weights from the ship to the ice is not recommended unless shelf ice conditions appear to be exceptionally good and no crevasses are observed between the ship and the barrier. When skidding is necessary, heavy cribbing made up of telegraph poles should be used to distribute the weight as far inboard on the ice as possible.

4. The ship must be kept ready for unmooring and getting under way. A quick breakup may call for that action. If more than one ship is tied up in the same vicinity, the situation may be more complicated, and no time should be wasted in getting clear.

5. Unless coming alongside to moor, a vessel should not steam too close to the barrier. Bergs frequently calve without warning.

6. Men should not be permitted to wander over the ice away from the ship until a careful check has been made for crevasses.

7. Material should be unloaded only as fast as it can be moved inland. Every effort must be made to reload material onto the ship if a breakup occurs.

In the Arctic, unloading cargo over the ice is considered practical over landfast or landlocked ice, but not over ice in open areas. Fast, smooth ice would present no difficulties for trucks or tractors provided it was thick enough to support the weight. Rough, hummocky ice would be more difficult but could probably be traversed by careful selection of route and use of a small bulldozer. In the event the ice is covered with soft snow and there are large amounts of cargo to be landed, a metal landing strip mat would serve very well to make a smooth roadway to the beach. In landlocked areas during the winter months, natural slips for unloading can easily be cut out by an icebreaker, thus entirely eliminating the problem of mooring lines or deadmen. It is recommended that landlocked areas be selected and checked by aerial observation before planning large-scale over-ice movements.

14.11. Single Ship in Ice. In general, ice is an obstacle to the progress of any vessel and is dangerous to vessels which, by their construction, were not intended for ice navigation. Nevertheless, it is possible for ordinary ships to navigate through open pack. The long periods of summer daylight in high latitudes greatly facilitate such operations, and the ability to see obstacles contributes markedly to the ease of shiphandling.

When a vessel encounters ice lying on her course, a decision must be made whether to attempt to penetrate the ice or to steam around it. If the boundaries of the ice are in sight, do not enter, but skirt to the windward. In the case of larger ice areas, unless they fill straits through which the vessel must pass or completely block access to her port of destination, time and fuel will be saved by taking the longer way around the ice zone.

When conditions make it necessary to enter the ice, the point of entry should be selected with great care. Make a thorough reconnaissance, using radar and helicopter (if so equipped), put an experienced ice pilot in the crow's-nest, and search for water sky. The following principles govern choice of the place of entry:

1. Consider the penetrability of the ice along the proposed course inside the edge of the ice field, with regard both to the thickness and the degree of consolidation.

2. Never enter ice where pressure exists as evidenced by tenting or rafting.

3. If possible, enter the ice upwind. The windward edge of an ice field is more compact than the leeward edge. If it is necessary to enter downwind, use great care to avoid damage to the hull of the vessel through collision with the ice cakes.

4. If the ice is thick and drifting rapidly, wait for a change in the direction of the ice movement which may be accompanied by an improvement in ice conditions. Take into account the time of ebb and flood; ice generally becomes more compact on the flood but begins to break up on the ebb.

5. The ice edge is usually not straight, but often has projecting tongues between bights. Enter at such a bight, for here the surge will be least.

6. Enter at the slowest possible speed to reduce the force of the initial impact on the stem. Once the bow is in the ice and is cutting or pushing ice aside, increase power to avoid losing headway and adjust revolutions thereafter in accordance with the state of the ice.

7. Always enter the ice on a course perpendicular to its edge. Failure to observe this precaution may result in a glancing blow which will likely damage the bow plating on the side toward the ice, and may swing the stern into the ice with resulting damage to rudder and propeller.

Some guiding principles of working in pack ice are:

1. Keep moving.
2. Work with the ice, not against it.
3. Do not rush the work.
4. Respect the ice; do not fear it.
5. Stay in open water or leads.
6. Watch the propeller(s).
7. Never hit a large piece of ice if you can go around it; if you must hit it, hit it head-on.

14.12. Convoying in Ice. An ice convoy consists of one or more ordinary slips, whether or not strengthened for ice navigation, accompanied by one or more icebreakers. Either the icebreakers or the escorted vessels may be naval ships.

There are three possible types of ice convoy:

1. Single ship convoy.
2. Simple convoy: one icebreaker escorting a group of ships.
3. Composite convoy: two or more icebreakers escorting several ships.

A simple convoy consists of several transports or other vessels and one leading icebreaker. The captain of the leading icebreaker decides upon the number

FIG. 14.8 ICEBREAKER MAKING A CHANNEL

of vessels he can take through. His decision depends on the type of ships which are to follow the icebreaker and the condition of the ice en route. If the ships to be convoyed are reinforced for ice navigation and have sufficiently powerful engines, an icebreaker can take an average of four of them through an ice coverage of 70 to 80 percent. If the conditions are more favorable, only 50 to 60 percent ice, the number of ships can be increased. If there is close pack with over 80 percent coverage, the number of ships must be limited to one or two. Figure 14.8 shows an icebreaker making a channel for ships to follow.

The arrangement of the convoy should be carefully worked out. The varying ice conditions in the area along the route and the variety of ships forming the convoy must be taken into consideration. The first factor to be considered is the power of the ships. The weakest, as a rule, are placed immediately after the icebreakers, so that they can avoid striking ice obstacles and be able to move in a comparatively clear channel. The most powerful and beamiest ships are placed in the convoy so that less powerful vessels can proceed in their wake. Consideration must also be given to whether a ship is loaded or in ballast. Finally, it is essential that one of the most powerful ships in the convoy be placed in the last position.

A composite convoy consists of two or three simple convoys. The number of ships to each icebreaker and their place in the column is determined in the same way as for a simple convoy. The difficulty of controlling from a position

FIG. 14.9 TWO ICEBREAKERS KEEPING A CHANNEL OPEN. *Official U.S. Navy Photograph*

in front is an important drawback to this type of convoy, which frequently stretches out over a distance of 1½ to 2 miles. The first icebreaker is designated the leader; the others are placed according to orders of the leader's captain, either in column or in line of bearing for breaking out.

The operating procedure is for the most powerful icebreaker to lead the convoy, breaking a channel in the ice without stopping to break out other ships. Following the leader at a distance decided upon by the leader's captain are two or three ships, the weakest in the entire convoy. The second icebreaker proceeds astern of the first group, followed by two or three ships, and so on. Figure 14.9 shows two icebreakers working in unison to keep a channel open for ships to proceed through and to moor. The assignment of the second icebreaker is to break out the ships ahead of her so that the leader will not have to return to them and thus delay the convoy. The second icebreaker, on receiving a signal "stuck" from any of the proceeding ships, increases speed, leaves the column, and breaks out the ship. When the latter is freed and moving, the icebreaker resumes her previous station in the column. The same action is taken by the second icebreaker upon hearing the same signal from one of the ships astern, provided there are no more icebreakers in the convoy. If there is a third icebreaker, she breaks out the ships following the second icebreaker. Ships must be broken out while proceeding, in order not to delay the progress of the entire convoy.

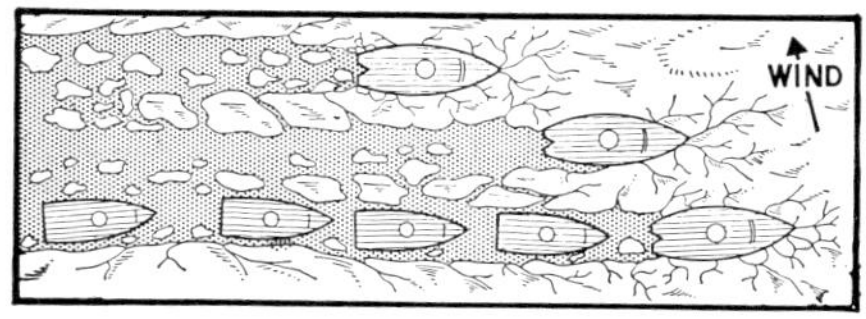

FIG. 14.10 LINE OF BEARING FOR BREAKING OUT

When several icebreakers are present in line of bearing for breaking out, they follow behind the leader at a set distance to leeward (Fig. 14.10) in such a way as to thin out the ice in the channel made by the leader, and remain always in readiness for breaking out or towing any ship that gets stuck or lags behind.

A final caution should be observed by all mariners in polar and subpolar waters. Radar, while invaluable in detecting icebergs, cannot distinguish brash or block ice in rough seas. The recent loss of a Danish steamer in the North Atlantic with all hands illustrates all too tragically the axiom that relatively small ice which cannot be detected in heavy weather by radar is a seaman's greatest danger in very cold oceans.

Part III

RULES OF THE ROAD

15

Rules of the Road—Principles and Application

15.1. History. The need for uniform Rules of the Road arose when steam vessels with higher speeds than sailing vessels appeared on the trade routes of the world. The modern Rules date from 1863 in which year Great Britain and France adopted uniform regulations for the prevention of collisions at sea.

The Congress of the United States approved a law on April 29, 1864, containing Rules similar to those already adopted by Great Britain and France. About 1885, Belgium, Denmark, Germany, Japan, and Norway accepted the same Rules as Great Britain, France, and the United States.

The President of the United States called in 1889 a Conference of all Maritime Nations to draw up Rules and Regulations for the safety of lives and property at sea. The Conference approved the International Rules of the Road, 1889, which have been accepted and used by all maritime nations until the International Rules of 1948 were adopted and put into effect.

In 1948 a Conference was held in London which approved changes in the 1889 Rules, which clarified and simplified them. More important, the 1948 Rules included the seaplane among seagoing vessels. The 1948 Rules are known as the International Regulations for Preventing Collisions at Sea, 1948. They were adopted by the United States by Public Law 172, 82nd Congress, approved October 11, 1951, and became effective January 1, 1954.

The most recent International Conference was held in the spring of 1960. It is therefore probable that further changes in the International Rules will occur before many years have elapsed.

The Inland Rules of the Road were approved by Congress on June 7, 1897, and made effective October 7, 1897. There have been some later additions and amendments.

The line of demarkation between the waters under the International and the Inland Rules was originally set forth by the Secretary of the Treasury under the Act of February 19, 1895. Some changes and additions have been made since that time. The Commandant of the U.S. Coast Guard is now charged with the determination of this boundary line which is shown on all charts.

The boundary line for seaplanes on the water is not the same as that for vessels. Public Law 172 states that the International Rules of the Road "shall not apply . . . with respect to aircraft, to any of the territorial waters of the United States." The Air Traffic Rules apply to seaplanes on the water in United States territorial waters.

The Statutory Rules for the Great Lakes and Western Rivers were enacted by Congress on *February 8, 1895,* and *May 21, 1948,* respectively.

The Pilot Rules (3) for Inland Waters, for the Great Lakes, and for the Western Rivers were originally issued by the Board of Supervising Inspectors for Steam Vessels, Department of Commerce, under the Acts of June 7, 1897, February 8, 1895, and R.S. 4412, respectively. The authority to issue Pilot Rules has now been transferred to the Commandant of the Coast Guard by Executive Order 9083 of February 28, 1942, effective March 1, 1942, and reaffirmed by subsequent Reorganization Plans. The three Pilot Rules supplement the Inland, Great Lakes and Western Rivers Rules, but they do not supersede those statutory Rules. The Pilot Rules have the force of statute, but they are not "judicially noticed." They must be proved in court. Courts can and have ruled certain Pilot Rules void because they conflicted with the statutory Rules.

15.2. Various Rules of the Road. All public and private vessels of the United States are required by law, regulation, and court decision to comply strictly with the following "Rules" for preventing collisions and related laws and regulations, namely:

1. International Rules of the Road, agreed to internationally and enacted by Congress. These Rules are applicable on the high seas, outboard of the boundary line, and in certain foreign waters.
2. Inland Rules of the Road enacted by Congress and used on Inland waters of the U.S., inboard of the boundary line, except on the Great Lakes and Western Rivers.
3. The Pilot Rules for Inland Waters prescribed by the Commandant of the U.S. Coast Guard for the above Inland Waters.
4. The Great Lakes Rules enacted by Congress for use on the Great Lakes and connecting and tributary waters as far east as Montreal.
5. The Pilot Rules for the Great Lakes issued by the Commandant of the Coast Guard.
6. The Western Rivers Rules, enacted by Congress and effective on the Red River of the North, Mississippi River, and waters tributary to the latter.
7. The Pilot Rules for Western Rivers, by the Commandant of the Coast Guard.
8. General Regulations of the Corps of Engineers, Department of the Army, applicable to the Great Lakes and Western Rivers.
9. The Motor Boat Act of 1940, applicable to every vessel propelled in whole or in part by machinery and not more than 65 feet in length,

except tugboats and towboats propelled by steam. It is effective in all U.S. waters.

10. Miscellaneous U.S. Statutes, i.e., Helm Order Act; Stand-by Act; Wreck Act; Death on High Seas Act; Naval Lights Act; Anchorage Act of March 3, 1899.
11. Panama Canal Rules.
12. Hawser Rules for tows in Inland Waters, prescribed by Commandant of the Coast Guard.
13. Anchorage Regulations issued by the Corps of Engineers, Department of the Army.
14. Statutes of the several states of the United States.
15. Decisions of U.S. Courts.
16. Customs accepted by the Courts.
17. Foreign Inland Rules in those foreign waters.

15.3. Basic Principles of International and Inland Rules. The International Regulations for the Prevention of Collisions at Sea, 1948, and the Inland Rules are based on the following principles, as interpreted by U.S. Courts:

a. Vessels affected.
b. Location.
c. When effective.
d. Mandatory.
e. Maneuverability.
f. Burdened and privileged vessels.
g. Obedience in time.
h. Shifting responsibility.
i. Assumptions.
j. Rule 27 (General Prudential Rule).
k. Customs.
l. Ferries, fire boats, vessels docking and undocking, getting under way.
m. Jurisdiction.
n. Legal personality.
o. Court decisions.
p. Unequal fault, equal responsibility.

15.4. Vessels Affected by the Rules. The law states that the Rules "shall be followed by all public and private vessels of the United States, and by all aircraft of United States registry. . . ." Hence the Rules must be obeyed by naval and merchant vessels, fishing vessels, ferries, tugs with tows, vessels being towed, yachts, power-driven and sailing vessels, pilot vessels, dredges, seaplanes on the water, in fact, by every vessel afloat.

15.5. Location. All of these vessels must comply with the Rules which apply to the place where the vessel is operating. The International Rules are effective on the high seas, beyond the boundary line separating the high seas from U.S. and foreign Inland waters. The Inland Rules and Inland Pilot Rules must be obeyed in Inland waters, except on the Great Lakes and Western Rivers. Seaplanes must obey the International Rules outside of territorial waters and the Air Traffic Rules within these waters.

15.6. When the Rules Apply. The exact time when the Rules must be applied is difficult to define. Mr. Justice Clifford, in the N.Y. AND LIVERPOOL CO. *v.* RUMBALL, said, "Rules of navigation . . . are obligatory upon vessels approaching each other, from the time the necessity for precaution begins and continue to be applicable as the vessels advance, so long as the means and opportunity to avoid danger remain. They . . . are equally inapplicable to vessels of every description, while they are yet so distant from each other that measures of precaution have not become necessary to avoid a collision." A British judge, Lord Esher, in the BANSHEE, said, "They [the Rules] only apply at a time when, if either of them [vessels] does anything contrary to the Regulations, it will cause danger of collision. None of the Regulations apply until that time has arrived."

15.7. Risk of Collision. The International Rules, 1948, and the Inland Rules prescribe certain "Steering and Sailing Rules" which must be followed when two vessels are approaching each other ". . . so as to involve risk of collision. . . ." Justice Longyear, in the MILWAUKEE, said, "Risk of collision begins the very moment when the two vessels have approached so near each other and upon such courses that by a departure from the rules of navigation . . . , a collision might be brought about."

Part C, Steering and Sailing Rules, International Rules, 1948, and Part IV, Inland Rules, state: "Risk of collision can, when circumstances permit, be ascertained by carefully watching the compass bearing of an approaching vessel. If the bearing does not appreciably change, such risk should be deemed to exist." This caution might be extended to the distance where the two approaching vessels are so far apart that either one or both could change course or speed without affecting the other vessel. It is reasonable to state that risk of collision is involved when "the bearing does not appreciably change" and when the approaching vessels are so close that the movements of one does affect the other and when "by a departure from the rules of navigation . . . a collision might be brought about."

15.8. Mandatory. The Rules are mandatory and are not optional. There is no choice of action until the collision is so imminent that both vessels must take action.

Justice Fuller, in BELDEN *v.* CHASE, stated that the Statutes and the Pilot Rules "are not merely prudential regulations, but binding enactments. . . . Obviously, they must be rigidly enforced, in order to attain the object for which they were framed, which could not be secured if the masters of vessels were permitted to indulge their discretion in respect to obeying or departing from them."

15.9. Ability to Maneuver. Both Rules are based, *in part,* on the premise that certain vessels are unable to maneuver as quickly and as easily as other types of vessels. The more maneuverable is required, therefore, to keep clear of the less maneuverable. An example is the rule which requires power-driven vessels to keep out of the way of sailing vessels except when the sailing vessel overtakes the power-driven vessel. There is a Rule for each case.

15.10. Burdened and Privileged Vessels. The terms *Burdened* and *Privileged* vessels are often used to indicate the vessel which must give way and the vessel which must hold its course and speed when there is risk of collision. The terms are expressive but misleading because each vessel is required by the Rules to act in a certain manner. For example: "Where two power-driven vessels are crossing, so as to involve risk of collision, the vessel which has the other on her own starboard side shall keep out of the way of the other." (Int. Rule 19.) "Where by any of these Rules, one of the two vessels is to keep out of the way, the other shall keep her course and speed. . . ." (Inland Article 21.)

15.11. Obedience in Time. It is important that the Rules be obeyed in time to avoid the immediate risk of collision, to give the other vessel an opportunity to understand the situation, and to take proper action. "The Rules must be obeyed when the vessels are far enough apart to adopt these maneuvers deliberately and safely." (The TRANSFER No. 10.)

15.12. Shifting Responsibility. The Courts have ruled, in many cases, that, when two vessels are approaching each other so as to involve risk of collision, the original responsibilities under the law cannot be changed by the subsequent movements of either vessel until the collision is so imminent that both must take appropriate action under Rule 27 or until the risk of collision exists no longer. In other words, no subsequent change in bearing or distance, after risk of collision is involved, will alter the fact that one of the vessels must keep clear and the other hold her course and speed. Of course, the time may come when there is no right of way and, therefore, each vessel must take measures to avoid the collision which is imminent. When the risk of collision exists no longer, the Rules apply no longer and a different situation may then arise. This principle is the reason that the old "port the helm" rule is bad seamanship and bad sea manners.

15.13. Assumptions. In order that a collision may be avoided after risk of collision exists, it is necessary that the movements of one vessel—the privileged one—must be known so that the other vessel—the burdened one—may change her course and speed, if necessary, to keep clear. A vessel has the right to assume that the other vessel will obey the Rules of the Road, will be navigated with care and attention, will keep to its own side of a channel, etc. However, the assumption does not hold when it is evident that the other vessel is not being navigated with care and attention. For example, a privileged vessel in a crossing situation should not hold its course and speed until collision results. Justice Longyear, in the MILWAUKEE, said, "It is true, prima facie, each has the right to assume that the other will obey the law. But this does not justify either in shutting his eyes to what the other may actually do, or in omitting to do what he can to avoid an accident, made imminent by the acts of the other."

15.14. Rule 27—General Prudential Rule. This Rule does not apply in every case where it suits the convenience of the vessel to use it. On the contrary, the other Steering and Sailing Rules should be strictly adhered to, in most cases. But the object of the Rules is to prevent and not cause collisions. There have been and will be many different situations. In order to cover *all* situations, this Rule (27) was added. It should be obeyed where collision is imminent and where both vessels must take further action to avoid collision. Until this point is reached, the other Steering and Sailing Rules should be applied.

15.15. Customs. Customs are not judicially noticed. They must be proved in each case; they must not conflict with the statutes and regulations; and they must be reasonable in view of a particular, permanent, local condition. Some customs have been accepted by the Courts, whereas some have not. "Moreover, each vessel has the right to assume that the other will conform to the requirements of an established usage, and must govern her own conduct accordingly"—Justice Wallace in the ALASKA.

15.16. Ferries, Fire Boats, Vessels, Docking, Undocking and Getting Underway. Ferries have a right to "make" and leave their slips. While they are doing so, if they are not on a steady course, the meeting, crossing, or overtaking rules do not apply and Rule 27 applies. Justice Ward, in the JOHN RUGGE, said, "The Steering and Sailing Rules apply to vessels navigating on steady courses. Where one of them is maneuvering merely, as, for instance, to get into or out of a dock, or, as in this case, winding around to get on her course, the situation is one of special circumstances, under Article 27 of the Inland Regulations which requires each vessel to act prudently."

In a like manner, a vessel has a right to dock or undock. Other vessels should not pass so close to the end of a pier as to restrict the movements of such vessels, entering or leaving a slip or dock. The regular Steering and Sailing Rules apply to such vessels if they are on a steady course, i.e., one where their future position can be ascertained. The Special Circumstances Rule (27) applies when they are not on a steady course.

A vessel getting underway is within her rights, too. Approaching vessels must apply Rule 27 until the former vessel has turned to a steady course. The meeting, crossing, or overtaking rules do not apply until that time.

A vessel backing out into a river or channel from a dock and turning preparatory to standing out must be allowed a reasonable amount of room to turn. Other approaching vessels should apply Rule 27 and not meeting, crossing, or overtaking rules.

A fire boat is subject to the Rules of the Road, but other vessels should recognize the urgency of her errands and make some allowance.

In all of these cases, the vessels concerned are subject to the Rules of the Road, and the only question is which Rule to apply. If the ferry entering or leaving her slip or the vessel getting underway, docking, or undocking is not on a steady course, Rule 27 applies. If they have steadied on one course, the other Steering and Sailing Rules apply, i.e., meeting, crossing, or overtaking.

15.17. Jurisdiction. When collision between vessels occurs in navigable waters used in interstate commerce, the case can be heard in the Federal Courts sitting as Courts in Admiralty. The District Courts are the trial courts, the Circuit Courts hear appeals, and the Supreme Court may hear the final appeal. If the collision occurs on waters wholly within a state, the state courts have sole jurisdiction. If the collision occurred on waters in or between two states which empty into the sea, there is concurrent jurisdiction. Either the Federal or state court may hear the case.

When a vessel runs into a pier or bridge through negligence or bad seamanship, the Admiralty Courts have no jurisdiction. On the other hand, if a bridge or draw damages a vessel due to improper construction or operation, the case can be heard by an Admiralty Court.

15.18. Legal Personality. A merchant ship is liable *in rem* (against a thing) for the faults of her master, officers, or men (perhaps a lookout) operating the ship. She can be sued in an Admiralty Court, attached, and sold to satisfy a judgment. The owners are not otherwise liable in such suits unless they have contributed to the fault by neglect, privity, or knowledge.

The Sovereign cannot be sued because one of her vessels, i.e., a naval vessel, has collided with a merchant vessel unless she has granted such privilege. Congress has approved such proceedings, but a naval vessel cannot be seized and thus placed out of commission or sold to satisfy a judgment.

15.19. Court Decisions. The Federal Courts in Admiralty determine the legal meaning of certain words in the Rules, such as "moderate speed," and interpret phrases such as "whose engines are going at full speed astern. . . ." (Inland Rule, Art. 28.)

15.20. Unequal Fault, Equal Responsibility. U.S. courts have held that where both vessels are in fault in a collision the liability of each vessel is one half of the total loss. However, if the fault of one is great and that of the other is minor, nonstatutory, and noncontributory, the courts may not inquire fully into the minor fault or they may disregard it and order the full costs to be paid by the vessel with the major fault.

15.21. Subdivision of International Rules. Read the International and Inland Rules in the following three chapters. It will be noted that the International Rules are divided into: Part A, Preliminary and Definitions; Part B, Lights and Shapes; Part C, Steering and Sailing Rules; and Part D, Miscellaneous. This arrangement differs from the 1889 Rules, in that Part B includes all the lights, shapes, and fog signals by which an approaching vessel is recognized and her position determined. The Steering and Sailing Rules, Part C, now contain only the Rules for maneuvering vessels to prevent collision.

A new Part D has been made which is truly miscellaneous. It includes: Rule 28, Whistle Signals; Rule 29, Seamanship Rule; Rule 30, Special Rules by Local Authority; Rule 31, Distress Signals; and Rule 32, Orders to Helmsmen.

15.22. Subdivision of the Inland Rules. The Inland Rules have not been changed. They also contain four parts: Part I, Enacting Clause, Scope, Penalty and Preliminary Definitions; Part II, Lights and so forth; Part III, Sound Signals for Fog, and so forth; and Part IV, Steering and Sailing Rules.

15.23. Structure of the International Rules. The student should notice that the International Rules contain two important parts, if the preliminary definitions and the miscellaneous Rules are not so considered. The first of these parts—Part B, "Lights and Shapes"—describes the lights, daytime shapes, and fog signals which various kinds of vessels show, carry, and sound in order that other vessels may recognize and locate them. The second part—Part C, Steering and Sailing Rules—prescribe the maneuvers which vessels shall make to avoid collision, having recognized and located the other vessel by her appearance, lights, shapes, or fog signals. These two parts, B and C, are essential parts of the Rules.

The next point to note is that vessels are divided generally into those at anchor, those under way, and those unable to maneuver easily and quickly. Vessels at anchor show anchor lights or, by day, an anchor ball. Vessels under way show (or flash, for sailing pilot vessels) side lights. Those unable to maneuver show special lights at night, usually red or white in a vertical line and shapes by day. The lights required on seaplanes are similar to those prescribed for other vessels.

To distinguish at night between power-driven and sailing vessels under way, the former carry a 20-point white light above the side lights. If the vessel is 150 feet or more in length, she must show a second 20-point white light, either forward and below or abaft and above the first (masthead) white light. Exceptions are power-driven vessels less than 40 tons and small, power-driven boats that are not required to carry the second white light (range light).

The range light is now a mandatory, instead of optional, requirement in the International Rules. It differs in character from the Inland Rules because it is a 20-point light, whereas the Inland Rules require an all-around white light for nonseagoing vessels.

Sailing vessels and any vessel being towed do not carry masthead or range lights.

15.24. Structure of the Inland Rules. The Inland Rules are assembled in the same general manner as the International Rules. The lights and shapes to aid recognition are found in Part II, but fog signals are placed in a separate Part III, although they are sounds used to aid recognition. Part IV contains the Steering and Sailing Rules and, in addition, miscellaneous Rules, i.e., whistle signals, seamanship, naval lights, distress signals, and orders to helmsmen. In other words, the subdivision of the Inland Rules is not as consistent and reasonable as the International Rules.

However, the Inland Rules divide vessels into three general classes in the same manner as the International Rules, i.e.: (a) vessels at anchor, (b) those under way, and (c) those unable to maneuver. The classes are distinguished in the same general manner, although the details vary. Seaplanes are not mentioned in the Inland Rules—a serious defect. The Rules for Aircraft on the water in U.S. waters will be found in the Air Traffic Rules of the Civil Aeronautics Board. Some of these Rules are quoted under the Inland Rules in the next three chapters.

Power-driven (called "steam" in the Inland Rules) vessels and sailing vessels are distinguished in the same manner as in the International Rules.

The differences and similarities between the two Rules will be pointed out in the following chapters.

15.25. Pertinent Statutes. *Anchorage Act*—A vessel has a right to anchor. The only question is: "Where can she anchor legally?" Certain water areas are designated by the Secretary of the Army as anchorages. (U.S. Act of March 4, 1915.) It is improper, although not unlawful, to anchor outside of these areas in restricted waters, save in an emergency. (The RICHMOND, 63 F 1020.) A vessel has a right to anchor in large, navigable water areas such as Chesapeake By or on soundings off the Coast. (LE LION, 84 F 1011.) The Anchorage Act of March 3, 1899, states: "It shall not be lawful to tie up or anchor vessels or other craft in navigable channels in such a manner as to prevent or obstruct the passage of other vessels or craft. . . ." The trend of judicial opinion has been to permit anchorage in a channel, "but to forbid them [vessels] from doing so in such a manner as to obstruct said channels or render their navigation difficult or dangerous." (JOB H. JACKSON, 144 F 896.) Channels are primarily for traffic and not for anchorage. It is sometimes necessary to anchor in a narrow channel—in a thick fog—but the anchorage should be shifted to an authorized anchorage ground as soon as possible. While the vessel remains in the narrow channel, every precaution to prevent collision should be taken such as: (a) use a short scope of chain, (b) keep power on the anchor engine, (c) station a watch in the engine room and keep steam up to the throttle, (d) station a watch on the bridge, (e) post extra lookouts.

Of course, if the channel is well lighted, marked, and wide and there is plenty of room for other vessels to pass, a vessel may remain at anchor there without violation of the statute. An example is the Hudson River off 96th Street, New York. Even so, she should anchor to one side of the channel, and be particularly careful that: (a) her anchor lights are showing, (b) a trained lookout is posted, (c) proper signals are sounded in a fog, and (d) the anchor ball is hoisted by day.

Attention is called to Article 11 of the Inland Rules which permits the Secretary of the Army to designate "special anchorage areas" where vessels not more than 65 feet in length may anchor and "not be required to carry or exhibit an anchor light." Such areas are usually near yacht clubs or marinas and are intended primarily for pleasure craft, secured for the night.

After the vessel has anchored in accordance with her legal rights, she must exhibit the proper lights at night, sound correct fog signals when necessary, and post a watch on deck. Sufficient trained lookouts must be stationed to inform the officer in charge of the ship of the approach of other vessels or objects which might require action.

An anchored vessel is presumed to be without fault if a moving vessel collides with her unless the moving vessel can prove that the collision "was the result of inevitable accident" (VIRGINIA EHRMAN, 97 U.S. 309) or that the anchored vessel was in fault (OTTO MARMET COAL CO. *v.* FIEGER CO., 259 F 435). The moving vessel must proceed lawfully in moving through an anchorage ground. Tides and current do not excuse her (STROUT *v.* FOSTER, 42 U.S. 89).

The anchored vessel may be found in fault if she has failed to anchor in a legal position, show the proper lights, or sound the proper fog signals. She may also be found in fault if she: (a) anchors without necessity in a narrow channel; (b) swings to the tide and obstructs the channel substantially; (c) anchors on a frequented compass course when a safer anchorage is available; (d) fails to move from a channel when possible; (e) fails to move when warned that her position is dangerous; (f) anchors where approaching vessels rounding a point sight her suddenly and belatedly; (g) fails to veer chain or use her helm when such action might prevent a collision, the moving vessel having done all in her power to avoid a collision; (h) anchors so close to another anchored vessel as to foul her when swinging (JUNIATA, 124 F 861); (i) fails to shift anchorage when dragging dangerously close to another anchored vessel. The vessel which anchored first should warn the one who anchored last that the berth chosen will foul the former's berth.

Vessels which drag should take every measure possible to stop dragging, such as dropping a second anchor (MARY FRASER, 26 F 872). Anchored vessels are not required to keep steam up unless weather, ice, or good seamanship demands it. An anchored vessel is not required to take unusual precautions to avoid collisions with moving vessels.

A moored vessel alongside a wharf may be found in fault if she projects into the channel and thereby obstructs it and contributes to a collision. A slight projection may not cause her to be found at fault if passing vessels can pass her safely using reasonable care. If the moored vessel unnecessarily obstructs the slip in which she is secured so that other vessels using the slip cannot evade collision by exercising reasonable care, she may be found at fault. A moored vessel alongside a wharf or slip is not required to show anchor lights, to sound fog signals, or to post a lookout. She must, however, house her anchors in the usual, secure manner so that passing vessels will not be damaged by them. In some areas she may be required by custom or local regulation to show a white light at each projection or extremity, and normally should of her own accord in congested waters.

Finally, anchorage may be regulated by state or local authorities, provided such ordinances or laws do not conflict with Federal statutes and maritime law as interpreted by U.S. courts.

Stand-by Act—Congress passed the "Stand-by Act" on September 4, 1890. This statute states, in part: "That in every case of collision between two vessels it shall be the duty of the master or person in charge of each vessel, if and so far as he can do so without serious danger to his own vessel, crew and passengers, to stay by the other vessel until he has ascertained that she has no need of further assistance, and to render to the other vessel, her master, crew and passengers such assistance as may be practicable and as may be necessary in order to save them from any danger caused by the collision, and also to give to the master or person in charge of the other vessel the name of his own vessel and her port of registry, or the port or place to which she belongs, and also the name of the ports and places from which and to which she is bound. . . ."

Reports of Collisions—U.S. vessels which have sustained or caused any accident involving loss of life, material loss of property, or any serious injury to any person, or have received any material damage affecting their seaworthiness or efficiency, are required to report the accident within five days or as soon thereafter as possible to the Commander of the Coast Guard District where the vessel belongs or where the accident took place. (Act June 20, 1874, as amended by Reorganization Plan 3.)

Helm Order Act—The Act of Congress of August 21, 1935, has been embodied in the Inland Rules (see Article 32 following). The International Rules, 1948, have changed the meaning of Article 32 of the International Rules, 1889. The New Rule will be found under Rule 32 in the third chapter following.

Motor Boat Act of 1940—The Act of April 25, 1940, was a revision of the Act of June 9, 1910. It applies to every vessel in U.S. waters propelled in whole or in part by machinery and not more than 65 feet in length, except tugboats and towboats propelled by steam. The Act will be quoted in a following chapter. The requirements of the Act are supplemented by Regulations issued by the Commandant of the Coast Guard, which may be found in 46 CFR Part 25.

Log Books—The Act of February 27, 1877, as revised, requires that: "Every vessel making voyages from a port in the United States to any foreign port, or, being of the burden of seventy-five tons or upward, from a port on the Atlantic to a port on the Pacific or vice versa, shall have an official log book; and every master of such vessel shall make or cause to be made therein, entries of the following matters, that is to say:

"First. . . .

"Twelfth. In every case of collision in which it is practicable to do so, the master shall, immediately after the occurrence, cause a statement thereof, and of the circumstances under which the same occurred to be entered in the official log book."

Wreck Act, March 3, 1899—The Act empowers the Secretary of the Army to remove any sunken boat, water craft, raft, or other similar obstruction, which has existed for a period longer than thirty days, from any river, lake, harbor, sound, bay, canal, or other navigable waters of the United States.

Death on the High Seas Act, March 30, 1920—The Act permits the personal representative of a decedent to maintain a suit for damages in the District Courts of the United States, in Admiralty, for the exclusive benefit of the decedent's wife, husband, parent, child, or dependent relative against the vessel, person, or corporation which is liable for the wrongful act, neglect, or default resulting in the death of the decedent.

Naval Vessels Lights Act—Section 2 of the Act of October 11, 1951, prescribing the new International Rules of the Road, 1948, and the Acts of March 5, 1948, and December 3, 1945, provide that the several Rules of the Road for International and Inland waters, the Great Lakes, and the Western Rivers shall not apply to any vessel of the Navy or of the Coast Guard, insofar as they pertain to lights, where the Secretary of the Navy, Secretary of the Treasury, or such official or officials as either may designate, shall find or certify that, by reason of special construction, it is not possible with respect to such vessels or class of vessels to comply with the statutory provisions as to the number, position, range of visibility, or arc of visibility of lights. The lights of any such exempt vessel or class of vessel shall, however, comply as closely to the requirements of the statutes as feasible. A notice of such finding or modification and of the character and position of the lights displayed must be published in the "Notice to Mariners."

Additional Whistle Signals—Attention is invited to Rule 28(c) of the International Rules, which provides: "Nothing in these Rules shall interfere with the operation of any special rules made by the government of any nation with respect to the use of additional whistle signals between ships of war or vessels sailing under convoy."

This paragraph does not permit a warship or vessel in a convoy to stop the sounding of fog signals in a fog, although war vessels often do, particularly in time of war. It merely legalizes the maneuvering whistle signals which have been used by naval vessels in formation to indicate or call attention to certain maneuvers.

Sea Manners—The expression is understood by seamen to mean a consideration for the other vessel and the exercise of common sense under certain conditions when vessels meet. A tug with a tow is difficult to maneuver. A large ship is more difficult to maneuver than a smaller one. A convoy or a formation of naval vessels is more difficult to maneuver than a single ship. All of these vessels are required to obey the Rules of the Road. No vessel is exempt. If a vessel disobeys the Rules, she is liable. Accordingly, seamen are not advised to disobey the Rules of the Road to show sea manners, but to obey them. The Rules of the Road apply when there is a risk of collision. Before that moment, there is enough time and plenty of opportunity for a single vessel to avoid a tug with a tow, a convoy, or a formation of naval ships. Small vessels can keep clear of large ones.

16

Lights and Shapes of Identification, General Rules (Articles) 1-6

INTERNATIONAL RULES

ENACTING CLAUSE, SCOPE, EXEMPTIONS, ETC.

Be it enacted by the Senate and House of Representatives of the United States of America in Congress assembled, That the President is authorized to proclaim the regulations set forth in section 6 of this Act for preventing collisions involving water-borne craft upon the high seas, and in all waters connected therewith. Such proclamation, together with the regulations, shall be published in the Federal Register, and, after the effective date specified in such proclamation, such regulations shall have effect as if enacted by statute and shall be followed by all public and private vessels of the United States, and by all aircraft of United States registry to the extent therein made applicable. Such regulations shall not apply to the harbars, rivers, and inland waters of the United States; to the Great Lakes of North America and their connecting and tributary waters as far east as the lower exit of the Lachine Canal in Montreal in the Province of Quebec, Canada; to the Red River of the North and the rivers emptying into the Gulf of Mexico and their tributaries; nor, with respect to air craft, to any territorial waters of the United States.

INLAND RULES

I. ENACTING CLAUSE, SCOPE, AND PENALTY

Whereas the provisions of chapter eight hundred and two of the laws of eighteen hundred and ninety, and the amendments thereto, adopting regulations for preventing collisions at sea [this act was replaced January 1, 1954, by the international rules of left-hand column], apply to all waters of the United States connected with the high seas navigable by seagoing vessels, except so far as the navigation of any harbor, river, or inland waters is regulated by special rules duly made by local authority; and

Whereas it is desirable that the regulations relating to the navigation of all harbors, rivers, and inland waters of the United States, except the Great Lakes and their connecting and tributary waters as far east as Montreal and the Red River of the North and rivers emptying into the Gulf of Mexico and their tributaries, shall be stated in one act: Therefore,

Be it enacted by the Senate and House of Representatives of the United States of America in Congress assembled, That the following regulations for preventing collisions shall be followed by all vessels upon the harbors, rivers, and other inland waters of the United States, except the Great Lakes and their connect-

INTERNATIONAL RULES | ***INLAND RULES***

ing and tributary waters as far east as Montreal, and the waters of the Mississippi River between its source and the Huey P. Long Bridge and all of its tributaries emptying thereinto and their tributaries, and that part of the Atchafalaya River above its junction with the Plaquemine-Morgan City alternate waterway, and the Red River of the North; and are hereby declared special rules duly made by local authority.

SEC. 2. Any requirements of such regulations in respect of the number, position, range of visibility, or arc of visibility of the lights required to be displayed by vessels shall not apply to any vessel of the Navy or of the Coast Guard whenever the Secretary of the Navy or the Secretary of the Treasury, in the case of Coast Guard vessels operating under the Treasury Department, or such official as either may designate, shall find or certify that, by reason of special construction, it is not possible for such vessel or class of vessels to comply with such regulations. The lights of any such exempted vessel or class of vessels, however, shall conform as closely to the requirements of the applicable regulations as the Secretary or such official shall find or certify to be feasible. Notice of such findings or certification and of the character and position of the lights prescribed to be displayed on such exempted vessel or class of vessels shall be published in the Federal Register and in the Notice to Mariners and, after the effective date specified in such notice, shall have effect as part of such regulations.

SEC. 3. Section 7(a) of the Air Commerce Act of 1926 (U.S.C., 1946 edition, title 49, sec. 177(a), is amended to read as follows:

"Except as specifically provided in the Act entitled 'An Act to authorize the President to proclaim regulations

SEC. 2. (*a*) *The Secretary of the Department in which the Coast Guard is operating (i.e., Commandant of) shall establish such rules to be observed, on the waters described in section 1 of this Act, by steam vessels in passing each other and as to the lights and day signals to be carried on such waters by ferryboats, by vessels and craft of all types when in tow of steam vessels or operating by hand power or horsepower or drifting with the current, and by any other vessels not otherwise provided for, not inconsistent with the provisions of this Act, as he from time to time may deem necessary for safety, which rules are hereby declared special rules duly made by local authority.*

(*b*) *Except in an emergency, before any rules or any alteration, amendment, or repeal thereof are established by the Secretary under the provisions of this section, the said Secretary shall publish the proposed rules, alterations, amendments, or repeals, and public hearings shall be held with respect thereto on such notice as the Secretary deems reasonable under the circumstances.*

SEC. 3. *Every licensed and unlicensed pilot, engineer, mate, or master of any vessel who violates the provisions of this Act or the regulations established pursuant hereto shall be liable to a penalty of not exceeding $500, and for all damages*

INTERNATIONAL RULES

for preventing collisions at sea,' the navigation and shipping laws of the United States, including any definition of 'vessel' or 'vehicle' found therein and including the rules for the prevention of collisions, shall not be construed to apply to seaplanes or other aircraft or to the navigation of vessels in relation to seaplanes or other aircraft."

SEC. 4. Section 610(a) of the Civil Aeronautics Act of 1938 (U.S.C., 1946 edition, title 49, sec. 560(a)), is amended by deleting the word "and" at the end of paragraph (4); by changing the period at the end of paragraph (5) to a semicolon and adding the word "and"; and by adding a new paragraph (6) reading as follows:

"(6) For any person to operate a seaplane or other aircraft of United States registry upon the high seas in contravention of the regulations proclaimed by the President pursuant to section 1 of the Act entitled 'An Act to authorize the President to proclaim regulations for preventing collisions at sea.'"

SEC. 5. After such regulations proclaimed under section 1 hereof shall have taken effect, all statutes, regulations, and rules in conflict therewith shall be of no further force and effect. Until such time as such regulations shall have been proclaimed and made effective pursuant to this Act, nothing herein shall in any way limit, supersede, or repeal any regulations for the prevention of collisions, which have heretofore been prescribed by statute, regulation, or rule.

SEC. 6. The regulations authorized to be proclaimed under section 1 hereof are the Regulations for Preventing Collisions at Sea, 1948, approved by the International Conference on Safety of Life at Sea, 1948, held at London from April 23 to June 10, 1948, as follows:

INLAND RULES

sustained by any passenger, in his person or baggage, as a result of such violation: Provided, *That nothing herein shall relieve any vessel, owner, or corporation from any liability incurred by such violation.*

SEC. 4. *Every vessel which is navigated in violation of any of the provisions of this Act or the regulations established pursuant hereto shall be liable to a penalty of $500, one-half to go to the informer, for which sum such vessel may be seized and proceeded against by action in any district court of the United States having jurisdiction of the offense.*

INTERNATIONAL RULES

INLAND RULES

Rule 1

(a) These Rules shall be followed by all vessels and seaplanes upon the high seas and in all waters connected therewith navigable by seagoing vessels, except as provided in Rule 30. Where, as a result of their special construction, it is not possible for seaplanes to comply fully with the provisions of Rules specifying the carrying of lights and shapes, these provisions shall be followed as closely as circumstances permit.

Rule 30

Reservation of Rules for Harbours and Inland Navigation

Nothing in these Rules shall interfere with the operation of a special rule duly made by local authority relative to the navigation of any harbour, river, lake, or inland water, including a reserved seaplane area.

AIR TRAFFIC RULES

60.1 Scope.—The air traffic rules in this part shall apply to aircraft operated anywhere in the United States, including the several States, the District of Columbia, and the several Territories and possessions of the United States, including the territorial waters and the overlying airspace thereof, except:

(a) Military aircraft of the United States armed forces when appropriate military authority determines that noncompliance with this part is required and prior notice thereof is given to the Administrator, and

(b) Aircraft engaged in special flight operations, requiring deviation from this part, which are conducted in accordance with the terms and conditions of a certificate of waiver issued by the Administrator.

PILOT RULES FOR INLAND WATERS

80.01 General instructions.—The regulations in this part apply to vessels navigating the harbors, rivers, and inland waters of the United States, except the Great Lakes and their connecting and tributary waters as far east as Montreal, the Red River of the North, the Mississippi River and its tributaries above Huey P. Long Bridge, and that part of the Atchafalaya River above its junction with the Plaquemine-Morgan City alternate waterway.

COMMENT: The boundary line between the high seas and Inland waters for vessels is prescribed by the Commandant of the U.S. Coast Guard.

The boundary line for seaplanes is the line which separates U.S. territorial waters from the high seas. See Sec. 1, International Rules, above.

The two boundary lines are not the same.

INTERNATIONAL RULES

Rule 1(b)

The Rules concerning lights shall be complied with in all weathers from sunset to sunrise, and during such times no other lights shall be exhibited, except such lights as cannot be mistaken for the prescribed lights or impair their visibility or distinctive character, or interfere with the keeping of a proper lookout.

INLAND RULES

ART. 1. *The rules concerning lights shall be complied with in all weathers from sunset to sunrise, and during such time no other lights which may be mistaken for the prescribed lights shall be exhibited.*

AIR TRAFFIC RULES

60.23 Aircraft lights.—Between sunset and sunrise:

(a) All aircraft in flight or operated on the ground or under way on the water shall display position lights,

(b) All aircraft parked or moved within or in dangerous proximity to that portion of any airport used for, or available to, night-flight operations shall be clearly illuminated or lighted, unless the aircraft are parked or moved in an area marked with obstruction lights,

(c) All aircraft at anchor shall display anchor lights, unless in an area within which lights are not required for vessels at anchor, and

(d) Within . . . Alaska the lights required in paragraphs (a), (b), and (c) of this section shall be displayed during those hours specified and published by the Administrator.

PILOT RULES

80.14 Lights; time for.—The following rules in this part concerning lights shall be complied with in all weathers from sunset to sunrise.

80.24(c) All floodlights or headlights which may interfere with the proper navigation of an approaching vessel shall be so shielded that the lights will not blind the pilot of such vessel.

80.34 Rule relating to the use of searchlights or other blinding lights.—Flashing the rays of a searchlight or other blinding light onto the bridge or into the pilothouse of any vessel under way is prohibited. Any person who shall flash or cause to be flashed the rays of a blinding light in violation of the above may be proceeded against in accordance with the provisions of R.S. 4450, as amended, looking to the revocation or suspension of his license or certificate.

80.36 Rule Prohibiting the carrying of unauthorized lights on vessels.—Any master, or pilot of any vessel who shall authorize or permit the carrying of any light, electric or otherwise, not required by law, that in any way will interfere with distinguishing the signal lights, may be proceeded against in accordance with the provisions of R.S. 4450, as amended, looking to a suspension or revocation of his license.

COMMENT: Navigation, identification, and position lights should be shown from sunset to sunrise and not from darkness to daylight.

INTERNATIONAL RULES	*INLAND RULES*
Rule 1(c)	
In the following Rules, except where the context otherwise requires:—	
(i) the word "vessel" includes every description of water craft, other than a seaplane on the water, used or capable of being used as a means of transportation on water;	*Not in Inland Rules.*
(ii) the word "seaplane" includes a flying boat and any other aircraft designed to manoeuvre on the water;	*Not in Inland Rules.*
(iii) the term "power-driven vessel" means any vessel propelled by machinery;	***Preliminary Definitions*** *The words "steam vessel" shall include any vessel propelled by machinery.*
(iv) every power-driven vessel which is under sail and not under power is to be considered a sailing vessel, and every vessel under power, whether under sail or not, is to be considered a power-driven vessel;	*In the following rules every steam vessel which is under sail and not under steam is to be considered a sailing vessel, and every vessel under steam, whether under sail or not, is to be considered a steam vessel.*
(v) a vessel or seaplane on the water is "under way" when she is not at anchor, or made fast to the shore, or aground;	*A vessel is "under way," within the meaning of these rules, when she is not at anchor, or made fast to the shore, or aground.*
(vi) the term "height above the hull" means height above the uppermost continuous deck;	*Not in Inland Rules.*
(vii) the length and breadth of a vessel shall be deemed to be the length and breadth appearing in her certificate of registry;	ART. 11. *The length of a vessel shall be deemed to be the length appearing in her certificate of registry.*
(viii) the length and span of a seaplane shall be its maximum length and span as shown in its certificate of airworthiness, or as determined by measurement in the absence of such certificate;	*Not in Inland Rules.*
(ix) the word "visible," when applied to lights, means visible on a dark night with a clear atmosphere;	*II. LIGHTS, AND SO FORTH* *The word "visible" in these rules, when applied to lights, shall mean visible on a dark night with a clear atmosphere.*

INTERNATIONAL RULES	*INLAND RULES*
(x) the term "short blast" means a blast of about one second's duration;	*Not in Inland Rules.*
(xi) the term "prolonged blast" means a blast of from four to six seconds' duration;	ART. 15. *The words "prolonged blast" used in this article shall mean a blast of from four to six seconds' duration.*
(xii) the word "whistle" means whistle or siren;	*Not in Inland Rules.*
(xiii) the word "tons" means gross tons.	*Not in Inland Rules.*

AIR TRAFFIC RULES

60.60 Definitions.—As used in this part, terms shall be defined as follows:

Aircraft.—Any contrivance used or designed for navigation of or flight in the air, except a parachute or other contrivance designed for such navigation but used primarily as safety equipment.

PILOT RULES

80.02 Definition of steam vessel and vessel under way; risk of collision.—In the rules in this part the words "steam vessel" shall include any vessel propelled by machinery. A vessel is under way, within the meaning of the rules in this part, when she is not at anchor, or made fast to the shore, or aground. Risk of collision can, when circumstances permit, be ascertained by carefully watching the compass bearing of an approaching vessel. If the bearing does not appreciably change, such risk should be deemed to exist.

80.24 Lights generally.—(a) All the lights required by §§ 80.18 to 80.23, inclusive, except as provided in §§ 80.18(b) and 80.21(b), shall be of such character as to be visible on a dark night with a clear atmosphere for a distance of at least two miles.

(b) The lights required by § 80.18(b) to be of the same character as the regular towing lights and the lights required by § 80.21(b) to be of the same character as the masthead light shall be of such character as to be visible on a dark night with a clear atmosphere for a distance of at least five miles.

SIGNALS

80.03 Signals.—The whistle signals provided in the rules in this part shall be sounded on an efficient whistle or siren sounded by steam or by some substitute for steam.

A short blast of the whistle shall mean a blast of about one second's duration.

A prolonged blast of the whistle shall mean a blast of from 4 to 6 seconds' duration.

COMMENT: A vessel is under way, among other conditions, when it is not secured to a buoy or when its anchor is aweigh. She is also under way when one anchor has been dropped "underfoot" to assist in maneuvering. (CITY OF ERIE, 250 F 259.) A vessel is liable if she anchors so suddenly in the path of a following vessel that collision cannot be avoided. (SOSUA, 271 F 772.) "A sailing vessel lying to in a fog, but having some of her sails up, is under way." (BURROWS *v.* GOWER, 119 F 616.)

A steam vessel (power-driven), motionless in the water, is under way. (NIMROD, 173 F 520.) A vessel drifting is also under way. (NYC, NO 18, 230 F 299.) A vessel in tow is also under way. (JOAQUIN MUMBRU—SCANDINAVIA, 11 F 2d, 542.)

Note that the definition of a vessel, under International Rule 1(c) (1) above, excepts the "seaplane on the water" and that, under International Rule 18(b), following, a seaplane on the water is deemed to be a vessel for the particular purposes of Rules 18-29 inclusive, i.e., for meeting, crossing, overtaking and narrow channel cases.

Notice, also, that "whistle" means whistle or siren (12). This definition will affect blowing fog signals, as given in Rule 15(b), and changes of course (see Rule 28). Naval vessels generally use the whistle for fog signals and changes of course.

INTERNATIONAL RULES

Rule 2 (Fig. 16.1)

(a) A power-driven vessel when under way shall carry:

(i) On or in front of the foremast, or if a vessel without a foremast then in the forepart of the vessel, a bright white light so constructed as to show an unbroken light over an arc of the horizon of 20 points of the compass (225 degrees), so fixed as to show the light 10 points (112½ degrees) on each side of the vessel, that is, from right ahead to 2 points (22½ degrees) abaft the beam on either side, and of such a character as to be visible at a distance of at least 5 miles.

(ii) Either forward of or abaft the white light mentioned in sub-section (i) a second white light similar in construction and character to that light. Vessels of less than 150 feet in length, and vessels engaged in towing, shall not be required to carry this second white light but may do so.

INLAND RULES

ART. 2. *A steam vessel when under way shall carry—(a) On or in the front of the foremast, or if a vessel without a foremast, then in the fore part of the vessel, a bright white light so constructed as to show an unbroken light over an arc of the horizon of twenty points of the compass, so fixed as to throw the light ten points on each side of the vessel, namely, from right ahead to two points abaft the beam on either side, and of such a character as to be visible at a distance of at least five miles.*

(*e*) *A seagoing steam vessel when under way may carry an additional white light similar in construction to the light mentioned in subdivision* (*a*). *These two lights shall be so placed in line with the keel that one shall be at least fifteen feet higher than the other, and in such a position with reference to each other that the*

INTERNATIONAL RULES

(iii) These two white lights shall be so placed in a line with and over the keel that one shall be at least 15 feet higher than the other and in such a position that the lower light shall be forward of the upper one. The horizontal distance between the two white lights shall be at least three times the vertical distance. The lower of these two white lights or, if only one is carried, then that light, shall be placed at a height above the hull of not less than 20 feet, and, if the breadth of the vessel exceeds 20 feet, then at a height above the hull not less than such breadth, so however that the light need not be placed at a greater height above the hull than 40 feet. In all circumstances the light or lights, as the case may be, shall be so placed as to be clear of and above all other lights and obstructing superstructures.

(iv) On the starboard side a green light so constructed as to show an unbroken light over an arc of the horizon of 10 points of the compass ($112\frac{1}{2}$ degrees), so fixed as to show the light from right ahead to 2 points ($22\frac{1}{2}$ degrees) abaft the beam on

INLAND RULES

lower light shall be forward of the upper one. The vertical distance between these lights shall be less than the horizontal distance.

(f) All steam vessels (except seagoing vessels and ferryboats), shall carry in addition to green and red lights required by article two (b), (c), and screens as required by article two (d), a central range of two white lights; the after light being carried at an elevation at least fifteen feet above the light at the head of the vessel. The headlight shall be so constructed as to show an unbroken light through twenty points of the compass, namely, from right ahead to two points abaft the beam on either side of the vessel, and the after light so as to show all around the horizon.

(b) On the starboard side a green light so constructed as to show an unbroken light over an arc of the horizon of ten points of the compass, so fixed as to throw the light from right ahead to two points abaft the beam on the starboard side, and of

FIG. 16.1 STEAMER WITH RANGE LIGHTS

INTERNATIONAL RULES

the starboard side, and of such a character as to be visible at a distance of at least 2 miles.

(v) On the port side a red light so constructed as to show an unbroken light over an arc of the horizon of 10 points of the compass (112½ degrees), so fixed as to show the light from right ahead to 2 points (22½ degrees) abaft the beam on the port side, and of such a character as to be visible at a distance of at least 2 miles.

(vi) The said green and red sidelights shall be fitted with inboard screens projecting at least 3 feet forward from the light, so as to prevent these lights from being seen across the bows.

(b) A seaplane under way on the water shall carry:

(i) In the forepart amidships where it can best be seen a bright white light, so constructed as to show an unbroken light over an arc of the horizon of 220 degrees of the compass, so fixed as to show the light 110 degrees on each side of the seaplane, namely, from right ahead to 20 degrees abaft the beam on either side, and of such a character as to be visible at a distance of at least 3 miles.

(ii) On the right or starboard wing tip a green light, so constructed as to show an unbroken light over an arc of the horizon of 110 degrees of the compass, so fixed as to show the light from right ahead to 20 degrees abaft the beam on the starboard side, and of such a character as to be visible at a distance of at least 2 miles.

(iii) On the left or port wing tip a red light, so constructed as to show an unbroken light over an arc of the horizon of 110 degrees of the compass, so fixed as to show the light from right ahead to 20 degrees abaft the beam on the port side, and of such a character as to be visible at a distance of at least 2 miles.

INLAND RULES

such a character as to be visible at a distance of at least two miles.

(c) On the port side a red light so constructed as to show an unbroken light over an arc of the horizon of ten points of the compass, so fixed as to throw the light from right ahead to two points abaft the beam on the port side, and of such a character as to be visible at a distance of at least two miles.

(d) The said green and red side lights shall be fitted with inboard screens projecting at least three feet forward from the light, so as to prevent these lights from being seen across the bow.

Not in Inland Rules.

CIVIL AIR REGULATIONS

3.700(b) Forward position lights.—Forward position lights shall consist of a red and a green light spaced laterally as far apart as practicable and installed forward on the airplane in such a location that, with the airplane in normal flying position, the red light is displayed on the left side and the green light is displayed on the right side. The individual lights shall be of an approved type.

3.701 Position light system dihedral angles.—The forward and rear position lights as installed on the airplane shall show unbroken light within dihedral angles specified in paragraphs (a) through (c) of this section.

(a) Dihedral angle L (left) shall be considered formed by two intersecting vertical planes, one parallel to the longitudinal axis of the airplane and the other at 110 degrees to the left of the first, when looking forward along the longitudinal axis.

(b) Dihedral angle R (right) shall be considered formed by two intersecting vertical planes, one parallel to the longitudinal axis of the airplane and the other at 110 degrees to the right of the first, when looking forward along the longitudinal axis.

PILOT RULES

80.15 Ferryboats.—(a) Ferryboats propelled by machinery and navigating the harbors, rivers, and other inland waters of the United States, except the Great Lakes and their connecting and tributary waters as far east as Montreal, the Red River of the North, the Mississippi River and its tributaries above Huey P. Long Bridge, and that part of the Atchafalaya River above its junction with the Plaquemine-Morgan City alternate waterway, shall carry the range lights and the colored side lights required by law to be carried on steam vessels navigating those waters, except that double-end ferryboats shall carry a central range of clear, bright, white lights, showing all around the horizon, placed at equal altitudes forward and aft, also on the starboard side a green light, and on the port side a red light, of such a character as to be visible on a dark night with a clear atmosphere at a distance of at least 2 miles, and so constructed as to show a uniform and unbroken light over an arc of the horizon of 10 points of the compass, and so fixed as to throw the light from right ahead to 2 points abaft the beam on their respective sides.

(b) The green and red lights shall be fitted with inboard screens projecting at least 3 feet forward from the lights, so as to prevent them from being seen across the bow.

(c) Officers in Charge, Marine Inspection, in districts having ferryboats shall, whenever the safety of navigation may require, designate for each line of such boats a certain light, white or colored, which will show all around the horizon, to designate and distinguish such lines from each other, which light shall be carried on a flagstaff amidships, 15 feet above the white range lights.

COMMENT: Power-driven vessels, 150 feet and more in length, are required by the International Rules to carry a 20-point, white range light, forward and below *or* abaft and above the masthead light. The Inland Rules require an additional white light, too, for all steam vessels except seagoing vessels and ferryboats. It must be an all-around white light. Seagoing vessels under Inland Rules may carry a 20-point range light and normally do, due to International Rule 2(a). Ferryboat lights are based on whether they are single- or double-ended.

The International and Inland Rules require the masthead and range lights to be placed "in line with and over the keel." This is not possible on aircraft carriers. Section 2, International Rules, above provides for the variation, as does the Act of December 3, 1945, as amended, which is similar, in Inland Waters.

The International Rules require the lower of the two range lights to be placed at a prescribed height above the hull. The Inland Rules do not prescribe the height.

Many vessels have been found "in fault" by the Courts because the side lights were

1. Not installed according to the Rules.
2. Not burning brightly.
3. Obscured.
4. Out.

The officer in charge of a vessel under way should require inspections and reports of the side, masthead, range, and stern lights to be made at regular (half-hour) intervals. He would be wise to inspect these lights himself frequently. If the vessel is not equipped with alternate, electrical side lights, he should require oil side lights to be cleaned, trimmed, lighted and ready on deck to replace the electrical ones without delay if necessary.

Under both Rules, the white range lights must be visible for 5 miles and the colored side lights, for 2 miles. White lights can be seen farther than colored lights. These two factors mean that the range lights are "picked up" more easily and quickly than the side lights. This is fortunate because the range lights give a more accurate estimate of the course of the approaching vessel than side lights.

Single-ended ferryboats are not encountered as frequently as double-ended ones. Single-ended ones carry the usual navigation lights, side lights, and two range lights. The double-ended ones, when in Inland Waters, carry a range of two all-around white lights at equal heights in addition to side lights. Thus the masthead light is replaced by an all-around light, and the after range light is at the same height above the hull as the forward one.

An occasional line of ferryboats is seen in Inland Waters, the boats of which carry a distinguishing blue light between and above the two white range lights.

The International Rules require each of the side lights of a seaplane to be visible through 110 degrees; this is true also of the Civil Air Regulations. The

International Rules require a visibility of 2 miles for side lights; the Civil Air Regulations do not prescribe the visibility distance.

INTERNATIONAL RULES

Rule 3 (Figs. 16.2 and 16.3)

(a) A power-driven vessel when towing or pushing another vessel or seaplane shall, in addition to her sidelights, carry two bright white lights in a vertical line one over the other, not less than 6 feet apart, and when towing more than one vessel shall carry an additional bright white light 6 feet above or below such lights, if the length of the tow, measuring from the stern of the towing vessel to the stern of the last vessel or seaplane towed, exceeds 600 feet. Each of these lights shall be of the same construction and character and one of them shall be carried in the same position as the white light mentioned in Rule 2(a) (i), except the additional light, which shall be carried at a height of not less than 14 feet above the hull. In a vessel with a single mast, such lights may be carried on the mast.

(b) The towing vessel shall also show either the stern light specified in Rule 10 or in lieu of that light a small white light abaft the funnel or aftermast for the tow to steer by, but such light shall not be visible forward of the beam. The carriage of the white light specified in Rule 2(a) (ii) is optional.

INLAND RULES

(Figs. 16.2 and 16.3)

ART. 3. *(a) A steam vessel when towing another vessel or vessels alongside or by pushing ahead shall, in addition to her side lights, carry two bright white lights in a vertical line, one over the other, not less than three feet apart, and when towing one or more vessels astern, regardless of the length of the tow, shall carry an additional bright white light three feet above or below such lights. Each of these lights shall be of the same construction and character, and shall be carried in the same position as the white light mentioned in article 2 (a) or the after range light mentioned in article 2 (f).*

(b) A steam vessel carrying towing lights the same as the white light mentioned in article 2 (a), when pushing another vessel or vessels ahead, shall also carry at or near the stern two bright amber lights in a vertical line, one over the other, not less than three feet apart; each of these lights shall be so constructed as to show an unbroken light over an arc of the horizon of twelve points of the compass, so fixed as to show the light six points from right aft on each side of the vessel, and of such a character as to be visible at a distance of at least two miles. A steam vessel carrying towing lights the same as the white light mentioned in article 2 (a) may also carry, irrespective of the position of the tow, the after range light mentioned in article 2 (f); however, if the after range light is carried by

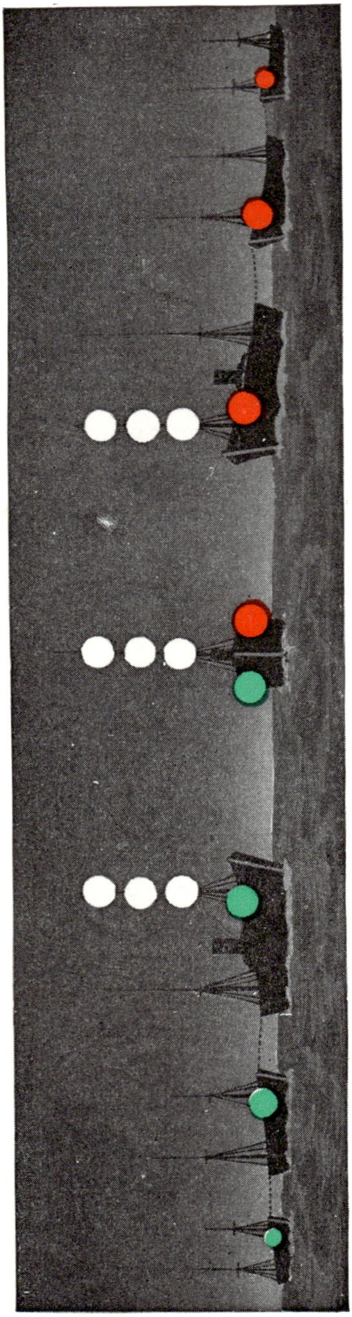

INTERNATIONAL

Towing More Than One Vessel (length of tow exceeding 600 feet)

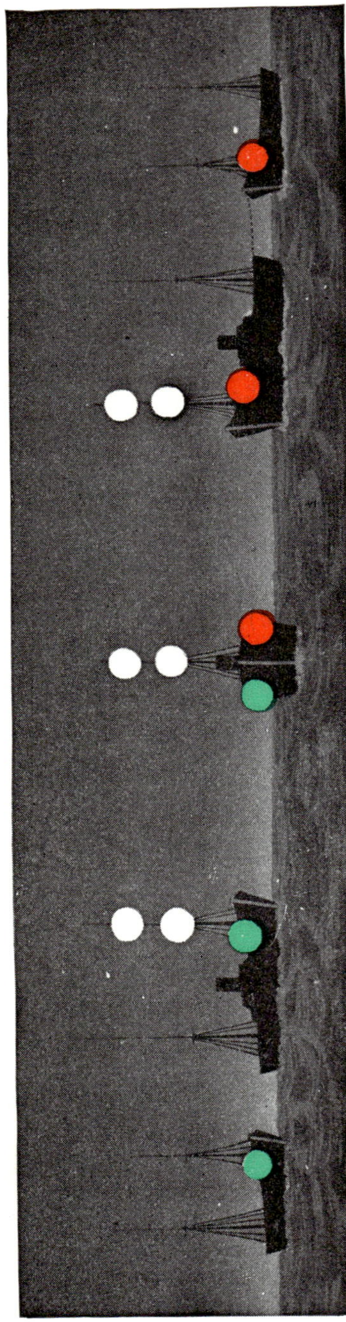

INTERNATIONAL

Towing One Vessel (or more than one if length not exceeding 600 feet)

FIG. 16.2 STEAMER TOWING WITHOUT RANGE LIGHTS

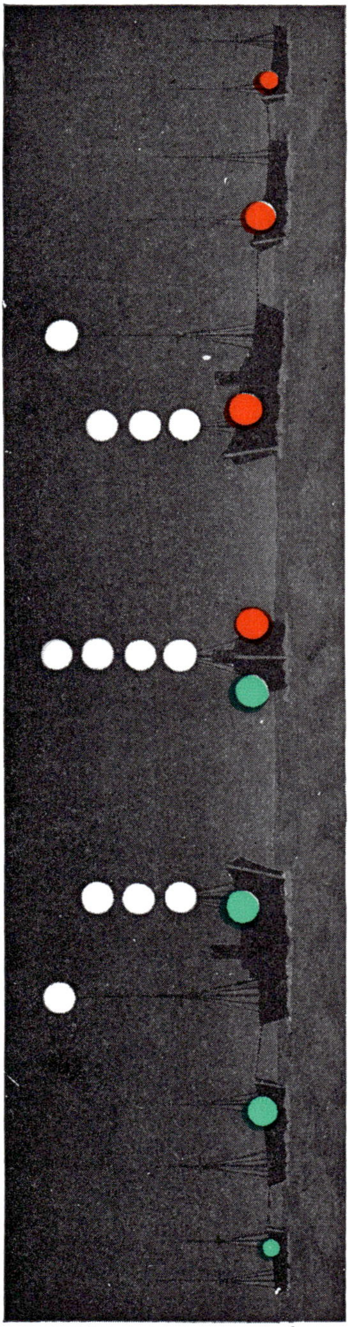

INTERNATIONAL AND INLAND

International: Towing More Than One Vessel (length of tow exceeding 600 feet). Inland: Towing one or more vessels (tow any length)

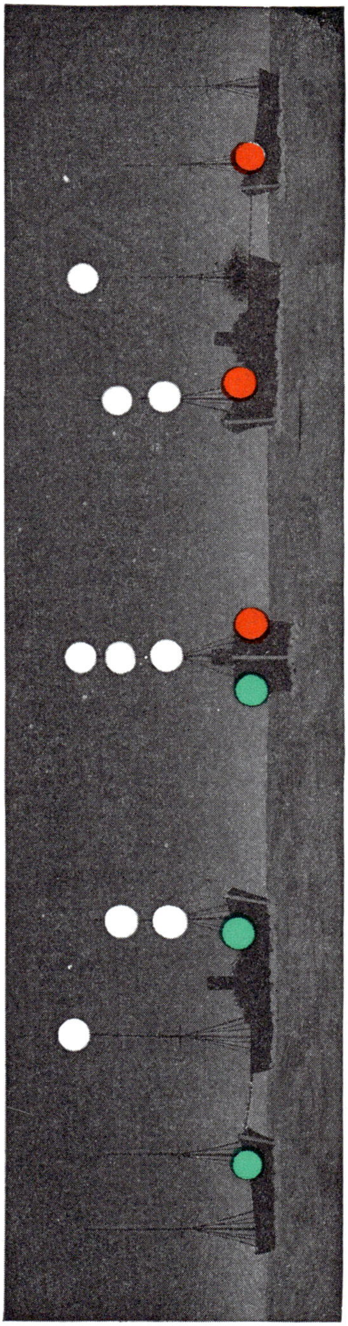

INTERNATIONAL

Towing One Vessel (or more than one if length not exceeding 600 feet)

FIG. 16.3 STEAMER TOWING WITH RANGE LIGHTS

INTERNATIONAL RULES

INLAND RULES

such a vessel when pushing another vessel or vessels ahead, the amber lights shall be carried in a vertical line with and at least three feet lower than the after range light. A steam vessel carrying towing lights the same as the white light mentioned in article 2 (a), when towing one or more vessels astern, may also carry, in lieu of the stern light specified in article 10, a small white light abaft the funnel or aftermast for the tow to steer by, but such light shall not be visible forward of the beam.

(c) A seaplane on the water, when towing one or more seaplanes or vessels, shall carry the lights prescribed in Rule 2(b) (i), (ii) and (iii); and, in addition, she shall carry a second white light of the same construction and character as the white light mentioned in Rule 2(b) (i), and in a vertical line at least 6 feet above or below such light.

Not in Inland Rules.

Air Traffic Rules do not provide for towing.

PILOT RULES

80.18 Signals to be displayed by a towing vessel when towing a submerged or partly submerged object upon a hawser when no signals can be displayed upon the object which is towed.—(a) The vessel having the submerged object in tow shall display by day, where they can best be seen, two shapes, one above the other, not less than six feet apart, the lower shape to be carried not less than 10 feet above the deck house. The shapes shall be in the form of a double frustum of a cone, base to base, not less than two feet in diameter at the center nor less than eight inches at the ends of the cones, and to be not less than four feet lengthwise from end to end, the upper shape to be painted in alternate horizontal stripes of black and white, eight inches in width, and the lower shape to be painted a solid bright red.

(b) By night the towing vessel shall display the regular side lights but in lieu of the regular white towing lights shall display four lights in a vertical position not less than three feet nor more than six feet apart, the upper and lower of such lights to be white, and the two middle lights to be red, all of such lights to be of the same character as the regular towing lights.

NOTE: The regulations in §§ 80.18 to 80.31a, inclusive, are applicable on the harbors, rivers, and inland waters along the Atlantic and Pacific coasts and the coast of the Gulf of Mexico. Similar Department of the Army regulations are applicable on the Great Lakes and their connecting and tributary waters as far east as Montreal (Great Lakes) and Western Rivers and the Red River of the North and are contained in §§ 201.1 to 201.16, inclusive, of Title 33, Code of Federal Regulations.

COMMENT: The International Rules do not mention towing alongside. Towing lights are carried by all steam- or power-driven vessels when towing. Sailing vessels, towing, show side lights only, plus a stern light. Towing lights are not required.

The towing lights are at least six feet apart under the International Rules and not less than 3 feet in the Inland Rules.

One of the towing lights is carried in the same position as the masthead light on the high seas. In Inland Waters, the towing lights may be carried in one of two positions:

(a) in place of the masthead light, in which case they are 20-point lights;

(b) in place of the after range light where they are all-around lights.

A masthead light is not required in Inland Waters if the towing lights are aft. A range light is not required but may be carried if the towing lights are forward. In International Waters, the after-range light may be shown.

The International Rules require a third towing light if both of the following conditions are met: (a) towing more than one vessel and (b) the length of the tow exceeds 600 feet. The Inland Rules require three towing lights when towing one or more vessels astern regardless of the length of the tow.

The towing vessel must show either a stern light or a white steering light aft in International Waters. In Inland Waters the towing vessel must show two 12-point amber lights visible from aft when pushing, if she is carrying 20-point towing lights. If towing astern with 20-point lights, she may show a white steering light or stern light aft or carry the all around after range light. Other towing vessels Inland may carry either the stern light or the all around after-range light or may omit both lights, depending on the arc of the towing lights. Note that the amber light requirement in Inland Rules took effect August 14, 1958. Its purpose is to inform overtaking vessels that the vessel ahead is pushing a tow.

A tug with a tow might show an approaching vessel one to four white lights, depending upon the following factors:

1. High seas or Inland waters.
2. Number of vessels in tow.
3. Length of the towline.
4. Bearing of approaching vessel.
5. Location of towing lights.

COURT DECISION: If the tow is manned, she is responsible that the proper lights on the tow are shown. The tug is responsible for her own lights and jointly for those on the tow. (EUGENE F. MORAN, 212 U.S. 466.) If the tow has no crew, the tug is responsible for showing proper lights on the tow. (LIZZIE M. WALKER, C.C.A.4, 3F Supp. 921.)

INTERNATIONAL RULES

Rule 4 (Figs. 16.4 to 16.6)

(a) A vessel which is not under command shall carry, where they can best be seen, and, if a power-driven vessel, in lieu of the lights required by Rule 2(a) (i) and (ii), two red lights in a vertical line one over the other not less than 6 feet apart, and of such a character as to be visible all around the horizon at a distance of at least 2 miles. By day, she shall carry in a vertical line one over the other not less than 6 feet apart, where they can best be seen, two black balls or shapes each not less than 2 feet in diameter.

(b) A seaplane on the water which is not under command may carry, where they can best be seen, two red lights in a vertical line, one over the other, not less than 3 feet apart, and of such a character as to be visible all around the horizon at a distance of at least 2 miles, and may by day carry in a vertical line one over the other not less than 3 feet apart, where they can best be seen two black balls or shapes, each not less than 2 feet in diameter.

(c) A vessel engaged in laying or in picking up a submarine cable or navigation mark, or a vessel engaged in surveying or underwater operations when from the nature of her work she is unable to get out of the way of approaching vessels, shall carry, in lieu of the lights specified in Rule 2(a) (i) and (ii), three lights in a vertical line one over the other not less than 6 feet apart. The highest and lowest of these lights shall be red, and the middle light shall be white, and they shall be of such a character as to be visible all around the horizon at a distance of at least

INLAND RULES

Not in Inland Rules.

INTERNATIONAL RULES

INLAND RULES

2 miles. By day, she shall carry in a vertical line one over the other not less than 6 feet apart, where they can best be seen, three shapes each not less than 2 feet in diameter, of which the highest and lowest shall be glob-

INTERNATIONAL

Carries Side Lights, if Making Way Through Water

VESSEL NOT UNDER COMMAND

Carries Side Lights, if Making Way Through Water

VESSEL WORKING WITH TELEGRAPH CABLE

FIG. 16.4 DAY MARKS AND LIGHTS. FISHING AND CABLE VESSELS

ular in shape and red in colour, and the middle one diamond in shape and white.

(d) The vessels and seaplanes referred to in this Rule, when not making way through the water, shall not carry the coloured sidelights, but when making way they shall carry them.

(e) The lights and shapes required to be shown by this Rule are to be

INTERNATIONAL RULES

taken by other vessels and seaplanes as signals that the vessel or seaplane showing them is not under command and cannot therefore get out of the way.

(f) These signals are not signals of vessels in distress and requiring assistance. Such signals are contained in Rule 31.

INLAND RULES

AIR TRAFFIC RULES

Not in these Rules

PILOT RULES

80.19 Steam vessels, derrick boats, lighters, or other types of vessels made fast alongside a wreck, or moored over a wreck which is on the bottom or partly submerged, or which may be drifting.—(a) Steam vessels, derrick boats, lighters, or other types of vessels made fast alongside a wreck, or moored over a wreck which is on the bottom or partly submerged, or which may be drifting, shall display by day two shapes of the same character and dimensions and displayed in the same manner as required by § 80.18(a), except that both shapes shall be painted a solid bright red, but where more than one vessel is working under the above conditions, the shapes need be displayed only from one vessel on each side of the wreck from which they can best be seen from all directions.

(b) By night this situation shall be indicated by the display of a white light from the bow and stern of each outside vessel or lighter not less than six feet above the deck, and in addition thereto there shall be displayed in a position where they can best be seen from all directions two red lights carried in a vertical line not less than three feet nor more than six feet apart, and not less than 15 feet above the deck.

80.20 Dredges held in stationary position by moorings or spuds.—(a) Dredges which are held in stationary position by moorings or spuds shall display by day two red balls not less than two feet in diameter and carried in a vertical line not less than three feet nor more than six feet apart, and at least 15 feet above the deck house and in such a position where they can best be seen from all directions.

(b) By night they shall display a white light at each corner, not less than six feet above the deck, and in addition thereto there shall be displayed in a position where they can best be seen from all directions two red lights carried in a vertical line not less than three feet nor more than six feet apart, and not less than 15 feet above the deck. When scows are moored alongside a dredge

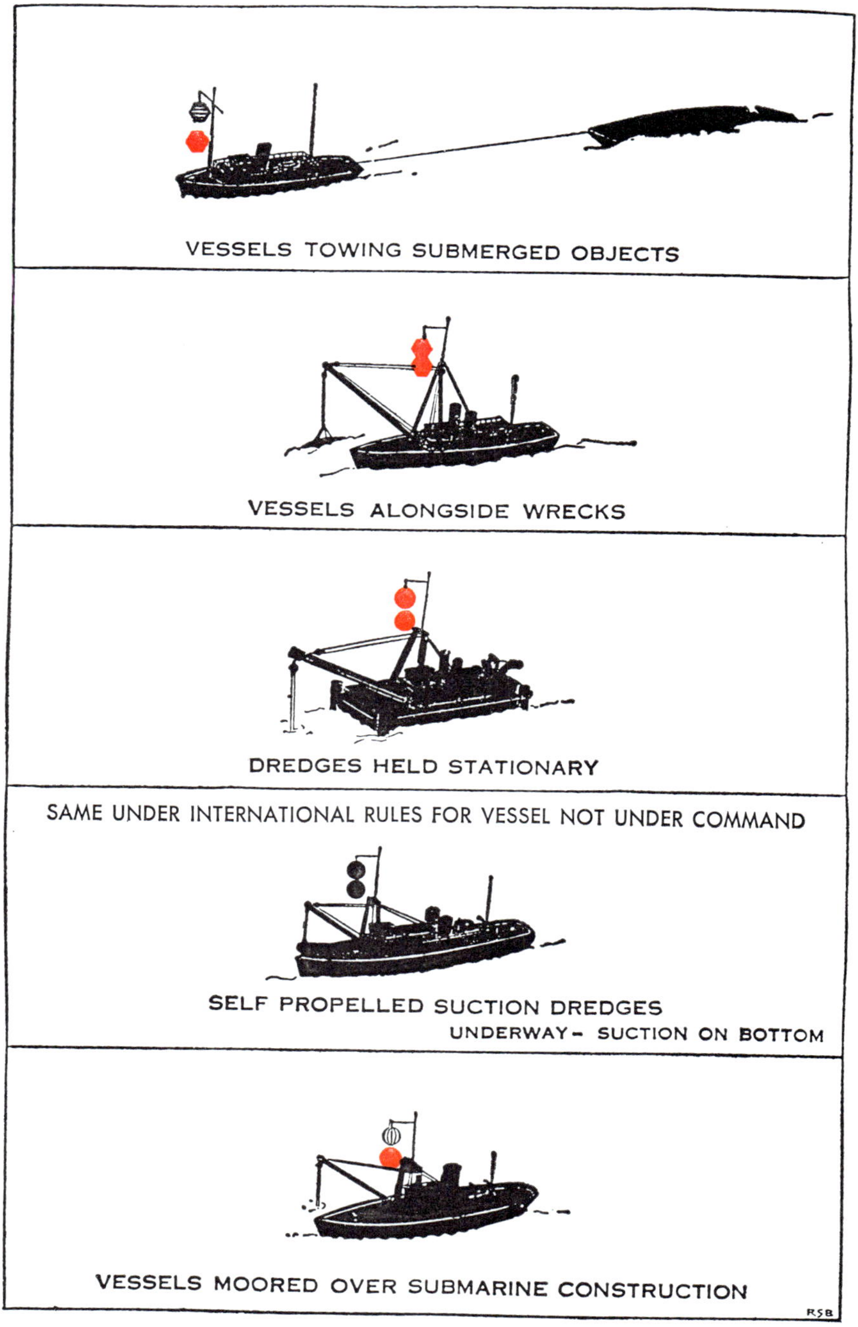

FIG. 16.5 DAY MARKS. Applicable to Inland Waters.

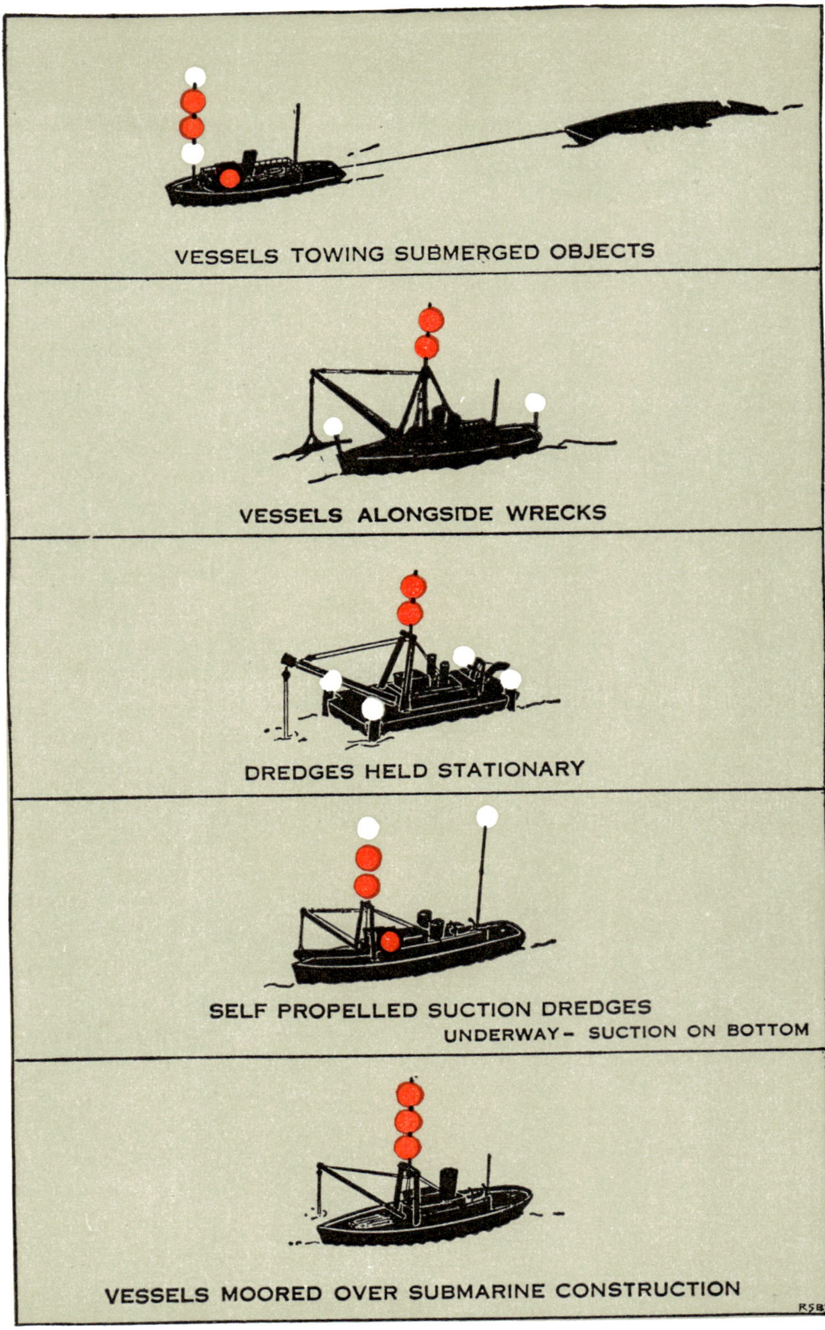

FIG. 16.6 LIGHTS ON VESSELS. Applicable to Inland Waters.

in the foregoing situation they shall display a white light on each outboard corner, not less than six feet above the deck.

80.21 Self-propelling suction dredges under way and engaged in dredging operations.—(a) Self-propelling suction dredges under way and engaged in dredging operations shall display by day two black balls not less than two feet in diameter and carried in a vertical line not less than 15 feet above the deck house, and where they can best be seen from all directions. The term "dredging operations" shall include maneuvering into or out of position at the dredging site but shall not include proceeding to or from the site.

(b) By night they shall carry, in addition to the regular running lights, two red lights of the same character as the white masthead light, and in the same vertical line beneath that light, the red lights to be not less than three feet nor more than six feet apart and the upper red light to be not less than four feet nor more than six feet below the masthead light, and on or near the stern two red lights in a vertical line not less than four feet nor more than six feet apart, to show through four points of the compass; that is, from right astern to two points on each quarter.

80.22 Vessels moored or anchored and engaged in laying cables or pipe, submarine construction, excavation, mat sinking, bank grading, dike construction, revetment, or other bank protection operations.—(a) Vessels which are moored or anchored and engaged in laying cables or pipe, submarine construction, excavation, mat sinking, bank grading, dike construction, revetment, or other bank protection operations, shall display by day, not less than 15 feet above the deck, where they can best be seen from all directions, two balls not less than two feet in diameter, in a vertical line not less than three feet nor more than six feet apart, the upper ball to be painted in alternate black and white vertical stripes six inches wide, and the lower ball to be painted a solid bright red.

(b) By night they shall display three red lights, carried in a vertical line not less than three feet nor more than six feet apart, in a position where they can best be seen from all directions, with the lowermost light not less than 15 feet above the deck.

(c) Where a stringout of moored vessels or barges is engaged in the operations, three red lights carried as prescribed in paragraph (b) of this section shall be displayed at the channelward end of the stringout. Where the stringout crosses the navigable channel and is to be opened for the passage of vessels, the three red lights shall be displayed at each side of the opening instead of at the outer end of the stringout. There shall also be displayed upon such stringout one horizontal row of amber lights not less than six feet above the deck, or above the deck house where the craft carries a deck house, in a position where they can best be seen from all directions, spaced not more than 50 feet apart so as to mark distinctly the entire length and course of the stringout.

80.23 Lights to be displayed on pipe lines.—Pipe lines attached to dredges, and either floating or supported on trestles, shall display by night one row of amber lights not less than 8 feet nor more than 12 feet above the water, about equally spaced and in such number as to mark distinctly the entire length and course of the line, the intervals between lights where the line crosses navigable channels to be not more than 30 feet. There shall also be displayed on the shore or discharge end of the line two red lights, three feet apart, in a vertical line with the lower light at least eight feet above the water, and if the line is to be opened at night for the passage of vessels, a similar arrangement of lights shall be displayed on each side of the opening.

80.24 Lights generally.—(a) All the lights required by §§ 80.18 to 80.23, inclusive, except as provided in §§ 80.18(b) and 80.21(b), shall be of such character as to be visible on a dark night with a clear atmosphere for a distance of at least two miles.

(b) The lights required by § 80.18(b) to be of the same character as the regular towing lights and the lights required by § 80.21(b) to be of the same character as the masthead light shall be of such character as to be visible on a dark night with a clear atmosphere for a distance of at least five miles.

(c) All floodlights or headlights which may interfere with the proper navigation of an approaching vessel shall be so shielded that the lights will not blind the pilot of such vessel.

80.29 Aids to navigation marking floating-plant moorings.—Breast, stern, and bow anchors of floating plant working in navigable channels shall be marked by barrel or other suitable buoys. By night approaching vessels shall be shown the location of adjacent buoys by throwing a suitable beam of light from the plant on the buoys until the approaching vessel has passed, or the buoys may be lighted by red lights, visible in all directions, of the same character as specified in § 80.24(a): *Provided,* That the foregoing provisions of this section shall not apply to the following waters of New York Harbor and adjacent waters: the East River, the North River (Battery to Spuyten Duyvil), the Harlem River and the New York and New Jersey Channels (from the Upper Bay through Kill Van Kull, Newark Bay, Arthur Kill, and Raritan Bay to the Lower Bay).

80.30 Obstruction of channel by floating plant.—Channels shall not be obstructed unnecessarily by any dredge or other floating plant. While vessels are passing such plant, all lines running therefrom across the channel on the passing side, which may interfere with or obstruct navigation, shall be slacked to the bottom of the channel.

80.31 Clearing of channels.—When special or temporary regulations have not been prescribed and action under the regulations contained in §§ 80.26 to 80.30, inclusive, will not afford clear passage, floating plant in narrow channels shall, upon notice, move out of the way of vessels a sufficient distance to allow them a clear passage. Vessels desiring passage shall, however, give the master of the floating plant ample notice in advance of the time they expect to pass.

NOTE: If it is necessary to prohibit or limit the anchorage or movement of vessels within certain areas in order to facilitate the work of improvement, application should be made through official channels for establishment by the Secretary of the Army of special or temporary regulations for this purpose.

80.31a Protection of marks placed for the guidance of floating plant.—Vessels shall not run over anchor buoys, or buoys, stakes, or other marks placed for the guidance of floating plant working in channels; and shall not anchor on the ranges of buoys, stakes, or other marks placed for the guidance of such plant.

80.33 Special signals for vessels employed in hydrographic surveying.—By day a surveying vessel of the Coast and Geodetic Survey, under way and employed in hydrographic surveying, may carry in a vertical line, one over the other not less than 6 feet apart where they can best be seen, three shapes not less than 2 feet in diameter of which the highest and lowest shall be globular in shape and green in color and the middle one diamond in shape and white.

(a) Vessels of the Coast and Geodetic Survey shall carry the above-prescribed marks while actually engaged in hydrographic surveying and under way, including drag work. Launches and other boats shall carry the prescribed marks when necessary.

(b) It must be distinctly understood that these special signals serve only to indicate the nature of the work upon which the vessel is engaged and in no way give the surveying vessel the right-of-way over other vessels or obviate the necessity for a strict observance of the rules for preventing collisions of vessels.

(c) By night a surveying vessel of the Coast and Geodetic Survey, under way and employed in hydrographic surveying, shall carry the regular lights prescribed by the rules of the road.

(d) A vessel of the Coast and Geodetic Survey, when at anchor in a fairway on surveying operations, shall display from the mast during the daytime two black balls in a vertical line and 6 feet apart. At night two red lights shall be displayed in the same manner. In the case of a small vessel the distance between the balls and between the lights may be reduced to 3 feet if necessary.

(e) Such vessels, when at anchor in a fairway on surveying operations, shall have at hand and show, if necessary, in order to attract attention, a flare-up light in addition to the lights which are, by this section, required to be carried.

80.33a Warning signals for Coast Guard vessels while handling or servicing aids to navigation.—(a) Coast Guard vessels while engaged in handling or servicing an aid to navigation during the daytime may display from the yard two orange and white vertically striped balls in a vertical line not less than three feet nor more than six feet apart, and during the nighttime may display, in a position where they may best be seen, two red lights in a vertical line not less than three feet nor more than six feet apart.

(b) Vessels, with or without tows, passing Coast Guard vessels displaying this signal, shall reduce their speed sufficiently to insure the safety of both vessels, and when passing within 200 feet of the Coast Guard vessel displaying this signal, their speed shall not exceed 5 miles per hour.

COMMENT: Notice particularly the difference under both Rules in the meaning of two red lights displayed in a vertical line. In the International Rules such a display calls attention to "A vessel which is not under command." Hence, a vessel which has suffered an accident affecting her maneuvering ability, one which has little or no steam pressure, a sailing vessel becalmed or in irons, all should show two red lights at night.

The Inland Rules do not provide special (two red) lights for vessels not under command. They must show the proper lights for vessels under way or at anchor.

Hence, commanding officers of naval vessels should realize that, if they show two red lights when broken down in Inland Waters, they may be held liable if a collision with a merchant ship occurs.

Two red lights in Inland Waters (Pilot Rules) indicates a vessel alongside a wreck (80.19), stationary dredges (80.20), self-propelled dredge (80.21), pipe lines attached to dredges (80.23), hydrographic survey vessel at work (80.33), Coast Guard vessel servicing aids to navigation (80.33a).

Under the International Rules, vessels and seaplanes, not under command, carry side lights when making way through the water. When not making way, they do not carry them.

A recent case was a carrier that rammed a destroyer and suffered serious damage to her (the carrier's) bow. She was forced to proceed stern first until the forward bulkheads were shored.

INTERNATIONAL RULES

Rule 5 (Figs. 16.7 to 16.12)

(a) A sailing vessel under way and any vessel or seaplane being towed shall carry the same lights as are prescribed by Rule 2 for a power-driven vessel or a seaplane under way, respectively, with the exception of the white lights specified therein, which they shall never carry. They shall also carry stern lights as specified in Rule 10, provided that vessels towed, except the last vessel of tow, may carry, in lieu of such stern light, a small white light as specified in Rule 3(b).

(b) A vessel being pushed ahead shall carry, at the forward end, on the starboard side a green light and on the port side a red light, which shall have the same characteristics

INLAND RULES

ART. 5. *A sailing vessel under way and any vessel being towed, except barges, canal boats, scows, and other vessels of nondescript type, when in tow of steam vessels, shall carry the same lights as are prescribed by article 2 for a steam vessel under way, with the exception of the white lights mentioned therein, which they shall never carry.*

ART. 10. (*a*) *A vessel when under way, if not otherwise required by these rules to carry one or more lights visible from aft, shall carry at her stern a white light, so constructed that it shall show an unbroken light over an arc of the horizon of twelve points of the compass, so fixed as to show the light six*

INTERNATIONAL RULES

as the lights described in Rule 2(a) (iv) and (v) and shall be screened as provided in Rule 2(a) (vi), provided that any number of vessels pushed ahead in a group shall be lighted as one vessel.

INLAND RULES

points from right aft on each side of the vessel, and of such character as to be visible at a distance of at least two miles. Such light shall be carried as nearly as practicable on the same level as the side lights.

FIG. 16.7 A SAILING VESSEL. In addition stern light not visible in these sketches shall be carried in both Inland and International Waters.

SEC. 2. (*a*) *The Secretary of the Department in which the Coast Guard is operating* (*i.e., Commandant of*) *shall establish such rules to be observed, on the waters described in section 1 of this Act . . . as to lights and day signals to be carried on such waters . . . by vessels and craft of all types when in tow of steam vessels . . . not inconsistent with the provisions of this Act, as he from time to time may deem necessary for safety. . . .*

PILOT RULES

80.16 Lights for barges, canal boats, scows and other nondescript vessels on certain inland waters on the Atlantic and Pacific Coasts.—(a) On the harbors, rivers, and other inland waters of the United States except the Great Lakes and their connecting and tributary waters as far east as Montreal, the Red River of the North, the Mississippi River and its tributaries above the Huey P. Long Bridge, and that part of the Atchafalaya River above its junction with the Plaquemine-Morgan City alternate waterway, and the waters described in §§ 80.16a and 80.17, barges, canal boats, scows, and other vessels of nondescript type not otherwise provided for, when being towed by steam vessels, shall carry lights as set forth in this section.

(b) Barges and canal boats towing astern of steam vessels, when towing singly, or what is known as tandem towing, shall each carry a green light on the starboard side and a red light on the port side, and a white light on the stern, except that the last vessel of such tow shall carry two lights on her stern, athwartship, horizontal to each other, not less than 5 feet apart, and not less than 4 feet above the deck house, and so placed as to show all around the horizon. A tow of one such vessel shall be lighted as the last vessel of a tow.

(c) When two or more boats are abreast, the colored lights shall be carried at the outer sides of the bows of the outside boats. Each of the outside boats in last tier of a hawser tow shall carry a white light on her stern.

(d) The white light required to be carried on stern of a barge or canal boat carrying red and green side lights except the last vessel in a tow shall be carried in a lantern so constructed that it shall show an unbroken light over an arc of the horizon of 12 points of the compass, namely, for 6 points from right aft on each side of the vessel, and shall be of such a character as to be visible on a dark night with a clear atmosphere at a distance of at least 2 miles.

(e) Barges, canal boats or scows towing alongside a steam vessel shall, if the deck, deck houses, or cargo of the barge, canal boat or scow be so high above water as to obscure the side lights of the towing steamer when being towed on the starboard side of the steamer, carry a green light upon the starboard side; and when towed on the port side of the steamer, a red light on the port side of the barge, canal boat, or scow; and if there is more than one barge, canal boat or scow abreast, the colored lights shall be displayed from the outer side of the outside barges, canal boats or scows.

(f) Barges, canal boats or scows shall, when being propelled by pushing ahead of a steam vessel, display a red light on the port bow and a green light on the starboard bow of the head barge, canal boat or scow, carried at a height sufficiently above the superstructure of the barge, canal boat or scow as to permit said side lights to be visible; and if there is more than one barge, canal boat or scow abreast, the colored lights shall be displayed from the outer side of the outside barges, canal boats or scows.

(g) The colored side lights referred to in this section shall be fitted with inboard screens so as to prevent them from being seen across the bow, and of such a character as to be visible on a dark night, with a clear atmosphere, at a distance of at least 2 miles, and so constructed as to show a uniform and unbroken light over an arc of the horizon of 10 points of the compass, and so fixed as to throw the light from right ahead to 2 points abaft the beam on either side. The minimum size of glass globes shall not be less than 6 inches in diameter and 5 inches high in the clear.

(h) Scows not otherwise provided for in this section on waters described in paragraph (a) of this section shall carry a white light at each end of each scow, except that when such scows are massed in tiers, two or more abreast, each of the outside scows shall carry a white light on its outer bow, and the outside scows in the last tier shall each carry, in addition, a white light on the

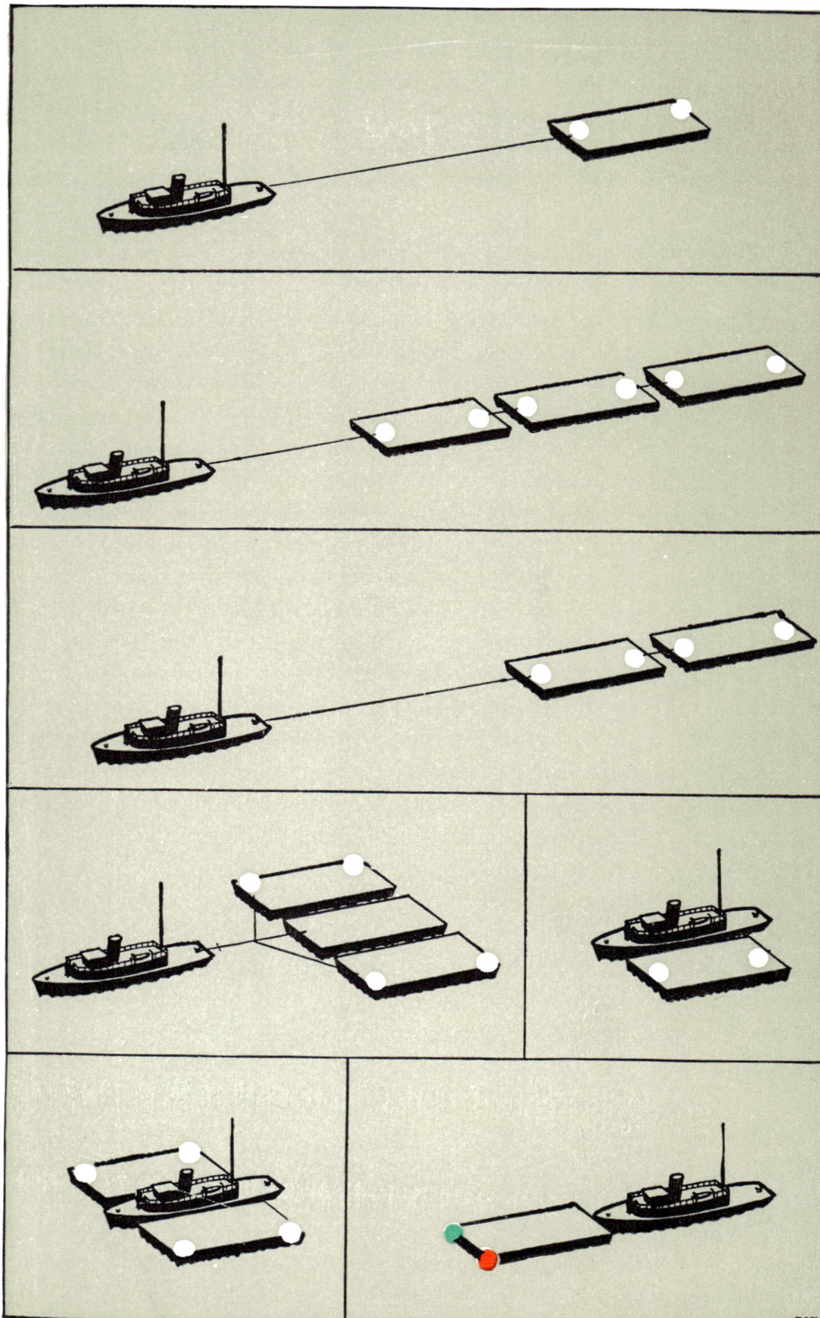

FIG. 16.8 INLAND WATER LIGHTS ON TOWED SCOWS (Hudson River area, Gulf Coast, and Gulf Intracoastal Waterway excepted)

outer part of the stern. The white light shall be carried not less than 8 feet above the surface of the water, and shall be so placed as to show an unbroken light all around the horizon, and shall be of such a character as to be visible on a dark night with a clear atmosphere at a distance of at least 5 miles.

(i) Other vessels of nondescript type not otherwise provided for in this section shall exhibit the same lights that are required to be exhibited by scows by this section.

NOTE: The regulations in §§ 80.16 to 80.17, inclusive, are not applicable to rafts. The requirements regarding lights for rafts are in § 80.32.

80.16a Lights for barges, canal boats, scows and other nondescript vessels on certain inland waters on the Gulf Coast and the Gulf Intracoastal Waterway.—(a) On the Gulf Intracoastal Waterway and on other inland waters connected therewith or with the Gulf of Mexico from the Rio Grande, Texas, to Cape Sable (East Cape), Florida, barges, canal boats, scows, and other vessels of nondescript type not otherwise provided for, when being towed by steam vessels shall carry lights as set forth in this section.

(b) When one or more barges, canal boats, scows, or other vessels of nondescript type not otherwise provided for, are being towed by pushing ahead of a steam vessel, such tow shall be lighted by an amber light at the extreme forward end of the tow, so placed as to be as nearly as practicable on the centerline of the tow, a green light on the starboard side of the tow, so placed as to mark the maximum projection of the tow to starboard, and a red light on the port side of the tow, so placed as to mark the maximum projection of the tow to port.

(c) When one or more barges, canal boats, scows, or other vessels of nondescript type not otherwise provided for, are being towed alongside a steam vessel, there shall be displayed a white light at each outboard corner of the tow. If the deck, deckhouse, or cargo of such barge, etc., obscures the side light of the towing vessel, such barge, etc., shall also carry a green light upon the starboard side when being towed on the starboard side of a steam vessel or shall carry a red light on the port side of the barge, etc., when being towed on the port side of the steam vessel. If there is more than one such barge, etc., being towed abreast, the appropriate colored side light shall be displayed from the outer side of the outside barge.

(d) When one barge, canal boat, scow or other vessel of nondescript type not otherwise provided for, is being towed singly behind a steam vessel, such vessel shall carry four white lights, one on each corner or outermost projection of the bow and one on each corner or outermost projection of the stern.

(e) When two or more barges, canal boats, scows, or other vessels of nondescript type not otherwise provided for, are being towed behind a steam vessel in tandem, with an intermediate hawser, such vessels shall carry white lights as follows:

(1) The first vessel in the tow shall carry three white lights, one on each corner or outermost projection of the bow and a white light at the stern amidships.

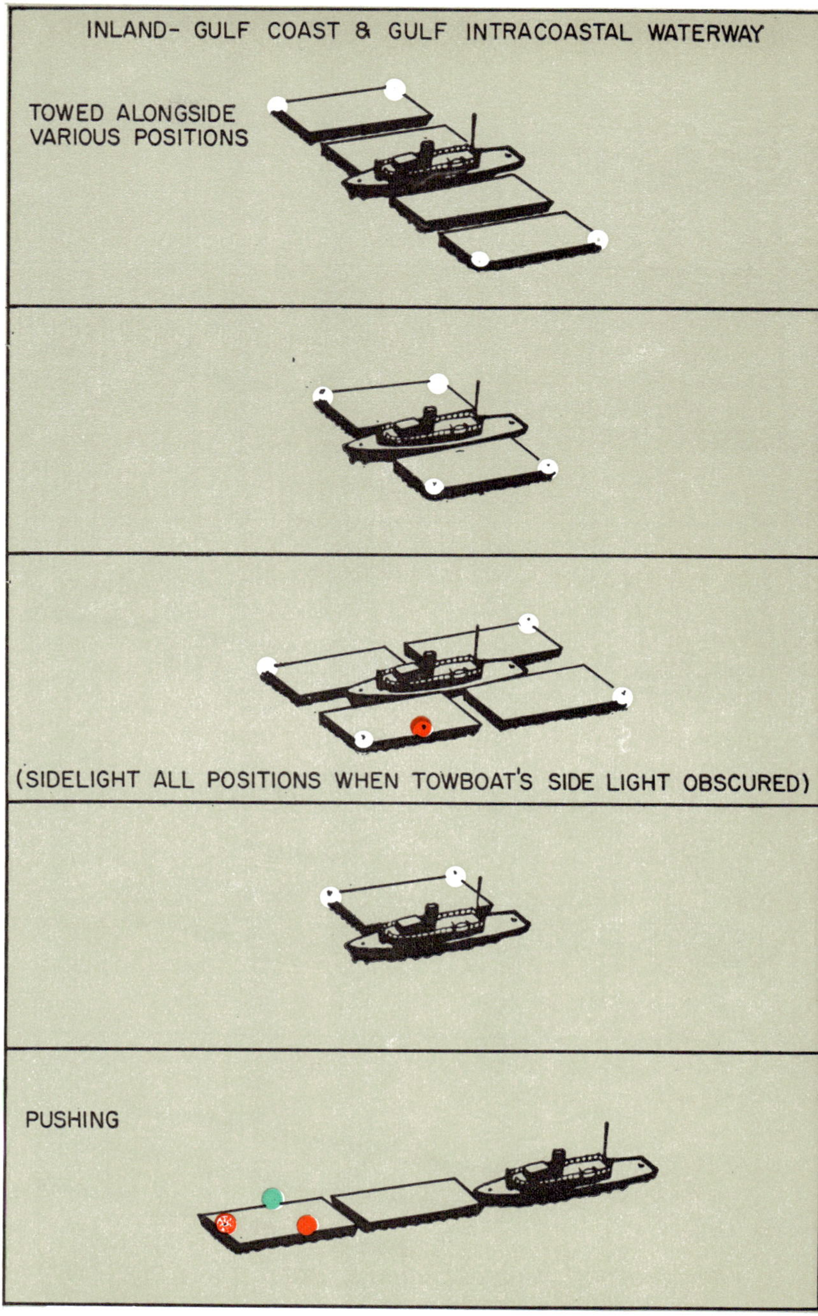

FIG. 16.9 LIGHTS ON NONDESCRIPT TOWED VESSELS

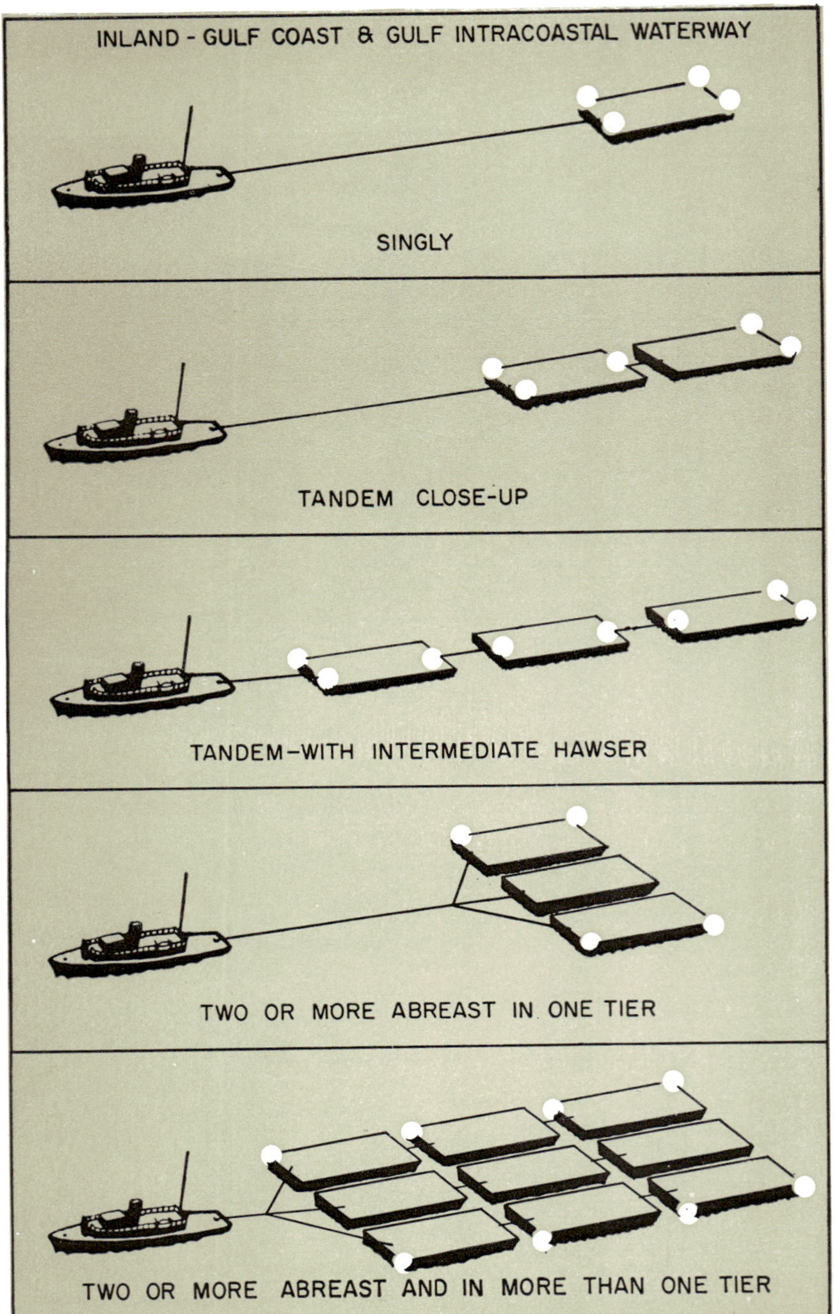

FIG. 16.10 LIGHTS ON NONDESCRIPT TOWED VESSELS

(2) Each intermediate vessel shall carry two white lights, one at each end amidships.

(3) The last vessel in the tow shall carry three white lights, one on each corner or outermost projection of the stern and a white light at the bow amidships.

(f) When two or more barges, canal boats, scows, or other vessels of nondescript type not otherwise provided for, are being towed behind a steam vessel in tandem, close-up, such vessels shall carry white lights as follows:

(1) The first vessel in the tow shall carry three white lights, one on each corner or outermost projection of the bow and a white light at the stern amidships.

(2) Each intermediate vessel shall carry a white light at the stern amidships.

(3) The last vessel in the tow shall carry two white lights, one on each corner or outermost projection of the stern.

(g) When two or more barges, canal boats, scows, or other vessels of nondescript type not otherwise provided for, are being towed behind a steam vessel two or more abreast, in one or more tiers, each of the outside vessels in each tier shall carry a white light on the outboard corner of the bow, and each of the outside vessels in the last tier shall carry, in addition, a white light on the outboard corner of the stern.

(h) When one or more barges, canal boats, scows, or other vessels of nondescript type not otherwise provided for, are moored to the bank or dock in or near a fairway, such tow shall carry two white lights not less than 4 feet above the surface of the water, as follows:

(1) On a single moored barge, canal boat, scow, or other vessel of nondescript type not otherwise provided for, a light at each outboard or channelward corner.

(2) On barges, canal boats, scows, or other vessels of nondescript type not otherwise provided for, when moored in a group formation, a light on the upstream outboard or channelward corner of the outer upstream boat and a light on the downstream outboard or channelward corner of the outer downstream boat; and in addition, any boat projecting toward or into the channel from such group formation shall have two white lights similarly placed on its outboard or channelward corners.

(i) The colored side lights shall be so constructed as to show a uniform and unbroken light over an arc of the horizon of 10 points of the compass, so fixed as to show the light from right ahead to 2 points abaft the beam on their respective sides, and of such a character as to be visible at a distance of at least 2 miles, and shall be fitted with inboard screens so as to prevent either light from being seen more than half a point across the centerline of the tow.

(j) The amber light shall be so constructed as to show a uniform and unbroken light over an arc of the horizon of 20 points of the compass, so fixed as to show the light 10 points on each side of the tow, namely, from right ahead to 2 points abaft the beam on either side, and of such a character as to be visible at a distance of at least 2 miles.

(k) The white lights shall be so constructed and so fixed as to show a clear, uniform, and unbroken light all around the horizon, and of such a character as to be visible at a distance of at least 2 miles.

(1) All the lights shall be carried at approximately the same height above the surface of the water and, except as provided in paragraph (h) of this section, shall be so placed with respect thereto as to be clear of and above all obstructions which might tend to interfere with the prescribed arc or distance of visibility.

80.16b Lights for barges, canal boats, scows, and other nondescript vessels temporarily operating on waters requiring different lights.—Nothing in §§ 80.16, 80.16a, or 80.17 shall be construed as compelling barges, canal boats, scows, or other vessels of nondescript type not otherwise provided for, being towed by steam vessels, when passing through any waters coming within the scope of any regulations where lights for such boats are different from those of the waters whereon such boats are usually employed, to change their lights from those required on the waters on which their trip begins or terminates; but should such boats engage in local employment on waters requiring different lights from those where they are customarily employed, they shall comply with the local rules where employed.

80.17 Lights for barges and canal boats in tow of steam vessels on the Hudson River and adjacent waters and Lake Champlain.—All nondescript vessels known as scows, car floats, lighters, and vessels of similar type, navigating the waters referred to in the following rules, shall carry the lights required to be carried by barges and canal boats in tow of steam vessels, as prescribed in such rules.

Barges and canal boats, when being towed by steam vessels on the waters of the Hudson River and its tributaries from Troy to the boundary lines of New York Harbor off Sandy Hook, as defined pursuant to section 2 of the act of Congress of February 19, 1895 (28 Stat. 672; 33 U.S.C. 151), the East River and Long Island Sound (and the waters entering thereon, and to the Atlantic Ocean), to and including Narragansett Bay, R. I., and tributaries, and Lake Champlain, shall carry lights as follows:

(a) Barges and canal boats being towed astern of steam vessels when towing singly shall carry a white light on the bow and a white light on the stern.

SINGLY

(b) When towing in tandem, "close up," each boat shall carry a white light on its stern and the first or hawser boat shall, in addition, carry a white light on its bow.

TANDEM—CLOSE UP

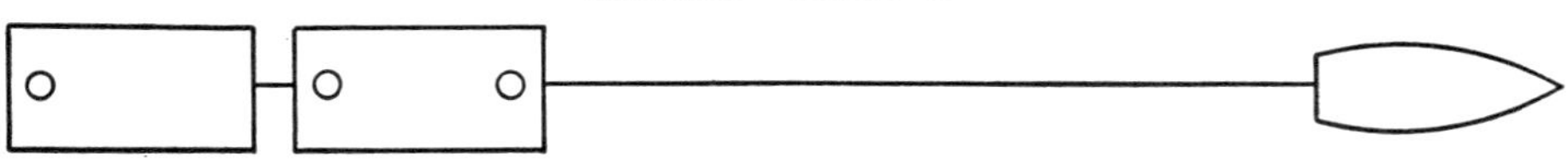

(c) When towing in tandem with intermediate hawser between the various boats in the tow, each boat shall carry a white light on the bow and a white light on the stern, except that the last vessel in the tow shall carry two white lights on her stern, athwartship, horizontal to each other, not less than 5 feet apart and not less than 4 feet above the deck house, and so placed as to show all around the horizon: *Provided,* That seagoing barges shall not be required to make any change in their seagoing lights (red and green) on waters coming within the scope of the rules of this section, except that the last vessel of the tow shall carry two white lights on her stern, athwartship, horizontal to each other, not less than 5 feet apart, and not less than 4 feet above the deck house, and so placed as to show all around the horizon.

TANDEM—WITH INTERMEDIATE HAWSER

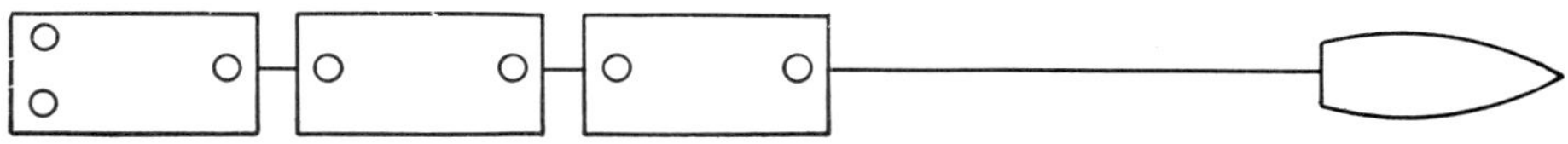

(d) Barges and canal boats when towed at a hawser, two or more abreast, when in one tier, shall each carry a white light on the stern and a white light on the bow of each of the outside boats.

TWO OR MORE ABREAST IN ONE TIER

(e) When in more than one tier, each boat shall carry a white light on its stern and the outside boats in the hawser or head tier shall each carry, in addition, a white light on the bow.

TWO OR MORE ABREAST AND IN MORE THAN ONE TIER

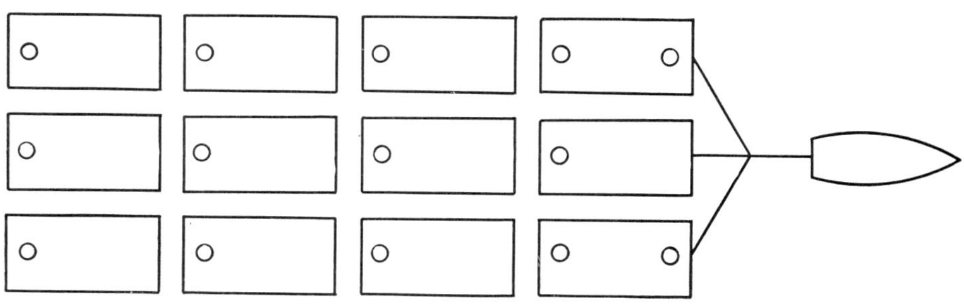

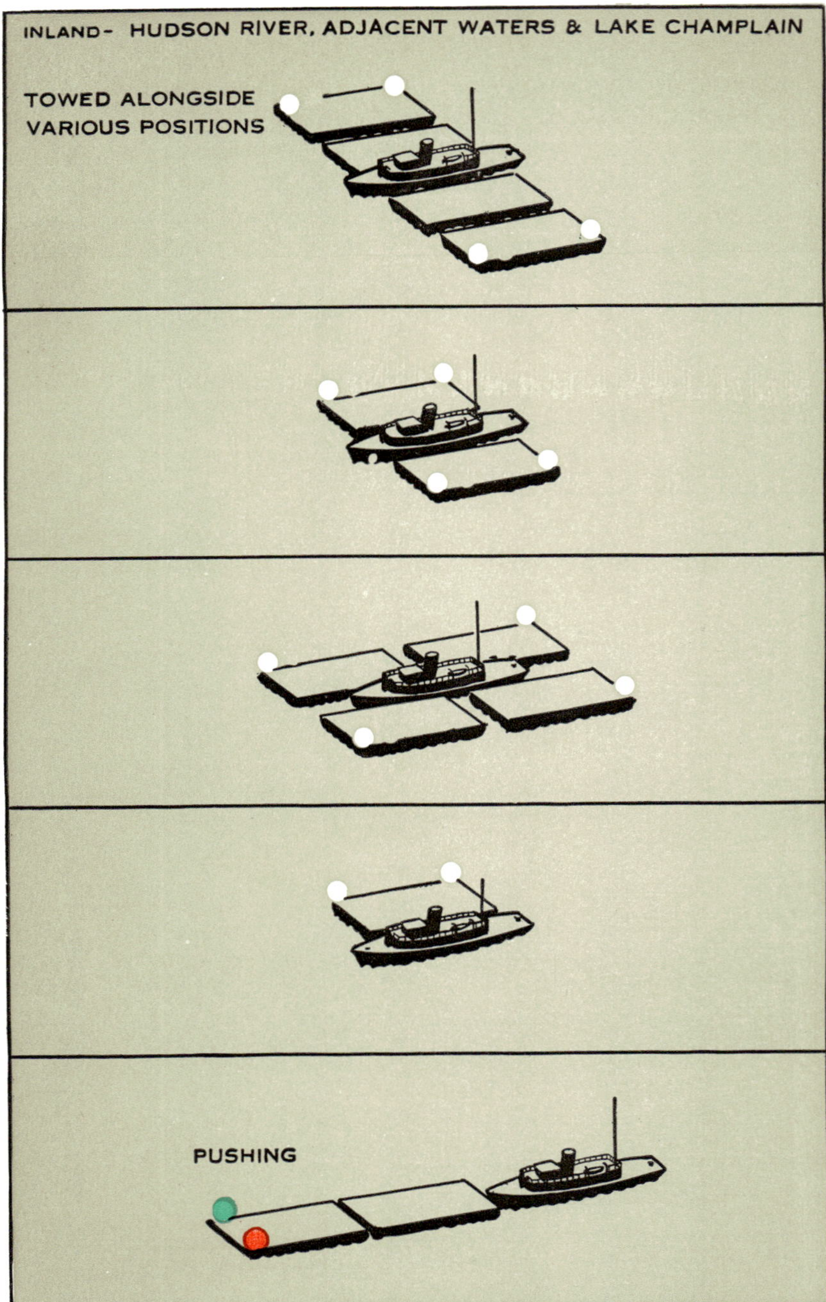

FIG. 16.11 LIGHTS ON NONDESCRIPT TOWED VESSELS

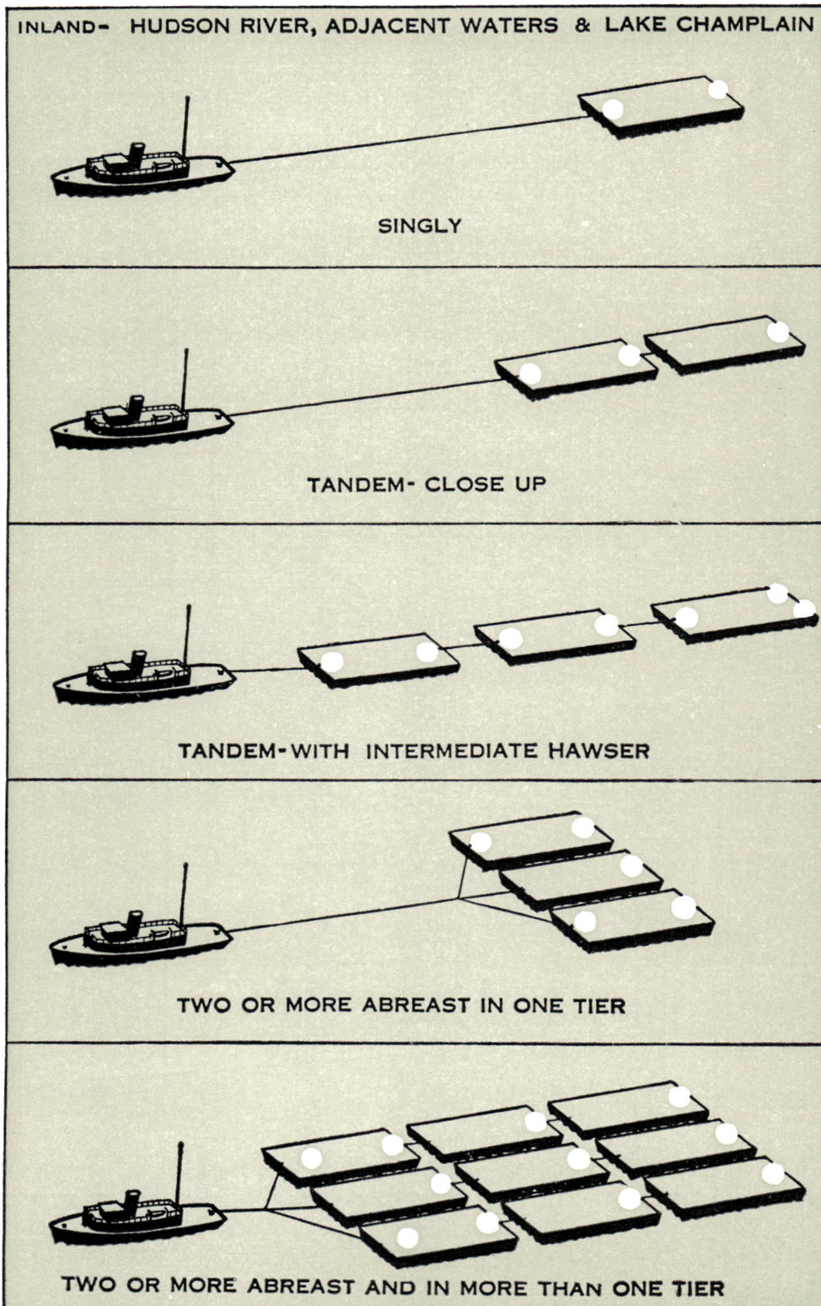

FIG. 16.12 LIGHTS ON NONDESCRIPT TOWED VESSELS

(f) The white bow lights for barges and canal boats referred to in the preceding rules shall be carried at least 10 feet and not more than 30 feet abaft the stem or extreme forward end of the vessel. On barges and canal boats required to carry a white bow light, the white light on bow and the white light on stern shall each be so placed above the hull or deck house as to show an unbroken light all around the horizon, and of such a character as to be visible on a dark night with a clear atmosphere at a distance of at least 2 miles.

(g) When nondescript vessels known as scows, car floats, lighters, barges or canal boats, and vessels of similar type, are towed alongside a steam vessel, there shall be displayed a white light at the outboard corners of the tow.

TOWED ALONGSIDE—VARIOUS POSITIONS

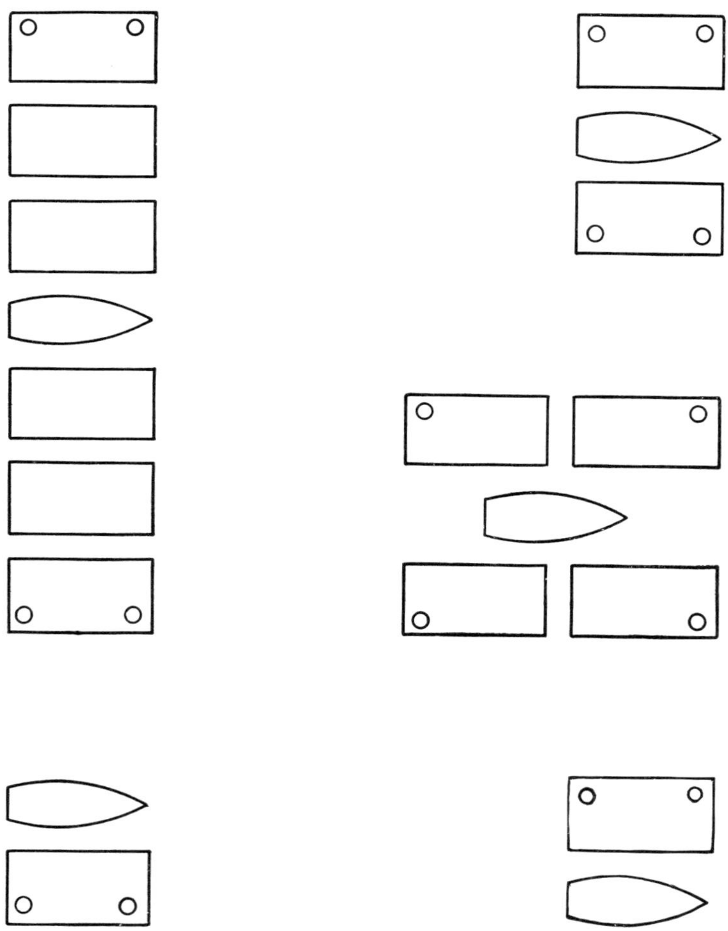

(h) When under way between the hours of sunset and sunrise there shall be displayed a red light on the port bow and a green light on the starboard bow of the head barge or barges, properly screened and so arranged that they may be visible through an arc of the horizon of 10 points of the compass; that is, from right ahead to 2 points abaft the beam on either side and visible on a dark night with a clear atmosphere at a distance of at least 2 miles, and be carried at a height sufficiently above the superstructure of the barge or barges pushed ahead as to permit said side lights to be visible.

PROPULSION OF BARGE OR BARGES BY PUSHING

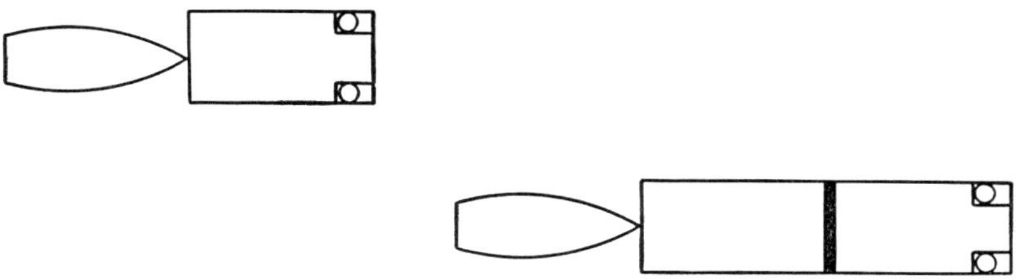

(i) Dump scows utilized for transportation and disposal of garbage, street sweepings, ashes, excavated material, dredging, etc., when navigating on the Hudson River or East River or the Waters tributary thereto between loading points on these waters and the dumping grounds established by competent authority outside the line dividing the high seas from the inland waters of New York Harbor, shall, when towing in tandem, carry, instead of the white lights previously required, red and green side lights on the respective and appropriate sides of the scow in addition to the white light required to be shown by an overtaken vessel.

The red and green lights herein prescribed shall be carried at an elevation of not less than 8 feet above the highest deck house, upon substantial uprights, the lights properly screened and so arranged as to show through an arc of the horizon of 10 points of the compass, that is, from right ahead to 2 points abaft the beam on either side and visible on a dark night with a clear atmosphere a distance of at least 2 miles.

Provided, That nothing in the rules of this section shall be construed as compelling barges or canal boats in tow of steam vessels, passing through any waters coming within the scope of said rules where lights for barges or canal boats are different from those of the waters whereon such vessels are usually employed, to change their lights from those required on the waters from which their trip begins or terminates; but should such vessels engage in local employment on waters requiring different lights from those where they are customarily employed, they shall comply with the local rules where employed.

COMMENT: The International Rules require side lights and a stern or steering light for "any vessel or seaplane being towed." The Inland Rules require side lights and a stern light "for any vessel towed." The Inland Rules do not cover seaplanes and the Rules except barges, canal boats, scows, and other vessels of nondescript type.

Hence, a steam- or power-driven vessel, being towed, should exhibit side lights and a stern or steering light, as appropriate, in both waters.

The Pilot Rules require two different kinds of lights for barges, canal boats, and scows, depending on the area in which they are being towed. Side lights are required in certain Inland waters, but in others only white bow and stern lights.

Such vessels are not required to change their type of lights when proceeding through an area where the other type is used unless they engage in local employment.

At present the Pilot Rules provide three separate systems of lights for barges, canal boats, scows, and other nondescript vessels—i.e., Inland Waters in general; Hudson River, adjacent waters, and Lake Champlain; Gulf Coast and Gulf Intracoastal Waterway. Those applicable to the Gulf Coast and Gulf Intracoastal Waterway were extensively revised August 9, 1958, and made effective January 1, 1959.

The Commandant of the Coast Guard (Exec. Order 9083, etc.) is charged with the power to make special rules to govern the lights to be carried by ferryboats, by barges, canal boats, etc., in tow of steam vessels, by dredges and vessels working on wrecks, and other nondescript vessels not provided for in the statutory rules.

The Act of August 8, 1917 (40 Stat. L 250) gives the Secretary of the Army the right to issue regulations for the use, administration, and navigation of navigable waters of the United States, "covering all matters not specifically designated by law to some other executive department." Regulations issued under the Act are usually published as Corps of Engineers regulations.

INTERNATIONAL RULES

Rule 6

(a) In small vessels, when it is not possible on account of bad weather or other sufficient cause to fix the green and red sidelights, these lights shall be kept at hand lighted and ready for immediate use, and shall, on the approach of or to other vessels, be exhibited on their respective sides in sufficient time to prevent collision, in such manner as to make them most visible, and so that the green light shall not be seen on the port side nor the red light on the starboard side, nor, if practicable, more than 2 points (22½ degrees) abaft the beam on their respective sides.

(b) To make the use of these port-

INLAND RULES

ART. 6. *Whenever, as in the case of vessels of less than ten gross tons under way during bad weather, the green and red side lights cannot be fixed, these lights shall be kept at hand, lighted and ready for use; and shall, on the approach of or to other vessels, be exhibited on their respective sides in sufficient time to prevent collision, in such manner as to make them most visible, and so that the green light shall not be seen on the port side nor the red light on the starboard side, nor, if practicable, more than two points abaft the beam on their respective sides. To make the use of these portable lights more certain and easy the lanterns con-*

INTERNATIONAL RULES

able lights more certain and easy, the lanterns containing them shall each be painted outside with the colour of the lights they respectively contain, and shall be provided with proper screens.

INLAND RULES

taining them shall each be painted outside with the color of the light they respectively contain, and shall be provided with proper screens.

COMMENT: The International Rule applies to "small" vessels, but the adjective "small" is not defined. The dividing line in Inland Waters is 10 gross tons.

17

Lights and Shapes of Identification, Specific Rules (Articles) 7–14

INTERNATIONAL RULES

Rule 7 (Fig. 17.1)

Power-driven vessels of less than 40 tons, vessels under oars or sails of less than 20 tons, and rowing boats, when under way shall not be required to carry the lights mentioned in Rule 2, but if they do not carry them they shall be provided with the following lights:—

(a) Power-driven vessels of less than 40 tons, except as provided in section (b), shall carry:—

(i) In the forepart of the vessel, where it can best be seen, and at a height above the gunwale of not less than 9 feet, a bright white light constructed and fixed as prescribed in Rule 2(a) (i) and of such a character as to be visible at a distance of at least 3 miles.

(ii) Green and red sidelights constructed and fixed as prescribed in Rule 2(a) (iv) and (v), and of such a character as to be visible at a distance of at least 1 mile, or a combined lantern showing a green light and a red light from right ahead to 2 points ($22\frac{1}{2}$ degrees) abaft the beam on their respective sides. Such lantern shall be carried not less than 3 feet below the white light.

(b) Small power-driven boats, such as are carried by seagoing vessels, may carry the white light at a less height than 9 feet above the gunwale, but it shall be carried above the sidelights or the combined lantern mentioned in subsection (a) (ii).

INLAND RULES

Article 7

Not in Inland Rules.

Not in Inland Rules.

INTERNATIONAL RULES

(c) Vessels of less than 20 tons, under oars or sails, except as provided in section (d), shall, if they do not carry the sidelights, carry where it can best be seen a lantern showing a green light on one side and a red light on the other, of such a character as to be visible at a distance of at least 1 mile, and so fixed that the green light shall not be seen on the port side, nor the red light on the starboard side. Where it is not possible to fix this light, it shall be kept ready for immediate use and shall be exhibited in sufficient time to prevent collision and so that the green light shall not be seen on the port side nor the red light on the starboard side.

(d) Small rowing boats, whether under oars or sail, shall only be required to have ready at hand an electric torch or a lighted lantern showing a white light which shall be exhibited in sufficient time to prevent collision.

(e) The vessels and boats referred to in this Rule shall not be required to carry the lights or shapes prescribed in Rules 4(a) and 11(e).

INLAND RULES

Not in Inland Rules.

ART. 7. *Rowing boats, whether under oars or sail, shall have ready at hand a lantern showing a white light which shall be temporarily exhibited in sufficient time to prevent collision.*

ART. 9. (*d*) *Rafts, or other water craft not herein provided for, navigating by hand power, horse power, or by the current of the river, shall carry one or more good white lights, which shall be placed in such manner as shall be prescribed by the Commandant of the Coast Guard.*

AIR TRAFFIC RULES

Not in Air Traffic Rules

PILOT RULES

80.32 Lights for rafts and other craft.—(a) Any vessel propelled by hand power, horse power, or by the current of the river, except rafts and rowboats, shall carry one white light forward not less than 8 feet above the surface of the water.

(b) Any raft while being propelled by hand power, by horse power, or by the current of the river, while being towed, or while anchored or moored in or near a channel or fairway, shall carry white lights as follows:

(1) A raft of one crib in width shall carry one white light at each end of the raft.

(2) A raft of more than one crib in width shall carry four white lights, one on each outside corner.

(3) An unstable log raft of one bag or boom in width shall carry at least two but not more than four white lights in a fore-and-aft line, one of which shall be at each end. The lights may be closely grouped clusters of not more than three white lights rather than single lights.

(4) An unstable log raft of more than one bag or boom in width shall carry four white lights, one on each outside corner. The lights may be closely grouped clusters of not more than three white lights rather than single lights.

(c) The white lights required by this section shall be carried from sunset to sunrise, in a lantern so fixed and constructed as to show a clear, uniform, and unbroken light, visible all around the horizon, and of such intensity as to be visible on a dark night with a clear atmosphere at a distance of at least one mile. The lights for rafts shall be suspended from poles of such height that the lights shall not be less than 8 feet above the surface of the water, except that the lights prescribed for unstable log rafts shall not be less than 4 feet above the water.

MOTORBOAT ACT OF APRIL 25, 1940; EXCERPTS FROM

AN ACT TO AMEND LAWS FOR PREVENTING COLLISIONS OF VESSELS, TO REGULATE THE EQUIPMENT OF CERTAIN MOTORBOATS ON THE NAVIGABLE WATERS OF THE UNITED STATES, AND FOR OTHER PURPOSES

Motorboat defined; inspection

Be it enacted by the Senate and House of Representatives of the United States of America in Congress assembled, That the word "motorboat" where used in this Act shall include every vessel propelled by machinery and not more than sixty-five feet in length except tugboats and towboats propelled by steam. The length shall be measured from end to end over the deck, excluding sheer: *Provided,* That the engine, boiler or other operating machinery shall be subject to inspection by the Coast Guard, and to their approval of the design thereof, on all said motorboats, which are more than forty feet in length, and which are propelled by machinery driven by steam.

Classes of motorboats

SEC. 2. Motorboats subject to the provisions of this Act shall be divided into four classes as follows:

Class A. Less than sixteen feet in length.

Class 1. Sixteen feet or over and less than twenty-six feet in length.

Class 2. Twenty-six feet or over and less than forty feet in length.

Class 3. Forty feet or over and not more than sixty-five feet in length.

Lights

SEC. 3. Every motorboat in all weathers from sunset to sunrise shall carry and exhibit the following lights when under way, and during such time no other lights which may be mistaken for those prescribed shall be exhibited:

(a) Every motorboat of classes A and 1 shall carry the following lights:

First. A bright white light aft to show all around the horizon.

Second. A combined lantern in the fore part of the vessel and lower than the white light aft, showing green to starboard and red to port, so fixed as to throw the light from right ahead to two points abaft the beam on their respective sides.

(b) Every motorboat of classes 2 and 3 shall carry the foliowing lights:

First. A bright white light in the fore part of the vessel as near the stem as practicable, so constructed as to show an unbroken light over an arc of the horizon of twenty points of the compass, so fixed as to throw the light ten points on each side of the vessel; namely, from right ahead to two points abaft the beam on either side.

Second. A bright white light aft to show all around the horizon and higher than the white light forward.

Third. On the starboard side a green light so constructed as to show an unbroken light over an arc of the horizon of ten points of the compass, so fixed as to throw the light from right ahead to two points abaft the beam on the starboard side. On the port side a red light so constructed as to show an unbroken light over an arc of the horizon of ten points of the compass, so fixed as to throw the light from right ahead to two points abaft the beam on the port side. The said side lights shall be fitted with inboard screens of sufficient height so set as to prevent these lights from being seen across the bow.

(c) Motorboats of classes A and 1 when propelled by sail alone shall carry the combination lantern, but not the white light aft, prescribed by this section. Motorboats of classes 2 and 3, when so propelled, shall carry the colored side lights, suitably screened, but not the white lights prescribed by this section. Motorboats of all classes, when so propelled, shall carry, ready at hand, a lantern or flashlight showing a white light which shall be exhibited in sufficient time to avert collision.

(d) Every white light prescribed by this section shall be of such character as to be visible at a distance of at least two miles. Every colored light prescribed by this section shall be of such character as to be visible at a distance of at least one mile. The word "visible" in this Act, when applied to lights, shall mean visible on a dark night with clear atmosphere.

(e) When propelled by sail and machinery any motorboat shall carry the lights required by this section for a motorboat propelled by machinery alone.

(f) Any motorboat may carry and exhibit the lights required by the Regu-

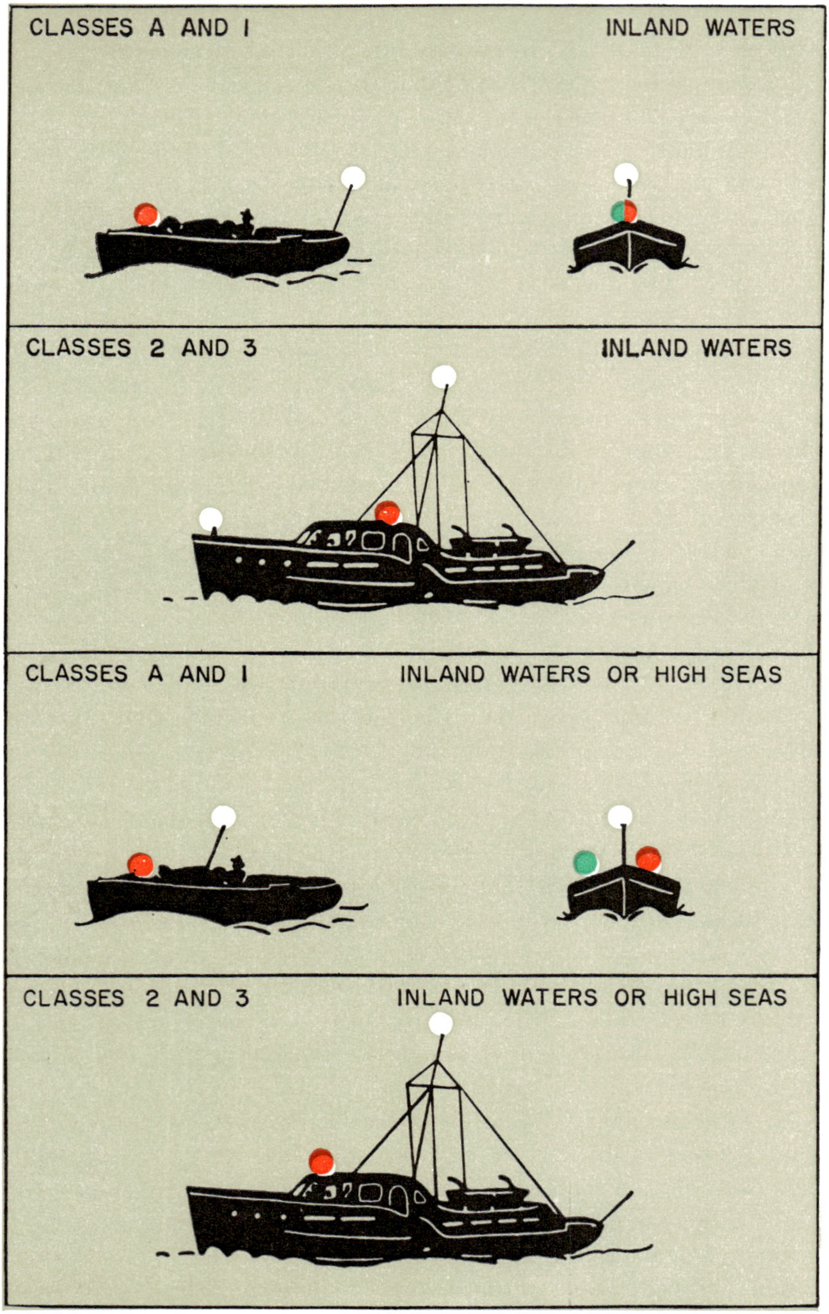

FIG. 17.1 MOTORBOATS. LIGHTS (For other combinations, see text)

lations for Preventing Collisions at Sea, 1948, Act of October 11, 1951 (65 Stat. 406–420), as amended, in lieu of the lights required by this section.

NOTE: On motorboats of classes A and 1 the aft white all around light or the 12 point white stern light may be located off the centerline.

Whistles

SEC. 4. Every motorboat of class 1, 2, or 3, shall be provided with an efficient whistle or other sound-producing mechanical appliance.

Bells

SEC. 5. Every motorboat of class 2 or 3 shall be provided with an efficient bell.

* * * * * * *

SEC. 9. The provisions of sections 4, 5, . . . of this Act shall not apply to motorboats propelled by outboard motors while competing in any race previously arranged and announced or, if such boats be designed and intended solely for racing, while engaged in such navigation as is incidental to the tuning up of the boats and engines for the race.

COMMENT: This International Rule prescribes lights for power-driven vessels of less than 40 tons, vessels under oars or sail of less than 20 tons and rowing boats.

The Inland (Pilot) Rules prescribe lights for rowing boats, rafts, or other water craft not provided for, navigated by hand power, horsepower, or the current.

The Inland Pilot Rules supplement the Inland Rules.

The International Rule divides power-driven vessels into two classes: (a) power-driven vessels of less than 40 tons and (b) small power-driven boats such as are carried by seagoing vessels. Motorboats of less than 40 gross tons are included in the latter.

From the point of view of an approaching vessel there is little difference in the lights carried. Both carry a masthead light and side lights, separated or in a combined lantern.

Vessels under oars or sail of less than 20 tons carry permanent or portable side lights, separate or combined.

The Motor Boat Act of 1940 applies to every vessel in U.S. waters propelled in whole or in part by machinery and not more than 65 feet in length except tug boats and tow boats propelled by steam. The Act applies to ship's boats and to sailing auxiliaries. The Act divides motorboats into two groups as far as navigation lights are concerned: (a) under 26 feet in length and (b) 26–65 feet in length, inclusive. Group (a) carries an all-around white light aft and combined side lights forward. Group (b) carries a 20-point white light near the stem, an all-around white light, above and abaft the forward white light, and separated side lights. Note, however, that by an amendment enacted June 4, 1956, any size motorboat may carry International Rules lights in Inland Waters rather than lights specified in the Motorboat Act itself.

Sections 3(a) and (b) apply to motorboats propelled by machinery. New Section 3(c) applies to motorboats propelled by sail. New Section 3(e) applies to motorboats propelled by both sail and machinery. Both changes were enacted June 4, 1956. The effect of the new sections is to require motorboats to show lights in conformance with their propulsion.

INTERNATIONAL RULES

Rule 8 (Fig. 17.2)

(a) (i) Sailing pilot-vessels, when engaged on their station on pilotage duty and not at anchor, shall not show the lights prescribed for other vessels, but shall carry a white light at the masthead visible all round the horizon at a distance of at least 3 miles, and shall also exhibit a flare-up light or flare-up lights at short intervals, which shall never exceed 10 minutes.

(ii) On the near approach of or to other vessels they shall have their sidelights lighted ready for use and shall flash or show them at short intervals, to indicate the direction in which they are heading, but the green light shall not be shown on the port side, nor the red light on the starboard side.

(iii) A sailing pilot-vessel of such a class as to be obliged to go alongside of a vessel to put a pilot on board may show the white light instead of carrying it at the masthead and may, instead of the sidelights above mentioned, have at hand ready for use a lantern with a green glass on the one side and a red glass on the other to be used as prescribed above.

(b) A power-driven pilot-vessel when engaged on her station on pilotage duty and not at anchor shall, in addition to the lights and flares required for sailing pilot-vessels, carry at a distance of 8 feet below her white masthead light a red light visible all around the horizon at a distance of at least 3 miles, and also

INLAND RULES

ART. 8. *Pilot vessels when engaged on their stations on pilotage duty shall not show the lights required for other vessels, but shall carry a white light at the masthead, visible all around the horizon, and shall also exhibit a flare-up light or flare-up lights at short intervals, which shall never exceed fifteen mnutes.*

On the near approach of or to other vessels they shall have their side lights lighted, ready for use, and shall flash or show them at short intervals, to indicate the direction in which they are heading, but the green light shall not be shown on the port side nor the red light on the starboard side.

A pilot vessel of such a class as to be obliged to go alongside of a vessel to put a pilot on board may show the white light instead of carrying it at the masthead, and may, instead of the colored lights above mentioned, have at hand, ready for use, a lantern with a green glass on the one side and a red glass on the other, to be used as prescribed above.

A steam pilot vessel, when engaged on her station on pilotage duty and in waters of the United States, and not at anchor, shall in addition to the lights required for all pilot boats, carry at a distance of eight feet below her white masthead light a red light, visible all around the horizon and of such a character as to be

INTERNATIONAL RULES

the sidelights required to be carried by vessels when under way. A bright intermittent all round white light may be used in place of a flare.

INLAND RULES

visible on a dark night with a clear atmosphere at a distance of at least two miles, and also the colored side lights required to be carried by vessels when under way.

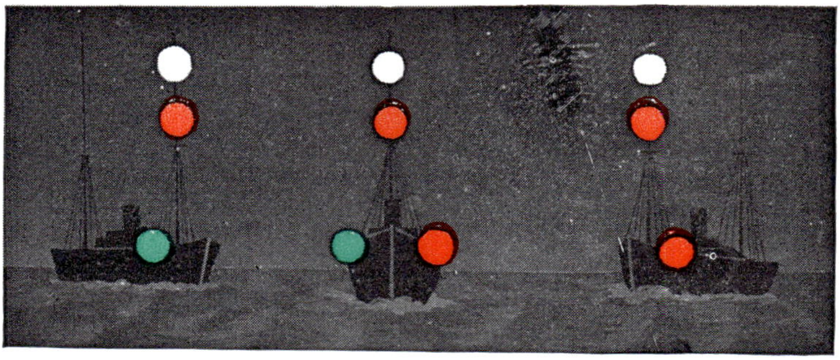

INTERNATIONAL AND INLAND

Power-Driven or Steam Pilot Vessel Under Way, Shows Flare-up or Intermittent Light

INTERNATIONAL AND INLAND

At Anchor	Under Way	At Anchor on Duty
	(Shows Flare-up at Intervals)	
(Shows Flare-up at Intervals)	(Shows Side Lights at Intervals)	(Shows Flare-up at Intervals)
Steam Pilot Vessel	Sailing Pilot Vessel	

All pilot vessels at anchor in International Waters shall also display anchor lights

FIG. 17.2

(c) All pilot-vessels, when engaged on their stations on pilotage duty and at anchor, shall carry the lights and show the flares prescribed in sections (a) and (b), except that the sidelights shall not be shown.

When engaged on her station on pilotage duty and in waters of the United States, and at anchor, she shall carry in addition to the lights required for all pilot boats the red light above mentioned, but not the

INTERNATIONAL RULES

They shall also carry the anchor light or lights prescribed in Rule 11.

(d) All pilot-vessels, whether at anchor or not at anchor, shall, when not engaged on their stations on pilotage duty, carry the same lights as other vessels of their class and tonnage.

INLAND RULES

colored side lights. When not engaged on her station on pilotage duty, she shall carry the same lights as other steam vessels.

Pilot vessels, when not engaged on their station on pilotage duty, shall carry lights similar to those of other vessels of their tonnage.

Not in Pilot Rules

COMMENT: The International Rules (1948) divide pilot vessels into the following classes:

1. Sailing pilot vessels.
2. Sailing pilot vessels which go alongside.
3. Power-driven pilot vessels.

The Inland Rules divide pilot vessels as follows:

1. All pilot vessels.
2. Pilot vessels which go alongside.
3. Steam pilot vessels.

These classes under way in both waters may be grouped as follows:

1. All pilot vessels carry or show a white all-around masthead light.
2. Power-driven or steam pilot vessels carry an all-around red light below the white masthead light.
3. All pilot vessels show steady or flashing side lights underway.
4. All pilot vessels show a flare-up or intermittent white light at intervals.

Pilot vessels on station at anchor show anchor lights in International Waters, but not in Inland Waters, the white masthead light, the red light below it if the vessel is steam or power-driven, and a flare-up or flashing white light.

All pilot vessels carry the same lights as other vessels of their tonnage and class when not on station.

The differences between the two Rules are: (a) the intervals between flares, (b) distances of visibility, (c) the use of an intermittent light in place of a flare on power-driven pilot vessels in International Waters, and (d) the classification of pilot vessels.

COURT DECISIONS: "When such pilot-boat is cruising (off the southern New Jersey coast), I think it clear that she is a pilot-vessel engaged on her station on pilotage duty." (THE HAVERTON, 31 F 563.)

INTERNATIONAL RULES

Rule 9 (Figs. 17.3 and 17.4)

(a) Fishing vessels when not fishing shall show the lights or shapes prescribed for similar vessels of their tonnage. When fishing they shall show only the lights or shapes prescribed by this Rule, which lights or shapes, except as otherwise provided, shall be visible at a distance of at least 2 miles.

(b) Vessels fishing with trolling (towing) lines, shall show only the lights prescribed for a power-driven or sailing vessel under way as may be appropriate.

(c) Vessels fishing with nets or lines, except trolling (towing) lines, extending from the vessel not more than 500 feet horizontally into the seaway shall show, where it can best be seen, one all round white light and in addition, on approaching or being approached by another vessel, shall show a second white light at least 6 feet below the first light and at a horizontal distance of at least 10 feet

INLAND RULES

Article 9 (Fig. 17.5)

ART. 9. (*a*) *Fishing vessels of less than ten gross tons, when under way and when not having their nets, trawls, dredges, or lines in the water, shall not be obliged to carry the colored side lights; but every such vessel shall, in lieu thereof, have ready at hand a lantern with a green glass on one side and a red glass on the other side, and on approaching to or being approached by another vessel such lantern shall be exhibited in sufficient time to prevent collision, so that the green light shall not be seen on the port side nor the red light on the starboard side.*

(*b*) *All fishing vessels and fishing boats of ten gross tons or upward, when under way and when not having their nets, trawls, dredges, or lines in the water, shall carry and show the same lights as other vessels under way.*

(*c*) *All vessels, when trawling, dredging, or fishing with any kind of drag nets or lines, shall exhibit, from some part of the vessel where they can be best seen, two lights. One of these lights shall be red and the other shall be white. The red light shall be above the white light, and shall be at a vertical distance from it of not less than six feet and not more than twelve feet; and the*

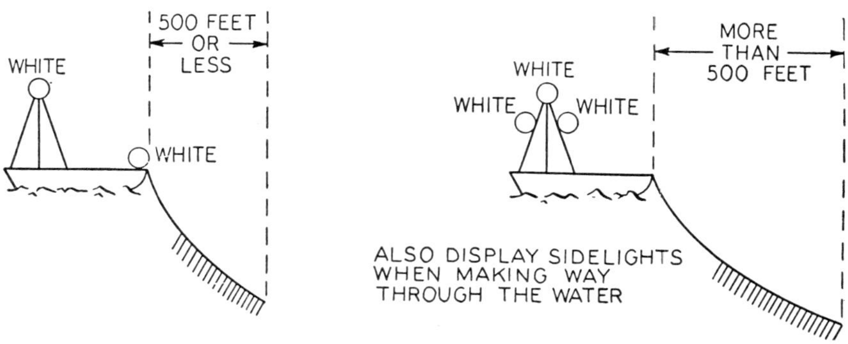

FIG. 17.3 VESSELS FISHING WITH NETS OR LINES EXCEPT TROLLING. INTERNATIONAL

INTERNATIONAL

Power-Driven Trawler

INTERNATIONAL

Line Fishing

Drift Net Fishing

Not more than 500 feet

INTERNATIONAL

Sailing Trawler or Dredge

Any Fishing Vessel Less Than 150′ in Length at Anchor and Attached to Fishing Gear

FIG. 17.4 VESSELS' LIGHTS

INTERNATIONAL RULES

away from it (6 feet in small open boats) in the direction in which the outlying gear is attached. By day such vessels shall indicate their occupation by displaying a basket where it can best be seen; and if they have their gear out while at anchor, they shall, on the approach of other vessels, show the same signal in the direction from the anchor ball towards the net or gear.

(d) Vessels fishing with nets or lines, except trolling (towing) lines, extending from the vessel more than 500 feet horizontally into the seaway shall show, where they can best be seen, three white lights at least 3 feet apart in a vertical triangle visible all around the horizon. When making way through the water, such vessels shall show the proper coloured sidelights but when not making way they shall not show them. By day they shall show a basket in the forepart of the vessel as near the stem as possible not less than 10 feet above the rail; and, in addition, where it can best be seen, one black conical shape, apex upwards. If they have their gear out while at anchor they shall, on the approach of other vessels, show the basket in the direction from the anchor ball towards the net or gear.

(e) Vessels when engaged in trawl-

INLAND RULES

horizontal distance between them, if any, shall not be more than ten feet. These two lights shall be of such a character and contained in lanterns of such construction as to be visible all around the horizon, the white light a distance of not less than three miles and the red light of not less than two miles.

FIG. 17.5

INTERNATIONAL RULES | ***INLAND RULES***

ing, by which is meant the dragging of a dredge net or other apparatus along or near the bottom of the sea, and not at anchor:

(i) If power-driven vessels, shall carry in the same position as the white light mentioned in Rule 2(a)(i) a tri-coloured lantern, so constructed and fixed as to show a white light from right ahead to 2 points (22½ degrees) on each bow, and a green light and a red light over an arc of the horizon from 2 points (22½ degrees) on each bow to 2 points (22½ degrees) abaft the beam on the starboard and port sides, respectively; and not less than 6 nor more than 12 feet below the tri-coloured lantern a white light in a lantern, so constructed as to show a clear, uniform, and unbroken light all round the horizon. They shall also show the stern light specified in Rule 10(a).

(ii) If sailing vessels, shall carry a white light in a lantern so constructed as to show a clear, uniform, and unbroken light all round the horizon, and shall also, on the approach of or to other vessels show, where it can best be seen, a white flare-up light in sufficient time to prevent collision.

Not in Inland Rules.

(iii) By day, each of the foregoing vessels shall show, where it can best be seen, a basket.

(f) In addition to the lights which they are by this Rule required to show vessels fishing may, if necessary in order to attract attention of approaching vessels, show a flare-up light. They may also use working lights.

Not in Inland Rules.

(g) Every vessel fishing, when at anchor, shall show the lights or shape specified in Rule 11(a), (b) or (c); and shall, on the approach of another vessel or vessels, show an additional white light at least 6 feet below the forward anchor light and at a hori-

Not in Inland Rules.

INTERNATIONAL RULES

zontal distance of at least 10 feet away from it in the direction of the outlying gear.

(h) If a vessel when fishing becomes fast by her gear to a rock or other obstruction she shall in daytime haul down the basket required by sections (c), (d) or (e) and show the signal specified in Rule 11(c). By night she shall show the light or lights specified in Rule 11(a) or (b). In fog, mist, falling snow, heavy rainstorms or any other condition similarly restricting visibility, whether by day or by night, she shall sound the signal prescribed by Rule 15(c) (v), which signal shall also be used, on the near approach of another vessel, in good visibility.

NOTE: For fog signals for fishing vessels, see Rule 15(c) (ix).

INLAND RULES

Not in Inland Rules.

PILOT RULES

80.32a Day marks for fishing vessels with gear out.—All vessels or boats fishing with nets or lines or trawls, when under way, shall in daytime indicate their occupation to an approaching vessel by displaying a basket where it can best be seen. If vessels or boats at anchor have their gear out, they shall, on the approach of other vessels, show the same signal in the direction from the anchor back towards the nets or gear.

COMMENT: *Inland Rules.* The Inland Rules have only one rule for fishing vessels, i.e., those which are trawling, dredging, or fishing at night with any kind of drag nets or lines. See Article 9(c). The remaining two sub-paragraphs, (a) and (b), in Article 9, prescribe lights for two sizes of fishing vessels when under way and not fishing, i.e., those of less than 10 gross tons and those of 10 gross tons or upward.

The Pilot Rules provide distinctive signals for vessels or boats fishing with nets or lines or trawls, under way and at anchor in the daytime. The Inland and Pilot Rules supplement each other.

International Rules for Fishing. The International Rules divide fishing vessels, under way and fishing, into five classes:

1. Vessels fishing with trolling lines (see 9(b)).
2. Vessels fishing with nets or lines, except trolling, extending from a vessel not more than 500 feet (see 9(c)).
3. The same type but extending more than 500 feet (see 9(d)).

4. Power-driven trawlers (9(e) (i)).
5. Sailing trawlers (9(e) (ii)).

Fishing vessels when not fishing are required to show the lights or shapes of similar vessels of their tonnage.

Fishing vessels whose gear has fouled on the bottom when fishing shall sound one prolonged blast followed by two short blasts on their whistle in restricted visibility and on the approach of another vessel in clear weather.

They are permitted to use working lights and may show a flare-up light to attract attention of approaching vessels.

Every vessel fishing at anchor must show the anchor ball or lights for vessels at anchor and, in addition, a basket or a white light below the anchor ball or forward anchor light and towards the outlying gear.

If the gear becomes fast to a rock or other obstruction, the vessel is required to haul down the usual fishing lights or shapes and hoist anchor lights or the anchor ball, as appropriate.

INTERNATIONAL RULES

Rule 10

(a) A vessel when under way shall carry at her stern a white light, so constructed that it shall show an unbroken light over an arc of the horizon of 12 points of the compass (135 degrees), so fixed as to show the light 6 points ($67\frac{1}{2}$ degrees) from right aft on each side of the vessel, and of such a character as to be visible at a distance of at least 2 miles. Such light shall be carried as nearly as practicable on the same level as the sidelights.

NOTE: For vessels engaged in towing or being towed, see Rules 3(b) and 5.

(b) In a small vessel, if it is not possible on account of bad weather or other sufficient cause for this light to be fixed, an electric torch or a lighted lantern shall be kept at hand ready for use and shall, on the approach of an overtaking vessel, be shown in sufficient time to prevent collision.

(c) A seaplane on the water when under way shall carry on her tail a white light, so constructed as to show an unbroken light over an arc of the

INLAND RULES

Article 10

ART. 10 (*a*) *A vessel when underway, if not otherwise required by these rules to carry one or more lights visible from aft, shall carry at her stern a white light, so constructed that it shall show an unbroken light over an arc of the horizon of twelve points of the compass, so fixed as to show the light six points from right aft on each side of the vessel, and of such a character as to be visible at a distance of at least two miles. Such light shall be carried as nearly as practicable on the same level as the side lights.*

(*b*) *In a small vessel, if it is not possible on account of bad weather or other sufficient cause for this light to be fixed, an electric torch or a lighted lantern shall be kept at hand ready for use and shall, on the approach of an overtaking vessel, be shown in sufficient time to prevent collision.*

Not in Inland Rules.

INTERNATIONAL RULES

horizon of 140 degrees of the compass, so fixed as to show the light 70 degrees from right aft on each side of the seaplane, and of such a character as to be visible at a distance of at least 2 miles.

INLAND RULES

CIVIL AIR REGULATIONS

3.700(c) Rear position light.—The rear position light shall be a white light mounted as far aft as practicable. The light shall be of an approved type.

3.701 Position light system dihedral angles.—The forward and rear position lights as installed on the airplane shall show unbroken light within dihedral angles specified in paragraphs (a) through (c) of this section.

(c) Dihedral angle A (aft) shall be considered formed by two intersecting vertical planes making angles of 70 degrees to the right and 70 degrees to the left, respectively, looking aft along the longitudinal axis, to a vertical plane passing through the longitudinal axis.

COMMENT: The International Rules now require a white stern light to be shown by all vessels when under way. Seaplanes on the water are also required to show a stern light by Air Traffic and International Rules. The latter Rules provide a partial exemption for small vessels.

Effective August 14, 1958, Inland Rules require "a vessel when under way, if not otherwise required by these rules to carry one or more lights visible from aft" to show the same stern light as required by the International Rules. Differences in language allow for special lights found in Inland Waters, such as the all-around after white range light.

The International Rules require an arc of visibility of 135 degrees for vessels, 140 degrees for seaplanes, and a visibility distance of 2 miles. The Inland Rules are identical for vessels. The Civil Air Regulations prescribe an arc of 140 degrees but no distance of visibility.

Another vessel should not see the stern light and the side lights of a seagoing vessel at the same time. But lights do "leak" across the limiting arc and it is possible that both may be seen. The side lights and the stern light of a seagoing vessel will probably be seen when about 2 points abaft the vessel's beam, unless so close "leaks" are on either side of the viewer.

INTERNATIONAL RULES

Rule 11 (Fig. 17.6)

(a) A vessel under 150 feet in length, when at anchor, shall carry in the forepart of the vessel, where it can best be seen, a white light in

INLAND RULES

Article 11 (Fig. 17.6)

ART. 11. *A vessel under one hundred and fifty feet in length when at anchor shall carry forward, where it can best be seen, but at a height not*

INTERNATIONAL RULES

a lantern so constructed as to show a clear, uniform, and unbroken light visible all round the horizon at a distance of at least 2 miles.

(b) A vessel of 150 feet or upwards in length, when at anchor, shall carry in the forepart of the vessel, at a height of not less than 20 feet above the hull, one such light, and at or near the stern of the vessel and at such a height that it shall be not less than 15 feet lower than the forward light, another such light. Both these lights shall be visible all round the horizon at a distance of at least 3 miles.

(c) Between sunrise and sunset every vessel when at anchor shall carry in the forepart of the vessel, where it can best be seen, one black ball not less than 2 feet in diameter.

INLAND RULES

exceeding twenty feet above the hull, a white light in a lantern so constructed as to show a clear, uniform, and unbroken light visible all around the horizon at a distance of at least one mile: Provided, That the Secretary of the Army may, after investigation, by rule, regulation, or order, designate such areas as he may deem proper as "special anchorage areas"; such special anchorage areas may from time to time be changed, or abolished, if after investigation the Secretary of the Army shall deem such change or abolition in the interest of navigation: Provided further, That vessels not more than sixty-five feet in length when at anchor in any such special anchorage area shall not be required to carry or exhibit the white light required by this article.

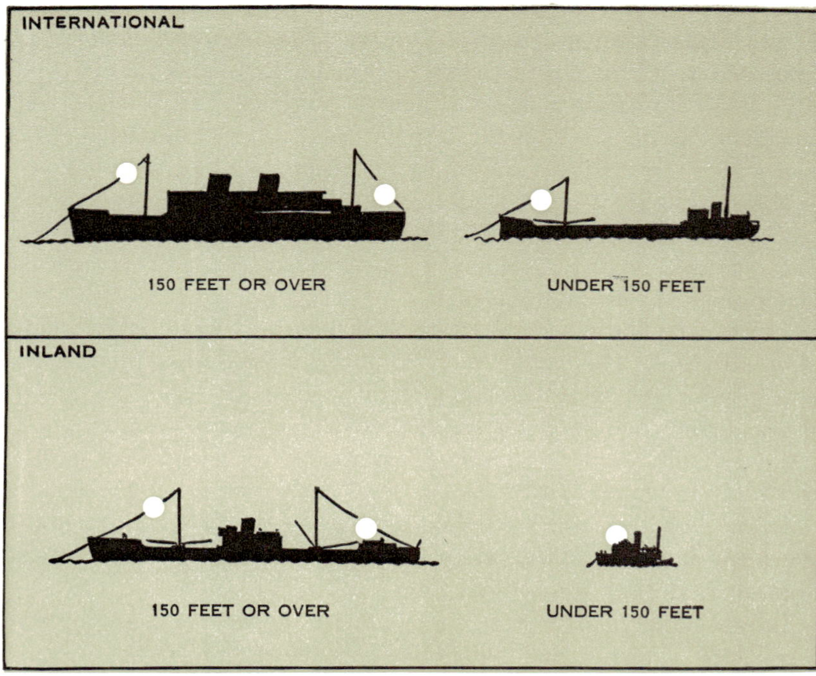

FIG. 17.6 LIGHTS ON VESSELS. Vessels at anchor.

INTERNATIONAL RULES

INLAND RULES

A vessel of one hundred and fifty feet or upward in length when at anchor, shall carry in the forward part of the vessel, at a height of not less than twenty and not exceeding forty feet above the hull, one such light, and at or near the stern of the vessel, and at such a height that it shall be not less than fifteen feet lower than the forward light, another such light.

The length of a vessel shall be deemed to be the length appearing in her certificate of registry.

(d) A vessel engaged in laying or in picking up a submarine cable or navigation mark, or a vessel engaged in surveying or underwater operations, when at anchor, shall carry the lights or shapes prescribed in Rule 4(c) in addition to those prescribed in the appropriate preceding sections of this Rule.

Not in Inland Rules.

(e) A vessel aground shall carry by night the light or lights prescribed in sections (a) or (b) and the two red lights prescribed in Rule 4(a). By day she shall carry, where they can best be seen, three black balls, each not less than 2 feet in diameter, placed in a vertical line one over the other, not less than 6 feet apart.

Not in Inland Rules.

(f) A seaplane on the water under 150 feet in length, when at anchor, shall carry, where it can best be seen, a white light, visible all round the horizon at a distance of at least 2 miles.

Not in Inland Rules.

(g) A seaplane on the water 150 feet or upwards in length, when at anchor, shall carry, where they can best be seen, a white light forward and a white light aft, both lights visible all round the horizon at a distance of at least 3 miles; and, in addition, if the seaplane is more than 150 feet in span, a white light on each side to indicate the maximum span, and visible, so far as practicable, all round the horizon at a distance of 1 mile.

Not in Inland Rules.

INTERNATIONAL RULES	*INLAND RULES*
(h) A seaplane aground shall carry an anchor light or lights as prescribed in sections (f) and (g), and in addition may carry two red lights in a vertical line, at least 3 feet apart, so placed as to be visible all round the horizon.	*Not in Inland Rules.*

AIR TRAFFIC RULE

3.704 Riding light.—(a) When a riding (anchor) light is required for a seaplane, flying boat, or amphibian, it shall be capable of showing a white light for at least 2 miles at night under clear atmospheric conditions.

(b) The riding light shall be installed to show the maximum unbroken light practicable when the airplane is moored or drifting on the water. Externally hung lights shall be acceptable.

CIVIL AIR REGULATIONS

60.23(c). All aircraft at anchor shall display anchor lights, unless in an area within which lights are not required for vessels at anchor. . . .

PILOT RULES

80.25 Vessels moored or at anchor.—Vessels of more than 65 feet in length when moored or anchored in a fairway or channel shall display between sunrise and sunset on the forward part of the vessel where it can best be seen from other vessels one black ball not less than two feet in diameter.

COMMENT: The Inland Rules allow the Secretary of the Army to designate "special anchorage areas" where vessels not more than 65 feet in length are not required to carry or exhibit an anchor light. This covers many small yachts at anchor in designated yacht anchorages in U.S. waters.

Aircraft carriers at anchor carry a white, all-around light on each bow and quarter just below the flight deck.

The International Rules prescribe two vertical red lights and the proper anchor lights for vessels aground. The location of the grounded vessel does not affect this requirement. The Inland Rules do not prescribe special lights for grounded vessels. They should show anchor lights.

The anchor ball is now required to be shown by a vessel at anchor in International and Inland waters, except yachts not more than 65 feet in length in U.S. yacht anchorages.

COURT DECISIONS: No rule required the launch (secured to a man-of-war's boat boom) to exhibit any additional light nor did the boom project an unusual or unreasonable distance (nearly 60 feet) for such a ship. . . ." (THE DIMITRI DONSKOI, 60 F 111.)

A vessel secured alongside another vessel at anchor must show the proper anchor lights. (THE PRUDENCE, 197 F 479.)

There are no rules, International or Inland, requiring a vessel moored to a wharf to show anchor lights. "It is not the practice to exhibit such (anchor) lights unless there are some special circumstances of danger, having reference to the ordinary navigation of other vessels." (HADDEN *v.* J. H. RUTTER, 35 F 365.) Anchor lights should be shown if custom requires it (SHIELDS *v.* THE MAYOR, 18 F 748), or if the vessel or any part of her projects near to the usual courses of passing vessels. (INDUSTRY, 27 F 767.)

If a vessel obstructs a slip by a line stretched across or by any other unusual obstruction, a light should be shown to warn entering vessels. (FULDA, 31 F 351.)

INTERNATIONAL RULES

Rule 12

Every vessel or seaplane on the water may, if necessary in order to attract attention, in addition to the lights which she is by these Rules required to carry, show a flare-up light or use a detonating or other efficient sound signal that cannot be mistaken for any signal authorized elsewhere under these Rules.

INLAND RULES

Article 12

ART. 12. *Every vessel may, if necessary, in order to attract attention, in addition to the lights which she is by these rules required to carry, show a flare-up light or use any detonating signal that cannot be mistaken for a distress signal.*

COMMENT: The International Rules provide three methods of attracting attention: (a) a flare-up, (b) a detonating signal, and (c) an efficient sound signal. The Inland Rules provide the first two methods but not the sound signal.

The use in International waters of efficient sound signal, i.e., a whistle, cannot consist of short, rapid blasts because such blasts show either a change in course or a doubt that sufficient action is being taken by the other vessel to avert collision (see Rule 28). The use of these signals is optional. (THE PACIFIC, 154 F 943.)

INTERNATIONAL RULES

Rule 13

(a) Nothing in these Rules shall interfere with the operation of any special rules made by the Government of any nation with respect to additional station and signal lights for ships of war, for vessels sailing under convoy, or for seaplanes on the water; or with the exhibition of recognition signals adopted by ship-

INLAND RULES

Article 13

ART. 13. *Nothing in these rules shall interfere with the operation of any special rules made by the Government of any nation with respect to additional station and signal lights for two or more ships of war or for vessels sailing under convoy, or with the exhibition of recognition signals adopted by shipowners, which have*

INTERNATIONAL RULES

owners, which have been authorized by their respective Governments and duly registered and published.

(b) Whenever the Government concerned shall have determined that a naval or other military vessel or water-borne seaplane of special construction or purpose cannot comply fully with the provisions of any of these Rules with respect to the number, position, range or arc of visibility of lights or shapes, without interfering with the military function of the vessel or seaplane, such vessel or seaplane shall comply with such other provisions in regard to the number, position, range or arc of visibility of lights or shapes as her Government shall have determined to be the closest possible compliance with these Rules in respect of that vessel or seaplane.

INLAND RULES

been authorized by their respective Governments, and duly registered and published.

Not in Inland Rules. (But, Act of December 3, 1945, as amended, which is applicable to Inland Waters, is similar with respect to lights.)

COMMENT: This International Rule allows naval vessels, vessels sailing under convoy, or seaplanes on the water to show additional station and signal lights. It does not exempt these vessels from showing the usual navigation lights. Subparagraph (b) of the International Rules has already been covered by Sec. 2 of Public Law 172 (see Chapter 16).

INTERNATIONAL RULES

Rule 14

A vessel proceeding under sail, when also being propelled by machinery, shall carry in the daytime forward, where it can best be seen, one black conical shape, point upwards, not less than 2 feet in diameter at its base.

INLAND RULES

Article 14

ART. 14. *A steam vessel proceeding under sail only, but having her funnel up, may carry in daytime, forward, where it can best be seen, one black ball or shape two feet in diameter.*

COMMENT: The International Rule deals with a vessel proceeding under sail when also being propelled by machinery. The Inland Rule affects a steam vessel proceeding under sail only but having her funnel up.

The distinguishing shape is a black cone, point up in International Waters, and a black ball or shape in Inland Waters.

18

Fog Signals, Rules 15, 16

INTERNATIONAL RULES

Rule 15 (Figs. 18.1 to 18.3)

(a) A power-driven vessel shall be provided with an efficient whistle, sounded by steam or by some substitute for steam, so placed that the sound may not be intercepted by any obstruction, and with an efficient fog-horn, to be sounded by mechanical means, and also with an efficient bell. A sailing vessel of 20 tons and upwards shall be provided with a similar fog-horn and bell.

(b) All signals prescribed by this Rule for vessels under way shall be given:

(i) by power-driven vessels on the whistle;

(ii) by sailing vessels on the fog-horn;

(iii) by vessels towed on the whistle or fog-horn.

(c) In fog, mist, falling snow, heavy rainstorms, or any other condition similarly restricting visibility, whether by day or night, the signals prescribed in this Rule shall be used as follows:

(i) A power-driven vessel making way through the water, shall sound at intervals of not more than 2 minutes a prolonged blast.

(ii) A power-driven vessel under way, but stopped and making no way through the water, shall sound at intervals of not more than 2 minutes two prolonged blasts, with an interval of about 1 second between them.

(iii) A sailing vessel under way shall sound, at intervals of not more

INLAND RULES

Article 15 (Figs. 18.1 to 18.3)

Art. 15. *All signals prescribed by this article for vessels under way shall be given:*

1. By "steam vessels" on the whistle or siren.

2. By "sailing vessels" and "vessels towed" on the fog horn.

The words "prolonged blast" used in this article shall mean a blast of from four to six seconds' duration.

A steam vessel shall be provided with an efficient whistle or siren, sounded by steam or by some substitute for steam, so placed that the sound may not be intercepted by any obstruction, and with an efficient fog horn; also with an efficient bell. A sailing vessel of twenty tons gross tonnage or upward shall be provided with a similar fog horn and bell.

In fog, mist, falling snow, or heavy rain storms, whether by day or night, the signals described in this article shall be used as follows, namely:

(*a*) *A steam vessel under way shall sound, at intervals of not more than one minute, a prolonged blast.*

(*b*) *Not in Inland Rules.*

(*c*) *A sailing vessel under way shall sound, at intervals of not more*

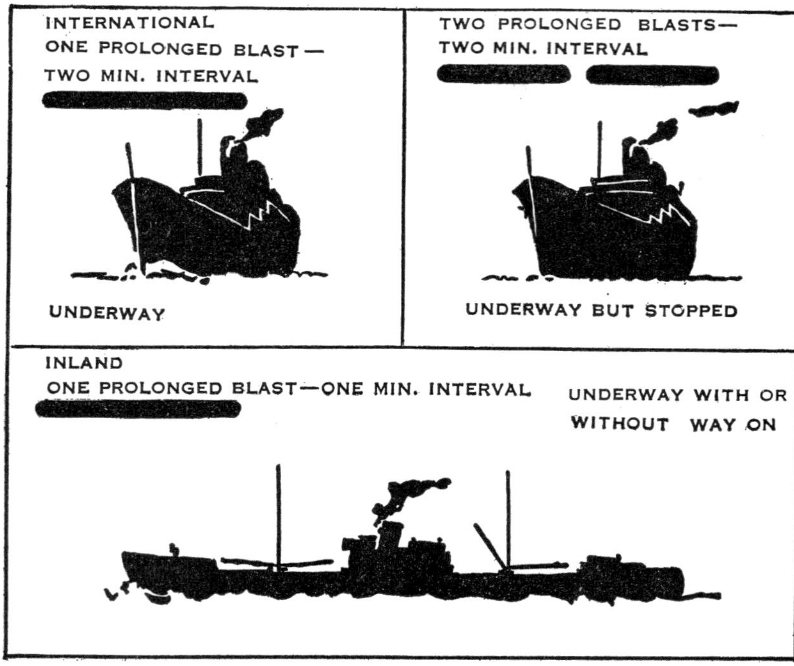

FIG. 18.1 SOUND SIGNALS IN THICK WEATHER. Steam vessels under way.

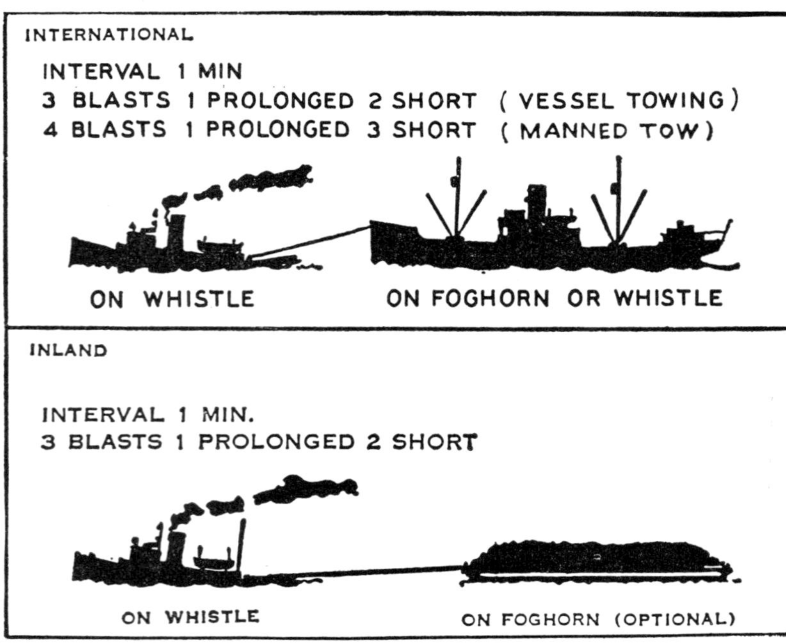

FIG. 18.2 SOUND SIGNALS IN THICK WEATHER. Vessels towed or towing.

INTERNATIONAL RULES	*INLAND RULES*
than 1 minute, when on the starboard tack one blast, when on the port tack two blasts in succession, and when with the wind abaft the beam three blasts in succession.	*than one minute, when on the starboard tack, one blast; when on the port tack, two blasts in succession, and when with the wind abaft the beam, three blasts in succession.*
(iv) A vessel when at anchor shall at intervals of not more than 1 minute ring the bell rapidly for about 5 seconds. In vessels of more than 350 feet in length the bell shall be sounded in the forepart of the vessel, and in addition there shall be sounded in the after part of the vessel, at intervals of not more than 1 minute for about 5 seconds, a gong or other instrument, the tone and sounding of which cannot be confused with that of the bell. Every vessel at anchor may in addition, in accordance with Rule 12, sound three blasts in succession, namely, one short, one prolonged, and one short blast, to give warning of her position	*(d) A vessel when at anchor shall, at intervals of not more than one minute, ring the bell rapidly for about five seconds.*

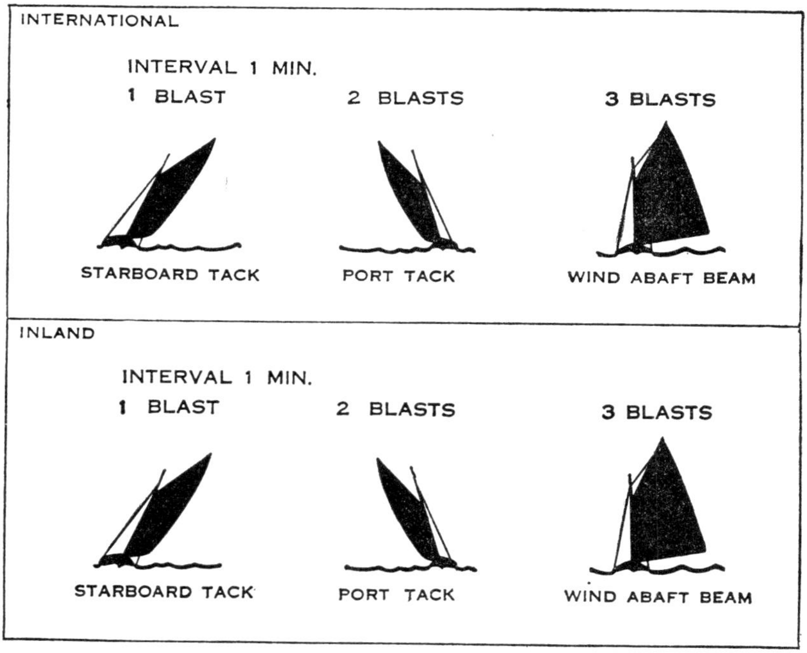

FIG. 18.3 SOUND SIGNALS IN THICK WEATHER. Sail vessels under way.

INTERNATIONAL RULES

and of the possibility of collision to an approaching vessel.

(v) A vessel when towing, a vessel engaged in laying or in picking up a submarine cable or navigation mark, and a vessel under way which is unable to get out of the way of an approaching vessel through being not under command or unable to manoeuvre as required by these Rules shall, instead of the signals prescribed in subsections (i), (ii) and (iii) sound, at intervals of not more than 1 minute, three blasts in succession, namely, one prolonged blast followed by two short blasts.

(vi) A vessel towed, or, if more than one vessel is towed, only the last vessel of the tow, if manned, shall, at intervals of not more than 1 minute, sound four blasts in succession, namely, one prolonged blast followed by three short blasts. When practicable, this signal shall be made immediately after the signal made by the towing vessel.

(vii) A vessel aground shall give the signal prescribed in subsection (iv) and shall, in addition, give three separate and distinct strokes on the bell immediately before and after each such signal.

(viii) A vessel of less than 20 tons, a rowing boat, or a seaplane on the water, shall not be obliged to give the above-mentioned signals, but if she does not, she shall make some other efficient sound signal at intervals of not more than 1 minute.

(ix) A vessel when fishing, if of 20 tons or upwards, shall at intervals of not more than 1 minute, sound a blast, such blast to be followed by ringing the bell; or she may sound, in lieu of these signals, a blast consisting of a series of several alternate notes of higher and lower pitch.

INLAND RULES

(e) A steam vessel when towing, shall, instead of the signals prescribed in subdivision (a) of this article, at intervals of not more than one minute, sound three blasts in succession, namely, one prolonged blast followed by two short blasts. A vessel towed may give this signal and she shall not give any other.

Not in Inland Rules.

(f) All rafts or other water craft, not herein provided for, navigating by hand power, horse power, or by the current of the river, shall sound a blast of the fog horn, or equivalent signal, at intervals of not more than one minute.

Not in Inland Rules.

PILOT RULES

80.12 Fog signals.—In fog, mist, falling snow, or heavy rainstorms, whether by day or night, signals shall be given as follows:

A steam vessel under way, except when towing other vessels or being towed, shall sound, at intervals of not more than 1 minute, on the whistle or siren, a prolonged blast.

A steam vessel when towing other vessels shall sound, at intervals of not more than 1 minute, on the whistle or siren, three blasts in succession, namely, one prolonged blast followed by two short blasts.

A vessel towed may give, at intervals of not more than 1 minute, on the fog horn, a signal of three blasts in succession, namely, one prolonged blast followed by two short blasts, and she shall not give any other.

A vessel when at anchor shall, at intervals of not more than 1 minute, ring the bell rapidly for about 5 seconds.

COMMENT: The International and Inland (Pilot) Rules require power-driven and steam vessels to sound fog signals on the whistle or siren. U.S. naval vessels use the whistle for fog signals.

Both Rules require sailing vessels to use a foghorn.

The International Rules require vessels towed to sound fog signals on the whistle or foghorn, but the Inland and Pilot Rules prescribe the foghorn.

The International Rules go a step farther than the Inland Rules about the conditions under which fog signals must be sounded. They state that fog signals shall be used not only in fog, mist, falling snow, heavy rainstorms, as required by Inland Rules, but also in "any other condition similarly restricting visibility." The Courts have ruled that statutory (Inland) rules apply to any obscuration of the air, such as smoke. (THE YOSEMITE, 28 F 2d, 939.) Fog signals should therefore be sounded in smoke or sandstorms in both waters if visibility is restricted.

Both the International and Inland Rules require a power-driven (steam) vessel in a fog to sound a prolonged blast. The interval is different under these two Rules—not more than 2 minutes in International Waters and not more than 1 minute in Inland Waters. The International Rules contains an additional paragraph, 15(c) (ii), which requires "a power-driven vessel under way but stopped and making no way through the water" to sound two prolonged blasts at intervals of not more than 2 minutes.

The International Rules prescribe the same fog signal for sailing vessels as the Inland Rules and at the same interval, not more than 1 minute.

The new International Rules provide an additional fog signal for vessels more than 350 feet in length when at anchor. They are required to sound a gong or other instrument, aft, in addition to the fog bell forward. The International Rules, 1948, also prescribe a new additional fog signal which may be sounded by a vessel at anchor in a fog to give warning of her position and of the possibility of collision—one short, one prolonged, and one short blast. The

Inland Rules neither recognize long vessels nor provide such a warning fog signal.

The International Rules have improved on the Inland Articles which prescribe a distinctive fog signal for steam vessels when towing, by applying the Rule to *all* vessels and by including "a vessel engaged in laying or in picking up a submarine cable or navigation mark, and a vessel under way which is unable to get out of the way of an approaching vessel through being not under command or unable to maneuver." Thus, a sailing vessel under way in International waters in a fog but unable to get out of the way of an approaching vessel because she is becalmed or is not under command should sound one prolonged blast followed by two short blasts. These additional steam vessels can only sound the prolonged blast or one, two or three blasts, if sailing vessels, under Inland Rules.

The International Rules require "a vessel towed or, if more than one vessel is towed, the last vessel of the tow, if manned, to sound a new, distinctive fog signal, one prolonged blast, followed by three short blasts." The Inland Rules permit but do not require towed vessels to sound three blasts in succession, one prolonged followed by *two* short blasts.

The International Rules contain a special rule for a vessel aground. She is required to sound the usual at anchor fog signals and "in addition, three separate and distinct strokes on the bell immediately before and after each such signal." The Inland Rules do not have a distinctive fog signal for a vessel aground. The Courts have ruled that distress signals, i.e., a continuous sounding with any fog-signaling apparatus, is proper for a vessel aground in Inland Waters. (THE LEVIATHAN, 286 F 745.)

Rule 15(c) (ix) contains the usual International fog signal for fishing vessels, if of 20 tons or upwards—a blast followed by ringing the bell. The blast and bell may be replaced by a blast consisting "of a series of several alternate notes of higher and lower pitch." This is a new type of fog signal.

In the case of the QUEVILLY, 253 F 415, a vessel was found at fault because she did not increase the frequency of her fog signal when proceeding through an anchorage.

LA BOYTEAUX suggests that fog signals should be started before the visibility has dropped to 2 miles which is the distance side lights must be visible.

Naval and some merchant vessels may use an automatic fog whistle. When no vessel is heard in the vicinity, it is a convenient, labor-saving device. After the fog whistle or horn of an approaching vessel is heard, the "automatic" should be shifted to "hand." Thereafter, the fog signal should be sounded more frequently but not immediately after the fog signal of the other vessel.

Both rules are so worded that vessels in or near fog shall sound only fog signals while they are not in sight of each other. There is one exception. Under the Inland and Pilot Rules the danger signal may be sounded when in doubt as to the course or intention of the other vessel, even though she is not in sight in the fog. (THE VIRGINIAN, 238 F 156.)

Fog distorts, suppresses, and changes the direction of fog signals at times.

The International and Inland Rules require power-driven (steam) vessels to be provided with an efficient whistle or siren, foghorn, and bell. Both Rules also require sailing vessels of 20 gross tons or upwards to be provided with an efficient foghorn and bell. Only the International Rules require these vessels to sound their foghorns by mechanical means.

COURT DECISIONS: The International Rules now require the last vessel of a tow, if manned, to sound a distinctive fog signal. The Inland Rules allow but do not require a vessel towed to sound the same signal as the towing steam vessel. A tow in Inland Waters need not sound a fog signal at all times in a fog, but it should do so when the circumstances make it desirable to indicate its location. The tow should sound it when it is in crowded waters or when the tow is long. (OPHELIA, 44 F 941.)

Naval vessels are not exempted from sounding fog signals even in war. They may not do so but, if a collision results, they may be found in fault (WATTS *v.* U.S., 123 F 105).

The Courts have not ruled the kinds of whistles and horns that are efficient, but they have ruled in definite cases that the fog-signaling apparatus was not efficient.

A steamer which left port with her whistle "out of commission" and which sounded a foghorn and not a whistle was held at fault. (THE MINNESOTA, 189 F 706.)

"A vessel is under obligation to observe the rule (sounding fog signals) not only when she is actually enveloped in the fog, but also when she is so near it that it is necessary that her position should be known to every vessel that may happen to be within it." (THE PERKIOMEN, 27 F 573.)

A vessel which is lying alongside a pier not projecting beyond the end of the pier into the stream is not required to sound fog signals. (THE EXPRESS, 48 F 323.)

A court has held recently that a radar-equipped vessel was in fault for entering a fog bank without using her radar. (THE MEDFORD, 65 F Supp. 622.)

When a tow is anchored, proper signals must be given by each vessel (at least by each vessel on the outside of a tier). The tug, if present, is responsible. (THE RALIEGH, 44 F 781.) If the tug is not present, the tow is responsible. (JERSEY CENTRAL, 221 F 625.)

The law requires every vessel at anchor in a group to sound a fog signal. (COHOCTON, 299 F 319.)

SUMMARY OF FOG SIGNALS

Type of vessel	International rules		Inland rules	
	Signal	Maximum interval	Signal	Maximum interval
Power-driven or steam vessel under way making way through water	—	2 min	—	1 min
Power-driven or steam vessel, under way but stopped and making no way through water	— —	2 min	—	1 min
Vessel towing	—··	1 min	—··	1 min
Vessel laying submarine cable or navigation mark	—··	1 min	—	1 min
Vessel not under command	—··	1 min	—	1 min
Vessel towed †	—···	1 min	—··	1 min
Vessel at anchor	*	1 min	*	1 min
Vessel aground	0 * 0	1 min	*	1 min
Sailing vessel				
On starboard tack	—	1 min	—	1 min
On port tack	— —	1 min	— —	1 min
Wind abaft beam	— — —	1 min	— — —	1 min
Vessels fishing, under way	—*	1 min	—	1 min
Vessels fishing, fast to a rock	—··	1 min	*	1 min

0 represents three strokes of bell.

* represents bell, forward, 5 seconds, except International Waters, where vessels over 350 feet sound gong aft also.

— represents a prolonged blast.

· represents a short blast.

† Optional, Inland. Single vessel, or last vessel of tow, International, *when* manned.

Sounding of gong aft applies only to vessels of more than 350 feet in length at anchor or aground in International waters. Vessels at anchor in International Waters may also sound a three-blast signal, one short, one prolonged, one short at unspecified intervals as a collision warning.

INTERNATIONAL RULES

Rule 16

(a) Every vessel, or seaplane when taxi-ing on the water, shall, in fog, mist, falling snow, heavy rainstorms or any other condition similarly restricting visibility, go at a moderate speed, having careful regard to the existing circumstances and conditions.

(b) A power-driven vessel hearing, apparently forward of her beam, the fog-signal of a vessel the position of which is not ascertained, shall, so far as the circumstances of the case admit, stop her engines, and then navigate with caution until danger of collision is over.

INLAND RULES

Article 16

ART. 16. *Every vessel shall, in a fog, mist, falling snow, or heavy rain storms, go at a moderate speed, having careful regard to the existing circumstances and conditions.*

A steam vessel hearing, apparently forward of her beam, the fog signal of a vessel the position of which is not ascertained shall, so far as the circumstances of the case admit, stop her engines, and then navigate with caution until danger of collision is over.

PILOT RULES

80.13 Speed in fog; posting of rules; diagrams—(a) Moderate speed in fog.—Every steam vessel shall, in a fog, mist, falling snow, or heavy rain storms, go at a moderate speed, having careful regard to the existing circumstances and conditions.

A steam vessel hearing, apparently forward of her beam, the fog signal of a vessel the position of which is not ascertained shall, so far as the circumstances of the case admit, stop her engines and then navigate with caution until danger of collision is over.

COMMENT: Masters of vessels in collision have pleaded a variety of excuses to prove that their speed was moderate under the existing conditions and circumstances. These excuses have included maneuverability, schedules, inability to slow down, and carrying U.S. mails. These excuses have seldom been accepted. Recently, they have pleaded radar, with no greater success. The courts have not been sympathetic to arguments that radar justifies a speed over that considered moderate for a vessel without radar or, for that matter, neglect, for that reason, of any other requirement set forth in the rules. The sense of the decisions has been that radar should be used with, not in lieu of, Rules (Arts.) 15 and 16, and that Rule (Art.) 29 should ever be borne in mind. The latter states: "Nothing in these Rules shall exonerate any vessel, or the owner, master or crew thereof, from the consequences of any neglect to carry lights or signals, or of any neglect to keep a proper lookout, or of the neglect of any precaution which may be required by the ordinary practice of seamen, or by the special circumstances of the case."

COURT DECISIONS: The courts have handed down many decisions defining moderate speed. They may be condensed into two specific rulings: ". . . a steamer is bound to use only such precautions as will enable her to stop in time to avoid a collision, after the approaching vessel comes in sight, providing such approaching vessel is herself going at a moderate speed required by law." (THE UMBRIA, 166 U.S. 404.)

Again, the Circuit Court of Appeal, 9th District, ruled in the CHICAGO, 94 F 2d 754, in 1937 that "one of the very long established principles of law in maritime navigation is that a vessel shall not proceed in a fog at a speed at which she cannot be stopped dead in the water in one half of the visibility before her."

The use of radar will not excuse immoderate speed as interpreted under Rule (Art.) 16, nor will it serve as an excuse for noncompliance with the Rules. (ARGENTINA—ANTINOUS, 1957 AMC 2356.)

"Radar is an aid, not a substitute, for prudent seamanship. . . ." (THE BUCENTAUR—THE WILSON VICTORY, 125 F. Supp 42, 1955 A.M.C. 142.)

"Prudent navigation involves taking advantage of all safety devices at hand. . . ." (THE HINDOO—THE AUSTRALIA STAR, 172 F 2d 472, 338 U.S. 823.)

The phrase in the Rule "having careful regard to the existing circumstances and conditions" comprises density of fog, nearby rocks and shoals, probability of encountering other vessels, vessels in company, known positions of anchored vessels, fog signals heard, channels, and many other conditions.

The Rule regarding moderate speed applies to sailing vessels as well as steamers. (ADAMS *v.* U.S., D.C. Mass. 1921.)

". . . These rules [Article 16] govern the navigation of a war vessel in time of war. . . ." (N.Y. AND CUBA MAIL S.S. CO. *v.* U.S., D.C. S.D. N.Y. 1924.)

The courts have ruled vessels at fault for leaving a safe berth in a fog. (THE GEORGIA, 208 F 635, 642.)

"In a dense fog it is the duty of a steam vessel to anchor in a permissible anchorage as soon as the circumstances permit; but there is no absolute duty to anchor in a thoroughfare. . . ." (THE MOHEGAN, 28 F 2d 795.)

U.S. Code Title 33, Sec. 409, forbids vessels to anchor "in navigable channels in such a manner as to prevent or obstruct the passage of other vessels. . . ." But there is no rule which forbids a vessel to anchor in a channel in a fog, unless she obstructs navigation. (THE QUIRIGUA, 93 F 2d 297.)

The court in the CITY OF NORFOLK, 266 F 641, ruled that a vessel caught in a dense fog and anchoring in the channel because in her master's best judgment it was safer than to attempt to reach another anchorage does not violate the statute and is not at fault.

In the case of THE GEORGIA, 208 F 635, the court ruled: "While it may be reasonable to say that the statute does not absolutely prohibit anchoring in navigable channels or make it a fault to anchor where controlling conditions make it absolutely necessary, yet the question whether a vessel is anchored in such a manner as to prevent or obstruct the passage of other vessels must be determined by looking not alone to the chart and to the geography of the situation but also to weather conditions and to the usual course of vessels using the thoroughfare." A vessel should anchor usually in designated anchorage areas in a fog. (THE RICHMOND, 63 F 1020.)

"The presumption, where a moving vessel comes into collision with one at anchor, in a fog, and where there is no evidence of negligence on the part of the anchored vessel, is, by well established rule, against the moving vessel." (THE CANANOVA, 297 F 658.)

"The INDRAKUALA should at least have slowed down upon approaching and entering the fog. . . ." (THE JULIA LUCKENBACH *v.* INDRAKUALA, 219 F 600.)

The courts have ruled many times that a steam vessel must stop her engines as far as the circumstances permit when she hears, forward of her beam, the fog signal of a vessel, the position of which is not ascertained. (THE SENECA, D.C. S.D. N.Y. 1908.) They have also ruled that a steamer should reduce her headway by reversing her engines in order to navigate with caution. (THE CITY OF ATLANTA, 26 F 456.)

"The notion that a ship, equipped with radar, may, once her navigation and range lights are bright, plunge through the seas at 15 knots in the hope that all other craft will keep clear of it cannot be accepted as a rule of safe and prudent navigation. . . ." (THE HINDOO—THE AUSTRALIA STAR, 172 F 2d 472, 338 U.S. 823.)

The apparent distance of the foghorn does not change the rule to stop the engines at once. Fog affects the sound of fog whistles and horns adversely. (LIE *v.* SAN FRANCISCO & PORTLAND S.S. CO., 243 U.S. 291.) The phrase "the position of which is not ascertained" has been ruled to include a knowledge of the position, i.e., the distance, direction, and course of the approaching vessel. (THE TREMONT, 160 F 1016.)

"It may be that proper observations on a PPI (i.e., radar scope) can 'ascertain' the position of a vessel. . . . They clearly did not do so in this case. . . ." (THE PRINS ALEXANDER, 2 Lloyds' List L.R.I.)

The courts have also ruled that the custom of some masters of porting the helm (turning right) when a fog signal is heard ahead is not navigating with caution. (THE COUNSELLOR, (1913) Prob. Div. 70.)

Sailing vessels are not required to stop when a fog signal is heard forward of the beam. They are required to proceed at a moderate speed in a fog. (THE CHATTAHOOCHEE, 173 U.S. 540.)

There is no right of way in a fog while the vessels are not in sight. The Steering and Sailing Rules do not apply until the vessels are in sight. If the vessels are then so close that both must take action to avoid collision, Article 27 applies. If there is room to maneuver in accordance with the Steering and Sailing Rules, the proper whistle signals should be sounded and the latter Rules obeyed.

"Scientific installations, and particularly radar, are potentially most valuable instruments for increasing safety at sea, but they only remain valuable if they are intelligently used, and if the officers responsible for working them work them and interpret them with intelligence. . . ." (THE ANNA SALEM, 1954, 1 Lloyds' List L.R. 475.)

19

Steering and Sailing Rules, Miscellaneous, Rules (Articles) 17–32

INTERNATIONAL RULES

PART C—STEERING AND SAILING RULES

Preliminary

1. In obeying and construing these Rules, any action taken should be positive, in ample time, and with due regard to the observance of good seamanship.

2. Risk of collision can, when circumstances permit, be ascertained by carefully watching the compass bearing of an approaching vessel. If the bearing does not appreciably change, such risk should be deemed to exist.

3. Mariners should bear in mind that seaplanes in the act of landing or taking off, or operating under adverse weather conditions, may be unable to change their intended action at the last moment.

INLAND RULES

PART IV—STEERING AND SAILING RULES

Preliminary—Risk of Collision

Not in Inland Rules.

Same as International Rules.

Not in Inland Rules.

PILOT RULES

80.02 Definition of steam vessel and vessel under way; risk of collision.—In the rules in this part the words "steam vessel" shall include any vessel propelled by machinery. A vessel is under way, within the meaning of the rules in this part, when she is not at anchor, or made fast to the shore, or aground. Risk of collision can, when circumstances permit, be ascertained by carefully watching the compass bearing of an approaching vessel. If the bearing does not appreciably change, such risk should be deemed to exist.

Comment: These are not compulsory rules, only suggestions that the risk of collision may be ascertained, with some certainty, by watching the compass bearing. They should be heeded, nevertheless. The bearing should be taken at adequate distance on a sharply defined point and repeated at frequent intervals until the vessels are safely past each other. The Rules require both vessels to take particular, prescribed actions when risk of collision is involved. A constant or nearly constant bearing is one indicator. It is obvious that the risk of collision does not exist when the two vessels are so far apart that a change in speed or course of either does not affect the other. However, such vessels may be on converging courses and speeds which will bring them dangerously close. One or both vessels may also change course or speed with the same result. When the two vessels have reached a point where a change in speed or course of one does affect the movements of the other and when the bearing is not changing, then a risk of collision is involved, and timely action must be taken. The Courts have handed down many rulings on this subject. One of these rulings has already been quoted, see Chapter 15, Risk of Collision. For further comment, see Rule 19.

Three final cautions are desirable:

1. Very gradually changing compass bearings should be watched suspiciously.
2. Comply with the Rules while there is yet time for each to understand the proposed movements of the other and to take the action prescribed by the Rules.
3. If a change of course is required by the Rules, it should be large enough to be noticeable by the other vessel.

INTERNATIONAL RULES

Rule 17 (Fig. 19.1)

When two sailing vessels are approaching one another, so as to involve risk of collision, one of them shall keep out of the way of the other, as follows:

(a) A vessel which is running free shall keep out of the way of a vessel which is close-hauled.

(b) A vessel which is close-hauled on the port tack shall keep out of the way of a vessel which is close-hauled on the starboard tack.

(c) When both are running free, with the wind on different sides, the vessel which has the wind on the port side shall keep out of the way of the other.

(d) When both are running free, with the wind on the same side, the vessel which is to windward shall

INLAND RULES

Article 17 (Fig. 19.1)

Same as International Rules.

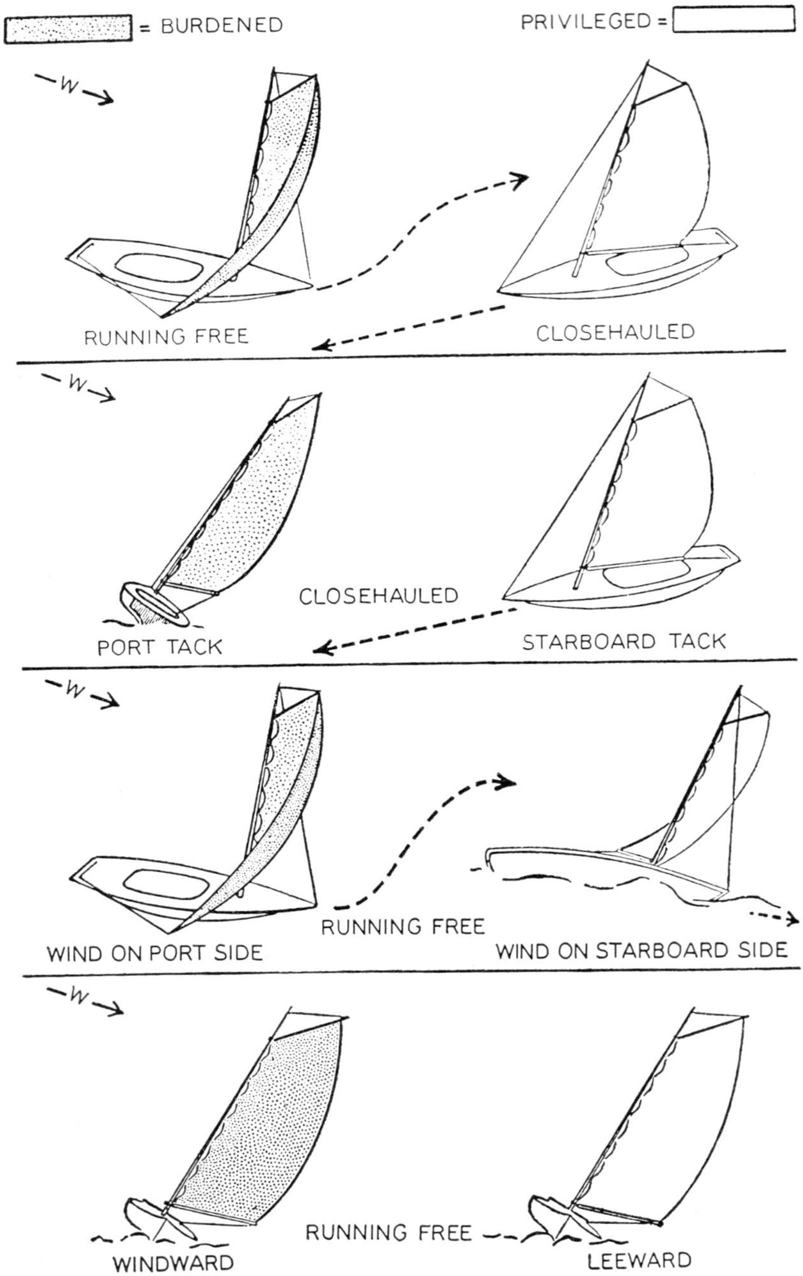

FIG. 19.1 SAIL VESSELS MEETING

INTERNATIONAL RULES

keep out of the way of the vessel which is to leeward.

(e) A vessel which has the wind aft shall keep out of the way of the other vessel.

INLAND RULES

COMMENT: The Rules apply only when risk of collision exists. The "burdened" vessel has one simple rule to observe—"keep out of the way of the other." The "privileged" vessel must hold her course. (ST. JOHN *v.* PAINE, 10 HOW (51 U.S.) 557.) The terms "burdened" and "privileged" depend on the course relative to the wind—the "tack" or whether the vessel is "running free" or "close-hauled." The "burdened" vessel is the one on the port tack, i.e., with the wind coming over the port side, or the one which is running free, or the one which is to windward.

The dividing line between "close-hauled" and "running free" has been the subject of a number of decisions, not always consistent, by U.S. and British Courts. The weight of opinion is that a vessel heading within but not including two points of being close-hauled, i.e., two points "free," is close-hauled within the rules. She is "running free" on other headings.

No specific provision is made in the Rules for a possible collision when both vessels are close-hauled on the same tack. The Overtaking Rule, Rule 24, would apply in such cases.

Sailing vessels are required to keep out of the way of vessels fishing with nets, lines, or trawls. See Rule 26 (Art. 26).

Steamers, i.e., power-driven vessels, must keep out of the way of sailing vessels. See Rule 20 (Art. 20).

There is an exception. Sailing vessels overtaking power-driven (steam) vessels must keep out of the way of the latter. See Rule 24 (Art. 24).

If a collision is imminent and it is apparent that the burdened sailing vessel cannot avoid a collision by her own acts, the privileged sailing vessel must take such seamanlike action as desirable to prevent the collision. (PIERRE CORNEILLE, 133 F 604.) A slight luffing is not a change of course. (THE MARMION, 1 ASP. 412.) If a vessel changes course into the wind by two points, it is a change of course under the Rules. (THE EARL WEMYS, 6 ASP. 407.) The change of course which is forced by a shoal or other well-known danger should be foreseen and expected by the burdened vessel. A vessel with the wind 1½ to 2½ points from dead astern was held sailing "with the wind aft." (TH' GOV. AMES, 187 F 40.)

INTERNATIONAL RULES

Rule 18 (Figs. 19.2, 19.3, and 19.4)

(a) When two power-driven vessels are meeting end on, or nearly

INLAND RULES

Article 18 (Figs. 19.2, 19.3, and 19.4)

ART. 18. RULE I. *When steam vessels are approaching each other head*

INTERNATIONAL RULES

end on, so as to involve risk of collision, each shall alter her course to starboard, so that each may pass on the port side of the other. This Rule only applies to cases where vessels are meeting end on, or nearly end on, in such a manner as to involve risk of collision, and does not apply to two vessels which must, if both keep on their respective courses, pass clear of each other. The only cases to which it does apply are when each of two vessels is end on, or nearly end on, to the other; in other words, to cases in which, by day, each vessel sees the masts of the other in a line, or nearly in a line, with her own; and by night, to cases in which

INLAND RULES

and head, that is, end on, or nearly so, it shall be the duty of each to pass on the port side of the other; and either vessel shall give, as a signal of her intention, one short and distinct blast of her whistle, which the other vessel shall answer promptly by a similar blast of her whistle, and thereupon such vessels shall pass on the port side of each other. But if the courses of such vessels are so far on the starboard of each other as not to be considered as meeting head and head, either vessel shall immediately give two short and distinct blasts of her whistle, which the other vessel shall answer promptly by two similar

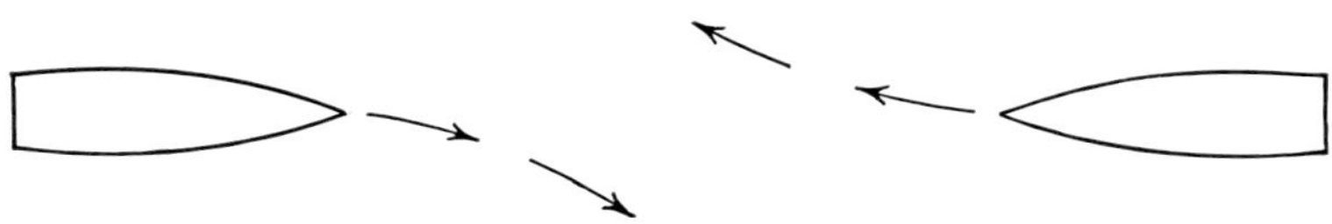

Here both ships change course to right and pass port side to port side, sounding one blast as course is changed. Same under International and Inland Rules.

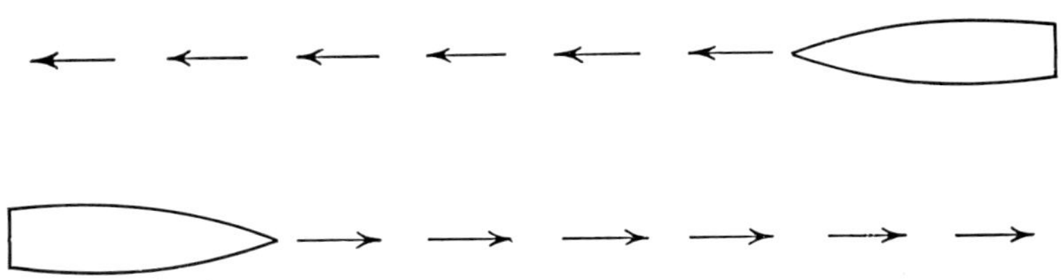

Both ships may hold course and pass port side to port side. No sound signal under International Rules unless course is changed. One blast given by either and answered by the other under Inland Rules.

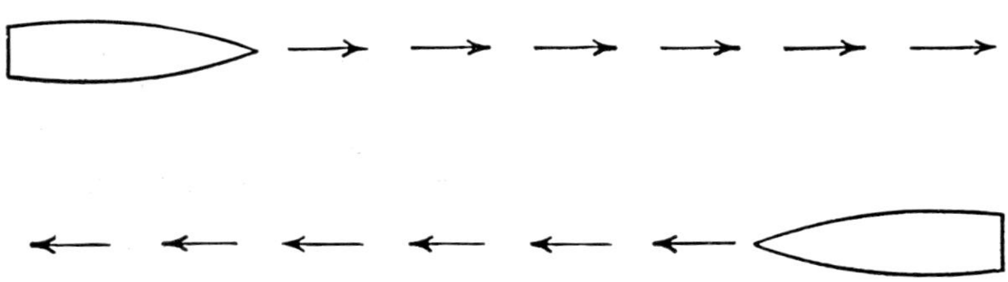

Both ships may hold course and pass starboard side to starboard side. No sound signal under International Rules unless course is changed. Two blasts given by either and answered by the other under Inland Rules.

FIG. 19.2 STEAM VESSELS MEETING HEAD-ON OR NEARLY HEAD-ON

INTERNATIONAL RULES

each vessel is in such a position as to see both the sidelights of the other. It does not apply, by day, to cases in which a vessel sees another ahead crossing her own course; or, by night, to cases where the red light of one vessel is opposed to the red light of the other or where the green light of one vessel is opposed to the green light of the other or where a red light without a green light or a green light without a red light is seen ahead, or where both green and red lights are seen anywhere but ahead.

(b) For the purposes of this Rule and Rules 19 to 29 inclusive, except

INLAND RULES

blasts of her whistle, and they shall pass on the starboard side of each other.

The foregoing only applies to cases where vessels are meeting end on, or nearly end on, in such a manner as to involve risk of collision; in other words, to cases in which, by day, each vessel sees the masts of the other in a line, or nearly in a line, with her own, and by night to cases in which each vessel is in such a position as to see both the sidelights of the other.

It does not apply by day to cases in which a vessel sees another ahead

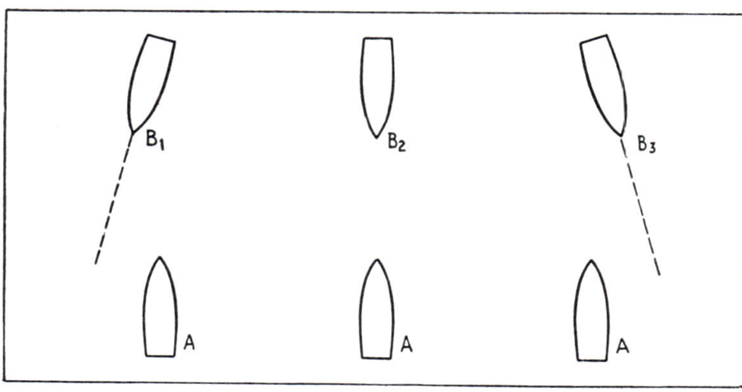

Showing "B" as Seen by "A"

Range Lights

FIG. 19.3 STEAMERS MEETING. (Sidelights should not show across bow, but sometimes do so)

INTERNATIONAL RULES

Rule 20(b), a seaplane on the water shall be deemed to be a vessel, and the expression "power-driven vessel" shall be construed accordingly.

INLAND RULES

crossing her own course, or by night to cases where the red light of one vessel is opposed to the red light of the other, or where the green light of one vessel is opposed to the green light of the other, or where a red light without a green light or a green light without a red light, is seen ahead, or where both green and red lights are seen anywhere but ahead.

Rule III. *If, when steam vessels are approaching each other, either vessel fails to understand the course or intention of the other, from any cause, the vessel so in doubt shall immediately signify the same by giving several short and rapid blasts, not less than four, of the steam whistle.*

Rule V. *Whenever a steam vessel is nearing a short bend or curve in the channel, where, from the height of the banks or other cause, a steam vessel approaching from the opposite direction can not be seen for a distance of half a mile, such steam vessel, when she shall have arrived within half a mile of such curve or bend, shall give a signal by one long blast of the steam whistle, which signal shall be answered by a similar blast given by any approaching steam vessel that may be within hearing. Should such signal be so answered by a steam vessel upon the*

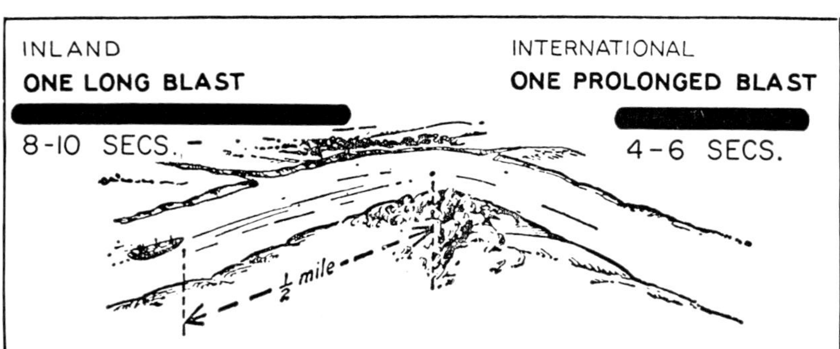

FIG. 19.4 STEAM VESSELS NEARING SHORT BEND

INTERNATIONAL RULES

INLAND RULES

farther side of such bend, then the usual signals for meeting and passing shall immediately be given and answered; but, if the first alarm signal of such vessel be not answered, she is to consider the channel clear and govern herself accordingly.

When steam vessels are moved from their docks or berths, and other boats are liable to pass from any direction toward them, they shall give the same signal as in the case of vessels meeting at a bend, but immediately after clearing the berths so as to be fully in sight they shall be governed by the steering and sailing rules.

Rule VIII. *When steam vessels are running in the same direction, and the vessel which is astern shall desire to pass on the right or starboard hand of the vessel ahead, she shall give one short blast of the steam whistle, as a signal of such desire, and if the vessel ahead answers with one blast, she shall direct her course to starboard; or if she shall desire to pass on the left or port side of the vessel ahead, she shall give two short blasts of the steam whistle as a signal of such desire, and if the vessel ahead answers with two blasts, shall direct her course to port; or if the vessel ahead does not think it safe for the vessel astern to attempt to pass at that point, she shall immediately signify the same by giving several short and rapid blasts of the steam whistle, not less than four, and under no circumstances shall the vessel astern attempt to pass the vessel ahead until such time as they have reached a point where it can be safely done, when said vessel ahead shall signify her willingness by blowing the proper signals. The vessel ahead shall in no case attempt to cross the bow or crowd upon the course of the passing vessel.*

INTERNATIONAL RULES

INLAND RULES

RULE IX. *The whistle signals provided in the rules under this article, for steam vessels meeting, passing, or overtaking, are never to be used except when steamers are in sight of each other, and the course and position of each can be determined in the day time by a sight of the vessel itself, or by night by seeing its signal lights. In fog, mist, falling snow or heavy rain storms, when vessels can not see each other, fog signals only must be given.*

AIR TRAFFIC RULES

60.22 Water operations.—An aircraft operated on the water shall, insofar as possible, keep clear of all vessels and avoid impeding their navigation. The following rules shall be observed with respect to other aircraft or vessels operated on the water:

(a) *Crossing.* The aircraft or vessel which has the other on its right shall give way so as to keep well clear;

(b) *Approaching head-on.* When aircraft, or an aircraft and vessel, approach head-on, or approximately so, each shall alter its course to the right to keep well clear;

(c) *Overtaking.* The aircraft or vessel which is being overtaken has the right-of-way, and the one overtaking shall alter its course to keep well clear;

(d) *Special circumstances.* When two aircraft, or an aircraft and vessel, approach so as to involve risk of collision each shall proceed with careful regard to existing circumstances and conditions including the limitations of the respective craft.

NOTE: The rules for operating aircraft on the surface of the water conform to marine rules for the operation of vessels. The "Special circumstances" rule is provided for situations wherein it may be impracticable or hazardous for a vessel or another aircraft to bear to the right because of depth of waterway, wind conditions, or other circumstances.

PILOT RULES

80.03 Signals.—The whistle signals provided in the rules in this part shall be sounded on an efficient whistle or siren sounded by steam or by some substitute for steam.

A short blast of the whistle shall mean a blast of about 1 second's duration.

A prolonged blast of the whistle shall mean a blast of from 4 to 6 seconds' duration.

One short blast of the whistle signifies intention to direct course to own starboard, except when two steam vessels are approaching each other at right angles or obliquely, when it signifies intention of steam vessel which is to starboard of the other to hold course and speed.

Two short blasts of the whistle signify intention to direct course to own port.

Three short blasts of the whistle shall mean, "My engines are going at full speed astern."

When vessels are in sight of one another a steam vessel under way whose engines are going at full speed astern shall indicate that fact by three short blasts on the whistle.

80.1 Danger signal.—If, when steam vessels are approaching each other, either vessel fails to understand the course or intention of the other, from any cause, the vessel so in doubt shall immediately signify the same by giving several short and rapid blasts, not less than four, of the steam whistle, the danger signal.

80.2 Cross signals.—Steam vessels are forbidden to use what has become technically known among pilots as "cross signals," that is, answering one whistle with two, and answering two whistles with one.

80.3 Vessels passing each other.—The signals for passing, by the blowing of the whistle, shall be given and answered by pilots, in compliance with the rules in this part, not only when meeting "head and head," or nearly so, but at all times when the steam vessels are in sight of each other, when passing or meeting at a distance within half a mile of each other, and whether passing to the starboard or port.

The whistle signals provided in the rules in this part for steam vessels meeting, passing, or overtaking are never to be used except when steam vessels are in sight of each other, and the course and position of each can be determined in the daytime by a sight of the vessel itself, or by night by seeing its signal lights. In fog, mist, falling snow, or heavy rainstorms, when vessels cannot so see each other, fog signals only must be given.

80.4.—Same as Inland Rule, Art. 18, Rule I.

80.5.—Same as Inland Rule, Art. 18, Rule V.

80.6 Vessels running in same direction; overtaking vessel.—When steam vessels are running in the same direction, and the vessel which is astern shall desire to pass on the right or starboard hand of the vessel ahead, she shall give one short blast of the steam whistle, as a signal of such desire, and if the vessel ahead answers with one blast, she shall direct her course to starboard; or if she shall desire to pass on the left or port side of the vessel ahead, she shall give two short blasts of the steam whistle as a signal of such desire, and if the vessel ahead answers with two blasts, shall direct her course to port; or if the vessel ahead does not think it safe for the vessel astern to attempt to pass at that point, she shall immediately signify the same by giving several short and rapid blasts of the steam whistle, not less than four, and under no circumstances shall the vessel astern attempt to pass the vessel ahead until such time as they have reached a point where it can be safely done, when said vessel ahead shall signify her willingness by blowing the proper signals. The vessel ahead shall in no case attempt to cross the bow or crowd upon the course of the passing vessel.

Every vessel coming up with another vessel from any direction more than two points abaft her beam, that is, in such a position with reference to the vessel which she is overtaking that at night she would be unable to see either of that vessel's side lights, shall be deemed to be an overtaking vessel; and no subsequent alteration of the bearing between the two vessels shall make the overtaking vessel a crossing vessel within the meaning of the rules in this part, or relieve her of the duty of keeping clear of the overtaken vessel until she is finally past and clear.

As by day the overtaking vessel cannot always know with certainty whether she is forward of or abaft this direction from the other vessel she should, if in doubt, assume that she is an overtaking vessel and keep out of the way.

80.13 (c) Diagrams.—The following diagrams are intended to illustrate the working of the system of colored lights and pilot rules.

FIRST SITUATION

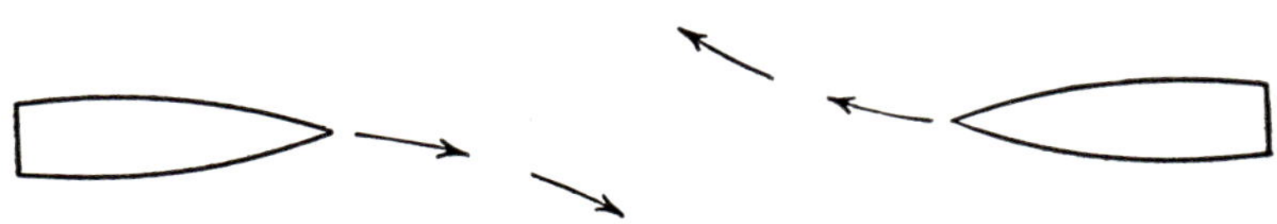

Here the two colored lights visible to each will indicate their direct approach "head and head" toward each other. In this situation it is a standing rule that both shall direct their courses to starboard and pass on the port side of each other, each having previously given one blast of the whistle.

SECOND SITUATION

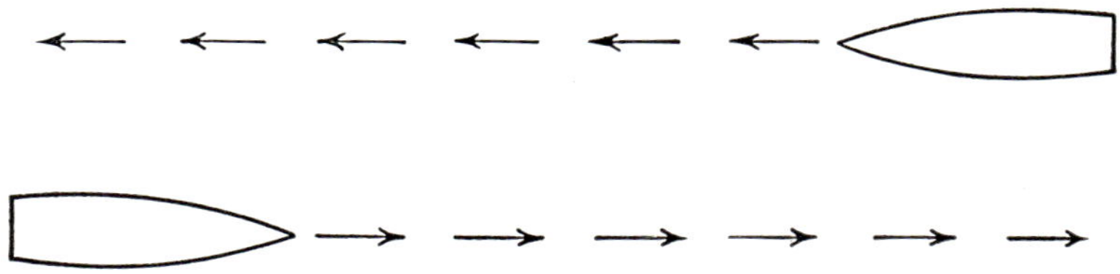

In this situation the red light only will be visible to each, the screens preventing the green lights from being seen. Both vessels are evidently passing to port of each other, which is rulable in this situation, each pilot having previously signified his intention by one blast of the whistle.

THIRD SITUATION

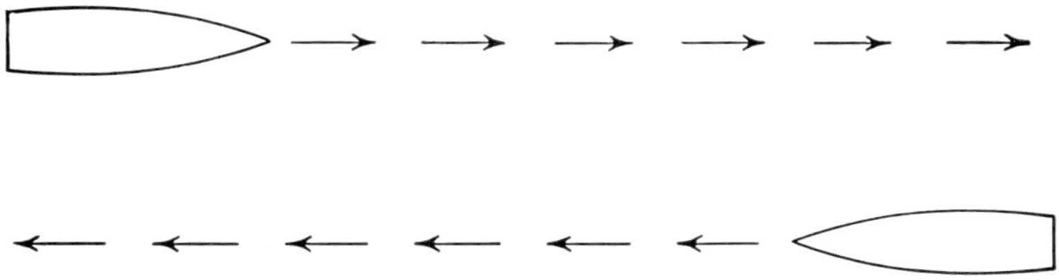

In this situation the green light only will be visible to each, the screens preventing the red light from being seen. They are therefore passing to starboard of each other, which is rulable in this situation, each pilot having previously signified his intention by two blasts of the whistle.

FOURTH SITUATION

In this situation one steam vessel is overtaking another steam vessel from some point within the angle of two points abaft the beam of the overtaken steam vessel. The overtaking steam vessel may pass on the starboard or port side of the steam vessel ahead after the necessary signals for passing have been given with assent of the overtaken steam vessel, as prescribed in § 80.6.

FIFTH SITUATION

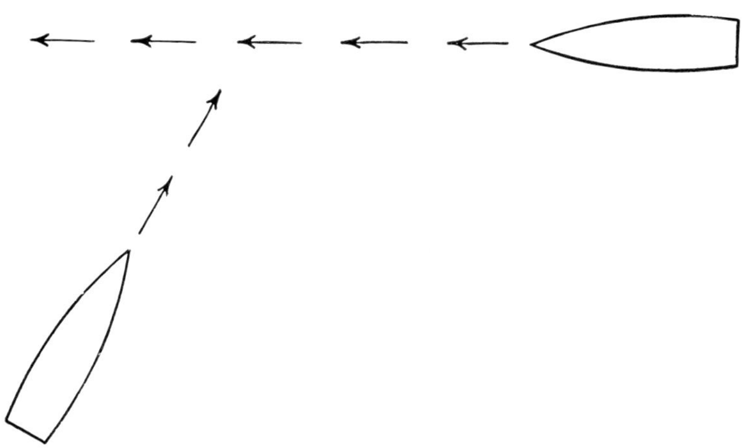

In this situation two steam vessels are approaching each other at right angles or obliquely in such manner as to involve risk of collision, other than where one steam vessel is overtaking another. The steam vessel which has the other on her own port side shall hold course and speed, and the other shall keep clear by crossing astern of the steam vessel that is holding course and speed, or, if necessary to do so, shall slacken her speed, stop, or reverse.

COMMENT: The "end on" meeting case is dangerous because the rate of approach of the two vessels is the sum of the speeds of both vessels. The rate is the greatest of any meeting situation. The time is shorter. Thus the need for a timely exchange of signals and a strict compliance with the Rules are greatest.

The dividing line between "end on" and "crossing" cases has not been determined with exactness because the Rules contain the phrases "end on or nearly so" and "end on or nearly end on." Related court decisions have held approaching vessels to be meeting when on courses diverging by as much as one or two points, rendering a stricter interpretation in open waters than in restricted waters, such as rivers and narrow channels. The true test is that contained in the rules, i.e., ability to see both side lights, essentially ahead.

There have been many instances of confusion between meeting and crossing cases near the bend of a narrow channel or river. The Courts have ruled that the present courses, when the two vessels are on opposite sides of a bend and in sight, do not determine whether the case is crossing or meeting. It is decided by the probable future courses when they actually pass each other. "The question, therefore, always turns on the reasonable inference to be drawn as to a vessel's future course from her position at a particular moment, and this greatly depends on the nature of the locality where she is at the moment." (THE PEKIN, APP. CAS. 532.) ". . . has reference to the position of the two vessels when they will be meeting and about to pass each other, and not to their position when signals are given, if at that time they are on a temporary course from which they will depart before they will nearly approach each other." (THE WILLIAM CHISHOM, 153 F 704.)

The custom of passing starboard to starboard in Hell Gate on the flood tide has been upheld by the Courts. The question of whether the custom prevails throughout the whole period of the flood tide or only at its greatest strength has not been decided. The custom is not accepted on the ebb tide. Generally speaking, a custom must be "a reasonable custom of navigation arising from the peculiarities of this locality." (TRANSFER NO. 21, 248 F 459.)

A vessel ascending a river against a current should stop, if necessary, to facilitate a safe passing because she can stop and maintain her position more readily than a vessel descending with the current. (THE GALATEA, 92 U.S. 439.)

A meeting vessel should slow, stop, or reverse if she considers the situation dangerous or uncertain and if her passing signal has not been assented to or answered. (THE VICTORY, 168 U.S. 410.) Under the Inland Rules, the danger signal should be sounded. In International Waters, the corresponding danger signal cannot be used, unless the sounding vessel is to keep her course and speed under the Rules, etc. She may, however, show a flare-up or use a detonating or other efficient sound signal (see Rule 12) to attract attention to the situation.

The Inland and International Rules require a port-to-port passing if the two vessels meet end on or nearly so. If one vessel proposes a starboard-to-starboard passing in violation of the rules, she must secure the assent of the other vessel before proceeding (THE CITY OF TOKIO, 77 F 2d 315), and she assumes the risk of passing (THE TITAN, 49 F 479). The assenting vessel will not be held in fault unless she is negligent (THE MARACAIBO, 14 F 2d 686) or fails to do her part (THE RICHMOND, 124 F 993).

The Inland Rules require sound signals as an evidence of intention. International Rule 18 does not prescribe sound signals. They are described in Rule 28 and mean an actual change in course.

The International Rules do not mention a starboard-to-starboard passing. The Inland Rules provide for such passing under certain conditions.

Inland Rule V will be discussed under International Rule 25.

Inland Rule VIII will also be discussed under International Rule 24.

Inland Rule IX prescribes the conditions under which whistle signals are to be used. The steamers must be in sight. In fog, mist, falling snow, or heavy rainstorms, when vessels cannot see each other, fog signals only must be given. The Courts have ruled that the alarm signal may be given in Inland Waters when the vessels are not in sight. (THE VIRGINIAN, 238 F 156.)

The second sentence of Article 18, Rule 1, Inland Rules, which begins "But if the courses . . . ," should be moved to the end of Rule 1, Article 18, in order that the third sentence, which begins "The foregoing only applies . . . ," shall refer to "end on" meetings and not to an "end on" meeting or a starboard-to-starboard meeting.

INTERNATIONAL RULES

Rule 19 (Figs. 19.5 and 19.6)

When two power-driven vessels are crossing, so as to involve risk of collision, the vessel which has the other on her own starboard side shall keep out of the way of the other.

INLAND RULES

Article 19 (Figs. 19.5 and 19.6)

Same as International Rule except "steam" is used in place of "power-driven."

AIR TRAFFIC RULES

60.22 Water operations.—An aircraft operated on the water shall, insofar as possible, keep clear of all vessels and avoid impeding their navigation. The following rules shall be observed with respect to other aircraft or vessels operated on the water:

(a) *Crossing.* The aircraft or vessel which has the other on its right shall give way so as to keep well clear;

(d) *Special circumstances.* When two aircraft, or an aircraft and vessel, approach so as to involve risk of collision each shall proceed with careful regard to existing circumstances and conditions including the limitations of the respective craft.

NOTE: The rules for operating aircraft on the surface of the water conform to marine rules for the operation of vessels. The "Special circumstances" rule is provided for situations wherein it may be impracticable or hazardous for a vessel or another aircraft to bear to the right because of depth of waterway, wind conditions, or other circumstances.

PILOT RULES

80.7 Vessels approaching each other at right angles or obliquely.— When two steam vessels are approaching each other at right angles or obliquely so as to involve risk of collision, other than when one steam vessel is overtaking another, the steam vessel which has the other on her own port side shall hold her course and speed; and the steam vessel which has the other on her own

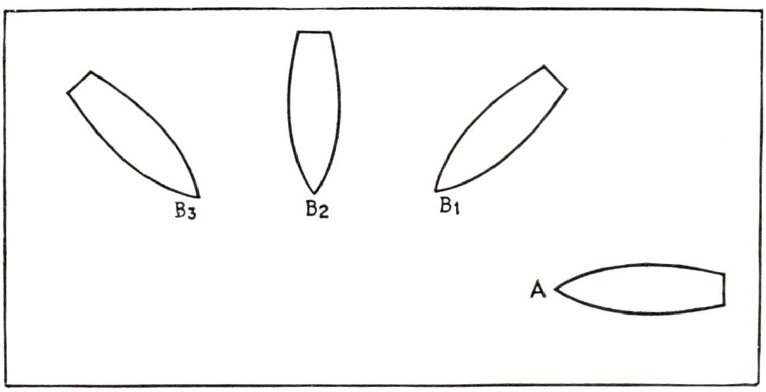

Crossing: "A" Sights "B" to Starboard

INTERNATIONAL AND INLAND

Range Lights

FIG. 19.5 STEAMERS WITH RANGE LIGHTS CROSSING FROM STARBOARD

starboard side shall keep out of the way of the other by directing her course to starboard so as to cross the stern of the other steam vessel, or, if necessary to do so, slacken her speed or stop or reverse.

If from any cause the conditions covered by this situation are such as to prevent immediate compliance with each other's signals, the misunderstanding or objection shall be at once made apparent by blowing the danger signal, and both steam vessels shall be stopped and backed if necessary, until signals for passing with safety are made and understood.

COMMENT: This rule applies when there is risk of collision. If a risk of collision exists, the subsequent changes of course or bearing do not change the original responsibility of "the vessel which has the other on her own starboard side" to keep clear. The burdened and privileged vessels are not changed until all danger of collision has passed.

Risk of collision exists "from the time the necessity for precaution begins. . . ." (MR. JUSTICE CLIFFORD, 1859.)

The Steering and Sailing Rules apply when a single vessel approaches a formation of warships. If the single vessel wishes to show good sea manners and good sense, she should change course, if necessary, to avoid the formation before the risk of collision exists.

Ferryboats are entitled to a reasonable opportunity to enter and leave their slips. Other vessels should not pass close to the slip. Rule 19 does not apply when ferryboats are maneuvering to enter or leave their slips. Rule (Article) 27 applies. After the ferryboats have "come to" a course, on leaving or while they are still on one course, straight or curved, when approaching the slip, Rule 19 does apply. The same instructions apply to a vessel getting under way from an anchorage or when about to anchor.

The exchange of a two-blast signal in a crossing case in Inland Waters does not, of itself, shift the responsibility of the burdened vessel to keep clear and of the privileged vessel to hold course and speed. Courts are divided as to the effect of such an exchange of signals, except for one point: agreement creates a situation of "special circumstances." In the absence of strict obedience to the rules, both vessels must be prepared to take action to avert collision. In International Waters, where one and two blast signals indicate rudder action, no signal is given, unless course is changed.

COURT DECISIONS: The failure of a privileged vessel to answer a crossing signal in Inland Waters, which violates the rule, is not an assent but is the same as a dissent. The burdened vessel should obey the rule, then. (THE ELDORADO, 89 F 1015.)

If the other vessel in Inland Waters ignores or does not hear the initiating vessel's blast or blasts, it is proper to repeat the original signal. (VICTORY, 168 U.S. 410.)

If the privileged vessel fails to answer the burdened vessel's two blasts, the latter vessel should not proceed to cross ahead under the impression that she has the initiative. (THE CYGNUS, 142 F 85.) "The right of way is not created by blowing a whistle. The law directs how a vessel shall pass." (THE HERMES, 21 F 2d 314.)

Comment has already been made on two vessels approaching a bend who seem, for the moment, to be crossing but who will be meeting when they pass. (THE PEKIN, APP. CAS. 532.)

When a vessel is meeting two others in Inland Waters, she should exchange whistles with the first encountered. After that exchange has occurred, she

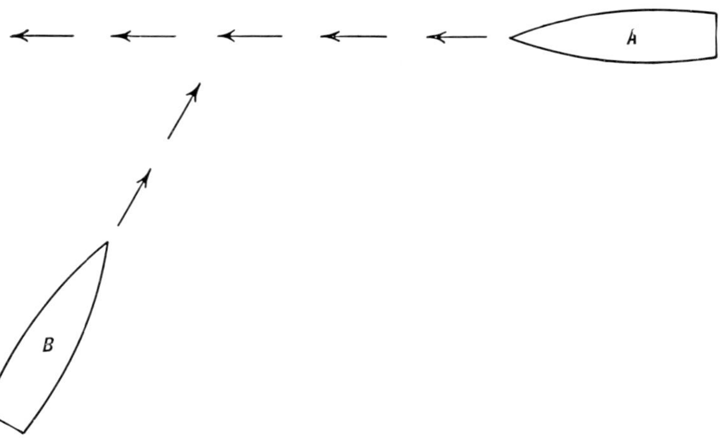

A holds course and speed.
No sound signal under International Rules.
One blast under Inland Rules.
B crosses astern of A, slowing, stopping, backing, or changing course as necessary.
Under International Rules B gives no sound signal unless she changes course or backs.
Under Inland Rules B sounds one blast to signify her intention of crossing astern of A whether she changes course or not.

FIG. 19.6 STEAM VESSELS IN CROSSING SITUATION

should then exchange signals with the second vessel encountered. (THE CITY OF TOKIO, 9 F SUPP. 715.)

The assenting vessel may be found in fault if she assents to a proposal which is dangerous. (THE ARTHUR M. PALMER, 115 F 417.)

When signals have been exchanged, each vessel must promptly act in accordance with the signal. (THE GALILEO, 24 F 386.)

Pilots exchange cross signals occasionally. These signals are forbidden in Inland Waters, see Pilot Rule par. 80.2 above. They are not mentioned in the International or Inland Rules.

"A course (for the privileged vessel) is not necessarily a straight course and may be a turning or a swinging one." (THE CRANFORD, 27 F 2d 710.) The giving way (burdened) vessel is not expected to foresee that the privileged vessel's course is not straight if such fact is not evident. (THE JOHN ENGLIS, 9 LL. L. REP. 400.)

INTERNATIONAL RULES

Rule 20

(a) When a power-driven vessel and a sailing vessel are proceeding in such directions as to involve risk of collision, except as provided in Rules 24 and 26, the power-driven vessel shall keep out of the way of the sailing vessel.

(b) A seaplane on the water shall, in general, keep well clear of all vessels and avoid impeding their navigation. In circumstances, however, where risk of collision exists, she shall comply with these Rules.

INLAND RULES

Article 20

ART. 20. *When a steam vessel and a sailing vessel are proceeding in such directions as to involve risk of collision, the steam vessel shall keep out of the way of the sailing vessel.*

Not in Inland Rules.

AIR TRAFFIC RULES

60.22 Water operations.—An aircraft operated on the water shall, insofar as possible, keep clear of all vessels and avoid impeding their navigation. The following rules shall be observed with respect to other aircraft or vessels operated on the water:

(a) *Crossing.* The aircraft or vessel which has the other on its right shall give way so as to keep well clear;

(b) *Approaching head-on.* When aircraft, or an aircraft and vessel, approach head-on, or approximately so, each shall alter its course to the right to keep well clear;

(c) *Overtaking.* The aircraft or vessel which is being overtaken has the right of way, and the one overtaking shall alter its course to keep well clear;

(d) *Special circumstances.* When two aircraft, or an aircraft and vessel, approach so as to involve risk of collision, each shall proceed with careful regard to existing circumstances and conditions including the limitations of the respective craft.

NOTE: The rules for operating aircraft on the surface of the water conform to marine rules for the operation of vessels. The "Special circumstances" rule is provided for situations wherein it may be impracticable or hazardous for a vessel or another aircraft to bear to the right because of depth of a waterway, wind conditions, or other circumstances.

PILOT RULES

80.8.—Same as Inland Article 20.

COMMENT: This rule applies to all cases where a power-driven (steam) vessel meets a sailing vessel except when the sailing vessel overtakes a steam vessel or meets one not under command or fishing with nets or lines or trawls. When a sailing vessel and a steam or power-driven vessel meet or cross, normally the sailing vessel should hold her course. A shoal, rock, or the bank may force the sailing vessel to tack. No fault can be ascribed to a sailing vessel for changing course or tacking if she does so when she must and no more than she needs to do. The sailing vessel should hold her course until there is an immediate danger of collision and then take the action which will prevent collision under Rule 27.

The second paragraph of Rule 20, International Rules, is new. There are four distinct points which should be made:

(a) A seaplane on the water is a power-driven vessel—see Rule 18(b).
(b) A seaplane on the water shall in general keep well clear of all vessels before risk of collision exists.
(c) However, after risk of collision exists, the seaplane and the other vessel shall obey the Rules.
(d) Masters of other vessels should realize that seaplanes cannot reverse their engines and that they do not maneuver readily in strong winds and choppy seas.

INTERNATIONAL RULES

Rule 21

Where by any of these Rules one of two vessels is to keep out of the way, the other shall keep her course and speed. When, from any cause, the latter vessel finds herself so close that collision cannot be avoided by the action of the giving-way vessel alone, she also shall take such action as will best aid to avert collision (see Rules 27 and 29).

INLAND RULES

Article 21

ART. 21. *Where, by any of these rules, one of the two vessels is to keep out of the way, the other shall keep her course and speed.*

[*See articles twenty-seven and twenty-nine.*]

COMMENT: This Rule requires the privileged vessel to keep the course and speed which may reasonably be expected under the circumstances so that the burdened vessel may know what changes of course or speed or both she must make "to keep out of the way." However, the privileged vessel must take such action as will best aid to avert collision when immediate collision cannot be avoided by the action of the burdened vessel alone. This point is a matter of judgment. The holding on vessel may be forced, eventually, to assume the responsibility of deciding when to depart from this Rule and to comply with Rule 27.

COURT DECISIONS: The Rule which requires the privileged vessel "to keep her course and speed" does not mean the actual compass course and speed at the time when signals are sounded but "in my judgment 'course and speed' in Article 21 mean course and speed in following the nautical maneuver in which, to the knowledge of the other vessel, the vessel is at the time engaged. . . ." (THE ROANOKE, PROB. DIV. 231.) The privileged vessel may be showing a flag requesting a pilot or she may be about to pick up or drop one. A shoal, bank, ledge, or other danger well known to the burdened vessel may force the privileged vessel to change her course and speed. The burdened vessel should still keep clear. The privileged vessel should not change her course or speed until it is necessary or more than necessary. (THE GEORGE DUMOIS, 153 F 833.) The Rule does not apply when vessels are not in sight in a fog. (THE D. S. GREGORY, D.C. N.Y. 1874.)

INTERNATIONAL RULES

Rule 22

Every vessel which is directed by these Rules to keep out of the way of another vessel shall, if the circumstances of the case admit, avoid crossing ahead of the other.

INLAND RULES

Article 22

Same as International Rules.

PILOT RULES

80.9.—Same as Inland Rule 22 except the Pilot Rule is applied to steam vessels only.

COMMENT: The International and Inland Rules apply to all vessels, but the Pilot Rule applies to steam vessels only. This limitation does not affect the validity of the Pilot Rule.

COURT DECISIONS: The Courts have frequently held that this rule should not be violated for convenience sake. (THE E. A. PACKER, 140 U.S. 360.) "Exceptions to the general rules of navigation are admitted with reluctance on the part of the Courts, and only when the adherence to such rules must almost necessarily result in a collision. . . ." (THE ALBERT DUOIS, 177 U.S 240.)

In some ports there is a custom among pilots and vessels customarily plying these waters to sound two blasts and cross ahead. "If there is a custom which permits Sound steamers to claim exemption from the operation of this Article . . . when approaching ferries in the East River on the ebb tide, such custom is opposed to law and cannot prevail." (THE PEQUOT, 30 F 839.)

INTERNATIONAL RULES

Rule 23

Every power-driven vessel which is directed by these Rules to keep out of the way of another vessel shall, on approaching her, if necessary, slacken her speed or stop or reverse.

INLAND RULES

Article 23

Same as International Rules except "steam" replaces "power-driven."

PILOT RULES

Not in Pilot Rules

COMMENT: The Rule does not prescribe the maneuvers which the master of the burdened ship shall make to keep out of the way. It merely states that the burdened vessel shall slacken her speed, stop, or reverse if it is necessary. The burdened vessel may change course, slow down, stop, or reverse, or any combination of these actions. There is only one course of action which is denied the burdened vessel, namely, she shall avoid crossing ahead if the circumstances of the case admit.

COURT DECISION: "The burdened vessel is to keep out of the way. How it shall do so, is not prescribed. . . . If she makes the attempt [to cross the bows of the privileged vessel] and thereby brings about a collision, she is in fault for not keeping out of the way of the privileged vessel." (THE GEORGE S. SHULTZ, 84 F 508.)

INTERNATIONAL RULES

Rule 24 (Fig. 19.7)

(a) Not withstanding anything contained in these Rules, every vessel overtaking any other shall keep out of the way of the overtaken vessel.

INLAND RULES

Article 24 (Fig. 19.7)

ART. 24. *Notwithstanding anything contained in these rules every vessel, overtaking any other, shall keep out of the way of the overtaken vessel.*

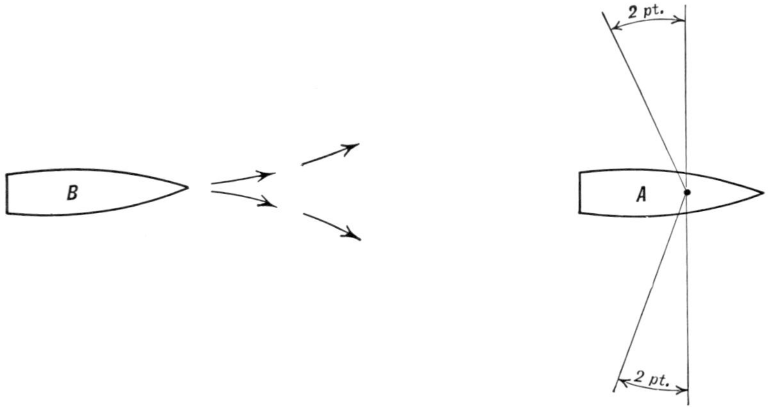

B is overtaking vessel if in arc 2 points abaft either beam of A. If in doubt B assumes she is overtaking. B must keep clear of A until finally past and clear.

International Rules: No sound signals are given unless either changes course or backs when appropriate signals are given.

Inland Rules: B gives one short blast if she wishes to pass on starboard hand and two short blasts if she wishes to pass on her port hand.

The same signal repeated by A gives permission for B to pass as requested.

FIG. 19.7 POWER-DRIVEN OR STEAM VESSELS IN OVERTAKING SITUATIONS

INTERNATIONAL RULES

(b) Every vessel coming up with another vessel from any direction more than 2 points ($22\frac{1}{2}$ degrees) abaft her beam, i.e., in such a position, with reference to the vessel which she is overtaking, that at night she would be unable to see either of that vessel's sidelights, shall be deemed to be an overtaking vessel; and no subsequent alteration of the bearing between the two vessels shall make the overtaking vessel a crossing vessel within the meaning of these Rules, or relieve her of the duty of keeping clear of the overtaken vessel until she is finally past and clear.

(c) If the overtaking vessel cannot determine with certainty whether she is forward of or abaft this direction from the other vessel, she shall assume that she is an overtaking vessel and keep out of the way.

INLAND RULES

Every vessel coming up with another vessel from any direction more than two points abaft her beam, that is, in such a position, with reference to the vessel which she is overtaking that at night she would be unable to see either of that vessel's side lights, shall be deemed to be an overtaking vessel; and no subsequent alteration of the bearing between the two vessels shall make the overtaking vessel a crossing vessel within the meaning of these rules, or relieve her of the duty of keeping clear of the overtaken vessel until she is finally past and clear.

As by day the overtaking vessel cannot always know with certainty whether she is forward of or abaft this direction from the other vessels she should, if in doubt, assume that she is an overtaking vessel and keep out of the way.

Article 18

RULE VIII. *When steam vessels are running in the same direction, and the vessel which is astern shall desire to pass on the right or starboard hand of the vessel ahead, she shall give one short blast of the steam whistle, as a signal of such desire, and if the vessel ahead answers with one blast, she shall direct her course to starboard; or if she shall desire to pass on the left or port side of the vessel ahead, she shall give two short blasts of the steam whistle as a signal of such desire, and if the vessel ahead answers with two blasts, shall direct her course to port; or if the vessel ahead does not think it safe for the vessel astern to attempt to pass at that point, she shall immediately signify the same by giving several short and rapid blasts of the steam whistle, not less than four, and under no circumstances shall the vessel astern attempt to pass the vessel ahead until*

INTERNATIONAL RULES | ***INLAND RULES***

such time as they have reached a point where it can be safely done, when said vessel ahead shall signify her willingness by blowing the proper signals. The vessel ahead shall in no case attempt to cross the bow or crowd upon the course of the passing vessel.

AIR TRAFFIC RULES

See Air Traffic Rule 60.22(c) under Rule 20.

PILOT RULES

80.6 Vessels running in same direction; overtaking vessel.—When steam vessels are running in the same direction, and the vessel which is astern shall desire to pass on the right or starboard hand of the vessel ahead, she shall give one short blast of the steam whistle, as a signal of such desire, and if the vessel ahead answers with one blast, she shall direct her course to starboard; or if she shall desire to pass on the left or port side of the vessel ahead, she shall give two short blasts of the steam whistle as a signal of such desire, and if the vessel ahead answers with two blasts, shall direct her course to port; or if the vessel ahead does not think it safe for the vessel astern to attempt to pass at that point, she shall immediately signify the same by giving several short and rapid blasts of the steam whistle, not less than four, and under no circumstances shall the vessel astern attempt to pass the vessel ahead until such time as they have reached a point where it can be safely done, when said vessel ahead shall signify her willingness by blowing the proper signals. The vessel ahead shall in no case attempt to cross the bow or crowd upon the course of the passing vessel.

Every vessel coming up with another vessel from any direction more than two points abaft her beam, that is, in such a position with reference to the vessel which she is overtaking that at night she would be unable to see either of that vessel's side lights, shall be deemed to be an overtaking vessel; and no subsequent alteration of the bearing between the two vessels shall make the overtaking vessel a crossing vessel within the meaning of the rules in this part, or relieve her of the duty of keeping clear of the overtaken vessel until she is finally past and clear.

As by day the overtaking vessel cannot always know with certainty whether she is forward of or abaft this direction from the other vessel she should, if in doubt, assume that she is an overtaking vessel and keep out of the way.

Comment: Neither the International nor the Inland Rules require the overtaking vessel to pass the overtaken vessel on a particular side. The Pilot Rule deals with steam vessels only. All three Rules allow the overtaking vessel to pass on either hand, provided, in Inland Waters, the proper signal has been sounded by the overtaking vessel and assented to by the other vessel. Good seamanship on confined waters suggests that the passing shall be to port if safe and practicable because the overtaken vessel is required by Article 25, Inland Rules, "to keep to that side of the fairway or mid-channel which lies on the starboard side of such vessel." The overtaken vessel may be forced to move to starboard if she meets another vessel, end on. Such a move to starboard might be awkward if not dangerous to the overtaking vessel trying to pass to starboard. The overtaken vessel must hold her course after the passing signal has been sounded. She can make such changes as might reasonably be expected under the conditions, i.e., change to avoid a vessel or danger.

The overtaking vessel must continue to keep out of the way until well past and clear where a change in her course or speed will not affect the other vessel.

COURT DECISIONS: The overtaken vessel should not assent to the passing unless it is safe to do so. The failure to sound the dissenting (danger) signal, in Inland Waters, when it is dangerous to pass, has been ruled a fault of the overtaken vessel. (THE JARANGER, C.C.A. Md, 1931.)

An overtaking vessel, in Inland Waters, cannot be denied her right to pass by the overtaken vessel which refuses to answer a passing signal. However, the overtaking vessel should not attempt to pass "unless in a clearly safe place for passing. . . ." She must not "attempt to pass in a place of doubtful safety without a signal of assent." (THE MESABA, 111 F 215.)

A great number of collisions in overtaking cases have been caused by suction. The overtaking vessel must allow ample room to pass the overtaken vessel. The overtaking vessel is liable for a collision caused by her failure to do so and by the sheering of the overtaken vessel due to suction or current. (THE FRED JANSEN, 49 F 254.)

The overtaking vessel must allow for changes of course of the overtaken vessel due to following the channel or to avoiding a third vessel. (THE HACKENSACK, 32 F 800.) (MORGAN'S LOUISIANA R.R. CO., 9 F 2d 174.)

INTERNATIONAL RULES

Rule 25 (See Fig. 19.4)

(a) In a narrow channel every power-driven vessel when proceeding along the course of the channel shall, when it is safe and practicable, keep to that side of the fairway or mid-channel which lies on the starboard side of such vessel.

(b) Whenever a power-driven vessel is nearing a bend in a channel where a power-driven vessel approaching from the other direction cannot be seen, such vessel, when

INLAND RULES

Article 25 (See Fig. 19.4)

ART. 25. *In narrow channels every steam vessel shall, when it is safe and practicable, keep to that side of the fairway or mid-channel which lies on the starboard side of such vessel.*

Article 18

RULE V. *Whenever a steam vessel is nearing a short bend or curve in the channel, where, from the height of the banks or other cause, a steam vessel approaching from the*

INTERNATIONAL RULES

she shall have arrived within one-half mile of the bend, shall give a signal by one prolonged blast of her whistle, which signal shall be answered by a similar blast given by any approaching power-driven vessel that may be within hearing around the bend. Regardless of whether an approaching vessel on the farther side of the bend is heard, such bend shall be rounded with alertness and caution.

INLAND RULES

opposite direction can not be seen for a distance of half a mile, such steam vessel, when she shall have arrived within half a mile of such curve or bend, shall give a signal by one long blast of the steam whistle, which signal shall be answered by a similar blast given by any approaching steam vessel that may be within hearing. Should such signal be so answered by a steam vessel upon the farther side of such bend, then the usual signals for meeting and passing shall immediately be given and answered; but, if the first alarm signal of such vessel be not answered, she is to consider the channel clear and govern herself accordingly.

When steam vessels are moved from their docks or berths, and other boats are liable to pass from any direction toward them, they shall give the same signal as in the case of vessels meeting at a bend, but immediately after clearing the berths so as to be fully in sight they shall be governed by the steering and sailing rules.

NOTE: A **long** blast is one of 8 to 10 seconds.

PILOT RULES

80.10.—Same as Inland Article 25.

COMMENT: The International Rule has three conditions, namely, the channel must be narrow, the vessel must be "proceeding along the course of the channel," and "it is safe and practicable." The first condition exempts wide bodies of navigable water, such as certain parts of Chesapeake Bay. The second has in mind the fact that vessels move across certain channels. The last is that obstructions may make navigation along one side of a channel unsafe.

The Inland Rules mention two limitations only, "narrow" and "safe and practicable."

Both Rules are not mandatory unless it is safe and practicable to obey them. The Rules apply to bodies of water which are narrow, carry two lanes of traffic, and are limited by banks or shoals or buoys or anchorage grounds. Certain channels which are much used, although surrounded by navigable water for some vessels, have been ruled "narrow."

The International and Inland Rules about power-driven (steam) vessels approaching a bend in a channel are fundamentally the same, except that the International Rules emphasize the need for alertness and caution in rounding the bed whether an answering blast is heard or not. The Inland Rules state the vessel is to consider the channel clear and govern herself accordingly if no answering blast is heard. If both power-driven (steam) vessels are within one-half mile of the bend and can see each other, no warning blasts are required.

After the vessels are in sight, the usual signals for meeting and passing should be given. The Inland Rules make definite provision for these signals. The situation is covered by Rules 18 and 28 in the International Rules.

The International Rules do not provide specifically for power-driven vessels moving from their docks or berths. The Inland Rules require such vessels to sound one long blast. This "slip whistle" or "change of status whistle" is prescribed so that an approaching vessel is warned that another vessel is about to emerge into the channel from the slip. Passing vessels should not proceed unnecessarily close to the end of the piers. The slip whistle should be continued or repeated until the emerging vessel is visible to other vessels. (THE DALZELLA, 109 F 2d 101.) A vessel which is backing out of a slip and must swing to proceed along a course is governed by Rule 27 until she reaches that course. (THE SERVIA, 149 U.S. 144.) If the vessel leaving encounters a vessel at close quarters, Rule 27 applies. If a vessel has emerged from a slip, bow first, and is on a recognized course, straight or curved, and if there is time to apply the crossing rule, it governs. (THE GULF OF SUEZ, PROB. DIV. 318.)

COURT DECISIONS: "Channels within the Rule are bodies of water navigated up and down in opposite directions, and that therefore harbor waters, with piers on each side where the necessities of commerce require navigation in every conceivable direction, up and down, across, and up and down between piers on the same side, cannot be considered narrow channels." (THE NO. 4, 161 F 847.)

The Courts have ruled that certain channels are narrow and others are not. (COMMONWEALTH *v.* DOMINION LINE, 258 F 707.)

"Rule 25 is not to be construed as prohibiting vessels from crossing such a channel at any convenient angle whenever the exigencies of their own navigation make it necessary or desirable for them to proceed from one to the other side of the channel; but when no exigency exists, they should keep to the proper side of the channel." (THE LA BRETAGNE, 179 F 286.)

The custom of violating this rule (narrow channels) to take advantage of counter currents "does not alter the fact that any vessel that goes up in that way [on the left side] violates the law and takes the risk, and, if there is a collision, is presumably at fault." (THE TRANSFER NO. 10, 137 F 666.)

"A vessel which approaches a narrow channel when another vessel is coming out, should keep to her own side of the channel." (THE ATLANTIC TRANSPORT CO., 15 F 2d 544.)

"Each of these vessels was entitled to presume that the other would act lawfully; would keep to her own side. . . ." (THE VICTORY, 168 U.S. 410.)

"When a vessel, located at rest in a slip, has cast off and is about to start out of the slip bow first or stern first, it is her duty to blow a single long blast, usually spoken of as a 'slip whistle.'" (THE JOHN ARBUCKLE, 185 F 240.)

INTERNATIONAL RULES

Rule 26

All vessels not engaged in fishing shall, when under way, keep out of the way of any vessels fishing with nets or lines or trawls. This Rule shall not give to any vessel engaged in fishing the right of obstructing a fairway used by vessels other than fishing vessels.

INLAND RULES

Article 26

ART. 26. *Sailing vessels under way shall keep out of the way of sailing vessels or boats fishing with nets, lines, or trawls. This rule shall not give to any vessel or boat engaged in fishing the right of obstructing a fairway used by vessels other than fishing vessels or boats.*

PILOT RULES

None

COMMENT: The International Rule applies to all vessels and the Inland Rule to sailing vessels only.

The Rule is based on the principle that vessels able to maneuver should keep out of the way of vessels whose movements are restricted or who cannot maneuver.

A fairway has been ruled as water on which vessels of commerce habitually move.

INTERNATIONAL RULES

Rule 27

In obeying and construing these Rules due regard shall be had to all dangers of navigation and collision, and to any special circumstances, including the limitations of the craft involved, which may render a departure from the above Rules necessary in order to avoid immediate danger.

INLAND RULES

Article 27

ART. 27. *In obeying and construing these rules due regard shall be had to all dangers of navigation and collision, and to any special circumstances which may render a departure from the above rules necessary in order to avoid immediate danger.*

PILOT RULES

80.11 Departure from rules.—In obeying and construing the rules in this part due regard shall be had to all dangers of navigation and collision, and to any special circumstances which may render a departure from said rules necessary in order to avoid immediate danger.

COMMENT: Note that the two Rules are the same except for the phrase "including the limitations of the craft involved" which has been inserted in the International Rules. It is apparent that the known limitations in the maneuverability of a seaplane on the water has been one of the reasons for adding this phrase. It applies, however, to many other cases.

The Courts have decided that this Rule (27) does not take precedence over the other steering and sailing rules. They should be obeyed in almost all situations, where risk of collision exists and there is time and room to apply them. However, the Rules were enacted to prevent collisions. The law cannot foresee the many different situations that could occur. Article 27 is binding, therefore, for the unusual, out-of-the-ordinary case where obedience to the other rules would probably cause a collision. The Courts have ruled that the following situations are governed by Rule 27:

1. A steamer backs out of a slip or a dock and is turning preparatory to proceeding farther.
2. A ferryboat is maneuvering to enter or to leave her slip.
3. A vessel has just weighed anchor and is "casting" to proceed.
4. A pilot is about to board or leave a vessel.

The common case is where two vessels are about to collide and the burdened vessel is not able to avoid a collision by her own actions.

A vessel cannot disregard the other rules and apply this General Prudential Rule when and as it suits her convenience.

COURT DECISIONS: "I am of the opinion that departure from Article 18 (and compliance with Rule 27) is justified when such departure is the one chance still left of avoiding danger which otherwise is inevitable." (THE BENARES, 5 ASP. M.C.)

"The duty is to avoid collision by observing the rules, primarily, by departing from them, if necessary, to avoid danger." (THE HERCULES, 51 F 452.)

"A vessel, coming out of her slip and maneuvering to get on her course or one maneuvering to get into her slip, is not navigating on any course, and the steering and sailing rules do not apply." (THE WILLIAM A. JAMISON, 241 F 950.)

INTERNATIONAL RULES

Rule 28

(a) When vessels are in sight of one another, a power-driven vessel under way, in taking any course authorised or required by these Rules, shall indicate that course by the following signals on her whistle, namely:

One short blast to mean "I am altering my course to starboard."

Two short blasts to mean "I am altering my course to port."

INLAND RULES

Article 28

ART. 28. *When vessels are in sight of one another a steam vessel under way whose engines are going at full speed astern shall indicate that fact by three short blasts on the whistle.*

See Article 18, Rules I, III, VIII, and IX.

INTERNATIONAL RULES	*INLAND RULES*
Three short blasts to mean "My engines are going astern."	
(b) Whenever a power-driven vessel which, under these Rules, is to keep her course and speed, is in sight of another vessel and is in doubt whether sufficient action is being taken by the other vessel to avert collision, she may indicate such doubt by giving at least five short and rapid blasts on the whistle. The giving of such a signal shall not relieve a vessel of her obligations under Rules 27 and 29 or any other Rule, or of her duty to indicate any action taken under these Rules by giving the appropriate sound signals laid down in this Rule.	
(c) Nothing in these Rules shall interfere with the operation of any special rules made by the Government of any nation with respect to the use of additional whistle signals between ships of war or vessels sailing under convoy.	*Not in Inland Rules.*

PILOT RULES

80.03 Signals.—The whistle signals provided in the rules in this part shall be sounded on an efficient whistle or siren sounded by steam or by some substitute for steam.

A short blast of the whistle shall mean a blast of about one second's duration.

A prolonged blast of the whistle shall mean a blast of from 4 to 6 seconds' duration.

One short blast of the whistle signifies intention to direct course to own starboard, except when two steam vessels are approaching each other at right angles or obliquely, when it signifies intention of steam vessel which is to starboard of the other to hold course and speed.

Two short blasts of the whistle signify intention to direct course to own port.

Three short blasts of the whistle shall mean, "My engines are going at full speed astern."

When vessels are in sight of one another a steam vessel under way whose engines are going at full speed astern shall indicate that fact by three short blasts on the whistle.

80.1 Danger signal.—If, when steam vessels are approaching each other, either vessel fails to understand the course or intention of the other, from any cause, the vessel so in doubt shall immediately signify the same by giving several short and rapid blasts, not less than four, of the steam whistle, the danger signal.

80.35 Rule prohibiting unnecessary sounding of the whistle.—Unnecessary sounding of the whistle is prohibited within any harbor limits of the United States. Whenever any licensed officer in charge of any vessel shall authorize or permit such unnecessary whistling, such officer may be proceeded against in accordance with the provisions of R.S. 4450, as amended, looking to a revocation or suspension of his license.

COMMENT: The basic difference between the International Rules and the Inland and Pilot Rules, as far as one-and-two blast whistle signals are concerned, is that the International Rules mean an actual change of course is being made and the Inland Rules signify an intention to pass a certain way, which may or may not include change of course. The Courts have ruled that, when a whistle signal, in Inland Waters, has been sounded and assented to, the two vessels must carry out promptly the maneuver proposed and accepted. "The steamer's delay in acting on her own signal was plainly at her own risk." (GALILEO, 24 F 386.)

Whistle signals are provided for power-driven (steam) vessels when vessels are in sight. The Rules require fog signals to be sounded in a fog when the vessels are not in sight. An exception is a Court ruling that permits the sounding of a danger signal in a fog in Inland Waters.

Comment has already been made that signals must be given in plenty of time for the other vessel to understand the maneuver and to act upon it.

The difference between the conditions under which a danger signal may be sounded in the International and Inland Waters should be noted. The International Rules lay down three conditions:

1. A power-driven vessel is to keep her course and speed. This eliminates meeting vessels, the burdened vessel in crossing cases, the overtaking vessel, either vessel when Rule 27 applies, i.e., when collision is imminent, any vessel navigating a bend in a channel.
2. When vessels are in sight of each other. This prevents the use of the danger signal in a fog before the vessels sight each other.
3. To indicate a doubt whether sufficient action is being taken by the other vessel to avert collision.

The Inland Rules, on the other hand, permit the danger signal to be blown when:

(a) Either vessel fails to understand the course or intention of the other from any cause.

(b) In a fog before vessels are in sight.

Rule 28(c) International Rules permits the use of additional whistle signals between ships of war or vessels sailing under convoy. The Rule does not sanction the failure to use the whistle in the fog or in meeting, crossing, or overtaking cases. The Rule merely legalizes an old custom in the U.S. Fleet of using the whistle to signal a simple maneuver in a formation of ships. Pilot Rule 80.35 does not prohibit such signalling, although care must be taken that such signals do not confuse merchant vessels.

COURT DECISION: The Courts have ruled that vessels backing at less than full speed or making sternway should sound three blasts. (THE SICILIAN PRINCE, 144 F 951.)

INTERNATIONAL RULES

Rule 29

Nothing in these Rules shall exonerate any vessel, or the owner, master or crew thereof, from the consequences of any neglect to carry lights or signals, or of any neglect to keep a proper look-out, or of the neglect of any precaution which may be required by the ordinary practice of seamen, or by the special circumstances of the case.

INLAND RULES

Article 29

Same as International Rules.

COMMENT: This Rule and number 27 are the ones which cover unusual situations and cases. They are sometimes called the "Articles of Good Seamanship." Collision cannot be avoided in every case by following a written set of rules. These two Rules require the master of a vessel to apply his knowledge of seamanship and to take appropriate action when adherence to the Rules would probably cause a collision.

A lookout must: (a) be trained, (b) have no other duties, (c) be stationed where he can see and hear best, (d) have good communication with the officer in charge of the vessel, and (e) be vigilant. Additional lookouts are necessary in restricted visibility. A station in the bow is considered by the Courts to be more desirable and useful than one in the foretop. A lookout aft is necessary when backing out of a slip. A lookout is also required at anchor. These are the proper lookouts and stations as handed down by the Courts.

The speed of a vessel is not prescribed by the Rules except:

1. In a fog (it must be moderate).
2. For a "privileged vessel" (she must keep her speed).
3. For a "burdened vessel" (she must slacken her speed or stop or reverse, if necessary).
4. Vessels passing near floating plants, working in a channel. (Passing vessels shall reduce their speed.)

The Courts, however, have ruled that vessels shall reduce speed when:

(a) In confined, busy waters.
(b) Approaching a bend when vessels on the other side cannot be seen.

(c) Passing close to another vessel which cuts off the view beyond.
(d) Whistle signals are not understood or agreed to.
(e) In doubt as to the intention or course of another vessel.
(f) Passing close to the ends of docks or near to ferry slips.
(g) Necessary "to do no injury to others that care and prudence may avoid." This includes reducing speed in harbors to prevent damage to other vessels, their mooring lines, or equipment.
(h) Cross signals are sounded.

Many of the requirements of good seamanship have already been discussed:

(a) Know the Rules and obey them.
(b) Know the differences between the International and Inland Rules.
(c) Carry the proper running lights.
(d) Make certain that these lights are burning brightly.
(e) If required to change course, make a noticeable change and in good time so that the other vessel will notice the change and have time to make her proper maneuver safely.
(f) Don't expect seaplanes on the water to maneuver as easily as a small boat.
(g) Give tugs with tows plenty of sea room before risk of collision exists.
(h) Observe and allow for the effect of the wind, tide, current, and suction on the tow.
(i) Pass dredges or vessels working under the water slowly, without leaving a dangerous wake.
(j) Give sailing vessels extra sea room. They have been known to luff or come about.
(k) Observe vessels narrowly approaching or leaving a pilot vessel.
(l) Don't run over fishing stakes or buoys.
(m) Navigate with care on well-known fishing grounds.
(n) Keep well clear of fishing vessels.
(o) Proceed at a "moderate" speed in a fog. You can make up the time later. The vessel and the lives of your men are your responsibility.
(p) Stop when a fog whistle is heard forward of your beam. Reduce speed still more, when the whistle sounds close.
(q) Reduce speed in restricted waters so that your wake will not damage other craft.
(r) Anchor in an authorized anchorage.
(s) Anchor outside of narrow channels.
(t) Take extra precautions if forced to anchor in a narrow channel.
(u) Anchor clear of lines of traffic.
(v) Select an anchorage which allows room to swing without fouling other vessels already anchored.
(w) Show good sea manners.

COURT DECISIONS: The Supreme Court in THE ARIADNE, in 1871, ruled that "For an officer to leave his vessel entirely without a lookout, especially when another vessel is known to be in the vicinity, is culpable negligence."

"A competent lookout, stationed upon the quarter of the vessel affording the best opportunity to see at a distance those meeting her, is indispensable to safe navigation and the neglect is chargeable as a fault in the navigation." (THE CATHARINE, 58 U.S. 170.)

"He [the lookout] has only one duty, that which its name implies, to keep a look-out." (THE ARLYN, 1930 A.M.C. 532.)

INTERNATIONAL RULES

Rule 30

Nothing in these Rules shall interfere with the operation of a special rule duly made by local authority relative to the navigation of any harbour, river, lake, or inland water, including a reserved seaplane area.

INLAND RULES

Article 30

ART. 30. *The exhibition of any light on board of a vessel of war of the United States or a Coast Guard cutter may be suspended whenever, in the opinion of the Secretary of the Navy, the commander in chief of a squadron, or the commander of a vessel acting singly, the special character of the service may require it.*

COMMENT: The Inland Rules, Great Lakes, and Western Rivers Rules, enacted by Congress, and the Pilot Rules, established by the Commandant of the Coast Guard, are "special rules" duly made by local authority. There are others. (See earlier chapters.)

Article 30, Inland Rules, has been discussed in earlier chapters.

INTERNATIONAL RULES

Rule 31

When a vessel or seaplane on the water is in distress and requires assistance from other vessels or from the shore, the following shall be the signals to be used or displayed by her, either together or separately, namely:

(a) A gun or other explosive signal fired at intervals of about a minute.

(b) A continuous sounding with any fog-signal apparatus.

(c) Rockets or shells, throwing red stars fired one at a time at short intervals.

(d) A signal made by radioteleg-

INLAND RULES

Article 31

ART. 31. *When a vessel is in distress and requires assistance from other vessels or from the shore the following shall be the signal to be used or displayed by her, either together or separately, namely:*

In the daytime—

A continuous sounding with any fog-signal apparatus, or firing a gun.

At night—

First. Flames on the vessel as from a burning tar barrel, oil barrel, and so forth.

INTERNATIONAL RULES

raphy or by any other signalling method consisting of the group ··· — — — ··· in the Morse Code.

(e) A signal sent by radiotelephony consisting of the spoken word "Mayday."

(f) The International Code Signal of distress indicated by N.C.

(g) A signal consisting of a square flag having above or below it a ball or anything resembling a ball.

(h) Flames on the vessel (as from a burning tar barrel, oil barrel, &c.).

(i) A rocket parachute flare showing a red light.

The use of any of the above signals, except for the purpose of indicating that a vessel or a seaplane is in distress, and the use of any signals which may be confused with any of the above signals, is prohibited.

NOTE: A radio signal has been provided for use by vessels in distress for the purpose of actuating the auto-alarms of other vessels and thus securing attention to distress calls or messages. The signal consists of a series of twelve dashes, sent in 1 minute, the duration of each dash being 4 seconds, and the duration of the interval between two consecutive dashes 1 second.

INLAND RULES

Second. A continuous sounding with any fog-signal apparatus, or firing a gun.

Rule 32

All orders to helmsmen shall be given in the following sense: right rudder or starboard to mean "put the vessel's rudder to starboard"; left rudder or port to mean "put the vessel's rudder to port."

Article 32

ART. 32. *All orders to helmsmen shall be given as follows:*

"Right Rudder" to mean "Direct the vessel's head to starboard."

"Left Rudder" to mean "Direct the vessel's head to port."

PILOT RULES FOR INLAND WATERS

80.26 Passing signals.—(a) Vessels intending to pass dredges or other types of floating plant working in navigable channels, when within a reasonable distance therefrom and not in any case over a mile, shall indicate such intention by one long blast of the whistle, and shall be directed to the proper side for passage by the sounding, by the dredge or other floating plant, of the signal prescribed in the local pilot rules for vessels under way and approaching each other from opposite directions, which shall be answered in the usual manner by the approaching vessel. If the channel is not clear, the floating plant shall sound the alarm or danger signal and the approaching vessel shall slow down or stop and await further signal from the plant.

(b) When the pipe line from a dredge crosses the channel in such a way that an approaching vessel cannot pass safely around the pipe line or dredge, there shall be sounded immediately from the dredge the alarm or danger signal and the approaching vessel shall slow down or stop and await further signal from the dredge. The pipe line shall then be opened and the channel cleared as soon as practicable; when the channel is clear for passage the dredge shall so indicate by sounding the usual passing signal as prescribed in paragraph (a) of this section. The approaching vessel shall answer with a corresponding signal and pass promptly.

(c) When any pipe line or swinging dredge shall have given an approaching vessel or tow the signal that the channel is clear, the dredge shall straighten out within the cut for the passage of the vessel or tow.

NOTE: The term "floating plant" as used in §§ 80.26 to 80.31a, inclusive, includes dredges, derrick boats, snag boats, drill boats, pile drivers, maneuver boats, hydraulic graders, survey boats, working barges, and mat sinking plant.

80.27 Speed of vessels passing floating plant working in channels.—Vessels, with or without tows, passing floating plant working in channels, shall reduce their speed sufficiently to insure the safety of both the plant and themselves, and when passing within 200 feet of the plant their speed shall not exceed five miles per hour. While passing over lines of the plant, propelling machinery shall be stopped.

80.28 Light-draft vessels passing floating plant.—Vessels whose draft permits shall keep outside of the buoys marking the ends of mooring lines of floating plant working in channels.

82.1 General basis and purpose of boundary lines.—By virtue of the authority vested in the Commandant of the Coast Guard under section 101 of Reorganization Plan No. 3 of 1946 (11 F.R. 7875), and section 2 of the act of February 19, 1895, as amended (28 Stat. 672, 33 U.S.C. 151), the regulations in this part are prescribed to establish the lines dividing the high seas from rivers, harbors, and inland waters in accordance with the intent of the statute and to obtain its correct and uniform administration. The waters inshore of the lines described in this part are "inland waters," and upon them the Inland Rules and pilot rules made in pursuance thereof apply. The waters outside of the lines described in this part are the high seas and upon them the International Rules apply. The regulations in this part do not apply to the Great Lakes or their connecting and tributary waters.

82.2 General rules for inland waters.—At all buoyed entrances from seaward to bays, sounds, rivers, or other estuaries for which specific lines are not described in this part, the waters inshore of a line approximately parallel with the general trend of the shore, drawn through the outermost buoy or other aid to navigation of any system of aids, are inland waters, and upon them the Inland Rules and Pilot Rules made in pursuance thereof apply, except that Pilot Rules for Western Rivers apply to the Red River of the North, the Mississippi River and its tributaries above Huey P. Long Bridge, and that part of the Atchafalaya River above its junction with the Plaquemine-Morgan City alternate waterway.

Authority of Master vs. Pilot

The Supreme Court in the OREGON, 158 U.S. 186, ruled: ". . . while the pilot doubtless supersedes the master for the time being in the command and navigation of the ship, and his orders must be obeyed in all matters connected with her navigation, the master is not wholly absolved from his duties while the pilot is on board, and may advise with him, and even displace him in case he is intoxicated or manifestly incompetent."

Not all pilots are competent.

Captains of naval vessels are reminded that a pilot is merely an adviser to the commanding officer. This officer is still responsible for the navigation and handling of the ship, except in the Panama Canal where the pilot assigned to a naval vessel has control of the navigation and movement of the vessel.

There have been cases where the pilot in the Panama Canal has relinquished control over a naval vessel to its usual commanding officer in moments of stress.

There are special orders about the responsibility of a commanding officer when the naval vessel under his command is moved, with or without her own power, in a naval station or in or out of a dry dock.

Part IV

WEATHER

20

Waves and Surf[1]

The seaman who must cope with the elements of nature during a lifetime on the sea should understand the fundamental forces which generate and develop waves and surf. Waves can be formed by a stone dropped in the water, a ship steaming along the surface, an earthquake, the wind, and other forces. We shall discuss wind waves only. Our knowledge is incomplete, but theory checks with observation fairly closely.

Waves on the surface of the sea may be defined as successive ridges with intervening troughs which, in the case of wind waves, advance in undulatory

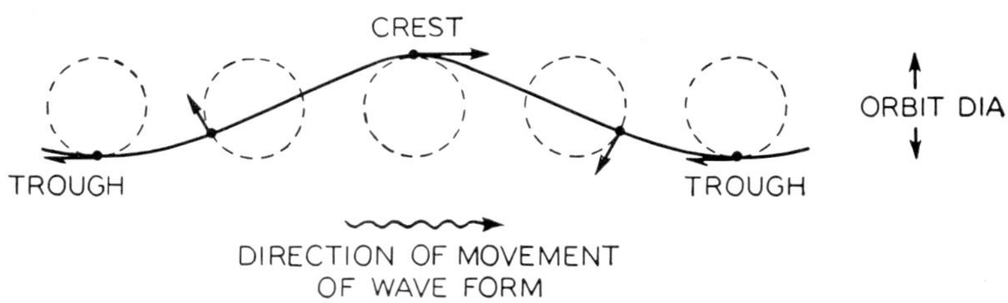

FIG. 20.1

motion. They range in size from ripples to storm waves. Flat calms are infrequent and of small area, since the wind generally blows and gentle breezes set up undulations in a short time.

The casual observer may think that the mass of water in each wave moves with the form of the wave. There is a very slight motion of the mass of water in the direction of the movement of the wave form. A particle of water, however, in a particular wave moves approximately in a circular orbit (Fig. 20.1) in deep water and in an ellipse in shallow water. The diameter of the orbit of particles on the surface is equal to the height of the wave from the crest to the trough. The diameter is independent of the length of the wave from crest to crest or of the speed of the wave's advance. The diameter decreases exponentially with depth (Fig. 20.2).

Even storm waves do not affect the water below 600 feet. Submarines at 90 feet seldom roll or pitch. A particle of water on the surface moves hori-

[1] Based on "Wind Waves at Sea, Breakers and Surf," H.O. 602, by Dr. Henry B. Bigelow and Dr. W. T. Edmondson. Courtesy of the Hydrographer, United States Navy, Scripps Institute of Oceanography and Woods Hole Oceanographic Institution.

zontally forward with the crest, downward after the crest has passed, to the rear in the trough, and upward on the windward side of the wave. In completing the circle it has also moved a short distance in the direction of the motion of the wave form. Ocean currents move large amounts of water in mass, and stormy winds blow the crests of waves before them or cause the crests to break. The velocity of a particle of water in a wave in its orbit is governed by (a) the length of the circumference of its orbit and (b) the time interval between two crests, i.e., the period of the wave.

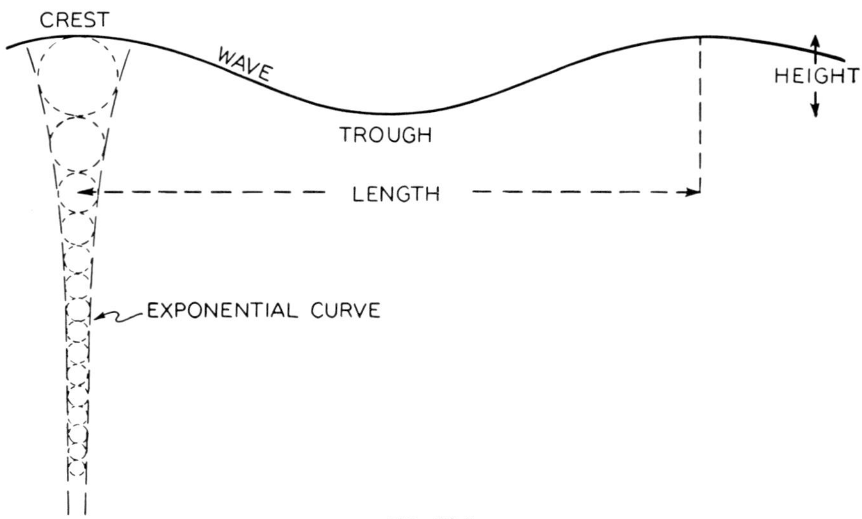

FIG. 20.2

Because the wave form must travel its length during its period, whereas water particles travel only around their orbits during the same period, the orbital velocity of a particle is always much less than the velocity at which the wave form is advancing. In long, low swells the orbital velocity may be $\frac{1}{100}$ to $\frac{1}{200}$ of the velocity of the wave form.

We have noted that the passage of a wave involves some progress of the particles of the wave in the direction of the wave motion. The effect on navigation is small for long, low waves, but it must be taken into account for high, steep waves. The mass transport of water particles is due, in part, to the difference in orbital velocity as the depth increases. Hence, a particle of water on the crest of its orbit is traveling faster than at the bottom of its orbit. Particles on the surface move more rapidly with the wind on the crest, and more slowly against the wind in the trough.

20.1. Growth and Decay of Waves. There has been a good deal of discussion of the causes of the generation of wind waves. Drs. H. Jeffreys, W. H. Munk, H. U. Sverdrup, and other scientists have presented a successful physical explanation of the process. Dr. Munk states that "the gustiness of the wind is responsible for generating waves of all lengths." Those shorter than

0.68 inch in length are called "capillary" waves; those longer, "gravity" waves. This dividing line has been chosen because it represents the length corresponding to the slowest possible velocity of waves. Shorter waves travel faster under the action of surface tension; longer waves travel faster under the action of gravity. Capillary waves advance more rapidly, the smaller they are, and subside quickly when the wind dies away. Gravity waves advance more rapidly, the longer they are, and continue to run long after the generating wind has ceased. Ripples are transformed into ordinary waves when they reach a length of 0.68 inch from crest to crest and have reached a velocity of 0.76 foot per second. A wind of one half to two knots is required to generate ripples, but a stronger wind will transform ripples into gravitational (ordinary) waves. Once formed, the wind pushes against the back slope of the wave and drags by frictional pull on the leeward side. The difference in velocity of the wind and wave naturally determines the amount of force applied.

The shape of the wave affects the movement because the wind has less effect on a long, low wave than on a steep one. Hence, if the wind velocity is greater than the wave velocity (and it can be smaller), the wind increases the speed of the particles on the crest now moving forward and decreases the speed of the particles in the trough which are moving backward. The particles do, therefore, acquire a small amount of movement in the direction of the motion of the wave form. The net result is that the waves gain in height and length until they reach the maximum height to which a wind of that strength can lift them. If they have already reached that height, they approach the limit in length.

The decay of waves is generally brought about by contrary winds or by the loss of energy due to internal friction and air resistance. Waves of oscillation would run for very long distances if unopposed. Such a condition is not met in nature because contrary winds may be encountered and, sometimes, a coast line intervenes. The velocity of steep waves is reduced more rapidly by the headwinds than that of long, low swells. A strong wind from a new direction generally flattens the waves with startling abruptness.

20.2. Wave Action in Depth. In deep water, the shape of the orbit of a particle of water is circular down to the depth affected by wave motion. On soundings the orbit becomes increasingly elliptical as the depth lessens.

The period deep under the surface is the same as on the surface, but, since the circumference of the orbit is shorter, the orbital velocity decreases with depth. Generally speaking, the wave action is negligible at depths greater than the wave length. Ripple marks on the bottom in deep water can be attributed to currents along the bottom.

20.3. Seismic Sea Waves (Often Called "Tidal Waves"). These waves are not ordinary waves of oscillation. They may be due to earthquakes, to superimposing one wave form on another, to the contracting shape of a bay, or to the movement of a high tide under pressure of winds of hurricane force.

The last two forces caused the inundation of Providence during the hurricane of 1938.

20.4. Dimensions of Waves. The dimensions of waves are:

1. *Height*—the elevation of the crest above the trough in feet.
2. *Length*—the distance in feet from one crest to another.
3. *Velocity*—speed in knots at which the wave form advances.
4. *Period*—length of time in seconds for two successive crests to pass one point.

20.5. Heights of Waves. There is much misinformation about the heights of waves, particularly in a storm. Heights are determined by the strength and duration of the wind and the "fetch," i.e., horizontal distance over which a constant and unveering wind has been blowing. The maximum heights of waves that winds of different strengths produce, given sufficient time and fetch, are shown in Table 20.1. The theoretical data agree fairly well with observational

TABLE 20.1. PROBABLE MAXIMUM HEIGHTS OF WAVES WITH WIND OF DIFFERENT STRENGTHS, COMBINED FROM VARIOUS OBSERVATIONS AT SEA WITH UNLIMITED FETCH

(Adapted from Krümmel)

Wind velocity (nautical miles per hour)	Wave height (feet)
8	2.6
12	4.6
16	7.9
19	11.5
27	19.7
31	24.6
35	29.9
39	36.0
43	39.4

data up to 40 knots. The effective fetch in the North Atlantic is often not more than 500 to 600 miles. In the Pacific, with prolonged gales and no change in wind direction, the waves may rise to 55 feet. Study data in Tables 20.2, 20.3, and 20.4.

Hurricanes in the Caribbean seldom blow long enough in one direction to give rise to the highest waves because the crests of storm waves break frequently, thus reducing their height. At any rate, waves do not continue to gain height at sea under a given wind. At first they grow rapidly and then slowly after they have attained 75 to 80 percent of their maximum height shown in Table 20.5.

The relative frequency of waves of different heights in different regions is given by Tables 20.6 and 20.7. The majority of waves are much lower than 7 to

TABLE 20.2. THE HEIGHTS OF WAVES, IN FEET, THEORETICALLY PRODUCED BY WINDS OF VARIOUS STRENGTHS BLOWING FOR DIFFERENT LENGTHS OF TIME

Wind velocity (nautical miles per hour)	Duration in hours						
	5	10	15	20	30	40	50
10	<2.0	2.0	2.0	2.0	2.0	2.0	2.0
15	3.5	4.0	4.5	5.0	5.0	5.0	5.0
20	5.0	7.0	8.0	8.0	8.5	9.0	9.0
30	9.0	13.0	15.5	17.0	18.0	19.0	19.0
40	14.0	21.0	25.0	28.0	31.0	32.5	33.0
50	18.5	29.0	36.0	40.0	45.0	48.0	50.0
60	23.5	37.0	47.0	54.0	62.0	67.0	69.0

TABLE 20.3. THE HEIGHTS OF WAVES, IN FEET, THEORETICALLY PRODUCED BY WINDS OF VARIOUS STRENGTHS BLOWING OVER DIFFERENT FETCHES

Wind velocity (nautical miles per hour)	Fetch (nautical miles)					
	10	50	100	300	500	1,000
10	<2	2.0	2.0	2.0	2.0	2.0
15	2.5	4.0	4.5	5.0	5.0	5.0
20	4.0	7.0	8.0	9.0	9.0	9.0
30	6.0	12.5	15.5	18.0	19.0	19.5
40	7.5	17.5	23.0	30.0	32.5	34.0
50	9.5	22.0	30.0	44.0	47.0	51.0

Based on a study of the basic energy relationships between wind and waves, by Dr. H. U. Sverdrup and Dr. W. H. Munk at the Scripps Institution of Oceanography. The calculated heights are the averages of about the highest 30 percent of the waves; these higher waves are of most practical significance, and the lower waves observed are likely to be of most recent origin.

TABLE 20.4. MINIMUM, MAXIMUM, AND AVERAGE HEIGHTS IN FEET OF WAVES FOR THE TRADE WIND BELTS

(After Krümmel, based on measurements by Paris)

Area	Minimum	Maximum	Average
Atlantic Trade Wind Belt	0	20	6
Indian Trade Wind Belt	3	16	9
Western Pacific, including the Trade Wind Belt	0	25	10

TABLE 20.5. MAXIMUM WAVE HEIGHTS THEORETICALLY POSSIBLE WITH VARIOUS WIND STRENGTHS, AND THE FETCHES AND DURATIONS REQUIRED TO PRODUCE WAVES 75 PERCENT AS HIGH AS THE MAXIMUM WITH EACH WIND VELOCITY

Wind velocity (nautical miles per hour)	Maximum wave height (feet)	75 percent of maximum height (feet)	Fetch for 75 percent of maximum height (nautical miles)	Duration for 75 percent of maximum height (hours)
10	2.0	1.5	13	5
20	9.0	6.8	36	8
30	19.0	14.3	70	11
40	34.0	25.5	140	16
50	51.0	38.3	200	18

12 feet, and waves greater than 20 feet are unusual anywhere. The stormier latitudes of all oceans experience about equally severe gales, and the waves in such storms are about the same in height.

Successive waves do differ in height. This is partly because longer and faster waves overrun slower ones, and the two kinds have probably come from different areas. Groups of waves greater than the average height are often encountered in a storm and even on a beach, usually the result of violent squalls. The rapidity with which the sea rises in a squall is not due solely to the strength of the wind at the moment but more probably to the violence *and* the *life* of the squall. If the squall occurs early in the life of the storm, it may flatten the

TABLE 20.6. RELATIVE FREQUENCY OF WAVES OF DIFFERENT HEIGHTS IN DIFFERENT REGIONS
(Adapted from a chart, based on 40,164 extracts from sailing ships' log books, in Schumacher, 1939)

Region	Height of waves (feet)					
	0-3 (percent)	3-4 (percent)	4-7 (percent)	7-12 (percent)	12-20 (percent)	>20 (percent)
North Atlantic, between Newfoundland and England	20	20	20	15	10	15
Mid-equatorial Atlantic	20	30	25	15	5	5
South Atlantic, latitude of southern Argentina	10	20	20	20	15	10
North Pacific, latitude of Oregon and south of Alaskan Peninsula	25	20	20	15	10	10
East equatorial Pacific	25	35	25	10	5	5
West Wind Belt of South Pacific, latitude of southern Chile	5	20	20	20	15	15
North Indian Ocean, Northeast monsoon season	55	25	10	5	0	0
North Indian Ocean, Southwest monsoon season	15	15	25	20	15	10
Southern Indian Ocean between Madagascar and northern Australia	35	25	20	15	5	5
West Wind Belt of southern Indian Ocean on route between Cape of Good Hope and southern Australia	10	20	20	20	15	15

TABLE 20.7. FREQUENCY DISTRIBUTION OF WAVES OF DIFFERENT HEIGHTS AT SOUTH BEACH, MARTHA'S VINEYARD, FROM OBSERVATIONS MADE BETWEEN NOVEMBER, 1943, AND APRIL, 1944. EACH CASE IS THE MEAN OF 20 CONSECUTIVE WAVES

Month	Mean height (feet)										Total cases	Monthly mean height (feet)
	0.0-1.0	1.1-2.0	2.1-3.0	3.1-4.0	4.1-5.0	5.1-6.0	6.1-7.0	7.1-8.0	8.1-9.0	9.1-10.0		
November	....	2	1	1	...	...	...	...	...	...	4	...
December	4	3	1	....	...	...	...	...	...	...	8	...
January	8	22	13	3	1	2	1	1	...	1	52	2.3
February	4	26	10	6	3	1	...	...	2	...	52	2.4
March	5	20	13	9	3	2	1	...	...	...	53	2.4
April	1	20	10	9	2	...	...	...	...	...	42	2.3
Total cases	22	93	48	28	9	5	2	1	2	1	211	...
Frequency (percent)	10.4	44.1	22.7	13.3	4.3	2.4	0.9	0.5	0.9	0.5	...	...

waves in its wake, but later it only adds to the height. The number of waves affected depends on the life and the area of the squall itself. A short, fierce gust may affect only one or two waves. Some squalls, traveling at the rate of 20 to 40 knots, advance unbroken for distances up to 1,000 miles or more, and trains of very large waves are not unusual in such circumstances.

20.6. The Length of Waves. The length of a wave is many times the height; therefore, the length must increase more rapidly than the height. If a 5-foot wave, 100 feet long, doubles in size, the height increases 5 feet, but the length increases 100 feet, 20 times as much as the height, provided that the ratio of 1:20 is maintained between the height and the length. The increase in length continues even after the height reaches its maximum value. Table 20.8 gives

TABLE 20.8. AVERAGE LENGTHS OF WAVES, OBSERVED AT SEA, ACCORDING TO THE STRENGTH OF THE WIND

(Adapted from Krümmel)

Beaufort scale	Wind description	Velocity (nautical miles per hour)	Waves, average length (feet)
2	Light breeze	11	52
4	Moderate breeze	20	124
6	Stiff breeze	30	261
8	Moderate gale	42	383
10	Strong gale	56	827

the average lengths of waves as observed at sea and shows that waves at sea are usually more than 100 feet in length. The length of storm waves is shown in Table 20.9. The longest storm waves are found in the "Roaring Forties." An

TABLE 20.9. LENGTHS OF STORM WAVES OBSERVED IN DIFFERENT OCEANS
(Adapted from Gaillard)

Ocean area	Wave length (feet)			Number of cases
	Maximum	Minimum	Average	
North Atlantic	559	115	303	15
South Atlantic	701	82	226	32
Pacific	765	80	242	14
Southern Indian	1,121	108	360	23
China Sea	261	160	197	3

average of 775 feet has been recorded for an entire day in this area with occasional waves 1,200 to 1,300 feet long.

The lengths quoted are for waves still developing because old swells may be much longer. Lengths of 1,320 feet have been observed in the Bay of Biscay, 1,481 feet in western Ireland, 2,594 on the south coast of England, and 2,719 feet in the equatorial Atlantic. Short seas are found in the China Sea and the Mediterranean due to the short fetches.

20.7. Steepness of Waves. Steepness can be expressed as the ratio of length to height, or vice versa. Steepness affects the strength and comfort of the ship.

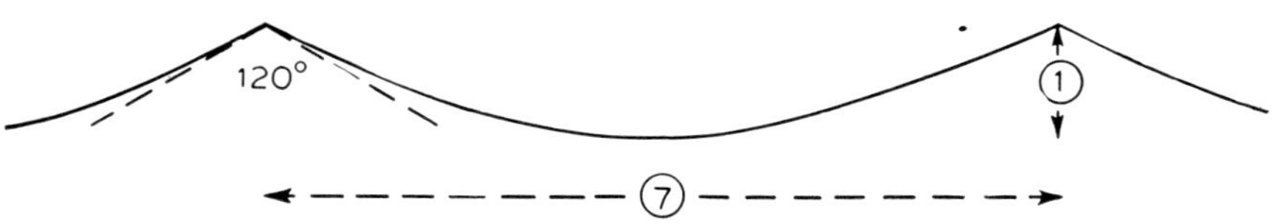

FIG. 20.3

Waves are steeper in the early stages of a storm than they are later. There is no general ratio of length to height because the age of the waves, as well as the velocity of the wind, affects the ratio. It is certain, however, that when

TABLE 20.10. CORRELATION BETWEEN THE AGE AND THE STEEPNESS OF GROWING WAVES
(Derived from an average curve fitted to empirical data, in a study by Sverdrup and Munk, Scripps Institution of Oceanography)

Age of wave, expressed as wind velocity/wave velocity	Steepness of wave, wave height/wave length	Age of wave, expressed as wind velocity/wave velocity	Steepness of wave, wave height/wave length
0.1-0.5	0.076	1.0	0.040
0.6	0.070	1.1	0.035
0.7	0.061	1.2	0.031
0.8	0.053	1.3	0.028
0.9	0.046	1.4	0.025

waves are as steep as indicated by a ratio of 7:1, they become unstable (Fig. 20.3). The usual ratio is between 7:1 and 25:1 while the waves are growing. After the waves have developed into swells, the ratio may be 25:1 to 100:1, and very old swells might be 1,000:1. The relation between age and steepness is given in Table 20.10. Seamen have often noticed that certain waves at sea get steeper until they break, and that this occurs again and again. So the ratio varies for the same wave.

20.8. Velocities and Periods of Single Waves. A careful observer will notice that waves continue to advance after the force that created them has disappeared, as demonstrated when a stone is dropped in a pool of water. Theoretically, the velocity of a *single* wave depends on its length. An old, low swell travels as fast as the higher wave from which it was formed, and it may travel even faster if its length increased while it was altering from a growing wave to a swell. The velocity (knots) of an ordinary (gravitational) wave, a free wave, is about 1.3 times the square root of the length in feet. High waves of a given length run faster than lower ones. The wave speed increases with increasing steepness, but the increase in speed never exceeds 12 percent. The steepness factor, however, can be ignored except in shoal water.

Because the velocity depends chiefly on length, it is misleading to calculate velocity from wind strength alone. The duration of the wind must be considered also, as is expressed in Table 20.11. Notice that the ultimate velocity of the

TABLE 20.11. THEORETICAL WAVE PERIODS (ITALIC), IN SECONDS, AND WAVE VELOCITIES (BOLDFACE), IN KNOTS, IN RELATION TO THE STRENGTH AND DURATION OF THE WIND

(Based on a study of the basic energy relationships between wind and waves by Sverdrup and Munk, Scripps Institution of Oceanography)

Wind velocity (nautical miles per hour)	Duration of wind (hours)				
	10	20	30	40	50
10	*3.0*	*3.8*	*4.6*	*4.8*	*4.8*
	9.1	**11.5**	**13.9**	**14.5**	**14.5**
20	*4.5*	*5.8*	*6.8*	*7.4*	*8.0*
	13.6	**17.6**	**20.6**	**22.4**	**24.2**
30	*5.6*	*7.4*	*8.6*	*9.5*	*10.4*
	17.0	**22.4**	**26.1**	**28.8**	**31.5**
40	*6.8*	*8.9*	*10.2*	*11.3*	*12.4*
	20.6	**27.0**	**30.9**	**34.2**	**37.6**

wave may be greater than that of the wind. Observers have noted that the wind has often a greater velocity than the waves while the storm is developing, and a smaller velocity as the storm dies away, or shall we say, moves away. Indeed, swells with a velocity of 30 to 60 knots have been seen in a flat calm. The general rule is that storm waves which are still rising travel at a lower velocity than the wind that produces them, and at a higher velocity than a dying wind. It is common knowledge that swells often warn mariners of an approaching hurricane. (See Chapters 21 to 24 on "Weather.") Hurricanes (and storms) move more slowly as a mass than the velocity of the winds involved and, hence, of the seas created. The violence of the approaching hurricane, however, cannot be determined solely by the velocity of the fore-running swells because fetch and age are factors in the velocity of a swell.

The velocity in knots can be roughly calculated by multiplying the period in seconds by 3. The length in feet is approximately equal to the square of the period in seconds multiplied by the factor 5.12. Table 20.12 shows how closely the theoretical and the observed values do agree. Periods of 26 to 30 seconds for wind waves are the longest recorded in print, but larger periods

TABLE 20.12. THE LENGTHS OF WAVES (AS OBSERVED AND AS COMPUTED FROM THE OBSERVED PERIODS) AND THE WAVE PERIODS (AS OBSERVED AND AS COMPUTED FROM THE OBSERVED LENGTHS) IN DIFFERENT PARTS OF THE OCEANS

(Adapted from Krümmel and from Sverdrup, Johnson, and Fleming)

Region	Length (feet)		Period (seconds)	
	Observed	Computed	Observed	Computed
Atlantic Trades	213	200	5.8	6.0
Indian Ocean Trades	315	341	7.6	7.3
South Atlantic Westerlies	436	535	9.5	8.6
Indian Ocean Westerlies	374	341	7.6	8.0
China Sea	259	282	6.9	6.6
Western Pacific	335	397	8.2	7.5

have been observed for waves created by earthquakes. The Hawaiian tidal waves of April 1, 1946, arrived at intervals of 15 to 17 minutes, a period corresponding to a length of about 100 nautical miles and a velocity of 430 knots. The slopes of such waves are so gentle that ships at sea would not be affected.

20.9. Velocity of a Group of Waves. The velocity of a group of waves is not always the same as that of a single wave. The velocity of a group which is still being formed by a constant wind is the same as that of a wave within the group. But after the group has parted company with the generating wind and after the waves in the group have changed into swells, the process of decay sets in. The leading waves eventually subside due to loss of energy by internal friction, i.e., setting undisturbed water in undulatory motion. This is in addition to the resistance of the air which causes a loss of energy in all waves of the group. The next waves then become the leading waves of the group. In this way the group, as a whole, travels at a slower speed than the individual waves in the group. Theoretically, the group has a velocity of about one half the velocity of the individual waves if the depth of water is greater than the wave lengths.

20.10. The Direction of Advance. Waves advance at right angles to the line of their crests and with the wind which made them. If the wind changes direction, the waves already formed continue in their original direction. Waves are not measurably affected by the earth's rotation on its axis. Winds are deflected by that rotation, to the right in the northern hemisphere and to the left in the southern hemisphere. This difference of effect occurs because the wave form advances and not the mass of water itself, whereas the whole mass of air moves. The result is that waves move with the generating winds when the isobars are long and nearly straight, but diverge when the isobars curve more rapidly and the winds veer. The wind must veer about 15 degrees from its original direction before a new train of waves is formed. Since the wind veers and waves do not change direction at sea, the wind direction and wave direction may be separated by as much as 60 to 100 degrees, depending on the observer's quadrant in the storm. This divergence often takes place after a storm. The difference between the direction of the wind and that of the waves cannot be reduced to a simple rule. "The wind bloweth where it listeth," but wave forms continue in their original direction.

20.11. Profiles and Contours of Waves. The theoretical profile of a free wave of oscillation (no longer driven by the wind) is very nearly the shape of a trochoid. In such a wave the curve of the crest is slightly steeper and narrower than in the trough. The length of a growing storm wave is usually 14 to 24 times its height. Its *average* slope from trough to crest, therefore, would range from 1:7 (8 degrees) to 1:12 (5 degrees). Actually, the trough is concave and the crest usually convex.

The curvature is gentle for long waves; hence the slope for a wave of 1:20 steepness is about 7 to 8 degrees along the steepest part near the crest. When the steepness ratio is 1:10, the steepest part of the profile is about 20 degrees. Thus, a small boat surmounting a short 1:10 wave would pitch three times as much as it would in riding over a long 1:20 wave. "Chop" is proverbially uncomfortable in small boats.

The crests become narrower and the troughs relatively longer if the height is large relative to the length; i.e., if the wave is steeper. When the ratio approaches 1:7, the angle at crest increases to about 120 degrees, and the wave is unstable (Fig. 20.3). Waves cannot continue to advance in this shape because they break and thus reduce the steepness and the height. Waves of this type appear frequently in stormy weather. Instability is often observed in small as well as in large waves, particularly where the wave is growing rapidly in height with a rising wind. The familiar "white caps" appear. Even after the wave stops rising, a strong wind may blow the crests forward to leeward, which is referred to by the phrase, "the sea is breaking." A really strong wind may blow the crests to leeward in masses of water. The pressure of the wind on the back of the crest is much stronger than on the front or lee side of the crest, and masses of water are thus driven on board ships and cause much damage. For example, the *Pittsburgh* lost her bow during a typhoon in the western

Pacific in 1945. The easiest course of the steamship in a storm should be determined with reference to the waves and not the wind because it is the waves and not the wind which cause the damage to, and try the strength of, the ship.

The discussion so far has dealt with a regular succession of waves moving in one direction. Actually, in a storm the seas consist of small waves superimposed on large ones and running at varying angles for varying lengths of time. Some waves are breaking. Some are not. The contour is still further complicated by peaks which rise suddenly when two waves from different directions come together and when gusts of wind and squalls from new directions confuse the seas. A strong wind given time and a sufficiently long fetch will generate waves moving in one direction, even though the surface may be confused by squalls and gusts.

20.12. Effect of Currents. The effect of currents, both "contrary" and "following," will be considered now. Briefly, a contrary current decreases the lengths of waves. Because the amount of energy in the wave is unchanged for the moment, a decrease in length will force an increase in height. The current does not alter the periods of the waves because it reduces their velocity in the same proportion that it decreases their lengths. The effect of following current is the reverse. The length is increased, the wave is not so steep, again without altering its period, and the amount of change depends on the relative velocities. The current is regarded as positive for "following" waves and negative for "contrary" waves.

If a wave, having a period of 4.2 seconds and, hence, 100 feet long in deep water and traveling at a rate of 13.4 knots, encounters a contrary current of 2 knots (relative ratio minus 0.15), the height would be increased by a factor of 1.39 and its length reduced to 67 feet. A general rule is that the stronger the current, the greater its effect on waves.

The effect of a current on the height of a wave is greater, the shorter the wave and the lower the wave velocity, because the force of the current meets less resistance. In the preceding example, if the wave were 50 feet long and had a period of 3 seconds, the ratio of the velocities would be about minus 0.22 and the height of the wave would be nearly doubled (factor 1.9)—if the wave did not break. In a like manner, the decrease in the height due to a following current is greater for a short wave of low velocity than for a long one of higher velocity.

The frictional drag of the wind on waves blowing in a constant direction for some time sets up a current, i.e., a wind drift. Drifts of this sort affect navigation and average 1.5 to 2 percent of the velocity of the wind. The drift caused by a 20-knot wind would average 0.3 to 0.4 knot. Wind drifts with velocities of 1.4 to 1.9 knots in 25 to 55 mile "blows" have been reported at Diamond Shoals. Tidal currents have no appreciable effect far at sea, but 1 to 1.6 knots are met off Georges Bank and 3 to 4 knots at the entrance to San Francisco. In such cases, a moderate sea and a contrary current can form waves which are dangerous to small craft, and such seas have swept destroyers in Pentland

Firth. Choppy seas are often encountered in the Gulf Stream when the current and the wave motion are contrary.

Waves which run counter to a strong current and pass through it lose height and become smoother rapidly. They drop to the size expected in the wind blowing at the time, and their lengths increase.

20.13. Alterations in Waves over Shoals. When waves leave deep water and reach "soundings," their lengths are gradually reduced. Their heights decrease at first but later may increase. The loss in height is far less than the loss in length. The lengths may decrease $\frac{1}{3}$ by the time the depth of water is $\frac{1}{10}$ the initial length and by $\frac{1}{2}$ when the depth is $\frac{1}{20}$ of that length. The steepness increases accordingly.

The length of the wave and the depth of the water determine the distance from shore that these changes take place. The longer the wave, the farther out from shore it will begin to steepen. The waves in the eastern North Sea, with westerly gales, are steep due to the wide shoals off the Dutch and Danish coasts.

20.14. Sizes of Waves in Shallow Water. The shapes, heights, and lengths of waves that are generated by strong winds in shoal water, such as Pamlico Sound, are of importance to small craft. Shoal water seems to limit the height of waves that do not break. The explanation lies, probably, in the effect of shoal water on length, although the relationship of depth of water to length is not definitely known. All that can be said is that waves in shallow water grow more slowly in length than they do in deep water. Hence they are steeper and break sooner. Waves may break when the depth is 0.8 of or equal to the wave height. If in shallow water waves do not meet the shore, they will continue to travel for considerable distances, breaking and losing height only to build up and break again.

Here are some rules derived from experience: (a) Waves in shoal water will never be much higher than the depth of the water and very likely much lower. (b) They will probably be steep and will break often if the wind is strong.

These rules also apply to waves forming and advancing along the shore. Their growth at the inshore ends will be retarded by the lack of depth and by refraction; therefore, there will be a decided difference in the heights of the same wave in deep water and along the shore. The degree of difference will depend, of course, on the length of the fetch, the shape of the bottom, and the strength of the wind.

20.15. Methods of Measuring Waves. The length can be measured easily if the ship is steaming at right angles to the crests by paying out a chip log until the chip is on the crest of one wave when the stern of the ship is on the crest of another. A traverse table will solve the triangle when steaming on other courses.

The velocity of a wave, calculated on a ship steaming with the sea, is the length of the wave divided by the time between passing crests minus the speed of the ship in feet per second. A simpler way is to note the time for a patch of foam or a piece of wood to move apparently from one crest to another.

The height can be measured by noting the height of the observer's eye above the water line when the crest of a wave is in line with the horizon, the ship being on even keel. Waves and breakers may be measured from piers by an ordinary sounding line or from the shore by observing the water level on a pile off shore.

20.16. Seas and Swells. Growing storm waves are often confused, but, when the development period is over and the waves have left or been left behind by the generating winds, the waves settle down to what is known as a "swell." This type of wave has a rounded crest, great length, and great width along the crest. The alteration in form is due to a loss of energy. The short, superimposed waves lose energy more rapidly than the longer ones and disappear first. The remaining crests decrease in height and assume a more rounded shape. The swells tend to approach the trochoidal form. The individual wave crests extend sideways until their width has been multiplied by 3 or 4 or even more. Observation seems to show that the lengths, velocities, and periods lengthen after the waves have become swells and have left their generating area, the fetch, behind. Sometimes the transformation into swells occurs rapidly—in a period of 2 hours—when the wind dies down quickly. This alteration in form may seem to be accomplished as rapidly for large as for small waves. However, the relative size of the two waves does affect the time required. In other words, small waves are assimilated by old swells more rapidly than larger waves.

20.17. Direction of Swells. The direction of the swells is the same as the winds which formed the original waves. Thus, a swell encountered at sea may be moving at any angle to the present wind. If a swell continues to come from one direction, it may be assumed that a storm is advancing directly toward or away from the observer, or it is stationary. Swells do outstrip storms and hurricanes. Swells from hurricanes sometimes reach an observer 24 hours before the arrival of the hurricane itself because tropical hurricanes, approaching the Caribbean, move about 12 mph in low latitudes and about 25 mph after recurving. The waves generated therein travel at 30 mph or more. A change in the direction of the swell may indicate a change in the direction of the storm center.

20.18. Persistence of Swells. Swells in the open sea may run for hundreds, even thousands, of miles unless they are beaten down by contrary winds or strike a coast. It is possible to predict the arrival of swells on a distant shore by studying the weather chart of a violent storm at sea. The French authorities at RABAT are successful in forecasting, in this manner, unusual swells and surf about 70 percent of the time.

We have already noted that, if the ocean were unlimited in extent, a swell could be flattened eventually by the energy lost in setting undisturbed water in motion (internal friction) and by the resistance of the air. Swells do run long distances. Roughly, swells lose one third of their height each time they travel a distance in miles equal to their length in feet. A rough curve could be plotted for the storm waves in the South Pacific which would emphasize the long distances that such waves could travel.

Breakers and Surf

20.19. Importance and Origin. When waves break after they have advanced into shoal water, they will be referred to as "breakers" to distinguish them from storm waves at sea whose crests curl over and break. When "breakers" develop into a more or less continuous line along the shore, they will be called "surf."

Any kind of a wave may break as it moves into shallow water, depending on various factors such as the slope of the bottom, the depth, shoals, ledges, etc. Swells that have come from a long distance and local storm waves break.

20.20. Energy of Surf. Breakers seen from seaward never seem as high or as dangerous as they really are because their backs are rounded and thus do not show the overhanging crest and towering concave form presented to the observer on the beach.

The energy of a wave may be divided into two approximately equal parts: (a) dynamic or kinetic, due to the momentum of its water particles; and (b) static or potential, due to the "head" or height of its center of gravity above the still water level. The force exerted by waves depends also on the shape of the object encountered, i.e., whether the object is broadside to or streamlined and end on.

The potential energy is periodic and moves with the wave form in deep water. In shallow water the kinetic energy is added to the potential, and large forces are thus released. It is true that the potential energy is reduced somewhat by changes in length and height as the wave moves into shoal water. Unfortunately for landing craft the total remaining energy lies above the still water level and moves forward with the wave form. This action brings about the destructive power of breakers, which can tear up concrete sea walls and wreck landing craft. (See Table 20.13.) Personnel in landing craft should take every precaution in the face of such forces.

Table 20.13. Pressures of Breakers on West Coast of Scotland, in 1845, as Recorded by Spring Dynamometers

(From Gaillard, after Stevenson)

Supposed height of of waves (feet)	Conditions of sea	Dynamometer readings (pounds/square feet)
6	Swell	3,041
10	Ground swell	3,041
20	Heavy sea	4,562
20	Strong gale, heavy sea	6,083

20.21. Causes of Surf. When the depth of the water has shoaled to about one half the length of the wave, the wave form begins to change, the length decreases, and the height may increase. Its crest is steepened, and, if the steepness increases, the wave will break. If the depth of the water does not continue to decrease, the altered wave may not break until it reaches the shore line. Thus, a wave advancing against a cliff with deep water alongside may just surge against the cliff without breaking, unless the irregular face of the cliff causes it to do so. Sand often accumulates against a sea wall, and the shallowing water may cause the waves to break heavily against the wall.

The relationship between length and velocity of waves at sea is altered by shallow water. Waves of different initial lengths tend to run at the same velocity as the breaker line is approached. Note the bottom line of Table 20.14 particularly. This table shows also that longer waves are slowed much more relatively than shorter ones. We have noticed that steep waves in deep water tend to travel a little faster (not more than 12 percent) than those with gentle contours. In shoal water the increase in height seems to oppose the tendency of the wave to get shorter with the result that they run a little faster than lower waves of the same wave length. The increase in velocity may be 10 percent, although usually it is less than this. The period of waves offshore is supposed to remain the same when the waves begin to alter their form "on soundings," but observation shows that the periods do change as much as 25 percent either way. This apparent change in period is probably due to the state of the sea, i.e., whether there is an underlying swell with waves superimposed upon it. The general rule is that the period of waves which preserve their identity, i.e., are not affected by overriding waves, is about the same when they break as in deep water. This is important to landing craft.

TABLE 20.14. DECREASE IN LENGTHS AND VELOCITIES OF WAVES OF DIFFERENT DIMENSIONS AS THEY ADVANCE OVER A SHOALING BOTTOM

Depth of water (feet)	Waves							
	Length (feet)	Velocity (knots)	Length (feet)	Velocity (knots)	Length (feet)	Velocity (knots)	Length (feet)	Velocity (knots)
Over 500	1,000	42.5	500	30.0	250	21.2	100	13.4
500	1,000	42.5	500	30.0	250	21.2	100	13.4
100	710	30.2	445	26.7	248	21.0	100	13.4
75	635	27.0	410	24.6	240	20.3	100	13.4
50	530	22.5	355	21.3	222	18.9	99	13.3
25	380	16.2	265	15.9	178	15.0	93	12.5
10	240	10.2	170	10.2	120	10.2	71	9.5

Seamen are quite aware of the fact that the height of the surf often increases shortly before the waves break. This increase in height, which can be considerable, is preceded by a small decrease in height at the time when the wave form begins to change. The wave regains its initial height when it reaches a depth 0.06 of its initial length. After this point, it may become higher.

It is the initial length, not the height, that determines the depth at which the form begins to change. A long wave suffers more change than a short one because the change continues for a longer time and starts earlier. The initial length, therefore, determines whether the height of the breaker will exceed the height of the original wave or not, and under most conditions on open coasts breakers are higher than the corresponding waves.

Imagine two waves in deep water, each 2 feet high, one 500 feet long and the other 100 feet. Alteration of the form of the 500-foot wave starts at a depth of 250 feet. Its height would increase to 2.1 feet at the 20-foot depth, 2.7 at the 7-foot line, and 3.5 feet at the 5.5-foot line where it might break. The shorter wave would be unaltered until it reaches the 40-foot depth. Its height would decrease to 1.8 feet at the 15-foot line and increase to 2 feet at the 5-foot line, and to 2.3 feet at the 3-foot mark where it might break.

TABLE 20.15. HEIGHTS (TO THE NEAREST FOOT) ATTAINED BY WAVES OF DIFFERENT INITIAL HEIGHTS AND PERIODS AT VARIOUS DEPTHS ON A GENTLY SLOPING BOTTOM

(Blank spaces indicate that the waves would, in all likelihood, have broken in deeper water, or that waves of the stated shapes could not exist. This table is based upon theoretical studies, substantiated by measurements of waves taken at the Scripps Institution of Oceanography and at the Woods Hole Oceanographic Institution)

Initial height of wave (feet)	Period of wave (seconds)	Depth of water (feet)					
		30	25	20	15	10	5
4	4	4	4	4	4	4	4
	6	4	4	4	4	4	5
	8	4	4	4	4	5	..
	10	4	4	4	5	5	..
	12	4	4	4	5	5	..
	14	4	4	5	5	6	..
	16	5	5	5	5	6	..
6	4	6	6	6	5	5	..
	6	5	4	6	6	6	..
	8	6	6	6	6	7	..
	10	6	6	6	7	7	..
	12	7	7	7	7	8	..
	14	7	7	7	8	..	..
	16	7	7	8	8	..	..
12	4	12	11	11	11	..	..
	6	11	11	11	11	..	..
	8	11	12	12	12	..	..
	10	12	12	13	..	..	..
	12	13	13	14	..	..	..
	14	13	14	15	..	..	..
	16	14	15	..	..	..	..
20	4	..	..	..	..	..	..
	6	18	18	..	..	..	..
	8	19	..	..	..	..	..
	10	20	..	..	..	..	..
	12	21	..	..	..	..	..
	14	..	..	..	..	..	..
	16	..	..	..	..	..	..

In general, in deep water waves whose steepness is less than 1:10 to 1:15 do not increase noticeably in height when they break, but longer waves do increase in height, sometimes considerably. Thus, old swells hardly perceptible even in calm weather may produce dangerous surf. Periods are easier to measure than lengths; hence, Table 20.15 gives the alteration in height of waves on a gently sloping bottom according to their periods. The height, attained when breaking, may be several times the initial height where the slope of the bottom steepens very abruptly.

When a wave advances into shoal water, the water particles suffer a change in velocity within their orbits and a change in the shape of the orbit itself. The circular orbits become elliptical. The elongation of these ellipses increases until finally the particles in contact with the bottom travel in straight lines. The orbital velocity, which has been uniform in deep water, is no longer so after the orbits become elliptical but is greatest at the crest and in the trough. This change in velocity is more pronounced as the wave advances into shoal water and is proportional to the length of the major axis. The velocity of the wave particles in the crest and trough grows greater and greater, although the height of the wave (i.e., the length of the minor axis) and the period may not change. The longer the wave and the greater the ratio, the higher is the final velocity. The water particles on the crest may be moving forward at the breaking point several times as fast as they were moving initially, which means the velocity in the trough has increased too. For this reason a boat on the crest of a wave will be carried forward rapidly. If the bow gets down in the trough, where the water is moving almost as fast as on the crest but in the opposite direction, there is good reason to expect that she will broach and swamp.

20.22. The Characteristics of Breakers. A wave may break at sea due to the pressure of the wind on the back of the crest, but "breakers" in shoal water are not caused primarily by the wind, although a strong "onshore" wind helps. Nor are breakers caused by the friction of the bottom but by progressive alterations of the wave form as the wave moves into shoal water. The wave just gets steeper until it breaks. There are three forms of breakers, "plunging," "spilling," and "surging." In the plunging type, the back of the wave continues to be well rounded, but the front becomes so concave that the crest of the wave suddenly falls forward and "breaks" with a roar. In the second type, both sides of the crest grow more concave as the wave advances, until the wave is so steep that it breaks down by "spilling." The wave form approaches a true "cycloid." The "surging" breaker neither plunges nor spills but surges into a peak up the beach face. All types may be seen in a single wave advancing along the shore. Long, gentle swells with an initial steepness ratio of greater than 1:200 (i.e., 1:300, 1:400, etc.) commonly produce the plunging breaker, whereas waves with ratios at sea less than 1:100 (i.e., 1:50, etc.) produce the "spilling" breaker. Some waves "break" several times in their advance up the shore. Other waves break well out, and the foaming mass of water is added to the underlying wave form and carried for long distances shoreward. Such waves are called waves of "translation." If the wave form is destroyed, there may be several lines of these waves. If the wave form is not destroyed, it will continue, carrying these foaming crests along until a secondary line of breakers occurs. This line is generally not so high as the first line.

20.23. The Characteristics of Surf under Different Conditions. The kind of surf that will develop for a given height and length (ratio) depends on the contour or grade of the bottom, the nature of the coast, obstructions offshore, the tide, currents, and the wind.

20.24. Height of the Surf. Generally the height of the breakers depends on the height and length of the waves in deep water. The steeper the waves in deep water, the less will be their proportionate increase in height before breaking. Read Table 20.16, but note that observations may vary 25 percent. Very

TABLE 20.16. RATIO OF BREAKER HEIGHT TO OFFSHORE HEIGHT FOR WAVES OF DIFFERENT DEGREES OF STEEPNESS IN DEEP WATER

Ratio between length and height of wave in deep water	Ratio between breaker height and height of wave in deep water
20:1	1.0:1
40:1	1.2:1
100:1	1.4:1
125:1	1.6:1

steep waves may break while the height is less than the original height. Waves 6 to 8 feet high at sea are only this same height when they break, if the ratio is small. Old, long swells may cause high breakers because they begin to change in deeper water and thus have time to increase their heights. It is not unusual for such old swells to double their heights.

Waves vary greatly in the ratios of height to length, and the height of surf varies in the same manner. The actual height of breakers at the highest point when breaking has been measured and estimated as 2 inches to 42 feet. Breakers have been seen in depths of 60 feet off the San Francisco and the Columbia River Bars. Breakers under such circumstances might be 30 to 45 feet high. In some places where the wind is the determining factor and old, long swells are unusual, the surf is highest with an "onshore" wind. Calm days and apparently calm seas are sometimes marked by surf. It is difficult to detect a long, low swell that can develop into bad surf. Cross seas may produce really large breakers and such seas are especially dangerous in small boat work. It is often possible to wait, when landing on a beach or launching a boat through the surf, until a series of smaller waves or breakers are sighted.

20.25. Depth of Water in Which Surf Develops. In deep water, waves of moderate steepness generally break when the depth is about 1.3 times their height. This ratio between the depth and the height when waves are breaking can vary widely under different conditions of wind, sea, and current, and the variation is often so great that all that can be said is that the ratio varies from 0.72 to 2.3 and maybe more. Observation seems to show that the steeper the bottom, the earlier the waves will break, but laboratory experiments do not support this observation. Perhaps it is well to leave the effect of this factor to future observation and experiment.

An "offshore" wind delays the breaking, and an "onshore" wind hastens the breaking. A general rule for a bottom slope of 1:40 can be stated as follows: (a) with an "onshore" wind, waves may break at a depth twice as great as their height; (b) with an "offshore" wind, at three quarters their height. Allowances must be made for the shape of the waves offshore. Long waves will break at greater depths than short ones of the same height. Attention is called to the old ratio between the height and length of the wave at breaking. The critical point is still 1:7, although some waves break when this ratio is larger. The reason that waves do not reach the critical ratio of 1:7 before breaking is that the forms of some waves change markedly as they pass into shoal water. The troughs become long and flat, and the crests steepen to instability while the ratio is still greater than 1:7.

20.26. The Effect of the Various Obstructions on Surf. Offshore ledges may act as breakwaters at low tide but affect approaching waves little at high tide. In such cases, it may be possible to land in their lee at low water but not at high. Bars are formed by tidal currents off river mouths, entrances to harbors, and even along some sections of the beach. At low water, when there is a bar off a beach, waves often will break along the bar, with little or no surf inshore. A bar off a river mouth or a harbor can cause breakers so large that access to the harbor will be denied vessels for long periods during stormy weather. The Columbia River entrance is an example. In deeper water, surf develops with more speed over ledges than it does along the shore. This action is probably due to the abrupt change in depth. If a long swell is running, the passage over a bar should be made with caution.

20.27. Types of Surf. The number of lines of breakers varies from one to six or even more. The variety in lengths and heights of succeeding waves as well as the slope of the bottom determines the number of lines of breakers. Long, high waves break farther out than short, low ones. Even regular swells may break more than once en route to the beach. Areas of surf over some outer bar may alternate with areas of unbroken water in the lee. Storm waves vary in height as do the breakers in their time of "breaking." The number of breakers differs along the same beach and even at one point on the beach.

The types of surf can be classified roughly as follows:

1. A single line of breakers due to abrupt, steep shores or ledges.
2. Breakers, close to the shore, with foaming water between them and the shore, found along steeply sloping beaches at high tide.
3. One line of large breakers with second line of lower breakers, inshore,

common in moderate weather with swells running in over a gently sloping bottom (Fig. 20.4).

4. Several lines of breakers of about the same height, same conditions as (3) above (Fig. 20.5).

FIG. 20.4 ONE LINE OF BREAKERS WITH SECOND LINE OF LOWER BREAKERS, INSHORE. *Courtesy W. H. Munk*

FIG. 20.5 SEVERAL LINES OF BREAKERS OF ABOUT THE SAME HEIGHT. *Official U.S. Navy Photograph*

5. Many lines of breakers, offshore and right up to the beach, characteristic of "onshore" storms.

20.28. Factors That Affect the Development of Surf. The development of breakers may be prevented where outlying boulders or ledges interfere with the movement of the wave form. If a ledge is close enough to the surface to cause the waves to break, it is quite effective in sheltering the shore beyond. A small island near the mouth of a bay shelters the water area inside and reduces the size of the waves and breakers within the bay. Outlying bars prevent the further movement of long ocean swells towards lee shores in such places as Georges Bank and Nantucket Shoals.

Tidal currents may cause waves to break far out or reduce their form in such a way as to act as a partial barrier to their further advance. Waves running against local currents, i.e., "rip" currents, along the shore, may break farther out at this point than elsewhere along the same shore.

Long marsh grass and large patches of seaweed provide shelter from storms for vessels which can lie behind the area of growth. Such growths are found all over the world and have often been used for shelter.

Hail tends to knock down the sea, although the swells continue in a smaller way thereafter. However, loose pack ice kills the sea and forms a real shelter from rough seas. Broken ice is more effective than oil because it increases the internal friction of the water and dissipates the energy of the wave forms.

The use of oil dates back to Pliny in A.D. 77 and probably before his time. Oil extinguishes the small waves and prevents the larger ones from breaking. Confused seas are smoothed and only the swell remains. Oil does not affect the development of surf to any great extent, but it can be used to prevent the breaking of waves before they reach the surf line proper.

20.29. Effect of the Wind on Persistence of the Surf. The regularity of the breakers, daily and seasonally, and the times of dangerous surf depend, in a great measure, on the regularity of local winds. Of course, swells from distant storms can and do affect this surf. Trade winds are more regular than other types of winds with the result that the surf on such windward shores is regular too. The windward side of the Lesser Antilles provides a good example, but even trade winds vary from one season to another, from night to day, and during different hours of the day because they are interrupted by storms. The surf of the Trade Wind belts, therefore, varies too. Surf that is caused by local winds dies rapidly with a change of wind, but surf that is generated by swells from a distance or by trade winds tends to persist and is not affected so promptly by a change in local winds. Of course, a contrary, strong, local wind will eventually flatten swells from a distance and stop the development of that surf.

20.30. Refraction of Waves. The angle at which breakers strike the shore depends upon the general line of advance of the waves offshore, the contour of the bottom, and the shape of the coast. Waves advancing parallel to the coast line may break on that coast at only a slight angle because the seaward ends of the waves are moving parallel to the coast at their usual speed, whereas the "inshore" ends are delayed by the shoaling bottom and by refraction. The amount of refraction has been calculated, given the angle of original approach of the waves in deep water, their length, height, period or velocity there, and the shape of the bottom. This is shown in Table 20.17.

TABLE 20.17. THE ANGLES WHICH BREAKERS MAKE WITH A STRAIGHT SHORE LINE, WHEN ALL BOTTOM CONTOURS ARE PARALLEL WITH THE BEACH, FOR WAVES OF DIFFERENT DEGREES OF STEEPNESS IN DEEP WATER APPROACHING THE SHORE LINE AT DIFFERENT INITIAL ANGLES

(It is assumed that the waves will break where the depth of water is 1.3 times the breaker height)

Steepness of wave in deep water (length: height)	Angle between wave in deep water and shore line						
	10°	20°	30°	40°	50°	60°	70°
10:1	9	16	23	30	36	41	46
20:1	6	10	15	20	23	27	30
40:1	5	9	13	17	20	23	25
100:1	3	7	10	12	15	16	17

In general, short waves are refracted less than long ones because the bottom affects them for a shorter time, and old swells are bent more than young storm waves. The table shows that waves break on the beach at greater angles than, perhaps, casual observation might show.

20.31. Loss of Wave Height by Refraction. When the "inshore" ends of waves are slowed and refracted and the "offshore" parts continue at their usual velocity, a reduction in height occurs due to the increasing width of the waves. They are stretched, as it were, and the wave energy is spread over a greater distance. There is reason to believe that the effect of refraction is different for waves of different steepness, although this needs further study. Theoretically, the decrease in height is inversely proportional to the amount of sidewise expansion. Hence, the greater the angle between the wave in deep water and

TABLE 20.18. PERCENTAGE DECREASE IN HEIGHT BETWEEN DEEP WATER AND THE BREAKER ZONE, FOR WAVES OF DIFFERENT INITIAL DEGREES OF STEEPNESS APPROACHING A STRAIGHT SHORE LINE (WITH STRAIGHT AND PARALLEL BOTTOM CONTOURS) AT DIFFERENT ANGLES

(It is assumed that the waves break where the depth of water is 1.3 times the breaker heights)

Steepness of wave in deep water (length:height)	Angle between wave in deep water and shore line					
	20° (percent)	30° (percent)	40° (percent)	50° (percent)	60° (percent)	70° (percent)
10:1	0	3	6	11	18	30
20:1	0	5	10	16	26	38
40:1	1	6	11	17	27	39
100:1	2	6	12	19	28	40

the coast, the more the waves tend to lose height. Table 20.18 shows that the decrease in height in surf due to refraction is negligible for waves that come in at angles smaller than 30 degrees, no matter what the steepness offshore. On the other hand, when the angle is greater than 60 degrees, the decrease in height may make a landing possible.

We have been considering straight coasts where the refraction will be the same all along a particular beach. If the coast is irregular, the amount of refraction will vary too.

20.32. Surf Around Shores of Bays (Fig. 20.6). If the bay is "open," waves advancing at right angles to the shore line of the center of the bay would

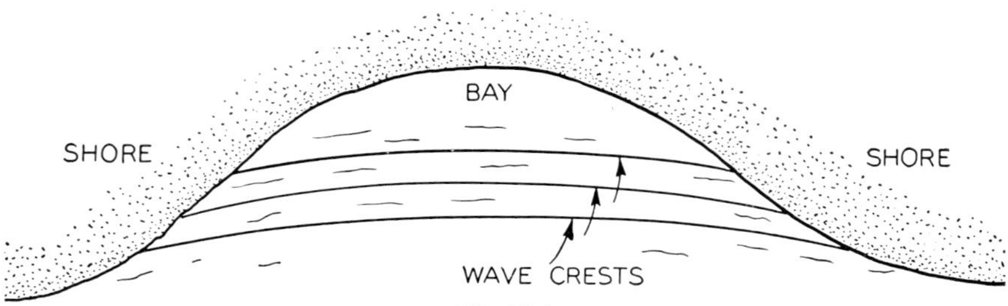

FIG. 20.6

suffer changes of height in accordance with the shoaling depths. The same waves whose line of advance might make an angle of 40 degrees with the flanks of the bay would have their heights reduced by 10 to 12 percent by refraction

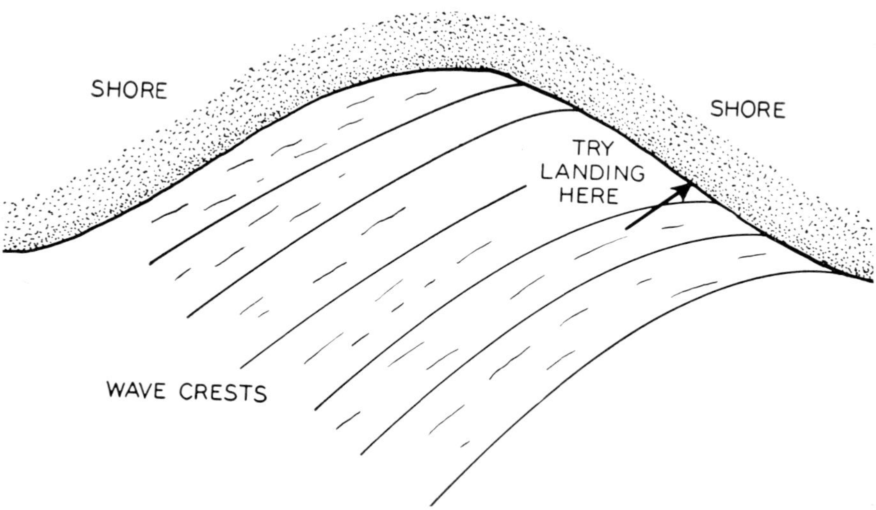

FIG. 20.7

on the flanks (Table 20.18). Of course, the tendency to lose height by refraction may be compensated by a gain in height for certain kinds of waves. In deep water, unless the waves are steeper than 1:25, the reduction by refraction would be more than counterbalanced by the gain in height. The general rule is that, when the wave crests are approximately parallel to the central part of an open bay, the surf will not vary enough along the flanks so that a landing can be attempted there in case a landing in the center would be dangerous. If the waves are advancing at a wide angle with the center of the bay shore, it may be possible to land on the lee shore when a landing in the center or on the exposed beach would be dangerous. (See Fig. 20.7.)

On the other hand, heavy swells moving into the mouth of a long, narrow bay (but not narrowing) will expend a large part of their energy by refraction and surf near the entrance. The head of the bay may be quiet with little or no surf. If the bay narrows, i.e., is funnel-shaped, the waves will be narrowed too. Their heights will increase because their energy is concentrated in ever-shortening lines. A landing at the head of the bay will be more difficult under this condition than at the entrance. A shallowing bay would only add to the heights of the surf at the head of the bay.

20.33. Surf Around Headlands. The surf on the tip of a headland may be very severe when the depths off the headland are shallow and the direction

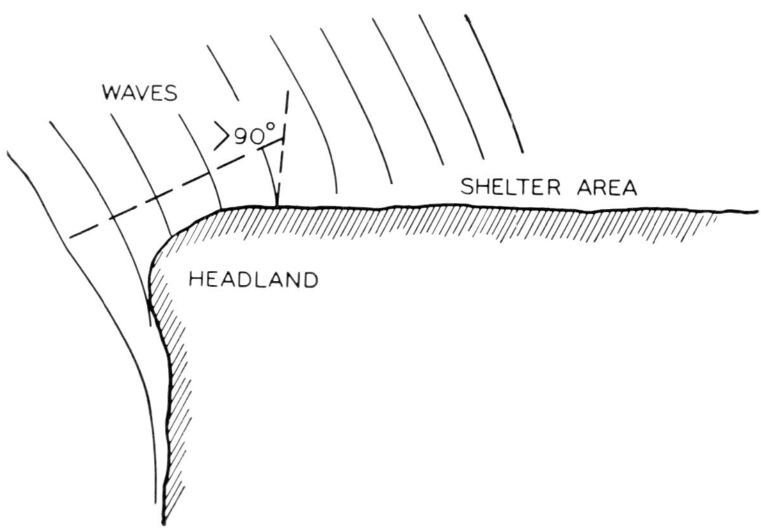

FIG. 20.8

of the waves is the same as the axis of the headland. If the water off the headland is deep, the swells may not break there but pass down the flanks of the headland with little or no change in direction. A heavy surf may be expected in stormy weather when the waves are higher and longer and break farther out.

The degree of protection offered by a headland against waves advancing at other angles depends on how abruptly the coast line changes around the headland and the direction of wave advance. The inshore ends of waves may be refracted right around a short headland with a broad round tip and thus be directed up the so-called sheltered side.

In cases where the shore line alters abruptly or the waves make an angle of more than 90 degrees with the sheltered side of the headland, the refracted waves may lose so much energy that the breakers on the lee side will decrease abruptly in height. Short waves may not be refracted at all, but may pass by the sheltered shore, leaving a calm area in the lee of the headland. Calm areas of this kind are frequently seen in the lee of steep headlands in moderate weather. Heavy weather does away with such calm areas because the waves, being longer, start refraction earlier and thus are enabled to turn around the headland. (See Fig. 20.8.)

The protection afforded by a headland is increased where the lee side is formed of smaller headlands and coves. If the protecting headland is long and narrow, like a breakwater, the lee side may be quite calm all the way to the end. Enclosed harbors with narrow entrances offer complete protection once the entrance is passed. Waves which get inside are so widened that their energy is dissipated completely.

20.34. Surf Around Islands. Waves advancing on an island with off lying shoals will be refracted around the island. The surf on the windward side may be very heavy and that to the leeward much less. Observations show that waves with an initial ratio of 1:20 to 1:100 height to length will break at angles of 16 to 35 degrees with the shore on the leeward sides, providing, of course, the depth is about 1.3 times their height when breaking. The height of the breakers here may be about one half what it was on the windward side.

Even so, there can be difficulties in landing on the leeward sides because the waves break at an angle with the shore line, and the bottom may be rocky or contain broken coral or coral heads. The surf may be confused if a landing is attempted where the refracted waves from both sides meet. On the whole, small, round islands are not very promising. Irregular shaped ones are better. A protected cove may be discovered to leeward or the lee of an abrupt headland used. The larger the island, the more lee it affords. Crescent-shaped islands with a concave side to leeward are excellent. Lagoons inside a reef with narrow entrances offer complete protection after the entrance has been passed because the energy of waves that do enter is soon expended in the widening of the wave crest line. The island lagoons in the Pacific are good examples.

A more complete description of waves and surf and methods of forecasting breakers and surf can be found in the several H.O. publications upon which this condensation is based.

21

The Atmosphere and Its Circulation

21.1. Introduction. Man lives at the bottom of a vast ocean of air known as the atmosphere that extends from the surface of the earth upward for many miles. He refers to the condition of the atmosphere at a given time and place as *weather,* whereas the term *climate* is used to denote the prevailing, or average, atmospheric characteristics which are to be found over a given area or region. The science which deals with the nature of the atmosphere, its changes, and the causes of the changes is known as *meteorology.* The term *aerology* is used by the U.S. Navy synonymously with *meteorology,* although by definition, the science of aerology is more restrictive than meteorology. *Climatology* is the science which is concerned with the study of climate.

Prior to the present century, books and articles dealing with weather were mainly descriptive in nature because there was only a limited understanding of the physical and mathematical laws which govern the weather and its changes. Rules for forecasting were developed empirically, and conditions in upper portions of the atmosphere, which have an important bearing on weather at the earth's surface, were known only in a general way.

During World War I weather forecasting was hampered because ships at sea did not make weather reports for reasons of security, and international cooperation in meteorology was at a minimum. This led to research for the purpose of gaining a better understanding of weather processes. The need for weather information on the part of the Scandinavian fishing industry and the lack of weather reports in that area stimulated Norwegian meteorologists and their colleagues to develop a better understanding of *air masses* and the *frontal* conditions along adjacent masses of air having different characteristics.

During the period between World Wars I and II the science of meteorology developed rapidly. The expansion of aviation during this period required detailed weather forecasts and information. A knowledge of conditions in upper portions of the atmosphere was needed for forecasting and for aircraft operation. New instruments and techniques were developed for probing upper air conditions, and the airplane itself was useful in gathering much information about the atmosphere. Charts, graphs, and formulas were developed in order to enable the meteorologist to interpret and forecast weather conditions more effectively than had been possible up to this time.

World War II again brought great demands on the part of the armed forces and of civil activities for weather information, but again the reports so vitally needed had to be curtailed. Those years saw many new developments for dealing with the weather. Radar equipment, the seismograph, and the airplane itself through reconnaissance operations became increasingly important in locating and tracking various storm conditions.

During the present century, meteorology has developed along sound scientific and mathematical lines so that today much is known not only about local conditions as they exist in many parts of the world but about the causes of weather as well. Experience has always been and still is necessary in dealing with and forecasting weather, but modern developments have done much to put the subject more nearly on a sound scientific basis. Meteorology has today developed into a specialized profession based on much training and experience. The armed forces and civil weather services of this and other nations now furnish frequent detailed weather information by radio. They maintain weather offices where one may call for information; well-trained personnel are available in some of these offices on a 24-hour basis. Many of the larger ships of the Navy carry Weather Service staffs.

The weather section of this volume is directed to those who, although not undertaking to be their own forecasters, should be able to understand the advice of the professional in order to make full use of existing weather facilities, to interpret conditions in the absence of information, to develop the ability to supplement official broadcasts with personal observations, and to make intelligent decisions on the basis of existing information. Presentation, therefore, has been made of some of the fundamental modern concepts of the physical processes which cause weather, along with descriptive weather information. This will afford a basis upon which to build further knowledge through observation and reference to literature on the subject. When a foundation has been laid, any further contacts with the subject should take on more meaning. Weather changes are interesting and sometimes quite fascinating. Many persons become quite able to interpret the weather signs. In the interests of safe, efficient, and pleasant operation of surface ships and aircraft, seamen and airmen should develop reasonable ability to do so. In many parts of the world official weather broadcasts are lacking or are insufficient. Even if this were not the case, no good mariner or aviator would want to rely solely on weather broadcasts over the radio for his meteorological guidance. At all times he not only wants to avoid poor weather conditions but to take advantage of favorable factors as well.

In the material which follows, emphasis has been placed on the applications of meteorology to the problems of the sea as they affect both the naval and merchant services. On first thought it might seem that weather entered into naval tactics and strategy more vitally in the days of sail a century or two ago than is the case in the present age of steam. Actually, however, weather as a factor is receiving increased attention. In the days of sail a commander always had two plans for battle based on weather conditions; if the enemy were to windward, one plan would be used; if to leeward, the other. Otherwise the weather did not particularly complicate the situation. Today with the many types of aircraft, surface ships, and activities of a campaign, weather enters into strategical and tactical problems in many ways, both in defensive and offensive operations. It should be realized that weather is friendly to him who knows how to take advantage of it.

For those who wish to become professional meteorologists there are many good elementary and advanced books on the subject. Several colleges and universities offer programs leading to science degrees with majors in the field of meteorology. The armed forces, with the exception of the U.S. Navy, uses these colleges to train their meteorologists. The U.S. Navy offers comprehensive graduate courses in theoretical and practical meteorology at the U.S. Naval Postgraduate School, Monterey, California. It is intended that the chapters here will afford a beginning toward a practical and working understanding of the subject.

The *United States Coast Pilots and the Sailing Directions* published by the Coast and Geodetic Survey and Navy Department, respectively, contain much descriptive material of the weather and climate to be found along our coasts and in many parts of the world. Such subject matter should be read by those operating in the areas concerned. Also the monthly *Pilot Charts* published by the Hydrographic Office of the Navy Department contain a mass of useful information concerning prevailing winds, fog, ocean currents, and average weather conditions. On the backs of many of these charts are special articles on such topics as hurricanes, water spouts, fog at sea, icebergs, and other phenomena.

The *U.S. Navy Marine Climatic Atlas of the World,* published by the direction of the Chief of Naval Operations, is a recent series of studies which presents meteorological information for the surface and upper-air, over the ocean areas of the world. Users will find this series of publications most helpful in weather studies of the ocean areas.

21.2. The Atmosphere. Atmosphere, or air, is the term which is applied to the mixture of gases which surround the earth. It is the vast air ocean at the base of which man makes his home. Its reaction with the surface of the earth results in the various kinds of weather and types of climate which exist from place to place. Scientists have reliable evidence that his gaseous envelope extends outward a distance of more than 600 miles from the earth's surface. Unlike the ocean, its outer limit is not sharply defined because of the air's thinness at a great distance. As the earth journeys around the sun and spins on its axis, the atmosphere is carried along.

21.3. Troposphere, Stratosphere, and Tropopause. From the standpoint of meteorology the atmosphere consists of two layers. These layers known as the troposphere and the stratosphere are quite different. The troposphere is the inner layer, and it reaches upward from the earth's surface to the base of the stratosphere which lies beyond. Tropopause is the name applied to the imaginary dividing surface between the troposphere and the stratosphere, and its height varies with the season, latitude, and weather conditions. It is higher over the equator than it is at the poles. At any given latitude it is higher in summer than in winter. It is higher over stormy areas than over regions of settled weather.

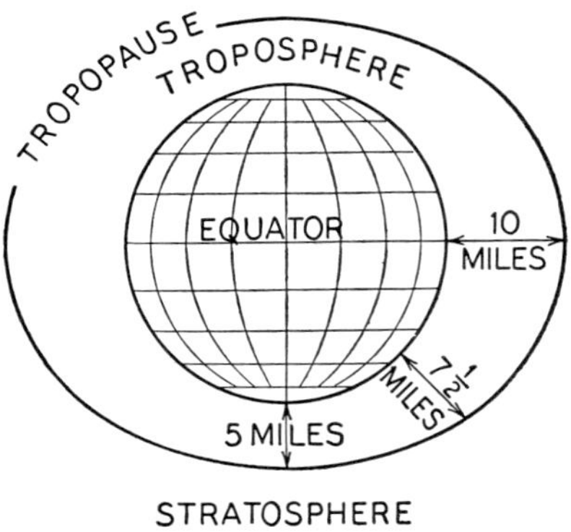

FIG. 21.1 THE EARTH AND ITS ATMOSPHERE, NOT SHOWN TO SCALE

Temperature distinguishes the troposphere from the stratosphere. Throughout the troposphere the temperature, on the average, is 1°F lower at intervals of 300 feet above the earth's surface. This decrease in temperature with elevation ends suddenly at the tropopause. Above the tropopause air temperature remains nearly constant at about −67°F for some distance upward into the stratosphere. Figure 21.1 shows the relation of the tropopause, troposphere, and stratosphere. For purposes of clarity the various portions of the figure are not shown to the same scale.

Other features of the troposphere are:

1. It has both horizontal and vertical air circulation; vertical motions of the atmosphere both upward and downward are known as air currents, while horizontal circulation of the air is called wind.
2. It is the region to which are confined such phenomena as clouds, storms, precipitation, and changing weather conditions.
3. About three fourths of the mass of the atmosphere is contained in the troposphere.
4. Its average upper limit is about 7½ miles above the earth's surface, but it varies from about 10 miles at the equator to about 5 miles at the poles.
5. The troposphere contains water vapor in varying amounts from less than 1 percent up to 5 percent by volume.
6. The troposphere is compressed and therefore quite dense as compared with the stratosphere. In the lower levels of the troposphere atmospheric pressure is approximately 1 inch of mercury less at each 1,000-foot interval above the earth's surface.
7. Flying conditions may be poor. Airplane icing, rough riding, poor visibility, cloud ceilings, and thunderstorm activity are common in the troposphere.

Essential characteristics of the stratosphere are:

1. A nearly constant temperature of −67°F exists for a considerable distance upward from the base of the stratosphere.

2. Vertical air motion is not in abundance. Wind in horizontal direction exists.
3. Very little if any water vapor is found in the stratosphere: clouds are virtually nonexistent.
4. Conditions favorable to flying prevail.

21.4. Composition of Air. Figure 21.2 shows the approximate percentages by volume of the principal constituents of the air. Throughout the troposphere air is composed of a mixture, not a chemical compound, of about 77 percent

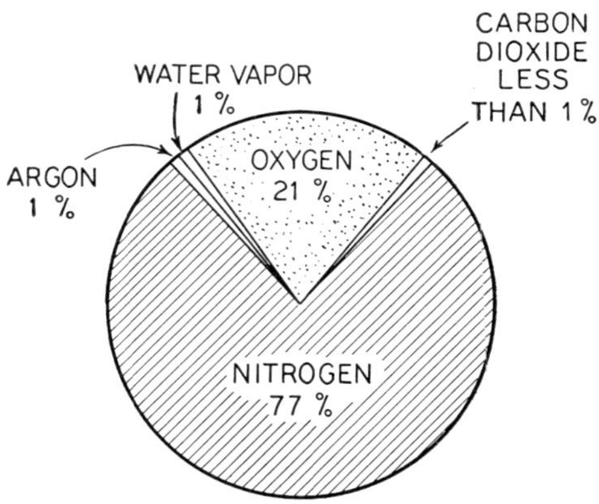

FIG. 21.2 AIR—A MIXTURE OF GASES. Nitrogen predominates; each has a function; but water vapor is the most important in weather and climate phenomena. The percentages by volume are indicated, but the percentage of water vapor may vary from less than 1 percent to about 5 percent.

nitrogen, 21 percent oxygen, 1 percent water vapor, 1 percent argon, and less than 1 percent carbon dioxide. There are also traces of a number of other gases, dust, smoke, and salt particles. Nitrogen seems to serve as a diluting agent for the oxygen. Oxygen is necessary for animal life and combustion. Carbon dioxide is essential to plant life. Weather changes, precipitation, visibility, rough air, and other phenomena are due, for the most part, to the relatively small amount of water vapor in the air. When water vapor condenses, it releases the energy which is found in violent, vertical currents and intense storms. Particles of dust, smoke, and salt in the air also play a major role in weather conditions.

Water vapor is an invisible gas. It can be seen only when it is precipitated as clouds, fog, dew, frost, etc. Although it usually composes only about 1 percent of the air, it may be present in an amount ranging anywhere from near nothing to 4 or 5 percent. Air temperature is the sole factor which determines the percentage of air which may be water vapor. For example, it might compose 5 percent of the volume of warm air, whereas the greatest amount which could exist in cold air is a much smaller percentage. If air contains the maximum amount of vapor possible at a given temperature, cooling of the air then would immediately result in the condensation of some of it. This means that some of the invisible water vapor would change to visible liquid or solid form. When air is cooled, the temperature at which condensation on a smooth polished surface begins to occur is known as the *dewpoint.*

As the percentage of water vapor in air varies, this causes the percentages of nitrogen and oxygen to vary. When the amount of vapor increases, the amounts of nitrogen and oxygen decrease. Water vapor is lighter, i.e., less dense, than nitrogen and oxygen. It would be expected then that of two parcels of air the one containing the higher percentage of vapor would be the lighter, other factors being equal.

There is always at least a small amount of vapor present in the atmosphere. When air contains the maximum amount possible at the existing temperature, the air is known as saturated. The term is rather misleading, as it will be recalled that saturated air, even at warm temperatures, never consists of more than a small percentage of water vapor by volume. Air which does not contain enough vapor to cause saturation at the existing temperature is known as unsaturated air. When dense fog exists, the air is not only saturated, but it contains condensed water vapor (fog particles) as well. The relative humidity then is 100 percent. *Relative humidity* is the ratio of the amount of water vapor in the air to the amount the air could hold at the existing temperature. When the air temperature increases 20°F, the capacity of the air for water vapor approximately doubles; therefore air at 80°F can hold 16 times as much as air at 0°F.

It is now believed that the gaseous constituents of the stratosphere and their percentages are the same as in the troposphere. The one exception is water vapor because this gas is virtually nonexistent throughout the vast region beyond the troposphere. In the troposphere the percentages of oxygen and nitrogen remain about constant because of the circulation and mixing of the air. Although there is no vertical motion of the air throughout the stratosphere, it is believed that the percentages of the various gases there remain constant because of molecular diffusion. If some such process were not in operation, the gases would tend to separate into layers with the heavier gases settling toward the lower portion of the stratosphere.

Air even at high levels and far inland contains large numbers of salt particles which have been carried away from the sea by the winds. These, together with other so-called hygroscopic particles, such as soot, smoke, etc., provide important nuclei necessary for the formation of raindrops. Such particles also affect the visibility and are a factor in sky coloring.

21.5. Heating and Cooling of the Atmosphere. Heat is transferred in three ways: by *radiation,* by *conduction,* and by *convection.* Virtually the only heat received at the earth's surface is from the sun. Heat from the sun is known as *insolation,* and it is transferred from the sun through space by means of short-wave radiation. Upon reaching the atmosphere a small amount of insolation is absorbed by the various gases, and some is reflected back into space. Of that which reaches the earth's surface, part is reflected, and the rest is absorbed by the land and water areas. All heat which is received from the sun by the earth and its atmosphere is eventually reradiated to space. If this were not true, the earth would gradually grow warmer and warmer, which, of course, is not the case. Heat reradiated by the earth is of long wave length rather than short wave length. Water vapor, which is the principal heat-absorbing constituent of the atmosphere, more readily absorbs the long wave radiation from the earth than the short wave incoming insolation from the sun. The

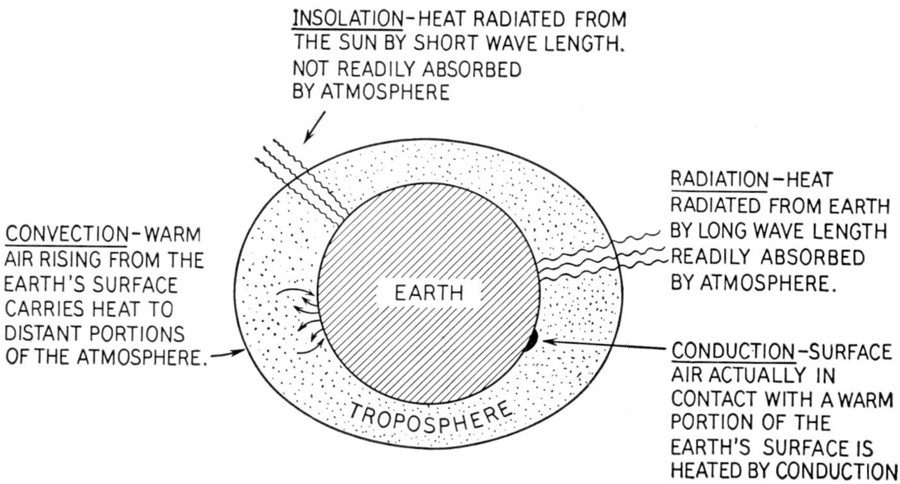

FIG. 21.3 MEANS BY WHICH THE ATMOSPHERE MAY RECEIVE HEAT: (1) Radiation, (2) conduction, and (3) convection

atmosphere then acts as a sort of trap, like the glass of a greenhouse. Incoming heat radiation is readily passed by the atmosphere, whereas outgoing radiation is impeded.

Radiation is not the only means by which the atmosphere is heated. Air in immediate contact with warmer water or land surfaces is heated by means of conduction. Conduction in itself, however, is effective in heating only that portion of the atmosphere which is actually adjacent to the earth. Convection can carry or circulate heated portions of surface air to upward and outlying portions of the troposphere. When a mass of air at the earth's surface becomes heated by conduction, it expands and becomes less dense or lighter per unit volume. It is then underrun by surrounding colder, denser, and heavier air and is pushed upward. It rises like a cork or a balloon. In that way its heat, acquired at the earth's surface, is carried to higher levels. The colder air from above, which replaces it at the surface, is heated in turn and eventually rises to upper levels. Examples of convection, conduction, and radiation are shown in Fig. 21.3.

21.6. Atmospheric Pressure. Air is very light, but it does have weight, and it, therefore, exerts pressure on the earth's surface. At sea level this pressure on the average amounts to 14.7 pounds per square inch. In other words, a column of air 1 inch in cross section and extending from sea level to the upper limit of the atmosphere weighs 14.7 pounds. At a height of 1,000 feet above sea level the air column would weigh less. This is because a portion of the column is below that elevation. At a height of 18,000 feet the column would weigh only about one half of the amount at sea level, or about 7.4 pounds. At an elevation of 36,000 feet, the average height of the upper limit of the troposphere, the air column would weigh only about one quarter as much as at sea level. Air is much lighter, i.e., less dense, in upper portions of the atmosphere than it is

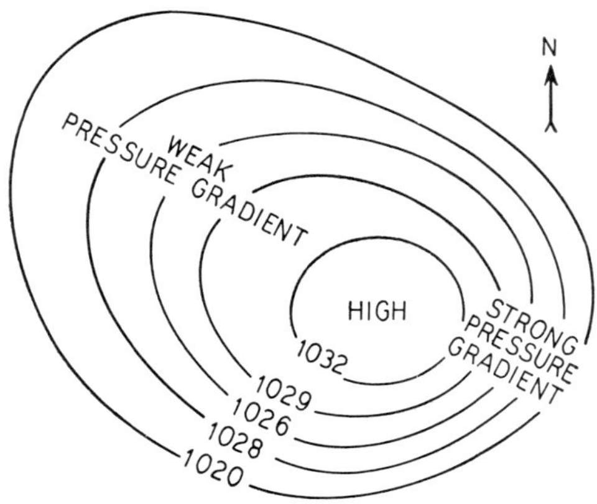

FIG. 21.4 AN AREA OF HIGH PRESSURE CENTERED AT "HIGH." In the southeast quadrant of this "High" the isobars are close together, and the pressure gradient is known as strong; at the center and to the northwest, the gradient is weak.

at sea level. Under standard conditions of temperature and pressure at sea level—i.e., at 59°F and 14.7 pounds per square inch—air weighs 1.22 ounces per cubic foot. Thirteen cubic feet weigh about 1 pound.

In meteorology, air weight, or atmospheric pressure, is usually expressed in terms of the length in inches of a mercury column or in millibar units. These units will be considered later in connection with barometers. Fourteen and seven-tenths pounds per square inch are equivalent to 29.92 inches of mercury or to 1,013.25 millibars.

Atmospheric pressure at any location constantly changes, and it varies from place to place. These changes in pressure values are due to changes in temperature. When air is warmed, it expands and therefore becomes less dense. The resulting atmospheric pressure is less than the average amount. On the other hand, when air becomes cold, it contracts and becomes more dense, or heavier. Areas having cold masses of air will record higher atmospheric pressure readings.

Lines which are drawn through points on the earth having the same atmospheric pressure are known as *isobars*. These lines of equal pressure enclose areas of high pressure and areas of low pressure. Horizontal *pressure gradient* refers to the decrease per unit distance in a horizontal direction perpendicular to the direction of the isobars. In Fig. 21.4 the isobars are seen to be spaced closer together in the southeast portion of the high pressure area than in the northwest section. When isobars are close together, the situation is known as a steep pressure gradient, but when they are far apart, the gradient is weak.

21.7. Wind Velocity and Direction—Causes. The velocity, or speed, of the wind is determined by the pressure gradient. Strong gradients cause strong winds, while weak gradients result in gentle winds. At times the pressure may

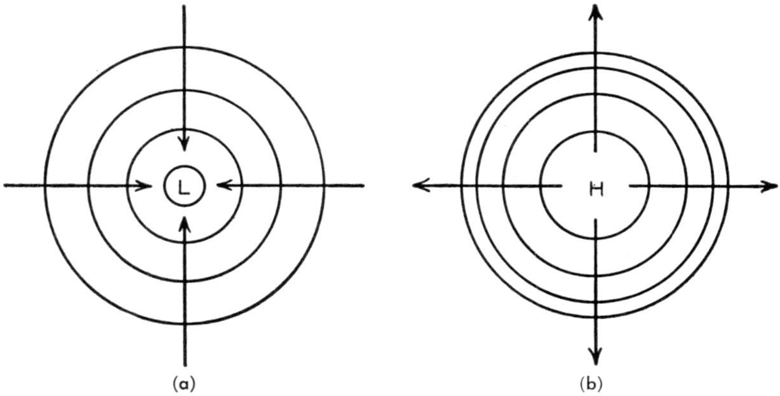

FIG. 21.5(a) AIR FLOWS TOWARD REGIONS OF LOW PRESSURE. Were it not for the apparent deflective force due to the earth's rotation, air would tend to flow directly toward points where pressures are the lowest.

FIG. 21.5(b) AIR FLOWS AWAY FROM REGIONS OF HIGH PRESSURE. Wind direction would be parallel to the gradient as shown, were it not for the earth's rotation.

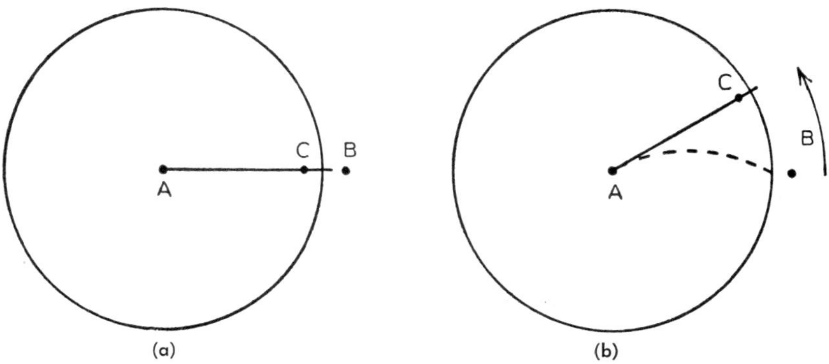

FIG. 21.6(a) Air starts to move from point *A*, located on a disk which is rotating counterclockwise, toward point *B* which is located just off the disk. As the air leaves point *A* it is also headed toward point C on the edge of the disk.

FIG. 21.6(b) The disk rotates through an angle of 30 degrees as the air moves across the disk from *A* to *B*. The air, because it is headed toward *B*, does not reach point C. Though the air moves directly from *A* to *B* its path appears as a curved line to an observer on the disk.

be about the same over a large area. A weather map showing such a condition is referred to as a *flat* map; the wind, if any, would be light. For any given pressure gradient the wind blows stronger over water areas than over land. This is due to less friction, which is usually greater over land than water. Hills, trees, buildings, and similar objects retard the speed of the wind more

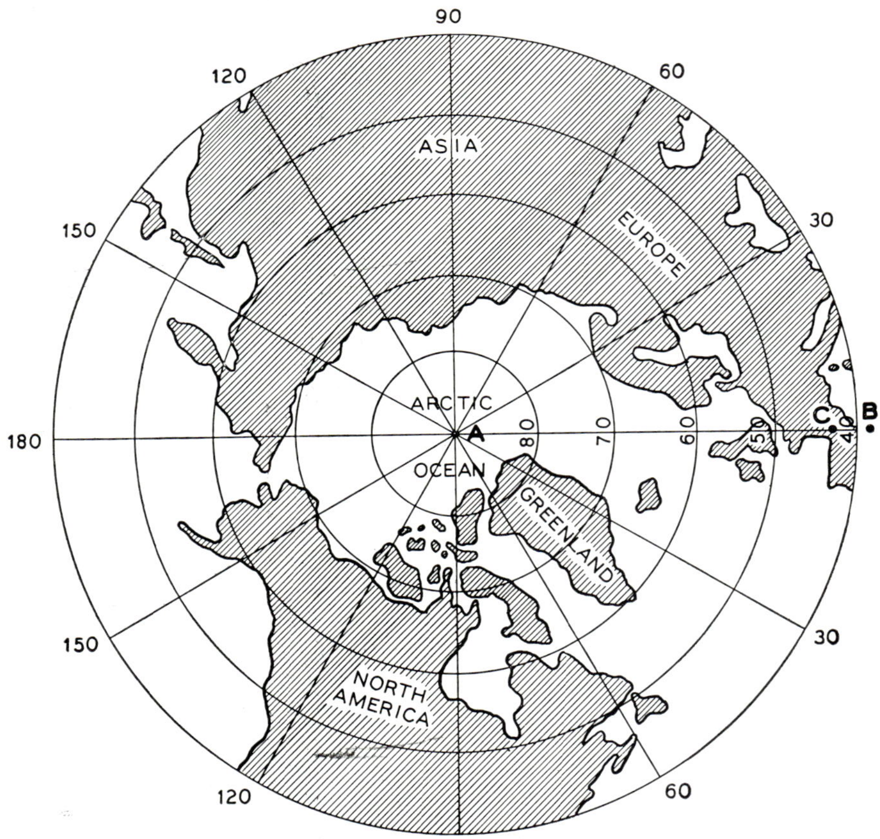

FIG. 21.7(a) As in FIG. 21.6, air starts to move from point *A* in the Arctic Ocean southward toward point *B* which is located in space.

than do water surfaces. The earth's frictional effect does not exist at elevations higher than about 1,500 feet above land and water surfaces. For this reason the wind in general blows stronger at elevations above the so-called surface frictional layer.

Wind direction depends chiefly upon the direction of the pressure gradient and the rotation of the earth. Let us first consider the effect of the pressure gradient. Figures 21.5(a) and (b) show the tendency of air to flow from a high pressure area to a section where the pressure is lower. This flow of air we know as wind, and it tends to blow parallel to the pressure gradient, i.e., directly across at right angles to the isobars. However, due to the rotation of the earth, wind does not blow directly from locations of high to low pressure, but it is deflected to the right in the northern hemisphere and to the left in the southern hemisphere.

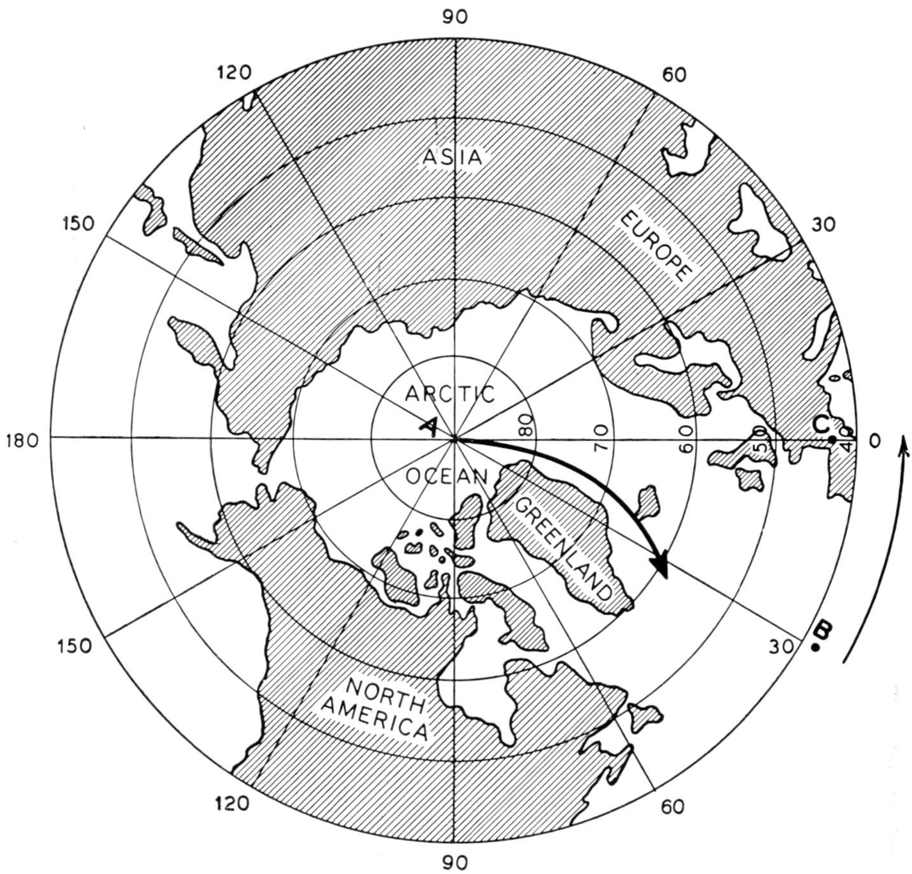

FIG. 21.7(b) As the earth rotates counterclockwise air appears to turn to the right as it moves toward point *B* in space.

Wind deflection caused by the earth's rotation is known as the *coriolis effect.* Refer to Figs. 21.6(a) and (b), and let us see what causes the coriolis effect: (a) shows a disk rotating to the left, counterclockwise. Let us assume that air starts to move in a straight line from *A* toward a point *B* which is located just off the disk. At the time the air starts to move from point *A* toward *B* it is also moving toward point *C* located at the edge of the disk. Let it also be assumed that point *C* on the disk rotates 30 degrees to the left during the time that it takes the air to move from *A* to *B*. Figure 21.6(b) shows the curved path which the air would take over the rotating disk to reach *B*. To see this clearly, cut a disk of paper and fasten it at the center to a board with a thumbtack.

Make an X on the board just off the disk. Place a pencil point at the center of the disk and draw a line slowly toward X, at the same time rotating the disk slowly in a counterclockwise direction. The pencil line will curve as in (b).

If the disk is rotated toward the right, a particle in motion on it would appear to be deflected toward the left.

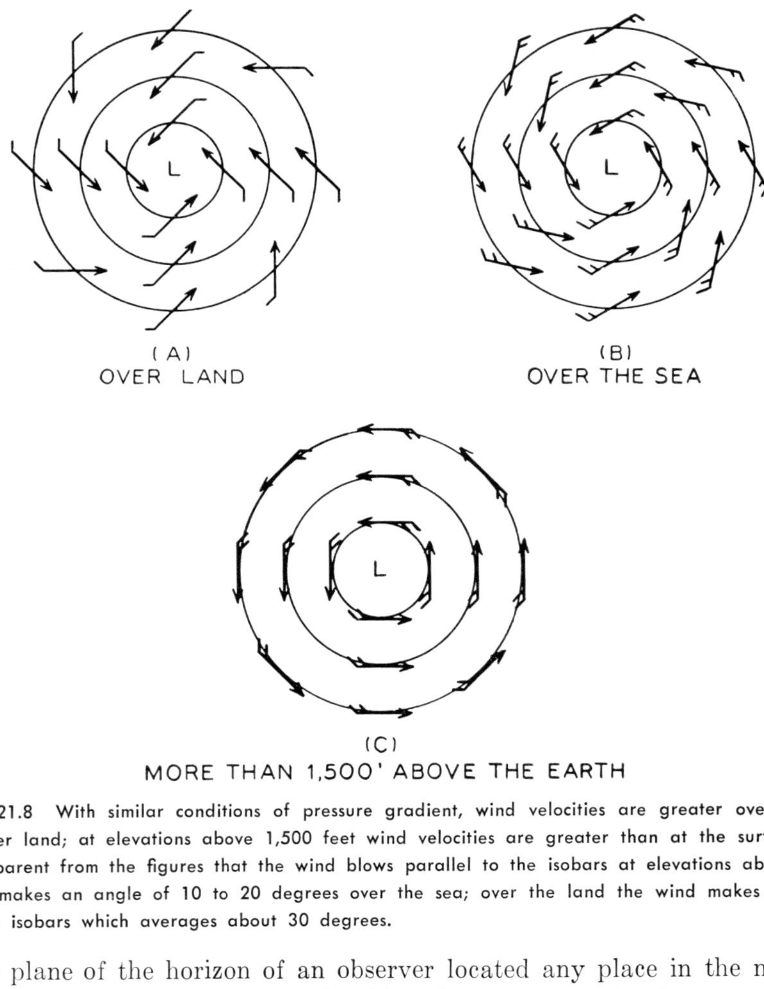

FIG. 21.8 With similar conditions of pressure gradient, wind velocities are greater over the sea than over land; at elevations above 1,500 feet wind velocities are greater than at the surface. It is also apparent from the figures that the wind blows parallel to the isobars at elevations above 1,500 feet; it makes an angle of 10 to 20 degrees over the sea; over the land the wind makes an angle with the isobars which averages about 30 degrees.

The plane of the horizon of an observer located any place in the northern hemisphere rotates toward the left with reference to a point in space. Therefore the plane of the horizon may be likened to the plane in (b), and air in motion in any direction on the earth north of the equator is deflected to the right. In the southern hemisphere wind from any direction is deflected toward the left. This is because the observer's horizon in the southern hemisphere rotates in a clockwise manner, or toward the right.

Figures 21.7(a) and (b) show why the polar northeasterlies blow from the northeast toward the southwest. Were it not for the coriolis effect, these winds would tend to blow from the north directly toward the south. Similar figures could be used to show the effect of the earth's rotation on the polar southeasterlies of the southern hemisphere and the trade winds of both hemispheres.

The coriolis effect is strongest in polar regions but is zero at the equator. At intermediate latitudes it varies directly as the sine of the latitude.

Air friction with the earth not only decreases wind velocities, but it decreases the coriolis effect as well. This means that the deflective effect of the earth's rotation is less over land surfaces than over ocean areas. The result is that under similar pressure gradient conditions the wind is deflected farther to the right over water than it is over land. Furthermore, at elevations above the frictional influence of the earth, levels more than about 1,500 feet above the earth, the coriolis effect is strong enough to pull the wind to the right so that it blows parallel to the isobars. Under those conditions low pressure is to the left, while high pressure is to the right, if a person stands with his back to the wind. This is known as *Buys Ballot's Law,* and it applies to the northern hemisphere. The opposite is true in the southern hemisphere. At or near the surface of the sea the lowest pressure would bear about 7 points on the port hand.

Figure 21.8 illustrates the results of earth friction on both the wind velocity and the coriolis effect.

Presently we will note the results of the coriolis effect in connection with the general circulation of the winds of the world, as well as in the cases of the many different types of local winds. Ocean currents such as the Gulf Stream are also affected by the deflective force of the earth's rotation.

21.8. General Winds of the Earth. Uneven heating of the earth's surface causes differences in atmospheric pressure which, in turn, cause winds. Equatorial regions of the earth receive considerably more heat than do the polar

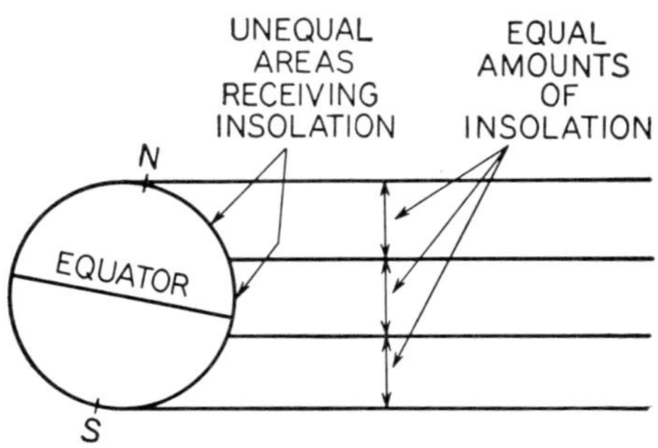

FIG. 21.9 The sun's rays reach the earth's surface more obliquely in polar regions than in the tropics. This causes unequal heating of the earth's surface. It will be noted that equal amounts of insolation affect unequal areas of the earth's surface.

areas. Figure 21.9 illustrates the effect of direct and oblique rays. This excess of heat at the equator is the basis of a definite world wind pattern. On a nonrotating globe of homogeneous surface the system would be simple. The atmosphere having been warmed and expanded over the hot equatorial belt would flow poleward at the higher levels of the troposphere. This would tend to increase polar surface atmospheric pressures. Air then would tend to flow away from the poles and along the earth's surface to the equatorial girdle of lower pressure. This simple circulation is shown in Fig. 21.10. Such a circulation is

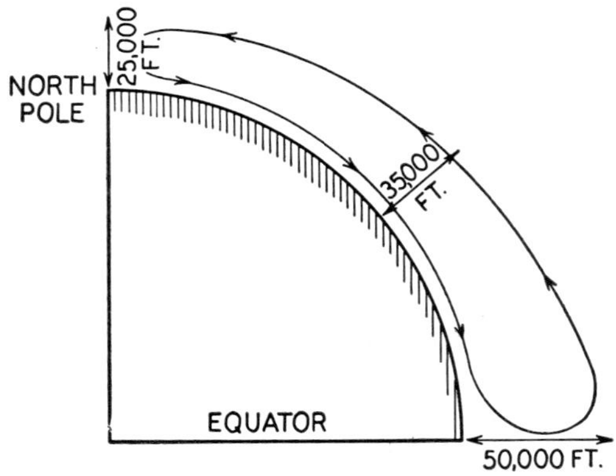

FIG. 21.10 THEORETICAL PATTERN OF WIND CIRCULATION DUE TO THE UNEQUAL HEATING OF THE EARTH'S SURFACE. Actually this scheme is considerably modified because of the rotation of the earth, the influence of oceans, continents, and other factors. (*After Rossby*)

impossible because of the influence of the earth's rotation. The world wind system is further complicated by the contrasting temperatures of continents and oceans and by many other local causes which will be considered in order. Refer now to Fig. 21.11.

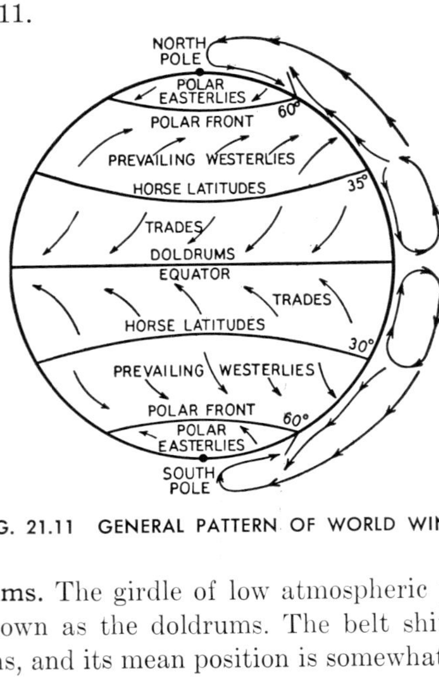

FIG. 21.11 GENERAL PATTERN OF WORLD WINDS

21.9. The Doldrums. The girdle of low atmospheric pressure in the region of the equator is known as the doldrums. The belt shifts slightly north and south with the seasons, and its mean position is somewhat north of the equator.

It is characterized by light and variable surface winds and frequent calm conditions. Warm temperatures and associated rising air currents are quite general. Cloud types are the bulging, piled-up cumulus, and cumulonimbus (thunderhead). The air is sultry, and showers and thunderstorms are frequent. The sky is often overcast.

21.10. The Horse Latitudes. In considering the world wind system we will first look at the conditions which exist in the northern hemisphere. Differences which are found in the southern hemisphere will then be pointed out.

Air rising over the doldrums flows poleward in the high levels of the troposphere, but it does not blow directly north. The coriolis effect causes it to be deflected to the right, and it becomes a southwest wind. In fact, at about latitude north 35 degrees it is supposed that the deflecting effect of the earth's rotation causes the wind at high levels to blow approximately from west to east, although the wind circulation at high levels is not nearly as well understood as conditions at the earth's surface. This deflection of the wind to the right causes the air to tend to pile up at about 35 degrees north latitude. The result is a ring of high pressure which extends around the earth at that latitude. The cooling of the air as it flows northward at high levels and its consequent shrinking and sinking contribute to the high-pressure belt. This region is characterized by descending air currents and cloudless skies. At the earth's surface the winds are light and variable. The weather is generally fine; air humidity is comparatively low. A marked contrast is quite apparent between weather conditions of these two zones, the doldrums and the horse latitudes. In the doldrums the air expands as it rises. This results in the cooling of the air and the formation of clouds and rain. With descending air currents in the horse latitudes no such process for the formation of clouds and rain is possible; rather, the relative humidity tends to become less in the descending currents of the horse latitudes.

21.11. The Trade Winds. With surface pressure conditions high at the horse latitudes and low in the doldrums we would expect to find wind blowing from the high- to the low-pressure region. This is precisely what happens. The trade winds blow from the horse latitudes to the doldrums and are quite dependable both in direction and force. Were it not for the rotation of the earth, the trade winds would blow directly from north to south. The deflective effect, however, turns them to the right so that they become the northeast trade winds. The trade wind belt shifts slightly in a north and south direction with the seasons. Off the northwest coast of Africa and the west coasts of Mexico and Central America the trades blow from a more northerly rather than northeasterly direction. The trades are not as well developed or as dependable in the southwest Pacific as in other oceans. Records from the Caribbean Sea on the other hand show a remarkable constancy in direction and velocity during most of the year.

21.12. The Prevailing Westerlies. Surface air also flows northward from the high-pressure region of the horse latitudes. The coriolis effect deflects it to the right so that it becomes a southwesterly wind. The prevailing southwesterlies of the temperate zone, or middle latitudes, are not nearly so consistent as the trades of lower latitudes. This is because the region of the prevailing westerlies includes the paths of many storms throughout the year; these storms are associated with winds from all points of the compass. Only occasionally are the trades interrupted by storms, the hurricanes which occur only during a portion of the year.

21.13. The Polar Cap of High-Pressure, and Prevailing Northeasterlies. Wind at the higher levels of the atmosphere flows poleward where it descends and tends to build up a region of high pressure at the earth's surface. Winds at surface levels therefore tend to blow southward but are deflected toward the right by the coriolis effect and become the prevailing northeasterlies of the polar regions. We now note a situation where the surface polar northeasterlies eventually encounter the southwesterlies of the temperate zone.

21.14. The Polar Front. The air masses of the converging polar northeasterlies and the southwesterlies of the middle latitudes do not readily mix. Instead, the cold northern air tends to underrun the lighter and warmer air from the south. The surface between these two currents is known as the *polar front.* The average position of this frontal surface is at about latitude north 60 degrees. It is important however to realize that the polar front shifts to positions which may vary from latitudes as far south as Florida to areas considerably north of the 60th parallel. For the most part the storms of the temperate zones have their origin along the so-called polar front, as we shall see in Chapter 24.

21.15. The Southern Hemisphere. It will be recalled that air in horizontal motion in the southern hemisphere is deflected toward the left. The trades south of the equator blow from the southeast rather than from the northeast; the prevailing westerlies are from the northwest rather than from the southwest. Wind from the cap of high pressure of southern polar regions blows from the southeast instead of the northeast, as is the case in the northern hemisphere. It will be noted that the northern hemisphere contains extensive land areas, whereas the southern hemisphere is predominantly a water surface. Inasmuch as the contrasting temperatures of land and water surfaces cause complications in the general wind circulation of the world, it follows that the prevailing winds of the southern hemisphere are much more constant in direction than is the case in the northern hemisphere. Also wind velocities of the southern hemisphere average higher than is the case north of the equator. This is accounted for by the fact that the air over water is less retarded by friction. The prevailing westerlies of the temperate zone of the southern hemisphere are known as the *Roaring Forties* because of their comparatively high velocities in the general area between latitudes south 40 and 50 degrees.

21.16. Upper Troposphere Wind Circulation. Some reference has been made to wind at high levels, but because of lack of data, a composite picture of the wind above 80,000 feet is not as well known as the wind at lower levels. During recent years, upper-air data to the height of 80,000 feet has been sufficient to meet most forecasting needs; however present methods of observing upper air above that level do not produce sufficient data for a complete understanding of the circulation patterns.

It is now known that there exists in each hemisphere a band of wind known as the "jet stream." These bands of high-speed winds meander around their respective hemispheres like rivers, never stationary but moving north or south, seemingly at will.

The jet stream was discovered during the latter days of World War II when American bombers, flying from bases in the Pacific to Japan, encountered headwinds which were almost equal to their ground speed. Investigation of this phenomena has established the fact that high-speed streams or jets of wind circle the earth in each hemisphere at the general level of 30,000 to 60,000 feet. The jets are relatively narrow and are generally not more than 2,000 to 3,000 feet thick in the vertical plane. Wind speed in the jet stream has been measured as high as 300 knots, but is more generally in the range of 100 to 150 knots. These jet streams seem to play an important role in the movement of surface storms. Meteorologists have determined that the main jet stream remains relatively distinguishable in areas of sufficient reports, but subjets or fingerjets appear to develop and dissipate over areas far removed from the primary jet stream. Investigations have shown that areas of clear air turbulence, abnormal temperature gradients, and erratic movement of cyclones (surface storms) are associated with the jet stream and its movement.

21.17. Heating and Cooling of Land and Water. The general pattern of global wind circulation at the earth's surface is considerably modified by the uneven heating of the continents and oceans. During the daytime, land areas are usually much warmer than they are at night. This is because the heat which is absorbed during the day penetrates only a short distance and is readily reradiated to open space. The balance between incoming and outgoing heat occurs at about 2 hours past noon. After that time land areas lose heat by radiation faster than they receive heat from the sun. Likewise there is a considerable annual variation in land temperatures; they are much colder during winter than in summer. The reason is similar to that of the daily temperature variations; i.e., there is a net loss of heat in winter and a net gain during the summer.

The effect of insolation on ocean surfaces is much different than on land surfaces. When the sun is shining diagonally on water surface, most of the heat rays are reflected back to space without heating the sea and its overlying air. When the sun is more nearly vertical, a considerable portion of the rays are absorbed, but the heat penetrates quite far into the water. The specific heat of water is great, which means that it requires more heat to raise the temperature of a given mass of water than an equal mass of land. Much of the heat absorbed by water surfaces is utilized to evaporate water at the surface rather than to raise its temperature. The sea of course is constantly in circulation so that the surface layers as they are warmed are carried to lower depths and replaced by cooler water from below. Circulation of this sort does not take place in the case of land heating. Observations show that the difference between day and night shaded air temperatures over the ocean is less than 1°F.

21.18. Permanent and Semipermanent High and Low Centers. In winter the continental land masses of North America, Asia, and Europe are much colder than the waters of the north Atlantic and Pacific oceans (Fig. 21.12). The result of this is the building up of high-pressure areas over the continents and low-pressure centers over the adjacent oceans. The *low* which lies between Canada and the Scandinavian peninsula is known as the *Icelandic low* because its center is located near Iceland. The low area in the Pacific is known as the *Aleutian low* because it is centered in the vicinity of the Aleutian Islands. These low areas are associated with much cloudiness, rain, drizzle, sleet, snow, fogginess, and strong winds. The stormy weather of these regions is not unlike stormy weather of any other section, but it is more widespread, persistent, and intense. Near land masses in the Aleutians, especially to leeward, there are regions of disturbed air currents known as *williwaws*. They are gusts of wind often reaching velocities of 100 knots or more. These winds are not only strong in a horizontal sense, but they are associated with violent vertical currents as well. Aleutian fog is especially thick and often extends to an elevation of 4,000 feet or more. Such fogs may persist in spite of winds as high as 60 knots.

The Icelandic and Aleutian lows actually are intensifications of the polar front ring of low pressure which lies between the region of prevailing westerlies and polar northeasterlies. During the summer they are less pronounced than in winter because of the warming of the continents in summer (Fig. 21.13).

It must be borne in mind that the Aleutian and Icelandic lows do not represent a continuation of one and the same low-pressure area. Rather, they are regions where low-pressure systems form or arrive from other places to remain for a time. Later the lows may move on or die out and are replaced by other lows. The Aleutian and Icelandic centers shift to various positions and are at times replaced by high-pressure areas.

Semipermanent high-pressure areas in the northern hemisphere are located in the Atlantic near the Azores and in the Pacific off the coast of California. A lesser center is found in the vicinity of Bermuda. In the southern hemisphere semipermanent high centers are located in the Pacific west of Chile, the Atlantic west of Africa, and the Indian Ocean. These high centers represent intensifications of the ring of high pressure which lies between the trades and the prevailing westerlies in both the northern and southern hemispheres.

The semipermanent lows and highs affect the general scheme of global wind circulation and have a decided effect on the weather in many parts of the world. They also have a direct relation to the direction and velocity of the currents of the oceans. Having considered the general pattern of world wind circulation, pressure, and heat distribution, we will now look at the seasonal winds.

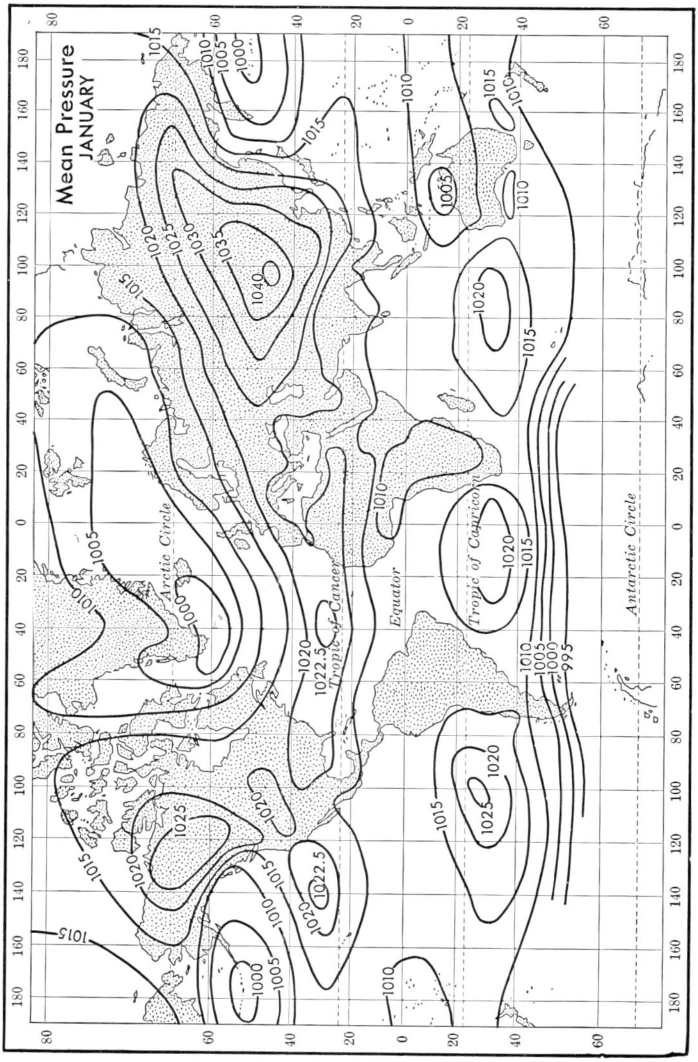

FIG. 21.12 ISOBARS OF MEAN PRESSURE FOR JANUARY

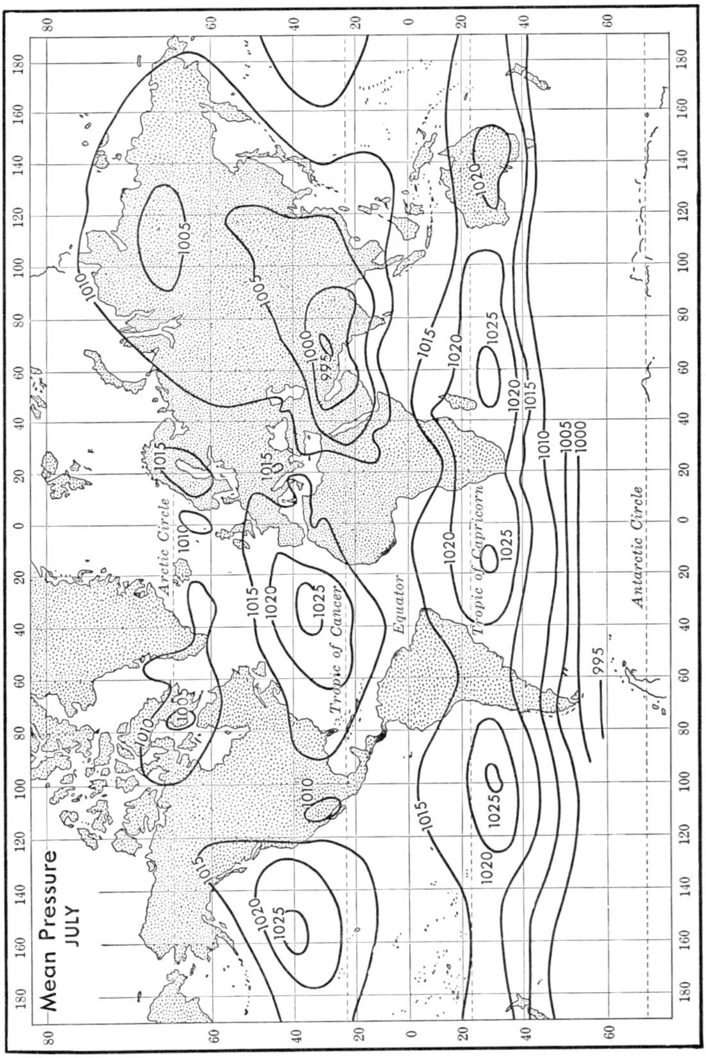

FIG. 21.13 ISOBARS OF MEAN PRESSURE FOR JULY

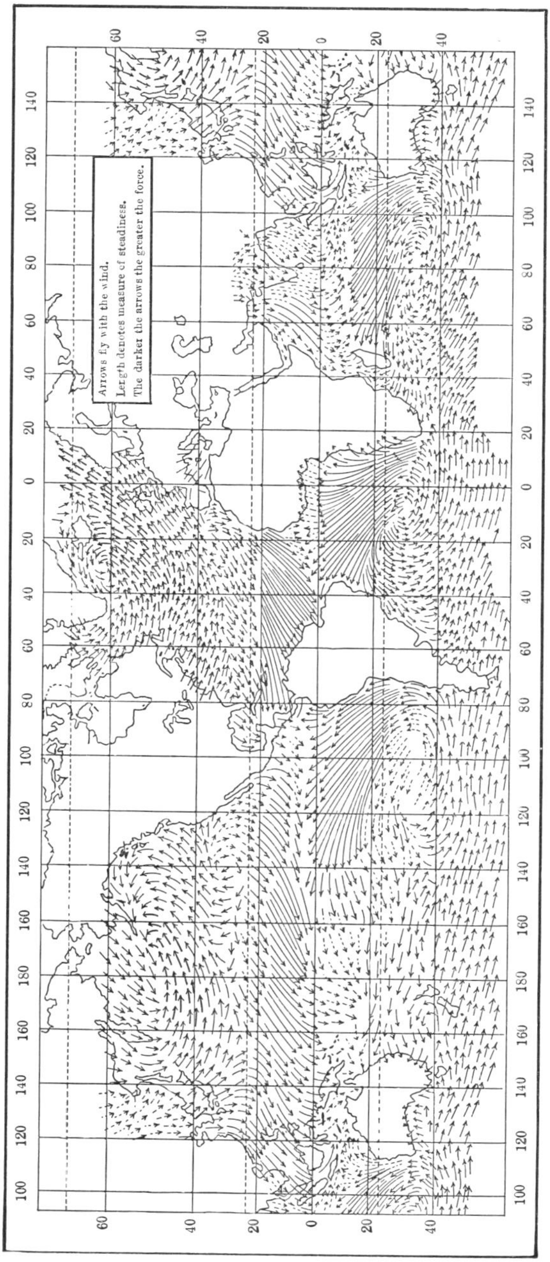

FIG. 21.14 OCEAN WINDS, JANUARY AND FEBRUARY. *Courtesy U.S. Weather Bureau*

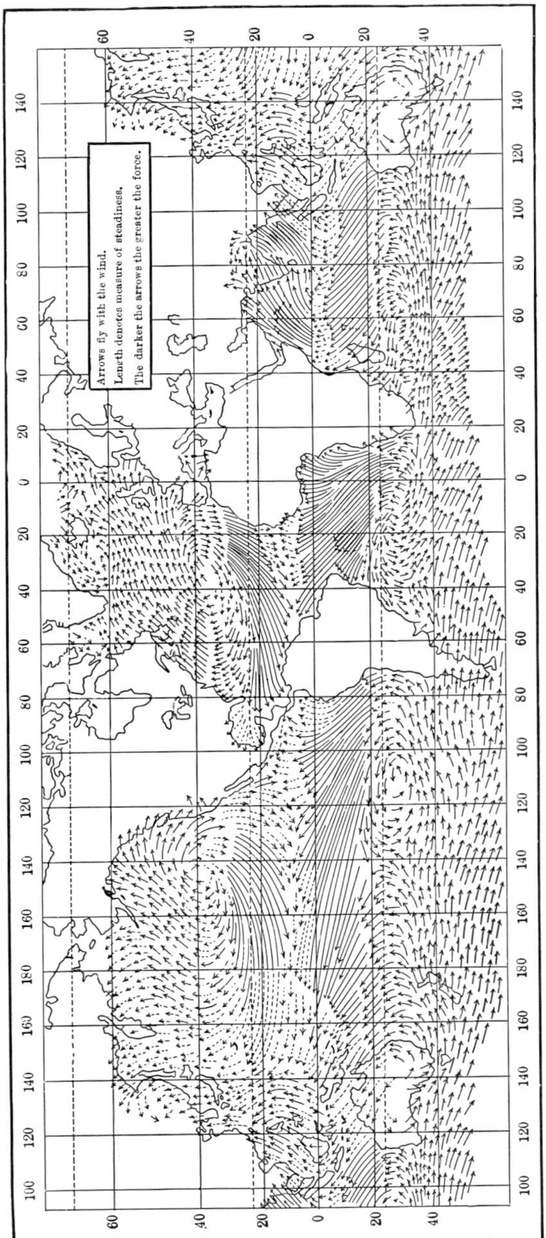

FIG. 21.15 OCEAN WINDS, JULY AND AUGUST. *Courtesy U.S. Weather Bureau*

21.19. The Monsoon Winds. It was seen that air over continents is colder in winter than the air of adjacent ocean regions, whereas in summer the reverse is true. These differences in temperature and therefore pressure cause semi-annual reversals in the wind direction in the areas affected. The results are quite marked in the Indian Ocean, China Sea, and south and southeastern Asia (Figs. 21.14 and 21.15). During the winter season air flows outward from the interior of the continent of Asia toward the regions of lower pressure which prevail over the warm waters of the Indian Ocean and Australia. In India and southeastern Asia the winds then prevail from the northeast and are dry because of their origin and their descent on the southern slopes of the east-west Himalaya range; it is the season of fine weather in that part of the world and extends from October to April. After the winds cross the equator, they are deflected toward the left and become northwest winds.

During the warmer portion of the year conditions are reversed. The southeast trades south of the equator cross to the northern hemisphere and are then deflected toward the northeast, becoming the winds of the southwest monsoon of south and southeast Asia. It is then that the wind flows from the relatively high-pressure area of Australia and the Indian Ocean to the area of low pressure which prevails over the continent of Asia. As the moist ocean winds reverse themselves and move inland over India and adjacent sections, heavy squalls, rains, and thunderstorms prevail. As the season advances, the squalls become less frequent, but the rainy season continues. The summer monsoon usually occurs from May to September. During this season there is considerable local variation in winds and rain, but, in general, it is the rainy season for that part of the world. In areas where the winds are deflected for considerable distances upward by the Himalayas, very heavy rainfall is reported.

Many other parts of the world have similar seasonal reversals in wind direction which are often associated with dry and rainy seasons. A mild monsoon wind reversal is noted in the states bordering the Gulf of Mexico.

21.20. Land and Sea Breezes. In the tropics and particularly during the warmer seasons of higher latitudes, the land during the day is commonly warmer than adjacent water. This applies not only to coastal sections but to inland lakes as well. Air overlying the land is heated, it expands, and is pushed upward by cooler air which flows onshore from the surfaces of adjacent water. Such sea breezes may penetrate inland for distances of 25 miles or more. On inland lakes the effect usually prevails for distances of only a few miles. Over land surfaces at night there is a net loss of heat because radiation exceeds insolation. Land surfaces then become cooler than water surfaces, and a reversal of the wind direction takes place. The contrast between land and water temperatures at night is not as great as during the day; therefore the nighttime land breeze is usually not as strong as the sea breezes of daytime.

Land and sea breezes extend to a height of only a few hundred feet. They are diurnal, i.e., daily, variations in wind direction due to differences in land and water temperatures during the day and night; monsoon winds, on the other hand, are also caused by differences in land and water temperatures, but they reverse in general only twice during a year. Monsoon winds are generated on a much larger scale than are land and sea breezes.

21.21. Mountain and Valley Breezes. During the daytime convectional currents tend to rise over mountains because of heating of mountainsides and summits. A general flow of air takes place up the valleys. At night radiation brings about chilling of the mountain slopes with the resultant chilling of the adjacent air. The cool, dense air then drains down the valleys. Winds of quite high velocity are sometimes noted, particularly in narrow canyons.

21.22. Gales from Land to Sea. During fall, winter, and sometimes spring months when strong high-pressure areas of cold, dense air build up over continental regions, adjacent coastal regions often experience winds of strong and sometimes gale force. When winds of this type are experienced along the American Atlantic Coast, they are usually preceded by low-pressure storm areas. With the passing of a storm to eastward, the pressure rises; and if pressure is sufficiently high east of the Rockies over the interior, strong northwest winds prevail for several hours or more. At times such winds are dangerous to coastal shipping. High pressure west of the Rocky Mountains, for example centered over Utah, is the cause of the *Santa Ana* of Southern California, which may endanger shipping in the harbors and coastal sections off Southern California. Strong winds of a similar nature are found at times in the Gulf of Mexico where they are known as *Northers*. On the west coast of Mexico and Central America they are known as *Tehuantepecers* and *Papagayoes*. In the Mediterranean along the north coast there are the *Black Northers*, *Boras*, and *Mistrals*.

21.23. Ocean Currents. Figure 21.16 shows the general pattern of ocean currents throughout the world. A glance at Figs. 21.14 and 21.15 will readily show how the direction of flow of ocean currents tends to coincide with the prevailing winds of the world. There are a number of other important factors, such as salinity, which contribute to the formation and maintenance of ocean currents.

One of the best-known ocean currents is the Gulf Stream. Originating in the Gulf of Mexico it flows northward off the Atlantic Coast of the United States to a point opposite Newfoundland, where it spreads out fanlike and drifts northeastward to Iceland, the British Isles, and the Scandinavian peninsula. It carries along warm water and warm air, and it affects profoundly the weather and climate of all places along its route. It is partly responsible for the comparatively mild temperatures of Iceland and of western Europe. Warm air from the Gulf Stream often flows out over adjacent colder water and land areas, and fog results. Noteworthy examples of this are the fogs which form during fall and winter months along the Atlantic seaboard and off Newfoundland where Gulf Stream air blows over the cold Labrador current. The Gulf Stream is known as a warm ocean current. Warm currents are producers of fog, rain, sleet, and snow. Let us now look at a cold ocean current.

In the Pacific Ocean south of the equator along the coasts of northern Chile and Peru is a cold ocean current known as the *Humboldt* current. Coming from

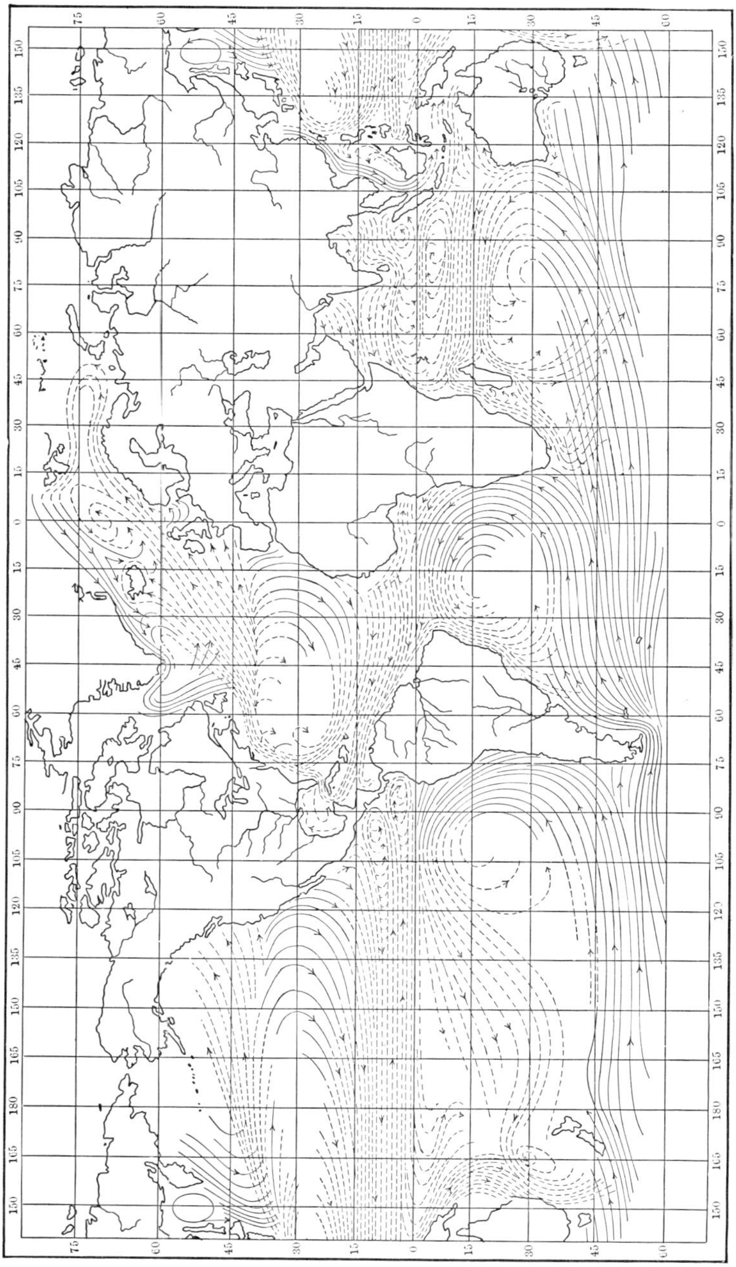

FIG. 21.16 OCEAN CURRENTS—COLD, SOLID LINES; WARM, DOTTED LINES. *Courtesy U.S. Weather Bureau*

far out in the Pacific it flows toward the east, then toward the northeast. At about latitude south 40 degrees it turns northward along the coasts of Chile and Peru. Before reaching the equator it turns toward the northwest, and then the west. The air over the Humboldt current is often warmer than the water surface. Three reasons for this are:

1. There is a decided upwelling of cold water along the coasts visited by this current.
2. The stream as it flows northward is moving into warmer latitudes.
3. Finally, any air which flows from adjacent water or land areas is likely to be warmer than this ocean current surface.

The result of warm air over cold water is fog or stratus clouds. That is precisely what exists along the northern coasts of Chile and Peru—frequent stratus cloudiness but little rainfall. Other coastal regions affected by cold ocean currents and little rainfall are the west coasts of Australia and the west coast of South Africa where cold currents flow northward. In the northern hemisphere cold currents flow southward off the northwest coast of Africa and the southern and Lower California coasts.

In order to appreciate the effect of cold ocean currents on climate, one has only to note the relatively small amounts of rainfall and lower temperatures which are recorded along the coasts and islands that lie in the paths of these currents.

22

Clouds, Fog, and Stability

22.1. Introduction. A knowledge of the various kinds of clouds, how they form, and what they mean is an indispensable tool to those at sea, in the air, or to anyone for that matter who must deal with the weather and its changing conditions. Because clouds offer visual evidence of conditions which exist in the atmosphere and of changes which are taking place, they also afford an indication of coming weather conditions, particularly if they are observed from time to time to note any changes from one kind to another which may be taking place. For example, one of the first indications of the existence of a tropical cyclone is the appearance of cirrus clouds. As the storm moves closer, cirrostratus clouds are noted. Finally, with the storm at hand, low, dark clouds are present.

Clouds are also useful in foretelling the approach of other storm types, such as the extra-tropical cyclones, or lows, of the temperate zones of both hemispheres, and of local disturbances, such as thunderstorms and tornadoes. Aviators soon learn to know the clouds which are associated with rough air and smooth air and the kinds which cause their planes to be coated with ice. We will therefore want to look into the identification of the various cloud types, their significance, and their causes.

As we look at the sky from time to time, there seems to be an endless variety of clouds. If we attempt to classify them, it is sometimes difficult to decide into which group some of them should be placed. Cloud identification, although at times difficult even for the experts, is important.

22.2. International Meteorological Conferences. The present scheme of cloud names, descriptions, and classification has, in general, been in effect for a great many years. It is based on the system proposed more than a century ago by the English scientist, Luke Howard. Cloud classification was revised, standardized, and brought up to date at an international conference at Copenhagen in 1929 and again in 1947 at a meeting of meteorologists in Washington. These actions at an international level were necessary so that the writings and conversations of meteorologists of all nations would not be confusing with regard to cloud descriptions. Since weather observers of most nations exchange synoptic weather reports in times of peace, it was also important that the kinds of clouds reported should be based on a classification known to all weather observers.

The classification agreed upon by the representatives of the meteorological services present at the international conferences consists of ten cloud genera grouped in four families. Each of the ten genera has a number of variations or cloud types. The four cloud families and ten cloud genera are as follows:

Family A: High Clouds. The three cloud genera of the high family are cirrus, cirrostratus, and cirrocumulus. The mean lower level of the base of the clouds in this family is 20,000 feet above the earth's surface. Because the temperature is normally about 1°F colder at each 300-foot interval above the earth's surface, temperatures at the heights of clouds of this family are always below freezing, and the particles of these clouds are crystals of ice. For example, when the temperature at the earth's surface is 100°F, cirrus clouds at an altitude of 20,000 feet or higher would have temperatures of 30°F or lower. It will be noted that *cirrus,* or *cirro,* is found only in the names of the three cloud genera of the family of high clouds.

Family B: Middle Clouds. The two cloud genera of the middle family are altocumulus and altostratus. Clouds of this family may have their bases anywhere from 6,500 feet to 20,000 feet above the surface of the earth. The names of both cloud genera of the middle family begin with *alto,* meaning high. These clouds, though high, are not so high as the cirrus genera, but they are higher than the clouds of the two low families which will be considered next.

Family C: Low Clouds. The three cloud genera of the low family are stratocumulus, stratus, and nimbostratus. Cloud bases of this family may be found quite near the earth's surface, or they may be as high as 6,500 feet. It will be noted that the names of none of the five cloud genera of the two low families of clouds contain *cirro* or *alto*. Hence, the absence of *cirro* or *alto* in the name of a cloud indicates that it is a member of one of the two low families, i.e., this family or the family which follows.

Family D: Clouds with Vertical Development. The two cloud genera of this family of low clouds are cumulus and cumulonimbus. Clouds in this family have bases which may occur anywhere from 1,600 to 6,500 feet above the surface of the earth. The tops of these clouds may reach an elevation of 20,000 feet or more. Unstable conditions of the atmosphere, which will be described later in this chapter, are responsible for the formation of these two cloud genera.

22.3. Description of Clouds—General. There are, in general, three cloud shapes: cirrus, stratus, and cumulus. Cirrus clouds, one of the members of the high family, are a detached kind of thin, silky-like cloud. They resemble somewhat a lock of hair, and because of their appearance are sometimes referred to as *mare's-tails.* Stratus clouds form a blanket or layer of a more or less uniform flat mass which in many cases covers the entire sky. Four of the ten cloud genera are of the stratus variety. Cumulus clouds are lumpy masses. Their bases are invariably flat and horizontal, but the tops are round and extend upward to various heights, depending upon atmospheric conditions. There are four cloud genera of the cumulus form. One of the ten cloud genera, stratocumulus, combines the appearances of both cumulus and stratus clouds.

22.4. Comparison of Strati-Form Clouds. Cirrostratus is a high cloud sheet through which the sun readily shines. Sometimes it is so thin that its presence is not suspected except for a milky appearance of the sky. Altostratus is lower, grayer, and thicker than cirrostratus. The sun's disk may be made out through altostratus, and rain or snow often falls from this type. Stratus is much lower and thicker than altostratus. It resembles fog, but its base does not rest on the ground. Nimbostratus is similar to stratus, except that precipitation, which may or may not reach the earth, usually falls from it. *Nimbo* means rain.

22.5. Comparison of Cumuli-Form Clouds. Cirrocumulus clouds never cover a large portion of the sky. They consist of a patch or patches of small, white flakes which may be arranged in rows. Among the most beautiful and rare of all the clouds, they are popularly referred to as *mackerel sky* because of their resemblance to the scales on the back of a mackerel or to a school of fish. Altocumulus are lower in the sky and larger than cirrocumulus. They may be found in groups and may be arranged in rows. Cumulus clouds are lower than altocumulus; they are the type quite common in summer. Their bases are flat, whereas the tops are rounded and may extend to great heights. Cumulonimbus clouds are similar to cumulus but are much larger. Their bases are flat and resemble nimbostratus. The tops extend to heights where temperatures are well below freezing. Showers and often thunder are associated with cumulonimbus clouds.

22.6. Stratocumulus Clouds. This type combines the appearance of both stratus and cumulus clouds and is often a low, lumpy layer which covers the entire sky. At other times it may consist of low rows or groups of cumuli-form clouds between which may appear patches of blue sky.

22.7. Specific Descriptions of the Ten Genera. *Cirrus* (Fig. 22.1) are detached clouds of delicate and fibrous appearance, without shading, generally white in color, and often of a silky appearance. Cirrus clouds appear in the most varied forms, such as isolated tufts, lines drawn across a blue sky, branching feather-like plumes, curved lines ending in tufts, etc. They are often arranged in bands which cross the sky like meridian lines and, owing to the effect of perspective, converge to a point on the horizon or to two opposite points. (Cirrostratus and cirrocumulus often take part in the formation of these bands.)

Cirrocumulus (Fig. 22.2) is a cirroform layer or patch composed of small, white flakes or of very small globular masses usually without shadows, which are arranged in groups or lines, or more often in ripples resembling those of the sand on the seashore.

Cirrostratus (Fig. 22.3) is a thin, whitish veil which does not blur the outlines of the sun or moon but usually gives rise to halos. Sometimes it is quite diffuse and merely gives the sky a milky look; sometimes it more or less distinctly shows a fibrous structure with disordered filaments.

FIG. 22.1 CIRRUS (Ci). *Official U.S. Navy Photograph*

FIG. 22.2 CIRROCUMULUS (Cc). *Official U.S. Navy Photograph*

FIG. 22.3 UPPER CLOUDS ARE CIRROSTRATUS: LOWER STRATA ARE CIRROCUMULUS. *Official U.S. Navy Photograph*

FIG. 22.4 ALTOCUMULUS, QUONSET POINT, RHODE ISLAND. *Official U.S. Navy Photograph*

Altocumulus (Fig. 22.4) is a layer (or patches) composed of laminae or rather flattened globular masses, the smallest elements of the regularly arranged layer being fairly small and thin, with or without shading. These elements are arranged in groups, in lines, or waves, following one or two directions, and are sometimes so close together that their edges join. The thin and translucent edges of the elements often show variations in coloring which are rather characteristic of this class of cloud.

Altostratus (Fig. 22.5) is a striated or fibrous veil more or less gray or bluish in color. This cloud is like thick cirrostratus but without halo phenomena; the

FIG. 22.5 ALTOSTRATUS. *Official U.S. Navy Photograph*

sun or moon shows vaguely with a faint gleam as though through ground glass. Sometimes the sheet is thin with forms intermediate with cirrostratus. At times it is very thick and dark, even completely hiding the sun or moon. In this case differences of thickness may cause relatively light patches between very dark parts; but the surface never shows real relief, and the striated or fibrous structure is always seen in places in the body of the cloud. Every form is observed between high altostratus and cirrostratus on the one hand, and low altostratus and nimbostratus on the other. Rain or snow may fall from altostratus (altostratus precipitans). When the rain is heavy, the cloud layer will have grown thicker and lower, becoming nimbostratus; but heavy snow may fall from a layer that is definitely altostratus.

Stratocumulus (Fig. 22.6) is a layer (or patches) composed of laminae, globular masses, or rolls; the smallest of the regularly arranged elements are fairly large; they are soft and gray with darker parts. These elements are arranged in groups, in lines, or waves, aligned in one or two directions. Very often the

FIG. 22.6 STRATOCUMULUS BELOW AND CUMULONIMBUS ABOVE. *Official U.S. Navy Photograph*

FIG. 22.7 STRATUS. A patch in the Cuyama Mountains near San Diego, California, as viewed from above. Note additional patches in background. *Official U.S. Navy Photograph*

rolls are so close that their edges join; and when they cover the whole sky, they have a wavy appearance.

Stratus (Fig. 22.7) is a low uniform layer of cloud resembling fog but not resting on the ground. When this very low layer is broken up into irregular shreds, it is designated *fractostratus.*

Nimbostratus (Fig. 22.8) is a low, amorphous, and rainy layer of a dark gray color usually nearly uniform; it is feebly illuminated, seemingly from

FIG. 22.8 NIMBOSTRATUS CLOUD, FOG AND RAIN OFF HATTERAS. *Official U.S. Navy Photograph*

inside. When it gives precipitation, it is in the form of continuous rain or snow. But precipitation alone is not a sufficient criterion to distinguish the cloud which should be called nimbostratus even when no rain or snow falls from it. There is often precipitation which does not reach the ground; in this case the base of the cloud is usually diffuse and looks wet on account of the general trailing precipitation, and it is not possible to determine the limit of its lower surface.

Cumulus (Fig. 22.9) are dense clouds with vertical development; the upper surface is dome-shaped and exhibits rounded protuberances, whereas the base is nearly horizontal. When the cloud is opposite the sun, the surfaces normal to the observer are brighter than the edges of the protuberances. When the light comes from the side, the clouds exhibit strong contrasts of light and shade; against the sun, on the other hand, they look dark with a bright edge. True cumulus is definitely limited above and below, and its surface often ap-

FIG. 22.9 CUMULUS CLOUDS NEAR SAN DIEGO, CALIFORNIA. *Official U.S. Navy Photograph*

FIG. 22.10 CUMULONIMBUS (Cb)

pears hard and clear-cut. But one may also observe a cloud resembling ragged cumulus in which the different parts show constant change; this cloud is designated *fractocumulus*.

Cumulonimbus (Fig. 22.10) are heavy masses of cloud with great vertical development whose cumuliform summits rise in the form of mountains or towers and the upper parts have a fibrous texture, often spreading out in the shape of an anvil. The base resembles nimbostratus, and one generally notices precipitation. This base has often a layer of very low, ragged clouds below it (fractostratus, fractocumulus). Cumulonimbus clouds generally produce showers of rain or snow and sometimes of hail and often thunderstorms. If the whole of the cloud cannot be seen, the fall of a real shower is enough to characterize the cloud as cumulonimbus.

22.8. What Do Clouds Mean? Cirrus clouds are the first signs of the approach of a storm area. They are especially significant in this respect if arranged in parallel bands across the sky. As the storm approaches, cirrostratus is observed. Rings around the sun and moon, i.e., solar and lunar halos, are quite common with cirrostratus skies because these clouds, like all high clouds, are composed of ice crystals. These crystals reflect and refract the rays of sun and moon to produce the rings of light known as halos. Cirrocumulus and altocumulus, like all cumulus cloud types, signify vertical currents in the atmosphere. They are not rain clouds.

The highest kind of cloud which is an active producer of rain or snow is altostratus. This cloud often succeeds cirrostratus clouds as a storm area moves closer. The kind of precipitation produced by altostratus clouds is a slow, steady, all-day rain or snow. Altostratus clouds are gradually replaced by nimbostratus clouds, and the precipitation continues with fog often present. Airplane flights are smooth with these cloud types, but the plane will ice up when air temperatures are near or below freezing. Nimbostratus clouds are associated with low ceilings and poor visibility. These clouds, therefore, offer much hindrance to flying, particularly in view of the fact that they generally cover quite an extensive area.

Stratocumulus clouds often form when cool or cold winds of moderate or strong velocities blow over surfaces which are moist and warmer than the wind. The air below the clouds is rough or turbulent, but it is smooth above the clouds. Stratocumulus clouds are not rain clouds.

Cumulus clouds mark the tops of rising air currents. Such currents are caused by temperature differences at the earth's surface. Air may also be started upward when deflected by a hill or mountain or by hitting a converging colder and denser air mass. Whether or not an air current continues to ascend depends upon its temperature as compared with the temperature of the surrounding air. Air stability is the technical term that the meteorologist uses when he refers to the relative temperature of air masses and their air currents. The air is rough underneath a cumulus cloud, but it is smooth above the cloud top. These clouds are typical of summer weather. At times they may extend to great heights, and of course this means that the air is turbulent or unstable up to the elevations indicated by the cloud tops. When cumulus clouds grow to heights of 20,000 feet or more, their tops become frozen because of the great distance from the earth's surface. The result then is a cumulonimbus cloud, or thunderhead. This type produces showers, often heavy but intermittent in character. Thunder, lightning, hail, and strong winds are also common. To the flier, however, the greatest hazard of a thunderstorm is the violence of the vertical currents. Planes in diving position have been carried upward thousands of feet with the pilot being helpless to do anything about it. The strain and stresses imposed upon the plane's structure by the vertical currents in a cumulonimbus cloud are often strong enough to cause structural damage.

22.9. How Clouds Form. Clouds result chiefly from ascending air currents. Air may be forced to higher levels in a number of ways, four of which will be referred to presently. When air rises, it encounters regions of lower pressure, and therefore it expands. If not saturated with water vapor, expanding air cools at the rate of 5.5°F for each 1,000 feet of rise above the earth. If a current of air rises sufficiently high, its temperature will eventually decrease to a value at which the air will be saturated with water vapor. Any further temperature decrease tends to cause some of the water vapor to be condensed into cloud particles.

22.10. Adiabatic and Pseudo-Adiabatic Rates. The cooling rate of 5.5°F per 1,000 feet for rising and expanding unsaturated air is known as the adiabatic, or dry, rate of cooling. The term *adiabatic* means that the air is cooling because it is expanding, and the drop in temperature does not represent a loss in heat to the regions surrounding the mass which is rising and cooling.

Should a volume of air grow colder than its surroundings, it contracts, becomes denser, and sinks in the atmosphere. Its descent is associated with warming at the adiabatic, or dry, rate of 5.5°F for each 1,000 feet of fall in elevation. The so-called dry adiabatic rate of warming or cooling of air applies only to air which is not saturated with water vapor, i.e., air which has a relative humidity of less than 100 percent.

If rising air is saturated with water vapor, i.e., has a relative humidity of 100 percent, it continues to expand, but its rate of cooling is only about one half the dry adiabatic rate. The reason for this is simple. When saturated air rises and expands, some of its water vapor condenses into liquid form, such as cloud particles. In changing from the gaseous to liquid state, the vapor releases heat of vaporization at the rate of about 2.8°F per 1,000 feet of elevation increase. Two processes, therefore, are in operation. The rising saturated air is cooling at the rate of 5.5°F per 1,000 feet and heating at the rate of about 2.8°F. The net result is a cooling rate of about 2.7°F for every 1,000 feet of ascent. This rate is known as the wet adiabatic, or pseudo-adiabatic, rate of cooling.

Descending air always heats at the dry adiabatic and never at the pseudo-adiabatic rate. Figure 22.11 shows adiabatic and pseudo-adiabatic rates of cooling for rising air columns.

Foehn, or Chinook Winds. These winds illustrate the effects of the adiabatic heating and cooling of air. A foehn is a warm, dry wind which blows down the side of a mountain range. When atmospheric pressure conditions are higher on one side of a range than on the other side, air is caused to flow upward, over, and down the range toward the area of lower pressure on the leeward

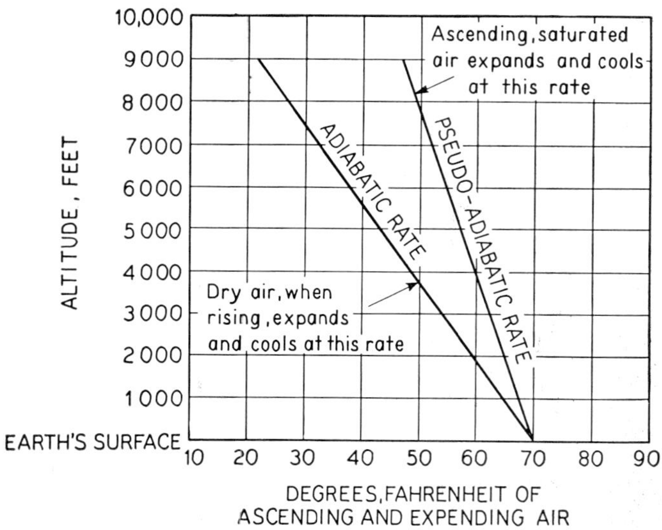

FIG. 22.11 GRAPHICAL REPRESENTATION OF THE ADIABATIC AND PSEUDO-ADIABATIC RATES OF COOLING. It will be noted that saturated air at 70°F will cool to 46°F if it ascends 9,000 feet, but air which is not saturated will cool to about 20°F if it ascends the same amount.

side. An example of these conditions is found in the Rockies when pressure is high over the north Pacific and low in the northern plains states. Fairly warm, moist air then flows inshore and ascends to higher ground. If at first the air is not saturated, it cools at the dry adiabatic rate of 5.5°F for each 1,000 feet rise in elevation. Once it becomes saturated, clouds form and the cooling rate decreases to about one half the dry adiabatic rate. Under these conditions rain or snow is very likely. Once the air reaches the region overlying the backbone of the Rocky Mountain range, it begins its descent at the dry adiabatic rate to the sections of eastern Montana and Wyoming, where it is known as a chinook wind and is much warmer and drier at any elevation than at a corresponding elevation on the western slope. This is, of course, due to the net cooling rate on the windward slope being less than the heating rate on the lee slope. The chinook wind is so pronounced at times that it has rapidly and completely melted and evaporated heavy blankets of snow east of the Rockies. Figure 22.12 illustrates the formation of a chinook wind.

In Europe this type of wind is known as a foehn. It is common on the north slope of the Alps. Warm, moist winds from the Mediterranean may be caused to ascend the southern slopes of the Alps when pressure is low over central Europe. As the wind descends to lower ground north of the Alps, it becomes the warm, dry *foehn.*

Four examples of cloud formation are as follows:

1. *Convection.* Let us consider the case of cumulus clouds which form over an island during the daytime. The land areas become warmer than the surrounding water surfaces. The warm land heats the overlying air and causes

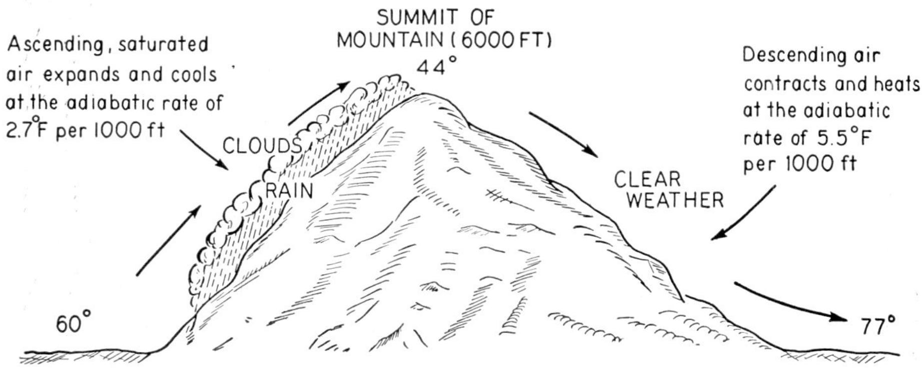

FIG. 22.12 THE FORMATION OF A CHINOOK WIND. Saturated air at a temperature of 60°F ascends a mountain range. Clouds and rain occur on the windward slopes. Upon reaching the summit the temperature of the air has decreased to 44°F. Descent on the leeward slopes results in the warming of the air so that its temperature is increased to 77°F, 17°F higher than at the corresponding level before ascent.

it to expand and become less dense. It is then pushed upward by cooler, denser air which flows inshore from the colder sea surfaces. If the rising air current continues to ascend, it will eventually cool to the dewpoint or temperature where its water vapor causes saturation of the air. The relative humidity then is 100 percent, and any further rising and cooling of the air tend to result in the condensation of some of the water vapor into visible cloud particles. Cumulus clouds have flat bases. This is apparent if they are viewed from an airplane or in a diagonal direction from the earth. The flat base marks the elevation at which the rising air current cooled to the dewpoint. Typical cumulus clouds have rounded, domelike, clear-cut tops. Their tops (Fig. 22.13) mark the height of the ascending air currents which caused the clouds. At elevations below the cumulus cloud base the air was cooling at the dry adiabatic rate. Throughout the cloud mass the rising air column cools at the wet, or pseudo-adiabatic, rate. Cumulus clouds, once formed, may be carried horizontally for considerable distances by the winds. Without the support of rising currents they eventually tend to sink and return to the invisible vapor form from which they had their genesis.

2. *Orographic Ascent of Air.* When the wind blows toward higher ground, such as a mountain range or barrier of hills, the air is forced upward. As it ascends, it expands and cools at the adiabatic rate. The cloud base forms at the height where the temperature of the rising and expanding air has decreased to the dewpoint. The tops of orographic-type clouds may extend to great heights above the terrain if the atmosphere is not stable. Stability is discussed later.

FIG. 22.13 CUMULUS (Cu). *Official U.S. Navy Photograph*

Regions of the world where prevailing moist winds are deflected upward by mountain ranges have heavy annual cloudiness and rainfall on the windward slopes. Sections to the leeward of mountains have much lower amounts of cloudiness and precipitation. India, lying south of the Himalaya range, records its heaviest rainfall during the portion of the year when the wind prevails from the south. In the winter when winds there prevail from north and are descending on the southern slope of the range, there is little cloudiness or rain.

3. *Overrunning Air Currents.* Let us consider what happens when a current of cool air from the east converges with a current of warm air moving from the south. As you look at Fig. 22.14, note that you are facing west with north to the right and south to the left. The current of cool air is moving away from you as indicated by the circle with a cross representing the tail of an arrow moving with the wind. The southern current is warmer and therefore lighter than the easterly current. When these currents meet, they do not tend to mix.

The warmer current is deflected upward by the heavier, denser air as indicated in the figure. The figure actually shows what is known as a warm front. Fronts, together with air masses, are discussed in Chapter 24.

Many types of clouds are formed when an air current overruns another, rises,

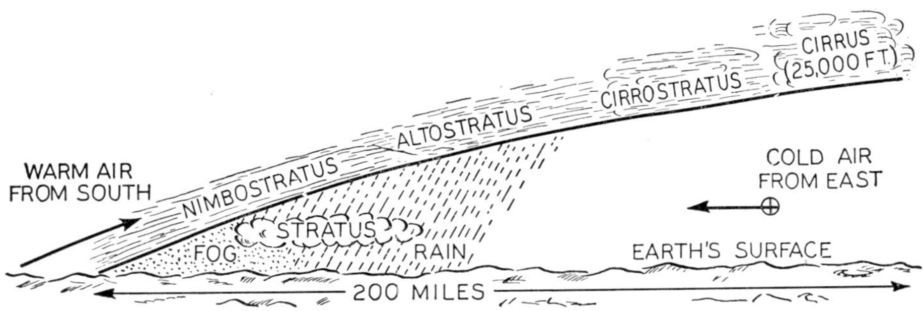

FIG. 22.14 A VERTICAL CROSS-SECTION THROUGH THE AIR LOOKING WEST. The cirrus clouds shown may be 25,000 feet above the earth's surface, and 200 miles north of the place where the warm current began its ascent over the cooler wedge of air.

expands, and cools adiabatically. Four types—cirrus, cirrostratus, altostratus, and nimbostratus—are shown in Fig. 22.14.

4. *Turbulence.* Figure 22.15 shows how clouds may be formed because of turbulence. Wind blowing over the earth's surface, particularly over rough

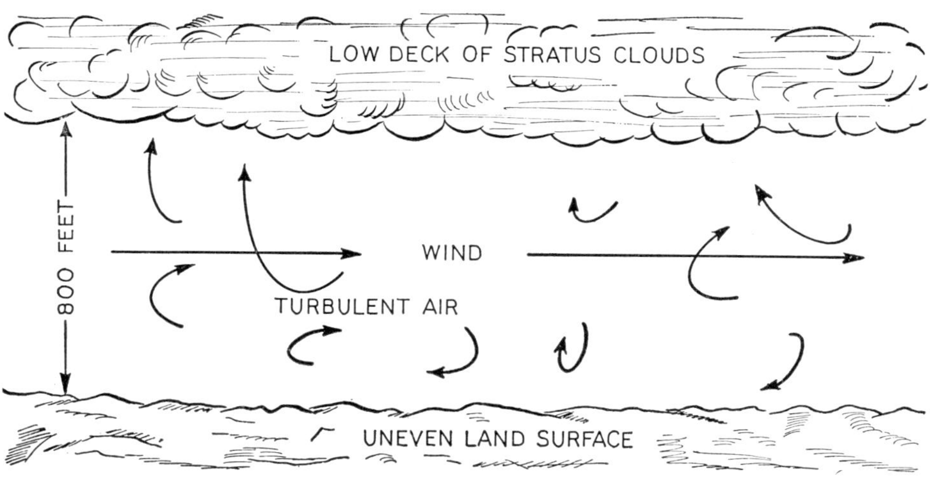

FIG. 22.15 STRATUS. These clouds may form when fresh to strong winds of high humidity blow over rough terrain. Friction with the ground surfaces causes upward air currents which expand, cool, and produce stratus or stratocumulus clouds.

terrain, causes upward and downward currents up to 1,500 to 2,000 feet above the surface and sometimes higher. Obviously if the air contains much water vapor, adiabatic cooling of the rising air currents may result in the air temperature reaching its dewpoint and the formation of clouds. The cloud types are either stratus or stratocumulus and are usually not rain-producing. The tops of these clouds ordinarily do not extend to elevations more than about 4,000 feet above the ground. Turbulence due to surface friction does not often exist above this level. If the air is unstable, the tops may extend to any height. Below and in turbulence-produced clouds the air is bumpy or rough; above the clouds it is smooth.

The value of the wind velocity often determines whether clouds or fog will form. As we shall see later, gentle or light winds when flowing over cold surfaces often produce fog. If the wind velocity is moderate to strong, friction of air and ground so stirs up the air that clouds are more likely to form than fog.

22.11. Stability, Instability, and Conditional Instability: Inversions. Why are cumulus cloud tops low and rather flat on some days, whereas at

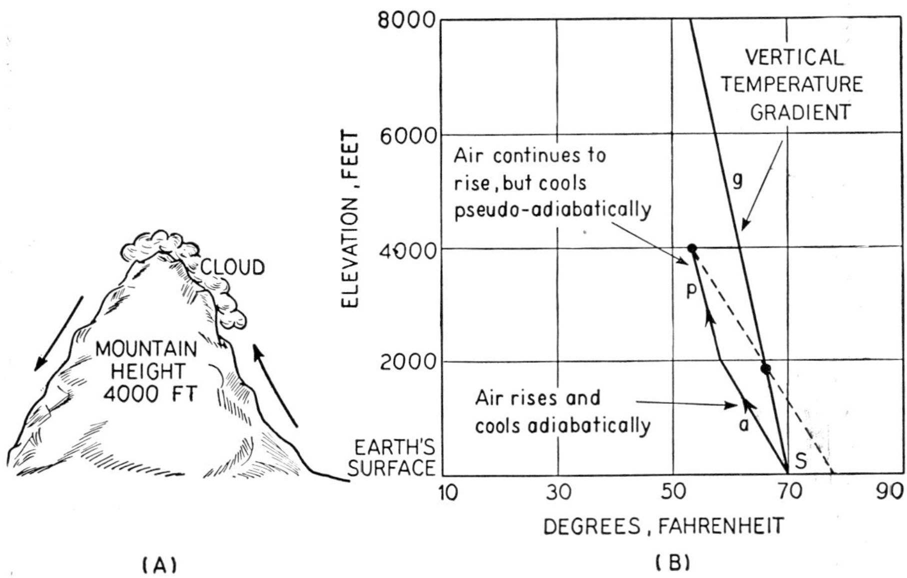

FIG. 22.16 STABILITY. (A) shows a current of air rising over a mountain. It expands and cools at the adiabatic rate (5.5°F/1,000 feet) until it reaches an elevation of 2,000 feet where the air becomes saturated. Rising from 2,000 to 4,000 feet it cools at the pseudo-adiabatic rate (about 2.7°F/1,000 feet). Now refer to graph (B). Let us assume that line *g* represents the existing vertical temperature gradient. Line *a* shows the rate of decrease in the temperature of the rising air current as it cools at the adiabatic rate. Line *p* shows the rate of decrease of the temperature of the rising air current as it cools at the pseudo-adiabatic rate. Looking back at (A) it will be noted that the mountain forces the air current only to a height of about 4,000 feet. Referring again to (B) it will be noted that at all heights the temperature of the rising column is less than its environment, and therefore stable. Hence upon reaching the leeward side of the mountain the air current will tend to descend.

other times the tops grow to great heights, sometimes building up to the great cumulonimbus cloud masses? The stability condition of the air is the answer to these questions, and the processes involved are as follows:

If a volume of air is caused to rise for any reason, such as in one of the ways described in the formation of clouds, it can do one of three things after the initial force which started it upward is removed. It may remain at the height to which it was forced; it may sink back to a lower level; or it may continue its journey upward. What it does depends on the stability of the air.

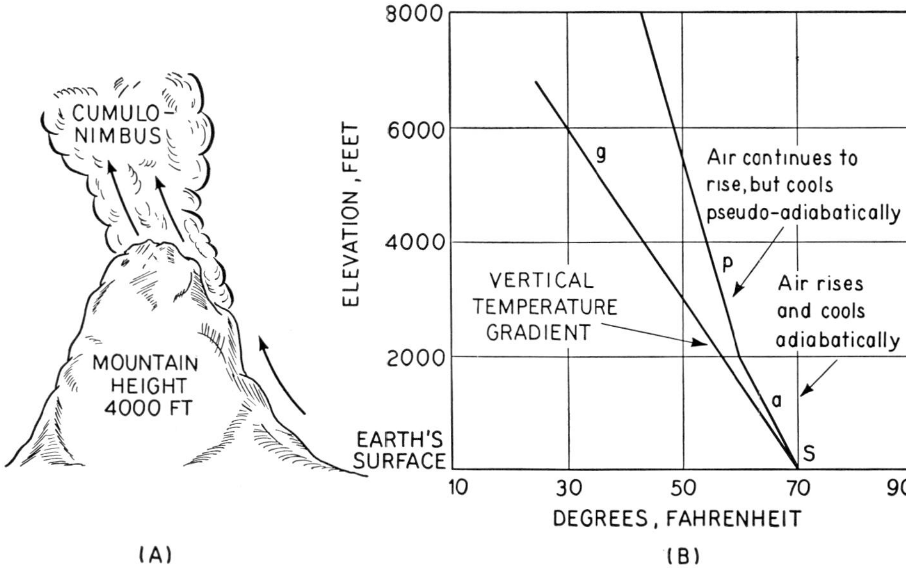

FIG. 22.17 INSTABILITY. Again, (A) shows a current of air moving from right to left and ascending a mountain slope. Expansion and cooling occur at the adiabatic rate up to 2,000 feet where the dewpoint is reached and a cloud base forms. Above 2,000 feet cooling takes place at the pseudo-adiabatic rate. Referring to (B) we note that the existing vertical temperature gradient is represented by line *g*. This temperature gradient is steeper than the one shown in Fig. 22.16. Lines *a* and *p* show the rate of temperature decrease while the air column is cooling at the adiabatic and pseudo-adiabatic rate, respectively. It will be noted that at any height the temperature of the rising air column (shown by line *a-p*) is greater than the air through which it is rising (shown by line *g*). Therefore, once the air begins to ascend the mountain slope it will tend to continue upward as long as it continues to be warmer than its environment.

If the air is in a condition of neutral stability, the particle remains at the elevation to which it was pushed; if the air is stable, it will tend to drop back; but if the air is unstable, it will continue its journey upward.

It will be recalled from Chapter 21 that the temperature of the atmosphere on the average is 1°F colder at each 300 feet of elevation above the earth. This is only an average condition and is known as the lapse rate, or vertical temperature gradient. The existing lapse rate at any place and time may be either greater or less than 1°F per 300 feet. Sometimes the temperature is warmer at increased elevations; such a condition is known as an *inversion.*

The existing lapse rate and the humidity of the air determine stability characteristics. If a volume of air is caused to move upward for any reason, such as those mentioned in cloud development, it will continue its upward journey after the initial impulse is removed, provided it finds itself surrounded by air colder than itself. The air then is known as unstable. On the other hand, if the air after being forced upward is colder than its environment, it will tend to sink back downward. This is a condition of stability.

Conditional instability is the term applied to air that is stable when cooling takes place at the adiabatic rate, but unstable when it occurs at the pseudo-

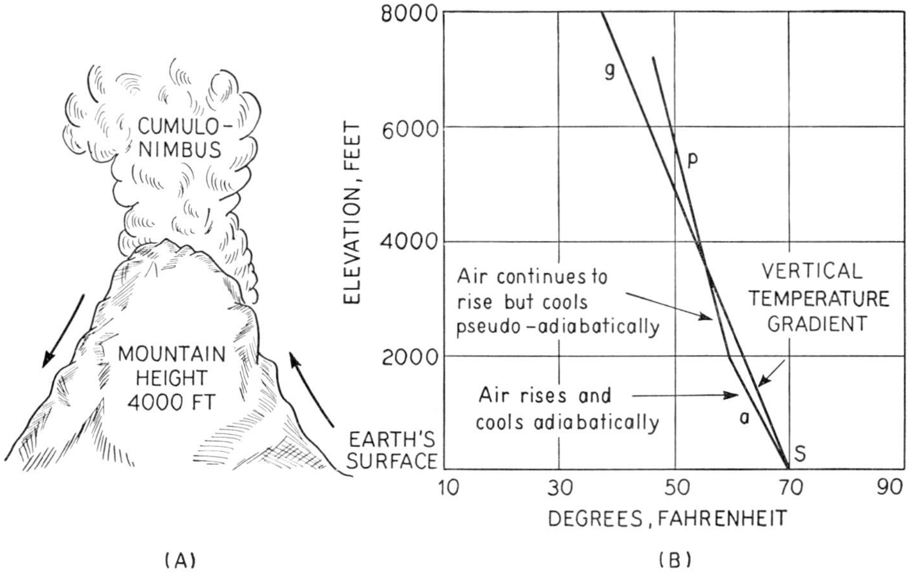

FIG. 22.18 CONDITIONAL INSTABILITY. Note that the existing vertical temperature gradient g is steeper than in 22.16, but not so steep as in 22.17. Air having this type of temperature gradient is stable for air columns rising and cooling at the pseudo-adiabatic rate. In this particular case air ascends the mountain slope from right to left. At 2,000 feet it has expanded and cooled to the dew-point where a cloud base forms. Between 2,000 and 4,000 feet the rising air column expands and cools at the pseudo-adiabatic rate. If the air column rises to just above 4,000 feet, it will be warmer than its environment and will continue upward. Had the air column contained less moisture it would have continued to a higher elevation than 2,000 feet at the adiabatic rate, and cumulonimbus might not have formed. Referring to (B), can you show why?

adiabatic. Conditional instability is quite common and is the basis of much of our cloudiness and precipitation.

Figures 22.16, 22.17, and 22.18 graphically illustrate stable, unstable, and conditionally unstable conditions; flat cumulus and cumulonimbus clouds are shown on the appropriate diagrams. In Fig. 22.16 it will be noted that the cloud top does not extend to a great height because the air column which forms the cloud does not tend to continue upward beyond the level to which it is forced by the mountain. In Fig. 22.18 the cloud top reaches high because its air column continues to rise after being forced to 4,000 feet by the mountain. In Fig. 22.17 any slight upward motion of the air at S will result in a rising air column; the column will continue to rise as long as it is warmer than its environment.

Figures 22.19A and B show low and high temperature inversions, respectively. Inversions at the earth's surface as in Fig. 22.19A, develop at night over land surfaces because of radiation of heat from the earth when skies are clear and winds are light. Ground fog is often the result. Inversions at levels

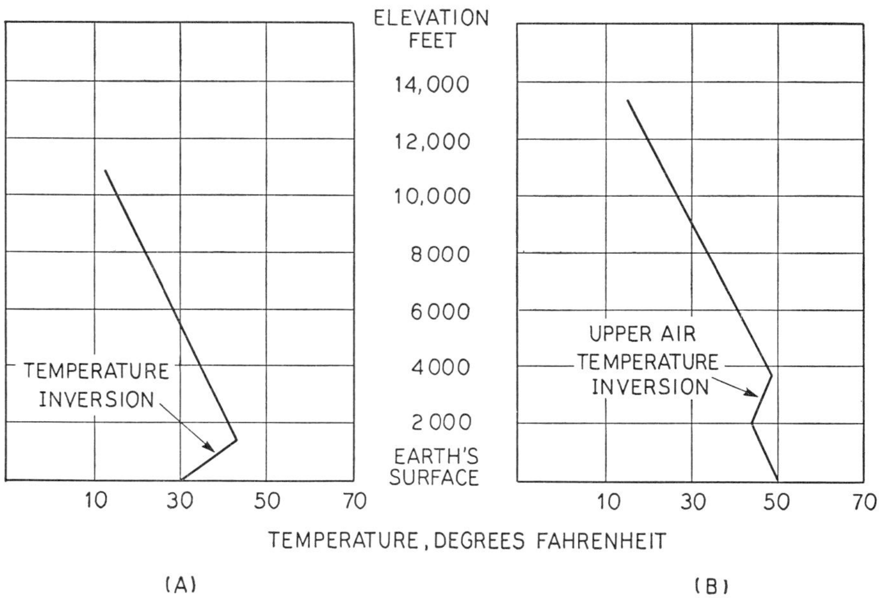

FIG. 22.19 Graph (A) shows a temperature inversion at the earth's surface. Surface inversions are characterized by stability and poor visibility; common at night, they are often associated with ground fog. Graph (B) shows an upper air inversion, the base of which is known as a "Lid." No air from below can penetrate the base of a lid; any smoke, fog, dust or haze present tends to spread out at the base of an upper air inversion.

above the earth's surface act as lids above which no air rising from below can penetrate. This condition causes a spreading out at the base of the upper air inversion of smoke and other restrictions to visibility.

22.12. Cloud Sequences. Cloud types in themselves are not so significant unless you consider their development, the types which preceded them, and the changes which are taking place.

If cumulus clouds form early on a summer morning, it means that the air is quite moist and convection has begun. This means that the clouds are likely to become more numerous by afternoon, build up to high levels, and by the middle of the afternoon thunderstorms are likely. If cumulus clouds do not appear until late in the morning, it means that the air has little water vapor because convection currents had to reach high levels before the air became saturated. The afternoon would have comparatively few clouds; the bases of these would be high, but their tops would not reach to the great heights of cumulonimbus clouds.

In the late afternoon if the air is stable, the afternoon cumulus clouds will gradually disappear, and by evening the sky will be clear. If the air, however, is unstable, the clouds will continue to build up to great heights even during the evening, and thunderstorms are likely then even at night.

Cirrus clouds mean little unless they increase in number and are succeeded by cirrostratus. Altostratus and nimbostratus clouds, rain, and stormy conditions in general are quite likely to succeed cirrostratus clouds. The thing to note is: Do the clouds seem to be replaced by thicker and lower cloud types?

To determine the end of stormy conditions watch for breaks in the low clouds, and watch for the appearance of blue patches of sky toward the west. As storms of middle latitudes move in general from west to east, their approach is indicated by clouds in the west, and their retreat is associated by clearing conditions which first appear to the west.

22.13. Fog. Fog is a stratus cloud whose base rests on the earth's surface. Although the substance of fog is the same as a cloud, the processes of cloud and fog formation are different. Clouds form chiefly because air rises, expands, and cools. Fog results from the cooling of air which remains at the earth's surface. This lowering of air temperature may occur in a number of ways, three of which are described as follows:

Warm Air Flowing over Cold Land or Water Surfaces. Fog may form when warm air sufficiently high in water vapor content flows over land or water surfaces cooler than the air. This type of fog may occur either during the day or at night, and the reason is not difficult to understand. The cool surface of the earth chills the overrunning air, and water vapor of the air tends to condense on particles of dust, smoke, or salt even before the relative humidity reaches 100 percent. The fog may be light, moderate, or dense, depending upon the amount of vapor which condenses. Dense fog is likely if the relative humidity value approaches 100 percent. The air in which fog forms does not tend to rise because it is being chilled. Because of this chilling it becomes more dense, heavy, and stable. However, if wind velocities are fresh to strong over land surfaces, turbulence prevents individual air particles from remaining long enough at the earth's surface to be chilled. It will be recalled that low stratus or stratocumulus clouds rather than fog form when wind velocities are other than light to moderate. Over the sea where there is less frictional turbulence than over land, fog may form even when wind velocities are quite strong.

Radiation Fog. This fog type is a nighttime phenomenon. During the daytime the earth's surface not only receives heat from the sun but radiates heat as well. At night the incoming heat ceases, but radiation continues, and the earth's surface therefore grows colder throughout the night. The low point is reached just prior to sunrise. Radiation fog is much more common over land than over water surfaces because the temperature of water surfaces varies little from day to night. Factors which favor the formation of radiation fog are light or gentle winds, sufficiently high humidity, and clear skies. Radiation fog is known to air pilots as ground fog. It is a shallow fog through which the sky is usually plainly visible, but it makes airplane landings and take-offs difficult; it obscures landmarks to some extent, thus complicating navigation. Sometimes it forms soon after sunset; at other times it may require a sustained drop in temperature throughout the night before forming.

FIG. 22.20 SEA SMOKE. *Official U.S. Navy Photograph*

Radiation fog burns off usually within an hour or so after sunrise. If mixed with smoke, which is particularly common in winter, it forms a greasy "smog" (smoke and fog) which may not be dissipated until later in the morning.

Fog in Air Changing Latitudes. Warm, moist air moving from low to higher latitudes may cool to such an extent that fog will form over large areas. This type is more common over the sea than over land. It may occur at sea during any part of the year, but over land areas it is most common during fall and winter months.

Two other fog processes may exist when the following conditions prevail:

1. Very cold stable air overlies a warm water surface or warm moist land surface. Fog formed under these conditions is known as steam fog or arctic sea smoke (Fig. 22.20). It is not common, but it is mentioned here so that if and when it is observed the reason for its presence will be understood. Steam

fog is caused by very rapid evaporation from a water surface or wet land surface which is overlaid by very cold stable air. The vapor is quickly condensed by the cold air and collects as a fog so long as the cold air is stable, i.e., not rising at the surface. It is most common in arctic regions where areas of both ice and open water exist. It is also of frequent occurrence in other regions, particularly over lakes and rivers when the first cold masses of air drift southward during fall and early winter. Another example of steam fog may be observed when roads steam in the sunshine after a rain.

2. Rain falls through cool air of high humidity. In this case the rain both chills and increases the water vapor content of the air through which it is falling. Thus the dewpoint of the air increases, and the air temperature decreases, which often produces a relative humidity sufficiently great to result in light, moderate, or dense fog. This type of fog is quite common in connection with the storms of the middle latitudes, over both water and land surfaces.

22.14. The Forecasting of Radiation Fog. In order to forecast radiation fog over land areas, temperature, sky, and wind conditions must be carefully studied. Also the forecaster must be familiar with local conditions, such as the terrain, proximity of smoke sources, location of nearby water areas, etc. Radiation fog is possible when the sky is expected to be clear or with only high, thin clouds present and the wind not more than force two. The temperature must drop to the dewpoint during the night. When the temperature drops to within a degree or two of the dewpoint, radiation fog may be expected soon if it has not already begun to form.

22.15. The Forecasting of Advection and Other Fog Types. The term *advection* is applied to fog formed because of air in horizontal motion. It must be borne in mind that fog may and often does form as a result of many processes operating at the same time. For example, let us consider the case where a warm, moist, easterly wind is blowing over Nebraska. First of all its elevation is increasing because of the gradual rise in the terrain from east to west. This results in expansion and adiabatic cooling of the air, which process was mentioned in connection with cloud formation. The air will also undergo cooling if it is passing over ground colder than itself. Then if rain is falling through the air, a third fog formation process is in operation. As in the forecasting of radiation fog, wind direction and velocity, temperature, and humidity conditions must be considered in forecasting advection and other fog types. Unlike radiation fog, the other types may form when skies are cloudy during the day or at night.

22.16. Fog at Sea. Fog is rare at the equator and in the trade-wind belt except along the coasts of California, Chile, and northwest and southwest Africa. On the other hand, it is a common phenomenon of middle and high latitudes, particularly in spring and early summer. The Newfoundland Banks is a region where the Gulf Stream and the Labrador current meet. When warm air from the Gulf Stream overruns the cold water of the Labrador current, dense fog banks result. Likewise, in the northwestern Pacific, fog is common off the coast of Asia where warm air of the Japanese current overruns the cold Kamchatka current. During spring and summer, warm, moist air currents which flow from land to sea produce fog over coastal water areas. A shift in wind direction tends to drift the fog back over adjacent land areas. During fall and early winter, air blowing from sea to land tends to produce fog over coastal sections. Such fog may drift to sea with a reversal of the wind direction.

In the north Pacific Ocean and Bering Sea wide areas of dense fog are common, and at times the fog may extend to elevations of 4,000 feet or more. It is caused by air moving northward from high-pressure areas which are centered in the Pacific Ocean at about latitude 35 to 40 degrees north. The air reaches its saturation temperature in passing over the colder waters to the north. Chilling of the air also increases its density so that there is no tendency of the air to rise and clear. Fog may persist in the Aleutian area when winds are quite strong. When fog-laden wind flows around and over land obstructions, clear spots may be found to leeward; the sun is not effective in burning off this type of fog. During fall and spring months, fog is at a minimum in the Aleutians; in winter, arctic sea smoke is common, and it may at times build up to elevations of several thousand feet above the surface.

23

Weather Elements, Instruments, and Reports

23.1. Introduction. In order to determine and describe the conditions of the atmosphere or weather at a given time and place, various weather elements must be observed and measured. This is done regularly at sea and on shore throughout most of the world, and the reports of the observations are quickly made available for the use of seamen, airmen, forecasters, and others concerned. The values of some elements are estimated or determined visually, while others are measured with the aid of instruments, some of which are carried aloft thousands of feet by free balloons in order to determine pressure, temperature, moisture, and wind conditions in the upper reaches of air masses.

A description of the weather elements and how they are measured and reported is given in this chapter. The nature of the information contained in weather broadcasts is given, and suggestions are made with regard to the preparation of weather maps at sea.

Weather elements that are estimated or determined by judgment without the aid of instruments are clouds, visibility, and the state of the weather. These are dealt with in the section which now follows. See Chapter 20 for a discussion of waves and surf.

23.2. Elements that Are Estimated—Clouds. Cloud observations include notations as to the types which are present, their direction of movement, and the proportion of the dome of the sky that they cover. Observers give particular attention to the clouds in the vicinity of the zenith, because types and movements are determined more reliably there than at the greater distances toward the horizon. Twenty-seven variations of the ten basic cloud genera can be recorded. For example, cumulus clouds may be of the relatively flat-topped type of fair, settled weather; or their tops may extend to a considerable height, thereby indicating unstable atmospheric conditions with the possibility of storminess. When cloud types and their variations are observed and reported adequately, many useful inferences may be drawn concerning present weather conditions and imminent changes.

The directions from which the high, middle, and low clouds move are recorded. High clouds move in general from west to east much of the time. Clouds of the low family move in a direction which is approximately that of the surface wind. Clouds of middle heights often move from unexpected directions.

They reveal converging and overrunning currents of air which may produce rain, particularly if they are traveling from a warm, moist, maritime source. In the absence of regular upper-air wind reports, information of middle-cloud directions offers reliable evidence of wind values to aviators who may be planning flights at those levels. For these and other reasons it is essential that middle-cloud directions be reported when known. Instructions for making weather reports specify that the direction of movement only of middle clouds be reported if known. If the direction of middle clouds cannot be determined, then the direction of motion of the upper clouds is reported. When neither middle- nor upper-cloud motions are known, the direction of low-cloud movement is reported. The direction and velocity of movement of some storm types are indicated by the motion of clouds at middle heights.

The amount of sky covered by clouds is reported in terms of eighths of the sky dome that is cloud-covered. For example, if clouds completely cover the sky, the condition of cloudiness is known as eight eighths or overcast; but if only one eighth of the visible dome is covered, cloudiness is known as one eighth. When the sky is "overcast but with openings" the condition is reported as seven eighths rather than eight eighths, even though the sizes and number of openings are almost negligible. The reason for this is that a condition of complete overcast followed by a report showing a break in the clouds often means the approach of clearing conditions, particularly if the breaks appear to the westward in the middle latitudes.

23.3. Visibility. By visibility is meant the greatest horizontal distance at the earth's surface that one can see with the unaided eye such large objects as ships, islands, hills, buildings, etc. The distinctness of the horizon also offers evidence of the prevailing degree of visibility.

There is a definite relationship between general weather conditions and the visibility. When the air is warmer than the water surface, it tends to be stable, and poor surface visibility due to fog, haze, smoke, drizzle, or dust may prevail. Poor visibility is often associated with clouds of the stratus type. If the water surface is warmer than the overlying air, horizontal visibility at the surface tends to be good. This is because particles that restrict visibility are carried upward by the surface air which expands and rises when heated by the water surface. Any clouds present would be the cumulus type. Fog and falling snow are two of the most serious restrictions to visibility, because the visibility can deteriorate so rapidly with the occurrence of either. Fog may be anticipated when the temperature and dewpoint values are approaching each other. This is especially true when the air has considerable foreign matter present, such as dust, smoke, and haze. Visibility is of prime importance in navigation on the sea and is equally important to landing aircraft.

23.4. State of the Weather. A weather report always includes a reference to the state of the weather. This, in general, refers to the existence, imminence, or absence of precipitation. It is possible to record and report 100 different states of the weather ranging from cloudless skies to heavy thunderstorms with hail. Other conditions which can be reported with existing weather codes include low fog, haze, distant lightning, smoke, and various degrees of fog, rain, snow, showers, drizzle, squalls, etc. Definitions of some of the more frequently reported state of weather elements follow.

Dust. This consists of minute earth particles which are picked up and carried along by the wind. Distant objects have a more or less grayish or tan appearance, depending on the amount of dust present. Although there is always some dust in the air, appreciable amounts are usually of local origin. Blowing dust is reported when sheets or clouds of it are carried along in a strong wind.

Haze. Haze is composed of very finely divided particles of matter from land areas or of salt particles from the sea. They are much more minute than dust particles. Haze gives distant objects a pale blue or sometimes yellow appearance. At a distance the details of objects disappear, and the objects stand out in silhouette fashion.

Smoke. Particles resulting from combustion are known as smoke. Except for its odor it might be confused with fog, dust, or haze, especially in light amounts. It is common in the vicinity of cities, especially to leeward of industrial areas. It gives a reddish tinge to the sun's disk at sunrise and sunset.

Hydrometeors are defined as water in solid or liquid form falling through the air. The following types are noteworthy in meteorology:

Drizzle. Drizzle is composed of minute liquid droplets that are so numerous they seem to fill the air. This form of precipitation originates in stratus clouds or fog. Drizzle particles seem to float in the air and appear to follow even slight motion of the air. Drizzle is characteristically associated with poor visibility. When the droplets instantly freeze to objects which they strike, they are known as freezing drizzle.

Rain. Rain is drops of water larger than drizzle falling from clouds. The drops are usually sparser than those of drizzle. When they freeze to objects, the condition is known as freezing rain.

Sleet. In the United States sleet is defined as frozen raindrops. In England and in the International Weather Code, as well as in popular press vernacular, a mixture of rain and snow is known as sleet.

Hail. Hail is almost exclusively a phenomenon of violent or prolonged thunderstorms. It is dangerous to aircraft and never occurs when temperatures at the earth's surface are below freezing. Hail consists of ice balls or stones with diameters ranging from $\frac{1}{5}$ inch to 2 inches or more, which either fall detached or fused in irregular lumps. They may be transparent or composed of alternate clear and opaque, snowlike layers.

Snow. This form of precipitation from clouds consists of white or translucent ice crystals mainly in branched hexagonal shapes often mixed with simple ice crystals.

Showers. This form of precipitation is associated with cumuliform clouds and is characterized by beginning and ending suddenly. Showers, usually of short duration, often occur in a series with periods of fair conditions between individual shower periods. Unstable atmospheric conditions are indicated by showers, and the form of precipitation may be either snow or rain. Snow showers are sometimes known as *snow flurries.*

23.5. Elements Measured by Instruments—Atmospheric, or Barometric, Pressure. The value of the atmospheric pressure is one of the most important elements in reporting and forecasting the weather. It is particularly significant when considered in connection with the direction and velocity of the wind and cloud types and sequences. Average pressure values reveal much regarding the prevailing or climatic characteristics of a region. Pressure variations are not for the most part perceptible to human senses. It is the one element that cannot be estimated; its value must be determined instrumentally. Pressure-measuring instruments in common use are the mercurial barometer, aneroid barometer, and the barograph.

The *mercurial barometer* is standard, but ships carry only the *aneroid* type. Mercurial barometers are direct reading, but corrections are applied for temperature, gravity, and the elevation above sea level. Also each instrument has an individual correction which must always be applied. The mercurial barometer (Fig. 23.1) is essentially a glass tube with one end closed. It is filled with mercury and inverted so that the open end projects into a well or cistern. The top of the mercury in the tube will then lower until the column just balances a column of the atmosphere of the same cross section as that of the mercury. The height of the mercury column as measured from the top of the

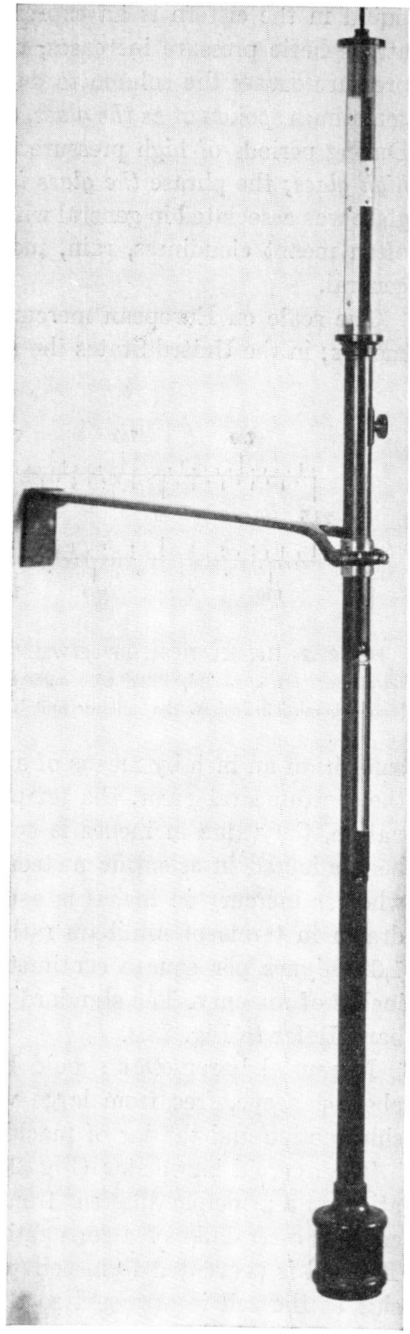

FIG. 23.1 MERCURIAL BAROMETER.
Official U.S. Navy Photograph

liquid in the cistern is an expression of the atmospheric pressure. When the atmospheric pressure increases, the mercury rises; a decrease in atmospheric pressure causes the column to decrease in height. The mercurial barometer is sometimes spoken of as *the glass,* especially in some of the earlier sea literature. During periods of high pressure the barometer reading would be known as *a high glass;* the phrase *the glass is low* refers to low pressure readings. A high glass was associated in general with fair weather conditions, whereas a low glass often meant cloudiness, rain, increased windiness, and stormy conditions in general.

The scale on European mercurial barometers is usually in terms of millimeters; in the United States the inch is used. Readings may be made to thou-

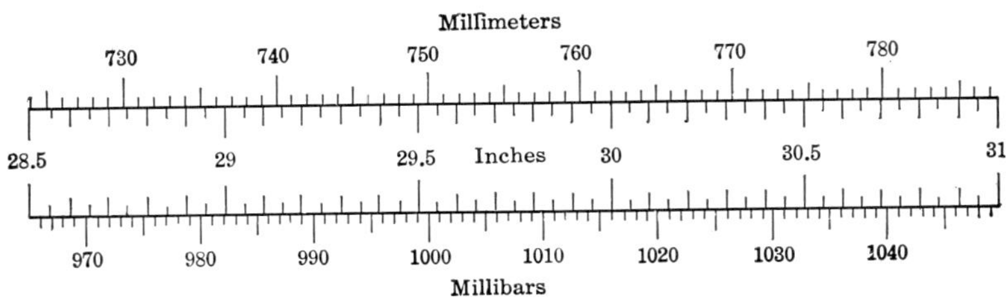

FIG. 23.2 THE RELATIONSHIP BETWEEN THE MILLIMETER, INCH, AND MILLIBAR SCALES. European barometers are commonly fitted with millimeter scales, while instruments in the United States and England are read in inches. The millibar unit is used on weather maps.

sandths of an inch by means of a vernier scale. After corrections are made for the instrumental error, the temperature at the barometer, gravity, and elevation, the value in inches is converted to units of millibars. This unit has been adopted in scientific meteorological work because it represents a force, whereas mercury in inches is a unit of length. Isobars on weather maps are drawn in terms of millibars rather than inches. A millibar is equivalent to 1,000 dynes per square centimeter; 1,000 millibars are equivalent to 29.53 inches of mercury. The standard pressure of 29.92 inches equals 1,013.2 millibars. Refer to Fig. 23.2.

Mercurial barometers should be mounted in a position convenient to the observer's eye, free from large variations in temperature, and as free from ship motion and the jar of machinery as possible.

The aneroid barometer (Fig. 23.3) is a small, convenient instrument operating on a principle different from that of the mercurial type. Without fluid it consists essentially of a corrugated metallic chamber or cell exhausted of air. The cell is prevented from collapsing by means of a strong steel spring. One side of the cell is arranged so that it responds to variations in atmospheric pressure by expanding and contracting. This motion is magnified by a system of levers and chains and transmitted to a hand on the dial of the instrument. Temperature differences are compensated for by means of a bimetallic link of brass and steel. It is necessary occasionally to test and compare aneroid barometers with standard mercurial instruments.

Barographs (Fig. 23.4) are instruments which afford a continuous record of atmospheric pressure and consist of two units. First, the motion of a series of

FIG. 23.3 ANEROID BAROMETER. *Courtesy Taylor Instrument Companies*

small aneroid cells is magnified and transmitted to a pen by means of a system of levers and chains. Second, a card mounted on a drum is rotated slowly on a vertical axis by means of a clock mechanism. The pen traces a continuous record of the barometric pressure on the card. The barograph is a most useful instrument because it not only shows the present reading but values at any prior moment as well. What is even more important is that it shows whether or not pressure values are rising or falling and the rate of changes. One of the most significant elements of a weather report is the so-called *pressure tendency* during the 3-hour period prior to the report. The tendency includes both the character of pressure change, i.e., whether the pressure is rising or falling, etc., and the net amount in tenths of millibars of such change. The barograph trace shows the pressure tendency at a glance. The barograph, like all aneroid barometers, must be compared frequently with standard mercurial instruments.

FIG. 23.4 MICROBAROGRAPH. *Courtesy Friez Instrument Div., Bendix Aviation Corp.*

The card on the drum may record the pressure for four or seven days, depending on the type of barograph in use. Figure 23.5 shows a typical barograph trace. It will be noted that pressure values are indicated along the vertical scale, whereas time units are marked off on the horizontal scale. The time lines are curved because the pen that records pressure values is pivoted at the aneroid unit.

23.6. Temperature. The temperature of the air near the earth's surface is measured with a thermometer graduated in degrees Fahrenheit, although some nations use the centigrade scale. In order to get good, representative shade temperatures, thermometers at land stations are exposed in shelters consisting of louvered sides and insulated roofs. The shelters are painted white, and they protect the instruments from rain and snow but permit the air to circulate freely through them.

Thermometer readings at sea are more representative of true air temperatures than readings taken at land stations. This is due to the fact that water surfaces are not subject to uneven heating and large variations in temperature, as are land surfaces. The proper exposure of thermometers on board ship is a

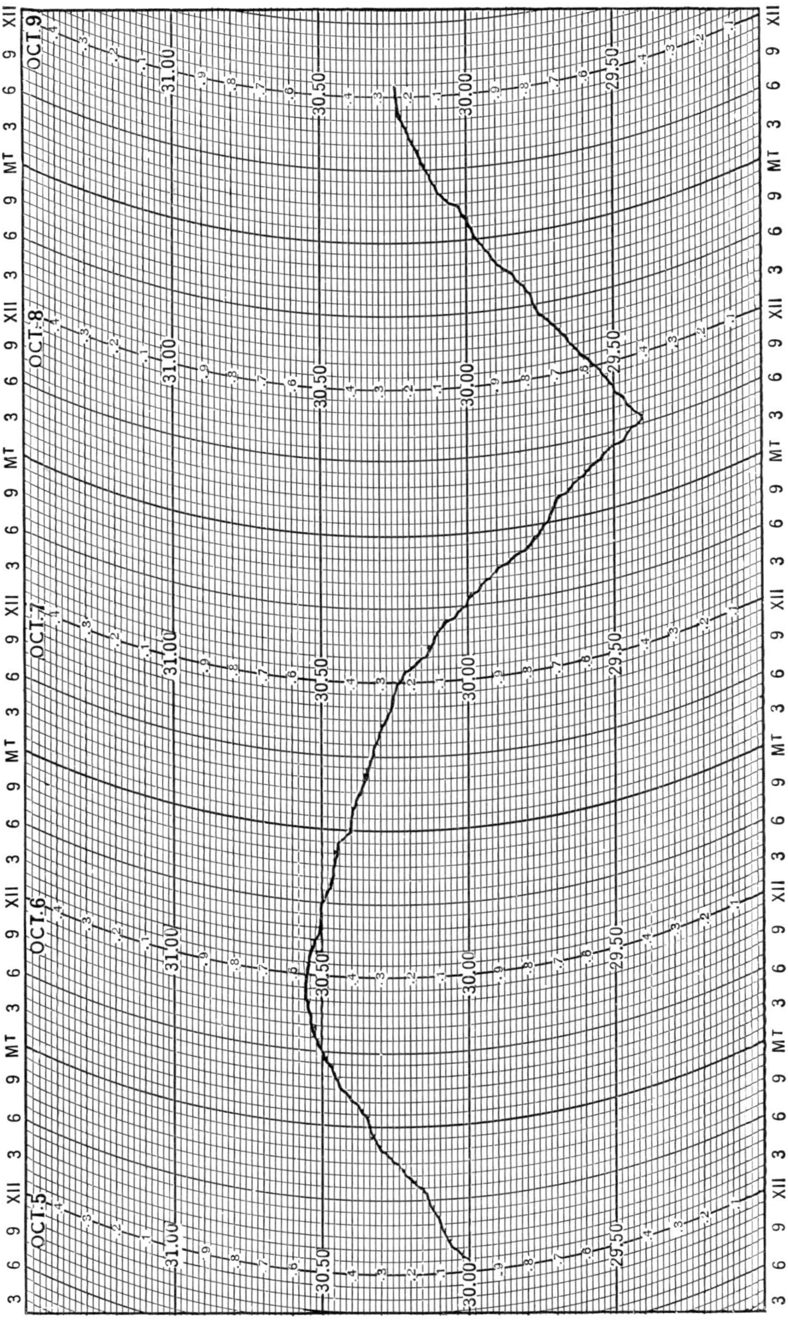

FIG. 23.5 TYPICAL BAROGRAPH TRACE

difficult problem. A shelter is not satisfactory because at any one location in the ship conditions are not always the same. For example, at one time heat may issue up from below decks to affect the thermometer; at other times the prevailing wind may carry the heat in the other direction. There is also the problem of temperature differences on leeward and weather sides. These difficulties are overcome to a large extent by means of the portable aspiration thermometer. It consists of a tubular shield and artificial ventilation. The latter is obtained by means of a mechanically driven fan attached to the top of the instrument. The *aspiration* thermometer may be taken to any part of the ship for readings, and it is not affected by the direct rays of the sun. It may be hung in the chart room when not in use, thus protecting it from the effects of fog, rain, and spray. An aspiration psychrometer is shown in Fig. 23.6.

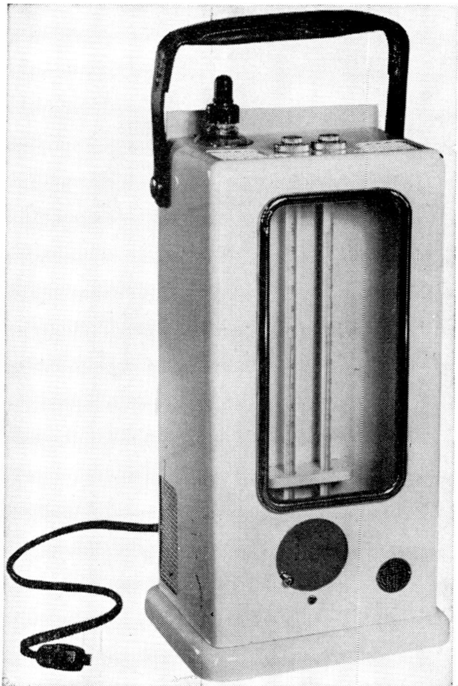

FIG. 23.6 PORTABLE ASPIRATION PSYCHROMETER. The electrical connection operates a small fan. *Courtesy Friez Instrument Div., Bendix Aviation Corp.*

On the Fahrenheit scale freezing is 32° and boiling is at 212°. Freezing on the centigrade scale is at 0° and boiling is 100°. Readings of one scale may be converted to the other by means of tables or formulas.

The *thermograph* (Fig. 23.7) is an instrument which affords a continuous record of the air temperature, just as the barograph gives continuous pressure values. In one type of instrument a flattened, curved, metal tube is filled with liquid, sealed, and fastened at one end. A change in the curvature of the tube is produced by the unequal expansion of metal and liquid. The tube actuates a pen which moves up and down on a card which is wrapped around a drum. The drum is driven slowly around a vertical axis by a clock mechanism. The values shown by such an instrument are subject to error, but

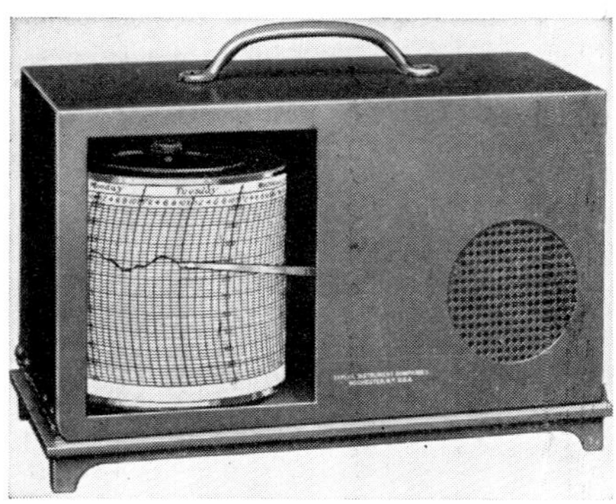

FIG. 23.7 THERMOGRAPH. *Courtesy Taylor Instrument Companies*

if the thermograph is compared frequently with a good mercurial thermometer, the errors may be determined and applied.

A *maximum* thermometer (Fig. 23.8) shows the highest temperature which occurred since the instrument was last read and set. This instrument is quite similar to the clinical, or fever, thermometer used by doctors. It is similar

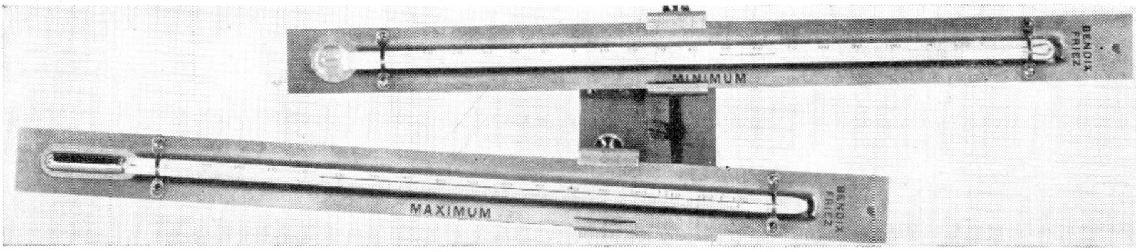

FIG. 23.8 Maximum and minimum thermometers show the highest and lowest temperature readings since the moment they were last read and set. They are mounted horizontally in a shelter. *Courtesy Friez Instrument Div., Bendix Aviation Corp.*

to an ordinary thermometer except that the bore has a constriction in it just above the bulb. When the mercury in the bulb warms and expands, mercury will flow past the constriction and rise in the bore, but mercury will not flow back into the bulb when the temperature decreases. To force the mercury back into the bulb it is necessary to whirl or shake the thermometer in a fashion similar to the way a doctor does.

A *minimum* thermometer (Fig. 23.8) shows the lowest temperature which occurred since it was last read and set. The fluid in this thermometer is alcohol, and a small index device is set at the top of the alcohol column in the stem. When the temperature drops, the alcohol contracts and carries the index downward in the stem. Expansion of the alcohol, however, does not affect the index, which remains at the lowest point to which it is carried. Maximum and minimum thermometers are mounted in a horizontal position.

23.7. Dewpoint and Relative Humidity. The water vapor content of the air may be obtained by means of *psychrometer* readings. The psychrometer is an instrument which consists of two thermometers, one of which is called the *dry-bulb* and the other is the *wet-bulb*. There is no difference in the two thermometers, but the bulb of one is fitted with a piece of cloth. The objection to a stationary thermometer shelter on board ship has been pointed out. Figure 23.9 shows the sling psychrometer. This instrument may be carried to any part of the ship so that suitable read-

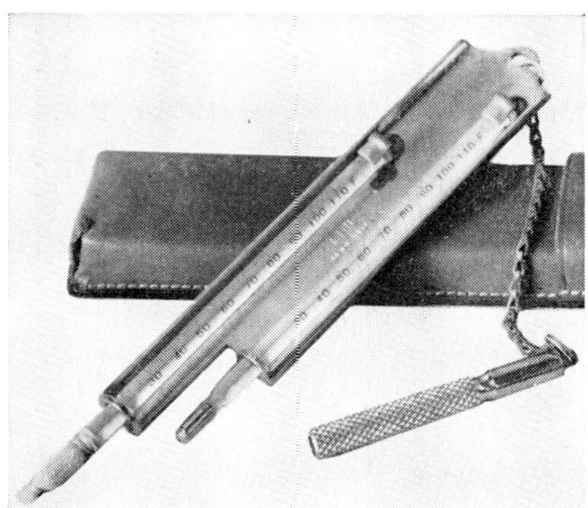

FIG. 23.9 SLING PSYCHROMETER. Swing in a vertical plane to obtain accurate wet and dry readings. *Courtesy Friez Instrument Div., Bendix Aviation Corp.*

ings may be had. When a reading is desired, the cloth on the wet-bulb thermometer is moistened. The psychrometer is then whirled. If the air is saturated with water vapor, such as it is during a dense fog, there will be no evaporation from the wet-bulb, and the two thermometers will read the same. When the air is not saturated, evaporation will take place from the wet-bulb. The evaporation is associated with cooling, and the reading of the wet-bulb will be lower than the reading of the dry-bulb thermometer. For any given air temperature, the greater the difference in the dry- and wet-bulb readings, the lower will be the dewpoint and relative humidity. In order to get the dewpoint and relative

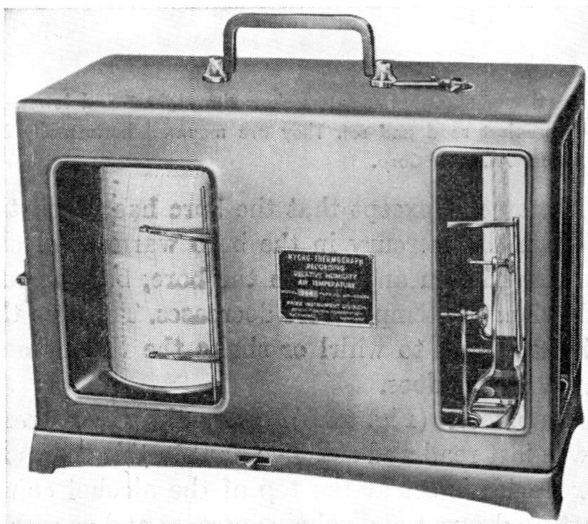

FIG. 23.10 HYDRO-THERMOGRAPH. *Courtesy Friez Instrument Div., Bendix Aviation Corp.*

humidity values it is necessary to look up their values in tables of dewpoint and humidity. The dry-bulb temperature and the difference between the dry- and wet-bulb readings are the two arguments which are used in entering the tables.

A continuous record of relative humidity is made by means of a *hygrograph* (Fig. 23.10). The recording pen is actuated by a unit of human hairs which expand and contract with changing humidity conditions. Values are recorded by the pen on a card which is mounted on a revolving drum driven by a clock mechanism. This instrument is occasionally compared with a psychrometer.

23.8. Wind Direction. The wind direction is the direction from which the wind blows. For example, a wind blowing from the northwest toward the southeast is known as a northwest wind. The direction may be estimated or it may be determined by means of a wind vane. *Wind vanes* (Fig. 23.11) point toward the direction from which the wind is blowing. Vanes may be wired to a disk in the weather office or some other location so that the resultant wind direction may be known at a glance. Some weather offices are equipped with meteorographs, a recording device that may be wired not only to the wind vane but to the anemometer, rain gauge, and sunshine recorder as well. Meteorographs have a card mounted on a drum driven by a clock mechanism. Wind direction and velocity, sunshine, and rain amounts may be recorded automatically on the card.

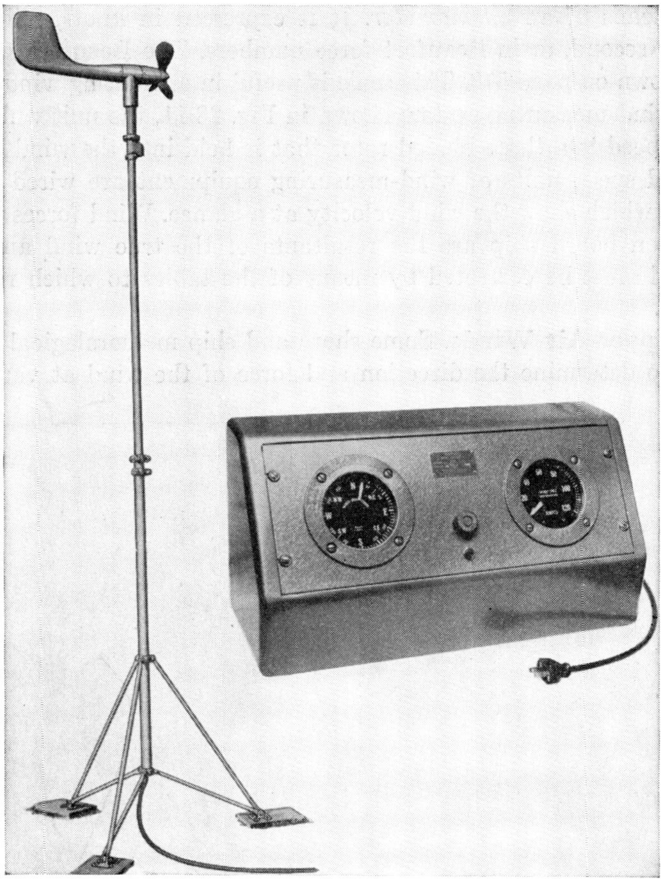

FIG. 23.11 THE AEROVANE WIND MEASURING SYSTEM. The roof installation consisting of the Aerovane Transmitter mounted on top of the mast transmits wind direction through 360 degrees and wind speed from 0 to 120 knots to the indicator assembly which is normally mounted in the office. The Aerovane is both indicating and recording and may be operated separately or in combination with additional Aerovane indicators—all controlled by a single Aerovane transmitter. The Aerovane Recorder not shown. *Courtesy Friez Instrument Div., Bendix Aviation Corp.*

In order to estimate the wind direction by the appearance of the sea, the crests of the small ripples are considered to be perpendicular to the wind direction. In strong winds the foam streaks reliably show the wind direction. At night and during heavy rains when the ripples cannot be made out, it may be necessary to note the apparent direction of the wind. This is the direction from which the wind seems to blow; it is the resultant of the true wind and the ship's movement. Wind vanes on board ship also register the apparent wind. Convenient tables are available for determining the true wind direction and velocity from the values of the apparent wind direction and force and the true course and speed of the vessel.

23.9. Wind Velocity. Wind velocity may be estimated or it may be measured by means of an *anemometer*. It is expressed in knots, miles per hour, meters per second, or in Beaufort force numbers. The Beaufort scale of wind force is shown on page 518. This scale is useful in estimating wind forces.

In the wind-measuring system shown in Fig. 23.11, the unit which measures the wind speed is a three-bladed rotor that is held into the wind by the wind vane. Wind speed units of wind-measuring equipment are wired to indicator assemblies which show the wind velocity at a glance. Wind forces measured or estimated on board ship are the resultants of the true wind and the ship's motion and must be corrected by means of the tables to which reference has been made.

23.10. Upper Air Winds. Some shore and ship meteorological stations are equipped to determine the direction and force of the wind at various heights

FIG. 23.12(A) THEODOLITE FOR PILOT BALLOON OBSERVATIONS. *Official U.S. Navy Photograph*

FIG. 23.12(B) PILOT BALLOON AND THEODOLITE FOR UPPER OBSERVATION. *Official U.S. Navy Photograph*

above the earth's surface. This information is important to airmen in working out their problems in navigation. Forecasters are also interested in upper air wind data because it indicates sources of air masses and other processes that determine cloudiness, precipitation, and other weather conditions.

Pibal observations are widely used to get upper air wind data. Small rubber balloons are inflated with hydrogen or helium so that they have a definite rate of ascension. After a balloon is released, it is watched through the telescope of a theodolite (Fig. 23.12), an instrument that resembles a surveyor's transit. By means of scales on the theodolite, the azimuth and the vertical angle of the balloon are determined at 1-minute intervals; a clock equipped with a buzzer is commonly used to indicate the time that the readings are to be made. Since the balloon has a predetermined rate of ascension, its height is known at all times; the height, together with the azimuth and vertical angle readings, is used to compute by means of trigonometry the wind force and direction at any altitude. The actual computations are made with the aid of a polar coordinate graph and tables.

For example, let us suppose that a 30-inch balloon is inflated so that it has an ascension rate of 600 feet per minute. If the balloon goes straight up after release, it is obvious that calm conditions prevail at the elevations corresponding to that portion of the balloon's journey. If after 2 minutes the balloon is observed to head south, we know that a north wind is blowing at an elevation of 1,200 feet. Any changes in the position of the balloon on the sky dome indicate variations in the wind direction and force at corresponding heights above the earth.

A pilot balloon observation terminates once the balloon gets out of sight because of clouds or distance. During World War II balloons were equipped with radar targets (reflectors), and tracking was done by radar. This practice is now common. This procedure allows upper air wind observations to be made in all kinds of weather because the path of the balloon does not have to be observed visually. Furthermore, upper wind force and direction may be determined by radio direction finding on the transmitted signal of a radiosonde (see Fig. 23.15).

23.11. Ceiling. The ceiling is the vertical distance between the earth's surface and the lowest layer of obscuring phenomena, such as clouds, that covers more than one half of the sky. The ceiling value is important in aviation, and it is often the principal item of interest in a weather report. Ceiling values may be estimated or they may be measured. The existing cloud types offer some indication in estimating ceilings. Previous ceiling values, weather trends, and conditions reported from surrounding areas should be considered. Many shore stations are equipped with ceilometers which are instruments that are used to determine ceiling conditions both at night and during daylight hours. A ceilometer is shown in Fig. 23.13. At the right in Fig. 23.14 will be seen a ceilometer projector that directs vertically an intense beam of modulated light to form a spot on the cloud base. At the left in the sketch is shown the ceilometer detector which receives the light from the spot the vertical angle of which is recorded. The ceiling is determined by triangulation. Meteorological stations not equipped with ceilometers depend upon the use of small rubber balloons to determine daytime ceiling values. Such balloons can be inflated so that they

FIG. 23.13 NOVALUX CEILOMETER. This instrument is used in connection with a projector and a recorder. (See FIG. 23.14) *Courtesy General Electric Co.*

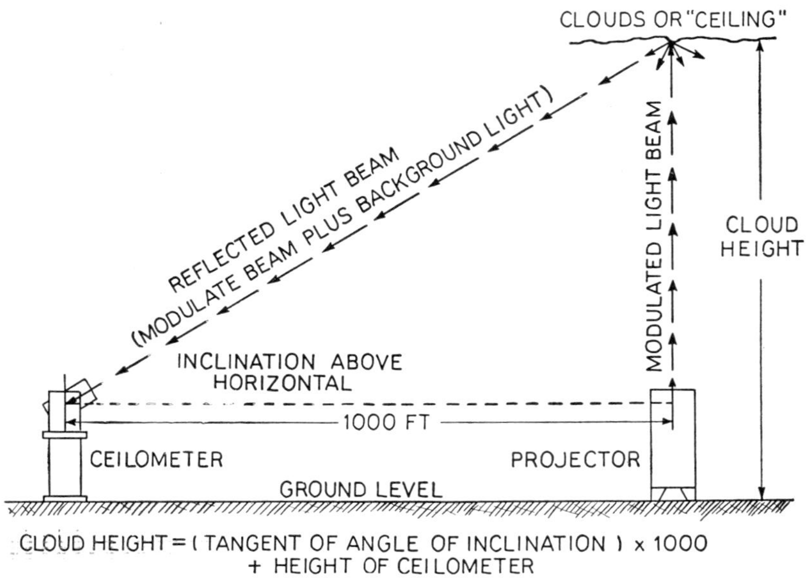

FIG. 23.14 Ceilings may be determined by the ceilometer at night and during daylight hours.

will ascend at a known rate. The elapsed time between the time of release and the time the balloon enters the cloud base is noted. This multiplied by the ascension rate gives the ceiling value. At night a vertical searchlight is trained on the cloud base, and the projector is located at some known distance from the weather office from 500 or 1,000 feet. The observer at the office measures the vertical angle of the spot on the cloud by means of a small hand instrument known as a *clinometer*. The tangent of the angle of elevation multiplied by the horizontal distance between the office and the light projector gives the ceiling height.

23.12. Ocean Surface Temperature. Observations from ships at sea include the temperature of the water at the surface. This may be obtained by hauling in a bucket of water and placing a thermometer in it at once, but usually a much more reliable and satisfactory method is a carefully installed thermometer in the condenser intake. Cooperation on the part of the engineer is, of course, essential. Water temperature is important in the forecasting of fog, clouds, and other phenomena.

23.13. Temperature, Humidity, and Pressure at High Levels. The *radiosonde* is an instrument which is used to obtain values of the pressure, temperature, and humidity at various levels above the earth's surface. Only a limited number of land stations and ships are equipped to make this type of weather observation known as a *RAOB*. The instrument consists of a box-like container which is equipped with elements for determining the pressure, temperature, and humidity of the upper air and with a battery-operated radio transmitter by means of which the measurements are transmitted to a receiver located at the meteorological station. The instrument is carried aloft by a small, free balloon to heights well into the stratosphere. When the balloon bursts, the radiosonde descends by means of a parachute. Weather element readings are transmitted during the ascent of the instrument.

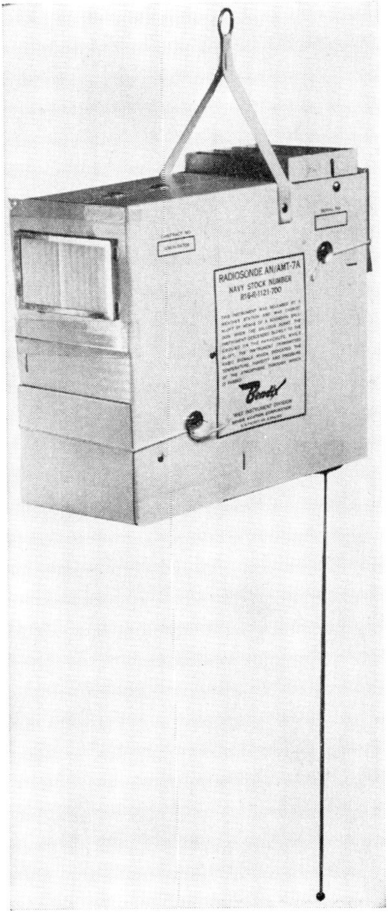

FIG. 23.15 RADIOSONDE. *Courtesy Friez Instrument Div., Bendix Aviation Corp.*

Information obtained by means of RAOBS is made use of immediately in

FIG. 23.16 THE BEGINNING OF A RADIOSONDE OBSERVATION OF THE WEATHER IN UPPER PORTIONS OF THE ATMOSPHERE. The radiosonde has just been released from the stern of a naval vessel. A radiosonde and a reflecting target may be seen below the balloon. *Official U.S. Navy Photograph*

the preparation of maps and charts that are valuable aids to weather forecasting. The information is also used by pilots in planning flights.

On 28 February, 1952, a radiosonde balloon reached an altitude of 155,000 feet above Manila, which was a world record. Such observations at great heights have added to the general knowledge of the atmosphere and its physical properties.

Weather reconnaissance flights often use a *dropsonde* to obtain upper air data. This instrument is similar to the radiosonde and is parachuted to the earth from the aircraft. The aircraft orbits within signal reading distance of the dropsonde and collects the data transmitted in a manner similar to that of the radiosonde receiver.

Recent developments in upper air measuring devices includes the *transosonde.* This device was developed by the U.S. Navy and is used to obtain pressure and wind data over the ocean areas between Japan and the United States. The transosonde is carried aloft by a large balloon to a predetermined height (usually around 30,000 feet) and is designed to maintain that altitude until the flight terminates. The flight is terminated once it reaches the United States or when a predetermined number of flight hours have elapsed. A standard parachute is attached to the instruments and carries the instruments to the earth in this manner.

Communication stations located throughout the Pacific and the United States are assigned the task of reporting the position of the transosonde and its altitude to the proper weather facilities.

23.14. Transmission of Weather Observations. On 1 January, 1949, by international agreement, the ships of all maritime nations began to use the same code for the transmission of weather reports by radio. The new code is slightly different from older codes, but its purpose is the same: it enables a report to be understood regardless of the nationality of the sender, and it requires less time to send than does a report in plain language. A coded weather message consists of a series of four, six, or more five-digit numbers, the number of groups depending upon whether a short, an abbreviated, or a full message is sent. Coded reports from ships differ slightly from land station messages because ships must report their position by latitude and longitude. Ship reports usually consist of the first six groups of a full weather message. This is known as an abbreviated message, and the symbols of these groups are as follows:

$YQL_aL_aL_a$	$L_oL_oL_oGG$	$Nddff$	$VVwwW$	$PPPTT$	$N_hC_LhC_MC_H$
20384	58718	61818	94100	11558	44620

The top line shows the symbols that stand for the data which is to be coded and transmitted by radio. The second line is the coded message ready to be transmitted as six numbers of five digits each. The meaning of the symbols and the digits shown in the sample report above is as follows:

Y The day of the week; 2 is the figure which codes Monday. Seven digits represent the seven days of the week.

Q The octant of the earth; 0 represents the octant in the northern hemisphere between 0 and 90 degrees west longitude. The digits 0–3 stand for the four octants in the northern hemisphere, while 5–8 are used for the octants of the southern hemisphere. It is always necessary to specify the octant in a report, because latitude is not indicated in reports as north or south; longitude is not designated east or west.

$L_aL_aL_a$ The latitude of the ship when the observation was made; 384 stands for 38.4 degrees.

$L_oL_oL_o$ The longitude; 587 is the code for 58.7 degrees.

GG The Greenwich civil time to the nearest hour, on the 24-hour scale. 18 is the code for 18^h G.C.T.

N Total amount of cloud; 6 means that six-eighths of the sky is covered by clouds.

dd The true direction, in 10's of degrees, from which the wind is blowing; 18 means that the wind is blowing from 180 degrees.

ff The wind speed in knots; the number 25 in the code stands for the wind speed in knots.

VV The visibility or horizontal distance at which objects can be seen in daylight or at which lights can be seen at night; 94 stands for ½ nautical mile (1,000 meters).

ww The present weather at the time of observation; the figure 10 stands for light fog, visibility 1,000 meters (1,100 yards) or more.

W Past weather; the digit 0 means that the weather within the six-hour period immediately preceding the observation was clear, or only a few clouds were present.

PPP The barometric pressure, in millibars and tenths (initial 9 or 10 omitted). The values refer to sea level and include all corrections for index errors, temperature and gravity; 115 stands for 1011.5 millibars.

TT The temperature of the air, in whole degrees Fahrenheit.

N_h The amount of low cloud, the height of which is reported by "h"; 4 means that the sky is four-eighths covered with low cloud.

C_L Clouds of types stratocumulus, stratus, cumulus, and cumulonimbus; 4 refers to stratocumulus formed by the spreading out of cumulus; cumulus also often present.

h The height of base of low cloud above the sea (or ground); 6 means 3,000 to 5,000 feet.

C_M Clouds of types altocumulus, altostratus, and nimbostratus; 2 stands for thick altostratus or nimbostratus.

C_H Clouds of types cirrus, cirrostratus, and cirrocumulus; 0 means that there are no high clouds present.

23.15. The Use of Weather Reports. Ship and land stations regardless of their locations observe and record weather conditions at the same time. The reports are transmitted to properly designated Weather Bureau Offices or Military Stations where they are relayed to the cognizant Weather Bureau Offices or Military Weather Centrals. These reports are used for the preparation of weather charts, from which forecasts are made. The forecasts are transmitted on assigned frequencies and at specified times to any party that can receive the broadcast. In addition to forecasts, a general description of the weather map is also broadcast, along with selected surface reports from which a ship may prepare its own weather map if it so desires. Recent advancements in radio facsimile have made it possible to broadcast facsimiles of the current weather maps in picture form. Many merchant ships and a great number of military ships are equipped to copy this form of transmission.

Complete weather broadcast schedules for various parts of the world are contained in the following publications: H.O. 206, *Radio Weather Aids,* issued by the Hydrographic Office of the Navy Department; *Weather Service for Merchant Shipping and Coastal Warning and Facilities Charts,* published by the U.S. Weather Bureau; *Information for Shipping,* Publication 9A published by the World Meteorological Organization.

23.16. Plotting Reports on the Weather Map. It has been noted how a large amount of useful weather information can be transmitted in a relatively short coded message. Persons who regularly receive and decode the reports soon learn to do so with very little reference to code tables. An element—for example, wind force—always occupies the same position in the same code group. If a weather element is missing for any reason, the observer who makes up the report at the point of origin always fills in with x's the space which the element would ordinarily occupy in the message.

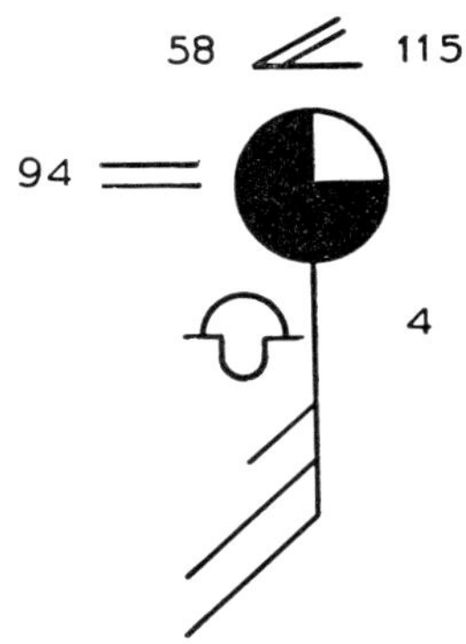

FIG. 23.17 ABBREVIATED WEATHER MESSAGE AS IT WOULD BE ENTERED ON A WEATHER MAP. The message described on page 515 would look like this when entered on a map.

Figure 23.17 shows the scheme used for plotting on a weather map the data of the ship's report appearing on page 515. First, the position of the ship is located on the chart, and a small circle is drawn there. The various weather elements of the report are then arranged at definite positions with reference to the circle. By using the same position for the same element each time there is then no question as to what element a figure represents. The circle is usually filled in first to represent the amount of the sky that is covered by clouds. Next, the arrow shaft is drawn to represent the wind direction. The shaft is always extended toward the direction from which the wind blows. The wind force is represented by feathers or barbs at the end of the shaft. One half of a barb stands for 5 knots, and one long barb stands for 10 knots. In Fig. 23.17 the wind force is 25 knots. Once the arrow shaft is drawn, the other elements may be recorded. If the arrow were to be drawn in last, it might extend through some of the other data with the danger of obliterating part of them.

Symbols are used to depict the wind direction and force, cloud forms, the present and past state of the weather, and the amount of clouds covering the sky. Coded figures represent the visibility and the amount of sky covered by low clouds. The elements represented directly by figures are the temperature and the pressure (tens and units of millibars).

BEAUFORT SCALE

Beaufort no.	Sea miles per hour (knots)	Seaman's description	Effect at sea
0	Less than 1	Calm	Sea like a mirror.
1	1-3	Light air	Ripples with the appearance of a scale are formed but without foam crests.
2	4-6	Light breeze	Small wavelets, still short but more pronounced; crests have a glassy appearance and do not break.
3	7-10	Gentle breeze	Large wavelets. Crests begin to break. Foam of glassy appearance. Perhaps scattered white horses.
4	11-16	Moderate breeze	Small waves, becoming longer; fairly frequent white horses.
5	17-21	Fresh breeze	Moderate waves, taking a more pronounced long form; many white horses are formed. (Chance of some spray.)
6	22-27	Strong breeze	Large waves begin to form; the white foam crests are more extensive everywhere. (Probably some spray.)
7	28-33	Moderate gale (high wind)	Sea heaps up and white foam from breaking waves begins to be blown in streaks along the direction of the wind. Spindrift begins.
8	34-40	Fresh gale	Moderately high waves of greater length; edges of crests break into spindrift. The foam is blown in well-marked streaks along the direction of the wind.
9	41-47	Strong gale	High waves. Dense streaks of foam along the direction of the wind. Sea begins to roll. Spray may affect visibility.
10	48-55	Whole gale	Very high waves with long overhanging crests. The resulting foam in great patches is blown in dense white streaks along the direction of the wind. On the whole the surface of the sea takes a white appearance. The rolling of the sea becomes heavy and shocklike. Visibility is affected.
11	56-66	Storm	Exceptionally high waves. (Small and medium-sized ships might for a long time be lost to view behind the waves.) The sea is completely covered with long white patches of foam lying along the direction of the wind. Everywhere the edges of the wave crests are blown into froth. Visibility affected.
12	Above 66	Hurricane	The air is filled with foam and spray. Sea completely white with driving spray; visibility very seriously affected.

In making a weather map it may be desirable to save time by entering only the data from the first five groups of the six-group abbreviated ship messages. The sixth group includes detailed cloud data; its elimination from the map would still leave the following elements recorded: total amount of cloudiness; wind direction and force; visibility; present and past weather; pressure and temperature.

On the other hand, should it be considered desirable to enter on the weather

OF WIND FORCE

Action of fishing smack	Effect ashore	Land miles per hour	International description	Beaufort no.
Makes no headway	Smoke rises vertically	Less than 1	Calm	0
Just has headway	Does not move wind vanes, but wind direction shown by smoke drift	1-3	Light air	1
Wind fills sails; makes up to 2 knots	Wind felt on face; leaves rustle; ordinary vane moved by wind	4-7	Light breeze	2
Heels slightly under full canvas; makes up to 3 knots	Leaves and small twigs in constant motion; wind extends light flag	8-12	Gentle breeze	3
Good working breeze; under all sail, heels considerably	Raises dust and loose paper; small branches are moved	13-18	Moderate breeze	4
Shortens sail	Small trees in leaf begin to sway; crested wavelets form on inland waters	19-24	Fresh breeze	5
Double reefs mainsail	Large branches in motion; whistling heard in telegraph wires; umbrellas used with difficulty	25-31	Strong breeze	6
Remains in harbor, or if at sea, lies to	Whole trees in motion; inconvenience felt in walking against wind	32-38	Moderate gale	7
Takes shelter if possible	Breaks twigs off trees; generally impedes progress	39-46	Fresh gale	8
	Slight structural damage occurs (chimney pots and slate removed)	47-54	Strong gale	9
	Seldom experienced inland; trees uprooted; considerable structural damage occurs	55-63	Whole gale	10
	Very rarely experienced; accompanied by widespread damage	64-75	Storm	11
		Above 75	Hurricane	12

map the data from a full weather message of eleven or more coded groups, this may be done in accordance with the procedure outlined in instructions issued by the Navy Department and the Weather Bureau.

23.17. Criteria for Wind Warnings. *Wind Warning Terminology in Common Usage by the U.S. Weather Bureau and the U.S. Navy.* Warnings of winds associated with closed cyclonic circulations of tropical origin are expressed in the following terms:

Type of Warning	*Corresponding Wind Speed*
Tropical depression	Winds up to 33 knots
Tropical storm	Winds between 34 and 63 knots
Hurricane/typhoon	Winds of 64 knots or greater

Warnings of winds associated with weather systems located in latitudes outside tropical regions, or by systems of tropical origin other than closed cyclonic circulations, will be expressed in the following terms:

Type of Warning	*Corresponding Wind Speed*
Small craft warning	Winds up to 33 knots (use in coastal and inland waters only)
Gale warning	Winds between 34 and 47 knots
Storm warning	Winds of 48 knots or greater

Wind Warning Terminology Used by the U.S. Weather Bureau for Coastal Warning Displays. Warnings of winds for coastal display purposes are issued by the U.S. Weather Bureau in accordance with the following criteria:

Type of Warning	*Corresponding Wind Speed*
Small craft warning	Winds up to 33 knots
Gale warning	Winds between 34 and 47 knots
Whole gale warning	Winds between 48 and 63 knots
Hurricane warning	Winds of 64 knots or greater

24

Weather of the Middle Latitudes

24.1. Introduction. The middle latitudes of the northern hemisphere, lying as they do between tropical and polar regions, have a temperate climate that is favorable to man. Large portions of the continents in this zone have been settled and developed, and the ocean areas, including the air above, are widely used in commerce. Because this portion of the world plays a highly important part in present-day civilization, this chapter is devoted to weather conditions to be found there. It should be recognized from the start that they are more complicated in many ways than in other parts of the world.

It will be recalled from Chapter 21 that weather conditions in the doldrums, the trade-wind belt, the horse latitudes, and the polar regions tend to conform to definite and more or less dependable patterns which are associated with those wind and pressure belts of the world. Such is not the case, however, in the middle latitudes. Weather here is varied, quite changeable, and it represents a continuous challenge to those who must contend with it at sea, on the land, and in the air.

It is in the region of the middle latitudes that masses of air moving northward from the tropics and southward from polar regions tend to converge and create various types of weather fronts. The reaction of these air masses with one another and with the land and water surfaces of the earth cause constant weather changes which characterize the middle latitudes of both hemispheres. Widespread areas of low- and high-pressure drift eastward in the prevailing westerlies of these latitudes, alternately bringing the changes in weather conditions that are associated with them.

This chapter deals with the air masses, fronts, and high- and low-pressure systems of the middle latitudes. The preparation and use of weather maps are described, and general principles are given for forecasting weather from surface weather maps and from local indications. In addition to the surface weather maps, professional meteorologists construct and use many other maps, charts, and graphs which show pressure, humidity, temperature, and wind conditions in the upper portions of the atmosphere. Those data are used in forecasting and in aircraft operation. The recent development of new instruments and techniques for gathering and interpreting atmospheric conditions at various elevations above the earth's surface has enabled the meteorologist to gain a better understanding of present weather conditions and to forecast changes more effectively than would be possible with only the use of the surface weather map.

24.2. The Meaning and Classification of Air Masses. When a large body of air remains for some time over a locality, it acquires the characteristics of that region. For example, air that sojourns for several days or weeks over northern Canada during the winter becomes cold. It will also lack moisture both because of its coldness and because of the absence of water at the earth's surface from which vapor could be received. It would be described as a cold,

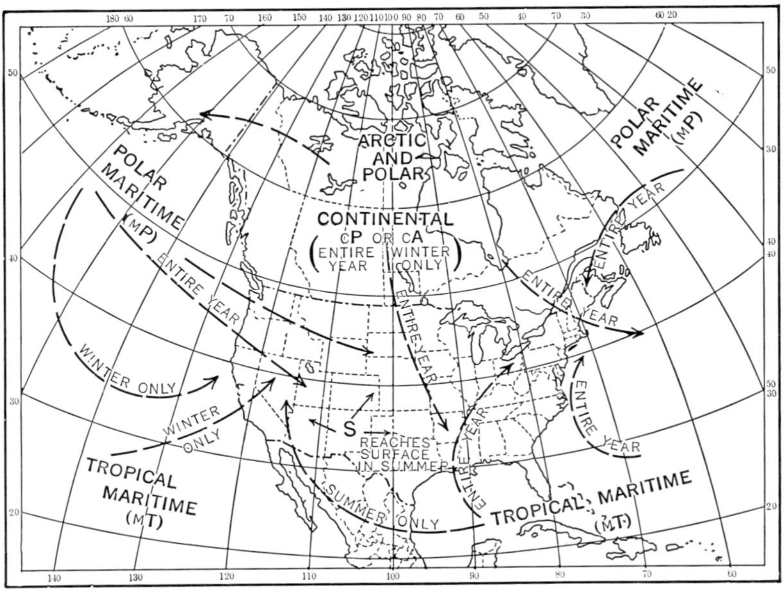

FIG. 24.1 AIR MASS TYPES THAT VISIT THE UNITED STATES. *Courtesy CAA*

dry mass of polar air, and the region where it acquired those characteristics is known as the source region (Fig. 24.1). On the other hand, a body of air which stagnates over the Gulf of Mexico for some time acquires the warmth of those waters, and it will acquire a large content of water vapor both because of the evaporation and also of the large capacity of warm air for vapor. Such a body of air would be known as a warm, moist mass of tropical air. In this case the source region is the Gulf of Mexico.

Air masses are classified according to their sources; they may be either polar or tropical. They are further classified as to their moisture content. Masses whose source regions are over the ocean are known as maritime air masses and are moist or high in water vapor content. Those originating over land areas are known as continental masses and are relatively dry or low in water vapor content. A further classification depends on whether an air mass, once it starts to move about, is warmer or colder than the surface over which it is moving. For example, a mass which moves northward over the Gulf states from the

Gulf of Mexico would be classed as a warm mass in winter because land surfaces of those states in winter are colder than the waters of the Gulf. However, in summer the mass would be classed as cold because then the land surfaces are warmer than the water surfaces. According to the foregoing classification (Bergeron) we have eight different air mass types, six of which may be found at times in the United States. The eight are as follows:

1. cPk—continental polar air, colder than the surface over which it is passing. (*k* is from the Norwegian word for cold, "kalt.")
2. cPw—continental polar air, warmer than the surface over which it is passing.
3. mPk—maritime polar air, colder than the surface over which it is passing.
4. mPw—maritime polar air, warmer than the surface over which it is passing.
5. cTk—continental tropical air, colder than the surface over which it is passing.
6. cTw—continental tropical air, warmer than the surface over which it is passing.
7. mTk—maritime tropical air, colder than the surface over which it is passing.
8. mTw—maritime tropical air, warmer than the surface over which it is passing.

Because there are no really widespread land masses immediately south of the United States, cTk and cTw air masses do not visit this country. Bergeron's classification also provides for four other types: cAw, cAk, mAk, and *S* air. *A* stands for air of Arctic origin, a region different from that which supplies the polar masses. Arctic masses do not invade the United States as frequently as do polar masses. *S* air refers to warm, dry air masses sometimes present at upper air levels.

24.3. Warm Air Masses. Because a warm air mass is one which is warmer than the land or water surface over which it is moving, it follows that warm air masses may exist during either summer or winter months. For example, in winter when maritime polar air from the Atlantic invades New England, the air may be considerably warmer than the land surfaces. Also, in summer when polar continental air flows from Canada westward out over the north Pacific, it is likely to be warmer than the ocean surface. Since the surface layer of a warm air mass is being chilled by the land or water underneath it, it follows that fog may result if the cooling of the air continues to the dewpoint. If surface wind velocities are moderate or stronger, stratus or stratocumulus clouds would form rather than fog; drizzle or rain might result.

24.4. Cold Air Masses. Just as warm air masses may occur during either winter or summer months, so cold air masses are found during any season. During summer months maritime tropical air from the Gulf of Mexico commonly moves northward to invade the Gulf Coast states. As this tropical air moves inland, it finds itself cooler than the land surfaces; convectional clouds are the result, and precipitation often occurs. Continental polar air which moves southward to the United States in winter is often much colder than the land surfaces which it overruns to the south. This air mass type acquires very cold temperatures during its stay in that part of the North American continent from the region of Hudson Bay to Alaska where it is segregated by mountain ranges from any possible warming by air from Pacific waters. As the surface layer of a cold air mass is warmed by the land or water surface over which it starts to move, vertical currents are set up. Any cloud types therefore would be of the cumuliform type, and precipitation would be in the form of showers. Visibility at surface levels would be good.

24.5. Fronts. Adjacent masses with different qualities as to temperature and humidity do not tend readily to mix; as the cold masses are heavy and the warm masses light, the warmer of two converging currents tends to overrun the colder. It will be recalled that the polar front is the surface between the converging southwesterly winds of middle latitudes and the prevailing northeasterlies of polar regions. The latter forms a wedge over which the winds from the south ascend. A front, then, is the surface between two masses of air of different density. In general, there are three kinds of fronts. They are the cold, warm, and occluded fronts. All types are watched closely on weather maps because it is along them that the poorest weather conditions occur, and changes from one type of weather to another take place in their vicinity. We will consider each type in order.

24.6. Warm Fronts. Figure 24.2 shows a west-east, vertical cross section through the atmosphere at some locality in the United States. To the right is

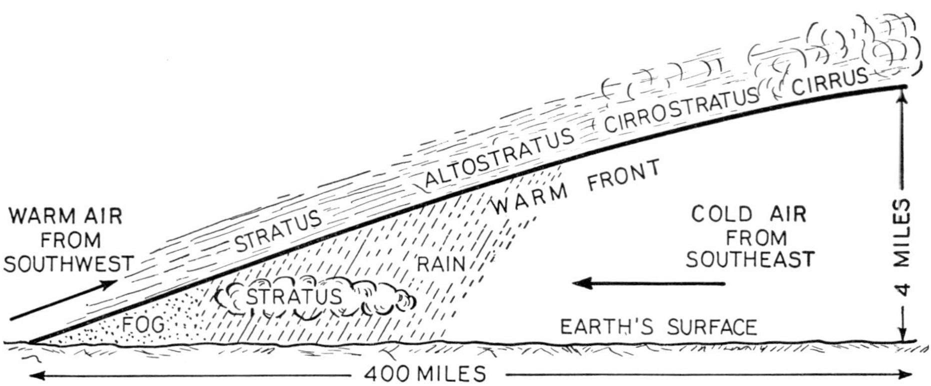

FIG. 24.2 TEMPERATURE, CLOUD, AND PRECIPITATION PHENOMENA

a mass of cool air which is flowing from the southeast; on the left is a warm air mass flowing from the southwest. As the currents are converging, the warm stream is forced to ascend the cool barrier. It is assumed that the system as a whole is drifting from west to east, as is customary for atmospheric combinations in the temperate zone of middle latitudes. It is further assumed that the warm current has a reasonably high relative humidity. The slope of the wedge is exaggerated in the figure in order to show clearly the processes involved. Actually the ratio of slopes of warm fronts averages about 1 mile in the vertical to 100 miles horizontally. As the warm air stream rises over the wedge of cold air, it expands and cools adiabatically, which results in the formation of the various cloud types shown. At the extreme right appear cirrus clouds at an elevation of about 4 miles. They are followed in order by cirrostratus, altostratus, and nimbostratus clouds. Precipitation in the form of steady rain is falling from the altostratus and nimbostratus types. In the rain curtain below the nimbostratus will be noted stratus clouds and fog. The figure shows a situation which is quite typical of warm fronts that occur in the United States and other portions of the middle latitudes. Warm front areas often cover hundreds of square miles—in fact, whole states or groups of states. Cloud ceilings are low because of the presence of nimbostratus clouds over wide areas, and visibilities may be poor due to the presence of rain or drizzle and fog. Where temperatures in the cloud and rain areas are near freezing, icing on aircraft is prevalent.

Although the air at all levels is likely to be smooth, considerable turbulence and thunderstorms may form above the front if the rising air becomes unstable in its upward glide over the cold wedge. In fact, any cloud type may appear along a warm front under various conditions. During late fall and early spring, frozen rain (sleet) may form in the rain curtain under the front when raindrops falling from above are frozen in the cold air of the wedge. Let us now take a look at a cold front.

24.7. Cold Fronts. A west-east, vertical cross section through the atmosphere showing a typical cold front of the middle latitudes appears in Fig. 24.3. At the left in the figure is shown a wedge of cold air advancing from the northwest. It is underrunning and forcing upward a stream of warm, moist air that

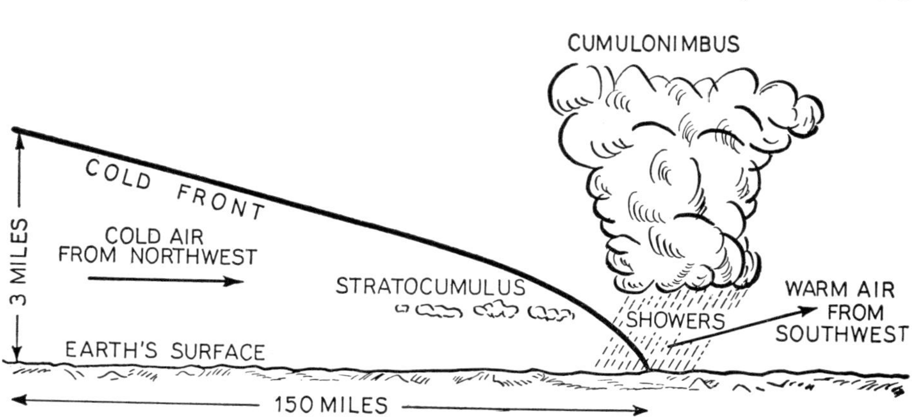

FIG. 24.3 COLD FRONT WIND, TEMPERATURE, CLOUD, AND PRECIPITATION PHENOMENA

flows from the southwest. The entire combination of masses is drifting toward the east. As in the case of the warm front the wedge is exaggerated so that the principles involved will stand out clearly. Cold fronts are steeper than warm fronts with a slant of about 1 to 50. As in the case of the warm front, weather is poor along the cold front, but it is a different type of weather. Warm air is nosed upward by the leading edge of the cold wedge, and cumulus rather than stratus clouds form. The cloud tops commonly reach quite high and often develop into cumulonimbus. Precipitation types are rain showers and often heavy hail or snow flurries. Cold-front cloud and precipitation areas are narrow horizontally as compared with those of the warm front. The air is usually rough up to an elevation of 6,000 feet, and it may be turbulent to heights of 20,000 feet or more when thunderstorms are present, which is quite common. At levels where the temperature is at or below freezing, aircraft icing will be encountered, and visibilities and ceilings are unfavorable for contact flight in the cloud and precipitation region.

In the cold air immediately behind the front, stratocumulus clouds often prevail. They form as the cold air moves rapidly over ground previously heated by the presence of the warm air ahead of the front.

Before considering occluded fronts let us examine an extra-tropical cyclone and note the relationships which exist between various kinds of air masses and fronts which may be found in an extra-tropical cyclone, or low.

24.8. Extra-Tropical Cyclones. Figure 24.4 shows the air masses, fronts, clouds, precipitation, and winds which are a part of an extra-tropical cyclone, also known as a depression, low-pressure area, or simply as a low. The figure represents an idealized model as visualized by the Norwegian meteorologist J. Bjerknes, and it shows the low in one particular stage of its development. The center sketch shows a plan view. At the top of the figure appears a vertical section of the low taken through the line *A–B*; the bottom diagram is a vertical section taken through the line *C–D*. Looking back at the plan view we note the warm front where warm air from the so-called warm sector is converging with, and ascending, cold air to the right. A cold front exists where cold air at the left is under-running air in the warm sector. A broken arrow pointing to the right at the middle of the figure indicates that the entire formation, that is the low, is moving toward the right, or east. The shaded portion shows where rain is occurring. Cloud types are not shown in the plan view. Warm fronts which appear on a surface map may be defined as regions where warm air is advancing to ground formerly held by cold air. The plan view of the figure clearly illustrates the definition. Similarly, a cold front is a region where cold air is advancing to ground formerly held by warm air. The figure also shows that situation.

The vertical section through *C–D*, which is shown on the lower portion of the figure, helps to clear up what is happening in the upper air. It will be noted that cumulonimbus clouds are shown at the cold front; altocumulus clouds also are often present there. In the cross section through *A–B* the warm air does not touch the earth's surface but is riding up over the cold air which is sweeping around toward the west. If we look back at the middle and lower views, the broad area of rain at the warm front and the narrow band of rain along the cold front are evident.

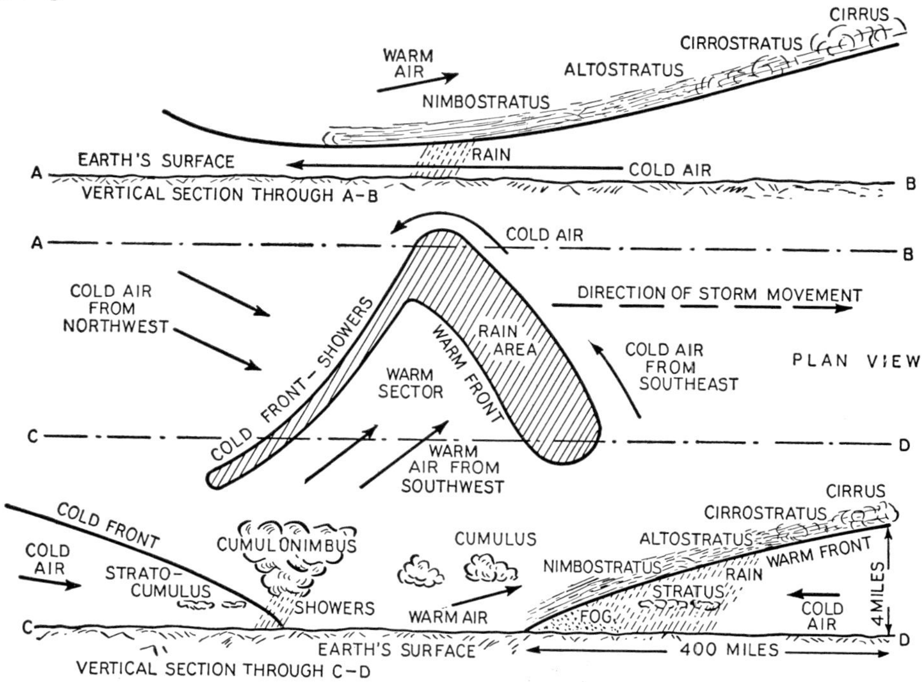

FIG. 24.4 A PLAN VIEW AND TWO VERTICAL SECTIONS OF AN EXTRA-TROPICAL CYCLONE. Hatched portions in the plan view show where rain is occurring. (*After Bjerknes*)

24.9. Formation and Occlusion of Extra-Tropical Cyclones. The life history of a depression is shown in Fig. 24.5. A stationary front such as the polar front is shown in *a* with cold air flowing toward the west while warm air flows toward the east. In *b* the front has ceased to be straight, possibly because pressure is being exerted at the left side of the figure by a mass of cold air to the north. A definite cold and warm frontal system has developed, and the arrows show that the wind has commenced to blow in a counterclockwise direction around a center of lower pressure. The hatched areas indicate that cloudiness and precipitation have begun. This development is known as a wave that has formed along the stationary front, and the wave will move from left to right along the front in much the same manner as an ocean wave. The cyclone has reached a normal stage of development in *c*, and the cold front is seen to be advancing faster than the warm front. In *d* the cold front has overtaken the warm front in the vicinity of the center of the depression; the dying-out or occlusion of the cyclone has begun. Further occlusion has taken place in *e* and *f;* under these conditions cloudiness and precipitation continue but in diminishing amounts. Occlusion, if it continues, results in the obliteration or filling up of the depression.

Two distinct types of occluded fronts are shown in *g* and *h*. When the cold front caught up with the warm front, the air in *g* to the left was colder than that at the right, and therefore it underran the latter as shown. The warm

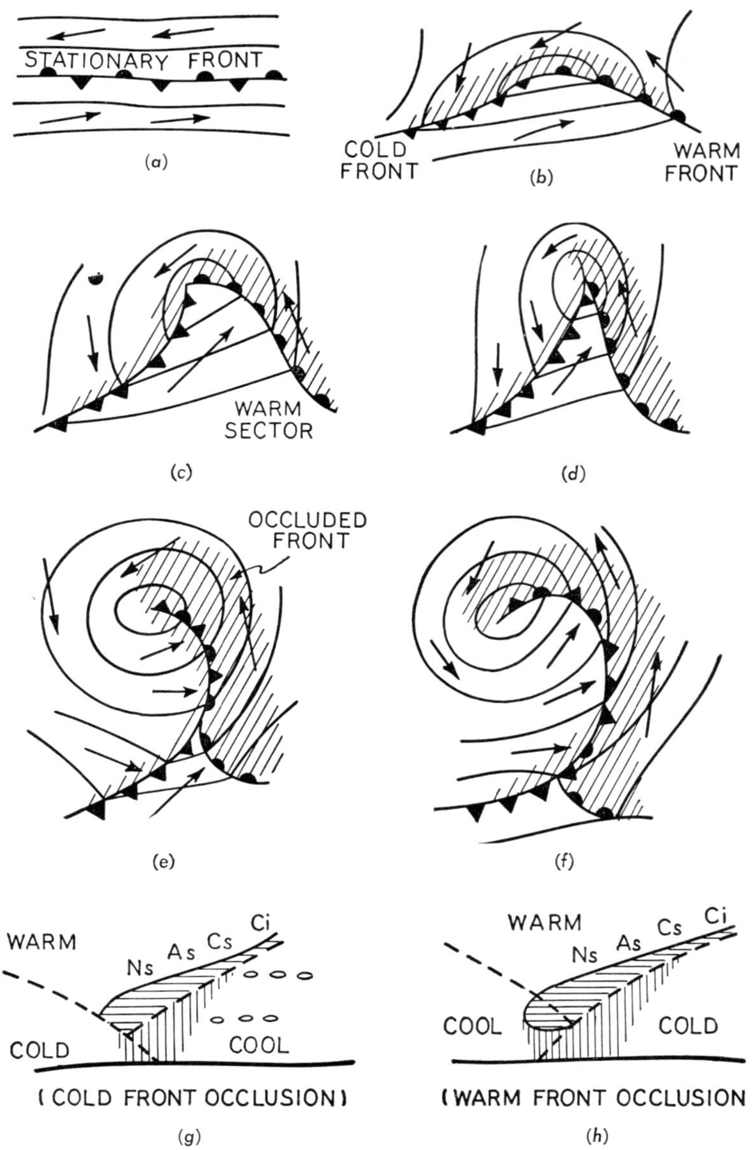

FIG. 24.5 DEVELOPMENT OF A DEPRESSION. Horizontal section, or plan view. Hatched areas are precipitation areas.

sector has been squeezed above the ground level. Precipitation continues at and behind the surface front, and it is chiefly that of the showery type such as is found along cold fronts. In *h*, which shows the warm-front type of occlusion, the air to the left is not as cold as that at the right. The cool air, therefore, is shown to be rising over the cold air, and precipitation continues at and ahead of the surface front; it is the steady rain or drizzle that is typical of warm fronts.

24.10. The Movement of Lows. Lows move in general from west to east across the United States. Many of them form in the general region of the semi-permanent low-pressure area of the Aleutian Islands and enter the north-western section of the United States. From there some continue across the northern part of the country to New England; others drift to the middle plains states where they recurve toward the northeast and leave the continent by way of the St. Lawrence valley. Some depressions form in the central and southern plains states from whence they move toward the east or northeast. The rate of movement in summer is about 500 miles per day; in winter it is somewhat faster, probably averaging 700 miles in 24 hours, but there are many variations. At times an extra-tropical cyclone may slow down and remain stationary over an area, while at other times a section of the country may be subjected to a series of lows which move along quickly one after the other. Lows vary in size from several hundred to 1,000 and sometimes 2,000 or more miles in diameter. They are more stormy and sharply defined in winter than in summer.

24.11. Anticyclones. Anticyclones (Fig. 24.6) are areas of high pressure; their name is due to the fact that the wind within them circulates in a manner

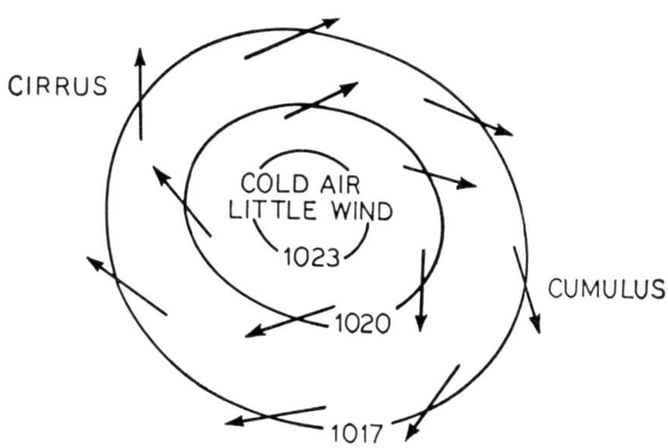

FIG. 24.6 AN ANTICYCLONE, HIGH, AND ASSOCIATED WEATHER ELEMENTS

opposite to that of cyclones. It has been noted that in extra-tropical cyclones of the northern hemisphere the wind circulates counterclockwise, spirally inward. In anticyclones the wind blows clockwise and outward. In general, anticyclones, or highs as they are most often called, alternate with lows in a sort of parade across the country. Highs commonly enter the country from the northern portion of the North American continent and tend to drift toward the southeast. The type of weather which they bring is chiefly cool or cold and fair, although under some circumstances there may be a considerable amount of cloudiness and rain. Their rate of movement and size are fairly comparable to that of lows. Many highs which invade the United States may be thought of as atmospheric mountains of cold, dense air which have broken away from their northern source regions to drift southward in order to lessen the pressure which builds up in polar areas.

24.12. Extra-Tropical Cyclonic Weather. Refer to Fig. 24.7 and let us see what sort of weather may occur when a typical low passes over your locality

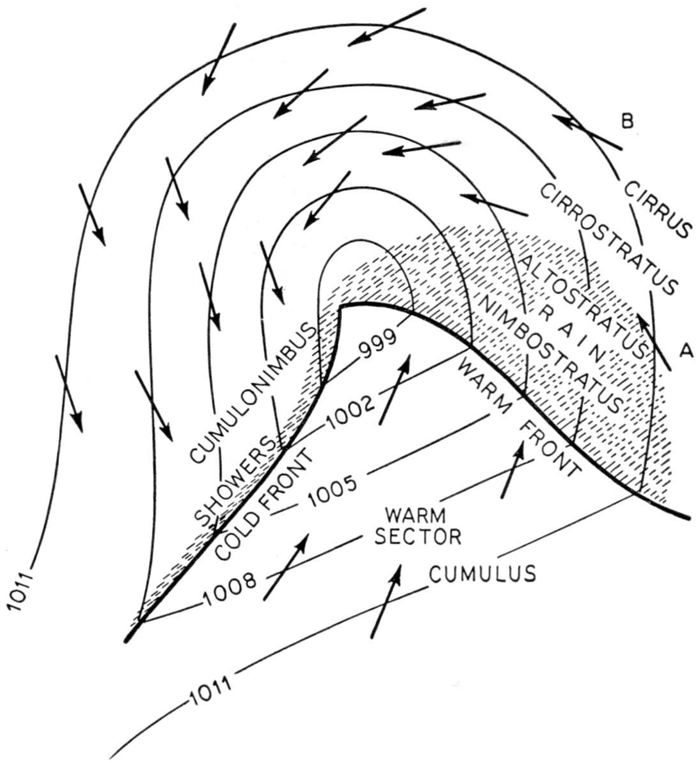

FIG. 24.7 EXTRA-TROPICAL CYCLONE. Rain or showers are occurring throughout the shaded area.

(point *A*). Assume that it is moving east-northeast at the rate of 600 miles per day, and that the center passes to the north of you. The approach of the low is indicated by the barometer, cloud types, and wind direction; the barometer begins to fall, cirrus clouds appear, and the wind sets in from the southeast. The temperature rises to some extent because presumably the preceding high-pressure area had brought cool or cold weather. Likewise, humidity values increase. A continued fall of the barometer is accompanied by the appearance of cirrostratus clouds, solar and lunar halos, and an increase in the velocity of the southeast wind. As the low draws closer, the cirrostratus clouds are replaced by altostratus clouds, and a light but steady rain begins. In due time nimbostratus and stratus clouds with fog appear, and the rain continues with increasing intensity. Presently the wind shifts to south-southwest, the barometer steadies, the rain ceases, and the sky clears except for some scattered cumulus clouds. The warm front has passed to the east, and we are now in the warm sector with temperatures and humidity high. With the approach of the cold front the typical roll, or squall, cloud (Fig. 24.8) looms on the western horizon,

FIG. 24.8 COLD FRONT CLOUD. *Official U.S. Navy Photograph*

and there may be cold-front thunderstorms with associated heavy showers of rain, gusty wind, and possibly hail. The cold-front storminess, however, is of brief duration as compared with that of the warm front. The wind soon shifts to northwest, the barometer rises sharply, and the temperature and humidity drop. Stratocumulus clouds may form back of the cold front, but the sky clears in due time, and the barometer continues rising.

It must be borne in mind that there have never been two lows exactly alike, and that cloudiness and rain types and amounts vary greatly; however, the behavior of wind, temperature, and pressure is more predictable.

Let us suppose now that we were located at point *B* and that the low center will pass to the south of us. A glance at Fig. 24.7 makes it obvious that the sequence of weather element values at our location will be different as the low moves along. The first indications of the approach of the low are cirrus clouds, a shift in wind direction to the east, and a downward trend of the barometer. As the low moves closer, cirrostratus clouds replace the cirrus, the wind continues from the east, and the barometer continues downward. Presently altostratus clouds take the place of the cirrostratus clouds, and steady, light rain begins; the wind continues from the east, and the barometer falls further. As the low progresses east-northeastward, the wind shifts to northeast and then to north with nimbostratus clouds replacing the altostratus. Eventually the wind shifts to northwest, the barometer begins to rise, the sky gradually clears, and the precipitation ceases. The low has passed to the eastward, and fine, clear weather prevails.

24.13. Preparation of Surface Weather Maps. Maps showing weather conditions at the earth's surface are prepared in meteorological offices every 6 hours and sometimes oftener (Fig. 24.9). Reports from a vast network of sta-

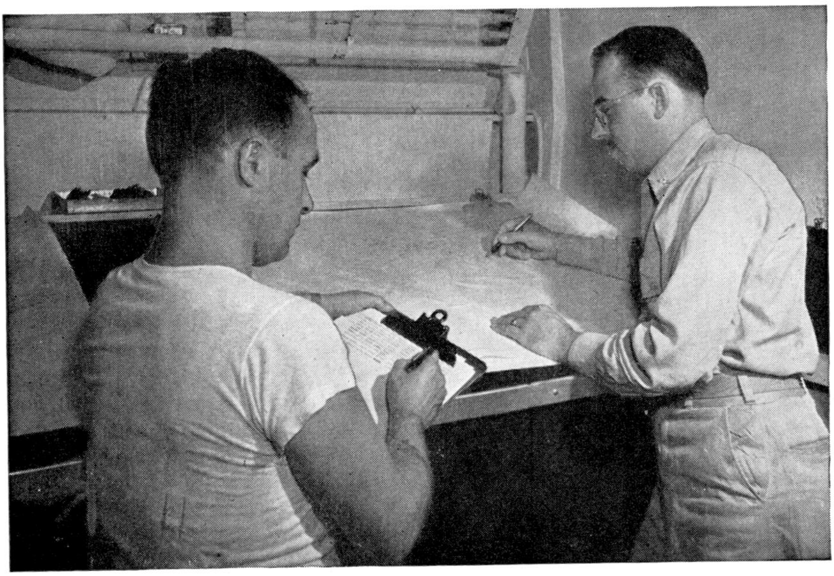

FIG. 24.9 CONSTRUCTING A WEATHER MAP ON BOARD AN AIRCRAFT CARRIER

tions are entered in ink on a base map in accordance with the scheme outlined in Chapter 23. Once the land and ship reports have been recorded, the meteorologist may then proceed to construct the isobars. It is important to have the preceding map in view in order better to construct the new map. The isobars are sketched in lightly, taking into consideration not only the reported pressure values, but the wind directions as well. It will be recalled that surface winds blow across isobars toward lows and away from highs. However, gentle to light winds, particularly over land areas, do not always obey this rule.

The next step is to identify the fronts and air masses. The positions which the fronts occupied on the preceding map are sketched in lightly with pencil as a guide to where they may be found on the current map. Cold fronts are drawn with blue pencil, whereas warm fronts are colored red; occluded fronts are purple. Cold fronts are easier to place than warm fronts because weather contrasts on either side of a cold front are greater than is the case with warm fronts. Wind directions are particularly significant in placing the positions of cold fronts; ahead of the front the wind commonly blows from the south or southwest, whereas behind the front it is likely to be from the west or northwest. Ahead of the front the 3-hour pressure characteristic values show falling barometers, but behind the front the pressure values are rising. Temperature and humidity readings are higher ahead than they are behind the front. Precipitation and cloud types are reliable guides. The position of a warm front often may be determined by noting the widespread area of clouds and precipitation which commonly precede it. After the fronts have been drawn, the air masses which are present should be labeled.

The meteorologist proceeds next to smooth the isobars and to shape them properly around the fronts, using black pencil. When isobars are originally sketched, there appear many meaningless irregularities of the lines due to errors of various kinds in the weather reports, some of which are unavoidable. Most errors may readily be spotted when compared with surrounding reports. The most common types of errors are those which appear frequently in reports from stations located at a considerable distance above sea level. The system used to convert station readings to corresponding sea level values involves a correction for temperature. No satisfactory method has been evolved to do this so that representative sea level values can be obtained.

Areas where precipitation was occurring when the observations were made are shaded green, and large green symbols are drawn to show at a glance the precipitation type, such as drizzle, rain, showers, snow, etc. Stations that reported fog conditions are underscored in red so that pilots and others who may consult the map may readily grasp such situations. Areas of high and low pressure are appropriately labeled. The date and time of the observations on which the map was based are entered in the legend.

Figure 24.10 shows a completed weather map that is typical of conditions during late spring over the central and eastern portions of the United States. On the particular day that this map was drawn, high pressure dominated the plains states. Areas of high pressure also prevailed east of New England and east of Florida. An area of low pressure that extended from the Mississippi Valley to the Atlantic seaboard dominated weather conditions throughout the entire eastern portion of the country. A cold front extended from northwestern Ohio to Louisiana, while a warm front ran through northern Ohio, southwestern Pennsylvania, and Virginia. A mass of cold continental polar air existed west of the cold front; cold maritime tropical air was in the warm sector of the low, i.e., between the cold and warm fronts; and maritime polar air was flowing in from the Atlantic Ocean east and north of the warm front. The highest pressure on the map was 1,025.1 millibars at Concordia, Kansas; the lowest was 1,011.9 millibars at Cleveland, Ohio.

A closer examination of the figure reveals that the high-pressure area in the middle west is associated with a clockwise circulation of wind that blows out-

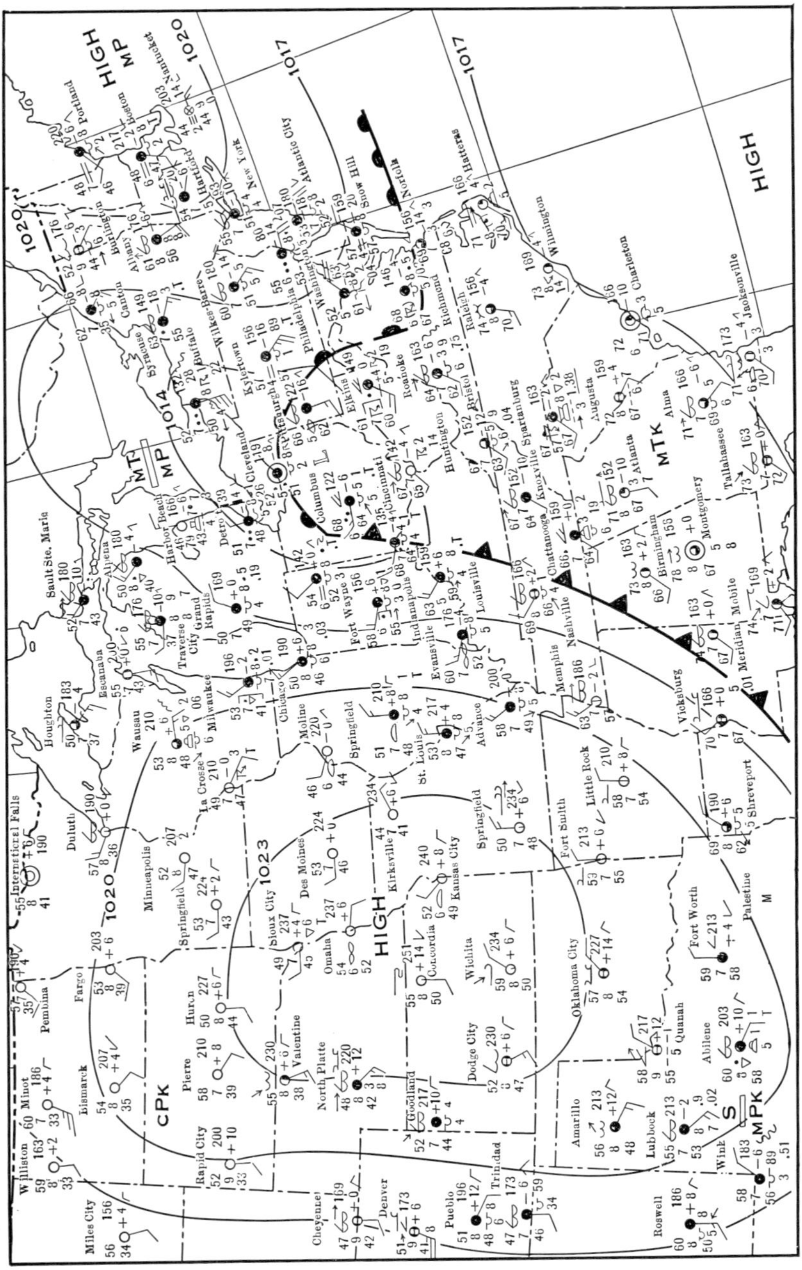

FIG. 24.10 A COMPLETED SURFACE WEATHER MAP

ward. Throughout much of the high, skies are clear and temperatures are cool. In the low-pressure area the wind is seen to be circulating in a counterclockwise manner and inward. Cloudy skies, precipitation, and higher temperatures prevail in the low. A broad area where it is raining is noted to the east and north of the warm front where maritime tropical air is converging with and

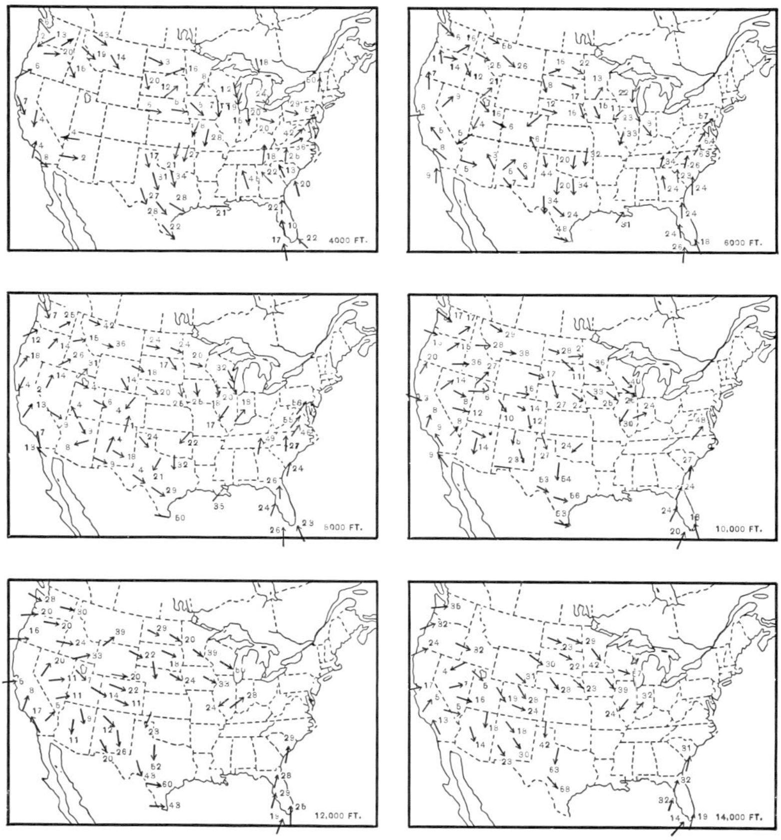

FIG. 24.11 CHART SHOWING WIND DIRECTIONS AND VELOCITIES AT SEVERAL HEIGHTS. *Courtesy of CAA*

ascending maritime polar air. The positions of these air masses, one above the other, are indicated by the air mass symbol found just east of Lake Huron.

It will be noted that in general the winds west of the cold front blow from the northwest, whereas those east of the cold front are from the southwest. Temperatures at places east of that front are, for the most part, higher than those to the west, and pressures to the east of it are falling while those to the west are rising. East and north of the warm front, pressures are falling, and the wind has an easterly component; south and west of the warm front, pressures are in general falling less rapidly, and winds blow from the south or southwest. Most places ahead of the warm front reported lower temperatures and dewpoints than those that prevailed behind the front.

It is apparent that the surface weather map does not show conditions that exist in the upper air, such as the direction and velocity of the wind at various heights, cloud levels, turbulence, regions where airplane icing may occur, air stability, temperature and humidity values, and other phenomena important to the forecaster and airplane pilot. Therefore, in addition to the surface map, auxiliary maps and diagrams are prepared. The nature and variety of these that are drawn depend on the needs of the forecaster and the size of the meteorology staff available. One of the most useful maps is known as the upper air wind chart. It may be drawn every 6 hours, and it shows the direction and velocity of the wind at the even-numbered thousands of feet above sea level. A glance at one of these charts (Fig. 24.11) reveals to forecasters the flow of air at different heights, which is important in the forecasting of cloudiness, precipitation, and other phenomena; to the airman it shows head and tail winds and their velocities at flight levels. Other diagrams may be made for the purpose of predicting thunderstorms, identifying air masses, and the solution of other problems of the forecaster and airman.

24.14. The Preparation of a Forecast. Upon the completion of the weather map the forecaster may see the conditions that prevail in all sections. If he desired to forecast what the weather would be at any place an hour later, the problem would not be difficult because values at most places would not radically change during that brief interval of time. Most forecasts, however, are made for periods of 8, 12, 24, 36, or more hours ahead. The conditions that a forecaster must take into consideration are:

1. Weather moves from west to east in general but at varying rates of speed and in uncertain specific directions.
2. Changes occur in the nature of the air masses, fronts, highs, and lows as they journey along.
3. New storm centers are continually forming while others die away.

Petterssen and other meteorologists have developed mathematical formulas for computing the speed and direction of movement of fronts and pressure centers. The successful use of these formulas, however, requires good judgment and considerable experience. At present no mathematical or physical formula for the solution of this problem is wholly reliable, but the systems so far developed are useful and no doubt point the way to more exact methods. Once the meteorologist has decided where the fronts and masses will move during the ensuing forecast period, he must then visualize what changes will occur in them. Finally, he must try to foresee any possible new developments.

All forecasters in time develop a number of empirical rules which help them to arrive at decisions. Some of those developed by Bowie are: east of the Rocky Mountains a storm which moves to the left of its normal track increases in intensity; abnormally high temperatures northwest of a storm indicate that it will either retrograde or remain stationary; and storms that start in the northwest and move southeastward do not gather great intensity until they begin to recurve northward. At the time of recurving they move slowly, as a rule, and care must be exercised in predicting clearing weather. Examples of rules which have been developed and based on Petterssen's formulas are: a high moves toward the area which shows the greatest rate of pressure increase; a new low will move in the direction of the warm sector isobars; that part of a cold front at a great distance from the center of a low moves slowly and is likely to die out.

If the new positions of fronts and pressure systems can be forecast, then wind directions and velocities may be determined, as the wind is definitely related to the pressure distribution. Once the winds are forecast, then future temperatures may be determined because wind directions determine temperatures. Precipitation is the most difficult and uncertain problem that the forecaster must confront. The times of beginning and ending and the amount that will fall, particularly for a specific local place, even when a clear-cut, energetic storm prevails, are often difficult to determine.

24.15. Aviation Meteorological Needs. Airmen are interested chiefly in forecasts for short periods ahead, especially regarding ceilings and visibilities. The forecasts must be specific, particularly for definite points along the route where landings may be made. The successful aviation forecaster must study the terrain and other peculiarities of each air terminal, as each locality varies with regard to the possibility of fog, low clouds, smoke, and other restrictions to visibility. Aviators before taking off and while in flight also want to know the conditions that exist at various points along their trip. Such information is possible because of the frequent reports that are made by the airport stations in this and other countries, from islands and by ships. Most airports report weather conditions hourly or oftener by means of radio or teletype. These reports, both of surface and upper air conditions, are collected and broadcast on regular schedules and by special request. Broadcast schedules may be obtained by applying to the Navy Department or the United States Weather Bureau.

24.16. Forecasting from Local Indications. Since the approach of a low is associated with a lowering barometer, winds from certain directions, a sequence of cloud types, and an increase in the temperature and humidity, it is possible to form a fair idea of future conditions by noting these elements, particularly their combinations. When cirrus appear and are followed by cirrostratus clouds which tend to grow more dense, and the barometer is dropping with wind from an easterly quarter, it is reasonable to assume that a low is approaching from the west. This means that the wind velocity may increase, that lower clouds and possibly fog may appear, and that the possibility of rain is good. On the other hand, clearing conditions are imminent when the wind shifts to a westerly quarter, the pressure consistently rises, and the cloud system breaks. Local indications vary in different parts of this country and from place to place throughout the world. A west wind in winter might mean rain in the state of Washington, warm, dry conditions in eastern Montana, but cold weather at Chicago. In summer in the southwestern portion of the United States a south wind is dry, but in India it is the wet monsoon that yields copious rainfall. In early summer at Milwaukee, Wisconsin, on the west shore of Lake Michigan, a southwest wind blowing from the central plains states may raise the temperature to near 100°F; but the same wind will be cooled as it continues across the lake, and temperatures on the east shore are likely to be 10 to 20 degrees lower. By keeping a record of cloud, barometer, wind, and other conditions, fairly reliable rules may be formulated. Forecasting from local indications is limited because there is no way of knowing how large or intense the approaching storm may be. Also the nature of the approaching fronts and air masses is unknown.

The following guide developed by the Weather Bureau is helpful in foreseeing through local pressure and wind indications the approach of an extratropical cyclone in the United States:

"When the wind sets in from points between south and southeast and the barometer falls steadily, a storm is approaching from the west or northwest, and its center will pass near or north of the observer within 12 to 24 hours, with wind shifting to northwest by way of south and southwest. When the wind sets in from points between east and northeast and the barometer falls steadily, a storm is approaching from the south or southwest, and its center will pass near or to the south of the observer within 12 to 24 hours, with wind shifting to northwest by way of north. The rapidity of the storm's approach and its intensity will be indicated by the rate and the amount of the fall in the barometer."

24.17. Weather Proverbs. Many weather sayings and proverbs that have been handed down through the years conform remarkably well with modern meteorological science. Those that are based on observations of elements such as wind, clouds, and pressure often work out well in sections where the sayings originated and in similar climatic areas. It is obvious, however, that many proverbs, while true in one locality, would be wholly inapplicable in foreign lands with different weather characteristics. The behavior of animals, birds, and fish is significant with respect to imminent weather changes, i.e., for periods of about 1 to 12 hours ahead. Their habits, however, are not related to the nature of a coming season. Such conditions as the storing of nuts and the thickness of fur are products of past seasons rather than indications of the weather of future months. Some proverbs have to do with determining future weather on the basis of conditions that prevail on certain days or during certain months. These are generally without scientific basis but should, in some cases, not be entirely disregarded, as they may be in agreement with certain observed weather and climatic trends and patterns. Comments on a few proverbs that apply in many parts of the temperate zone, particularly the United States, are as follows:

Every wind has its weather.—Bacon. It will be recalled that easterly winds are associated with rain or snow, and heavy falls may occur with winds from the northeast; south winds are warm and humid; while those from the west bring fair weather. Even the strength of winds determines such elements as fog, clouds, and thunderstorms.

Mackerel scales and mare's-tails make lofty ships carry low sails. The scales and tails refer to cirrocumulus and cirrus clouds, respectively, which often precede storm areas with accompanying high winds that necessitate shortening sail.

Enough blue sky in the northwest to make a Scotchman a jacket is a sign of approaching clear weather. As storms move from west to east, clearing conditions in the west are significant.

Clouds flying against wind indicate rain. This refers to a condition such as is found ahead of warm fronts. Surface winds are often from the southeast, but altocumulus clouds in the overflowing warm current a mile or two above might well be moving from the southwest.

When the locks turn damp in the scalp house surely it will rain.—American Indians. The retreat of a high and the approach of a low is accompanied by an increase in the humidity.

The moon, if in house be, cloud it will, rain will come soon.—Zuñi Indians. The house means a halo around the moon. Halos form when cirrostratus clouds are present, and this cloud type may be succeeded by lower clouds and rain.

When it is evening, ye say, It will be fair weather: for the sky is red.—Matthew xvi, 2. At sunset the sun's rays reaching an observer have traveled diagonally through a considerably long path in the atmosphere, and even when little cloudiness and vapor are in the air, only the longer waves toward the red end of the solar spectrum succeed in getting through. When a storm area with its attendant high humidity and cloudiness is west of the observer—often out of sight because of the curvature of the earth—even the red rays are absorbed, and the sunset will be gray.

25

The Tropical Cyclone—Thunderstorms—Tornadoes

The Tropical Cyclone

25.1. Introduction. Tropical cyclones are the most destructive of all the storms of the world. They are not as large in area as the extra-tropical cyclones, or lows, of the temperate zones, but their winds are stronger and the rainfall which they bring is much more intense. These destructive storms of the tropics have caused the loss of many ships at sea. In some cases where cyclones have approached coast lines, the associated wind and tide conditions have destroyed cities never to be rebuilt. Upon moving inland many of the storms have caused extensive damage to cities and rural sections.

Ships at sea should make every effort to avoid a tropical cyclone. As we shall presently see, it is only occasionally that one would be encountered. In passing through such a storm, wind and sea are most certain to bring about some structural damage. Even a well-found ship in some cases may be in danger of foundering. Masts and superstructures are vulnerable because of the extreme violence of wind and sea. Personnel may be lost overboard or injured by objects adrift ordinarily considered secured. Lifeboats, airplanes, and other exposed objects are most certain to be carried away by the wind and sea. In view of all these unpleasant circumstances an astute seaman finds it worth while to study the nature of the tropical cyclone and to avoid it if possible. Many ships at sea have never encountered a tropical cyclone; if ordinary precautions are used, most ships should never have to pass through a violent one.

25.2. Areas Affected. Regions in which tropical cyclones form and move are shown in Fig. 25.1. Tracks are shown in the Pacific and Indian oceans both north and south of the equator and in the north Atlantic Ocean. It will be noted that in the Atlantic Ocean south of the equator, the south Atlantic, the tropical cyclone is entirely absent. These storms are unknown in the Arctic and Antarctic oceans.

Tropical cyclones invariably form over the ocean near but not at the equator. Once formed they may travel distances of hundreds or even thousands of miles before losing their force and finally dissipating. Ocean areas and, of course, islands in the paths of these storms are affected. Occasionally tropical cyclones move near to or cross over continental coastal areas, but South

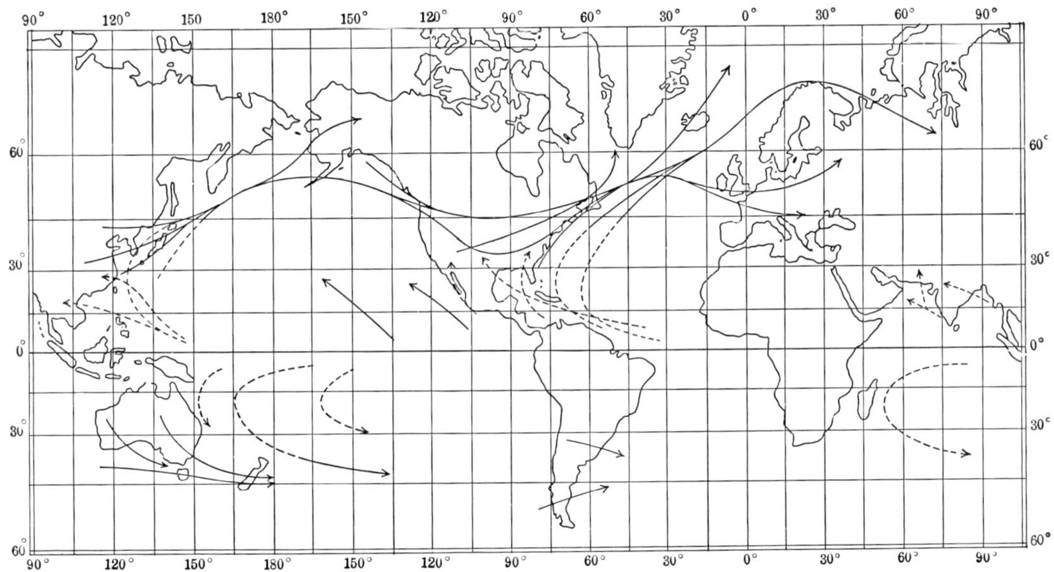

FIG. 25.1 PRINCIPAL STORM TRACKS OF THE WORLD. Solid lines, extra-tropical cyclones; dashed lines, tropical cyclones. Note that when tropical cyclones reach the higher latitudes in the North and South Pacific, and North Atlantic, they become extra-tropical cyclones as indicated by solid lines. (*From Pilot Chart of the Central American Waters*, October, 1946)

America, Europe, and Africa are free from such visitations. In North America the east and west coasts of Mexico and the Central American countries and the states of the Gulf and Atlantic coasts may be affected. China, Japan, India, and the northwestern and northeastern portions of Australia are other regions subject to the tropical cyclone. *Tropical cyclone* is the general or scientific term applied to this type of storm wherever found, but in the Atlantic and in the Pacific waters near Mexico, Central America, and Australia they are known as *hurricanes*. They are referred to as *baguios* in the Philippines, *typhoons* in China, and *cyclones* in India. It should be noted that the term *hurricane* has two distinct meanings. It refers to any wind reaching or exceeding 75 miles an hour (64 knots) and is the name given to tropical storms which originate in the areas adjacent to Mexico, Central America, and Australia. When used in the latter sense, it refers to the storm as a whole.

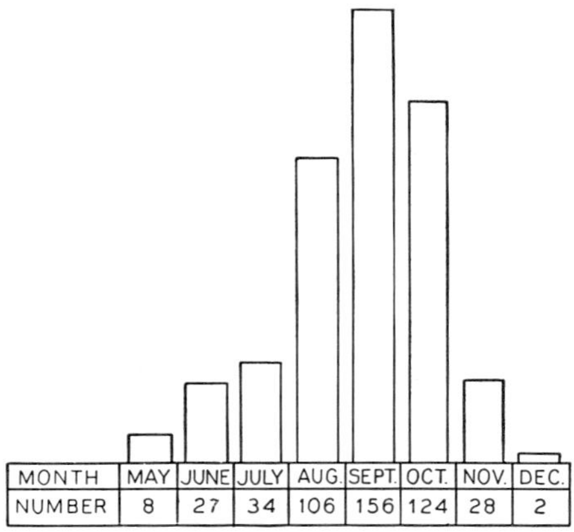

FIG. 25.2 The total number of tropical storms in the North Atlantic is shown by months for the 65-year period 1887–1951. The grand total was 485.

25.3. Frequency of Tropical Storms. Tropical storms in the Atlantic occur most frequently in August, September, and October (Fig. 25.2). Since 1951 a total of 46 tropical storms has been reported, 30 of which reached hurricane intensity, i.e., wind of 64 knots or greater. The distribution of occurrence of these storms is as follows:

1952–1957	*Jan.*	*May*	*June*	*July*	*Aug.*	*Sept.*	*Oct.*	*Nov.*	*Dec.*
Hurricane *			1		9	16	3	1	
Tropical Storm	1	2	1	3		8	2		1

* Winds of hurricane force, i.e., 64 knots.

It is interesting to note the occurrence of the tropical storm in January 1955. Although this storm did not reach hurricane intensity, it is noteworthy because heretofore a tropical storm in January had not been reported. The number of storms that occur each year is erratic (Fig. 25.3). Since 1951, the greatest number of storms to occur in any one year was 10 in 1955.

Tropical cyclones of the eastern North Pacific may be encountered from May through December off the western coasts of Mexico and Central America. Hurricanes of this region are as violent but not usually so large as those of the North Atlantic.

In the western North Pacific the tropical cyclone may occur during any month of the year, though the months of greatest frequency for this storm in that region are July, August, September, and October.

Tropical cyclones of the Bay of Bengal and the Arabian Sea are more likely to be encountered during May and October than in other months, whereas the season for tropical cyclones of the South Pacific and South Indian Ocean extends from September to May with the months of January, February, and March having the greatest frequencies.

25.4. Formation of the Hurricane. These storms form almost invariably in the region of the doldrums or equatorial belt of calms. This belt, migrating slightly north and south with the seasons, is farthest north during the month of September, the month of greatest hurricane frequency in the Atlantic. The belt is farthest south during March, but at no time during the year does it extend south of the equator in the Atlantic. These facts tend to show that the position of the doldrums with reference to the equator has a direct bearing on the possibility of hurricane genesis.

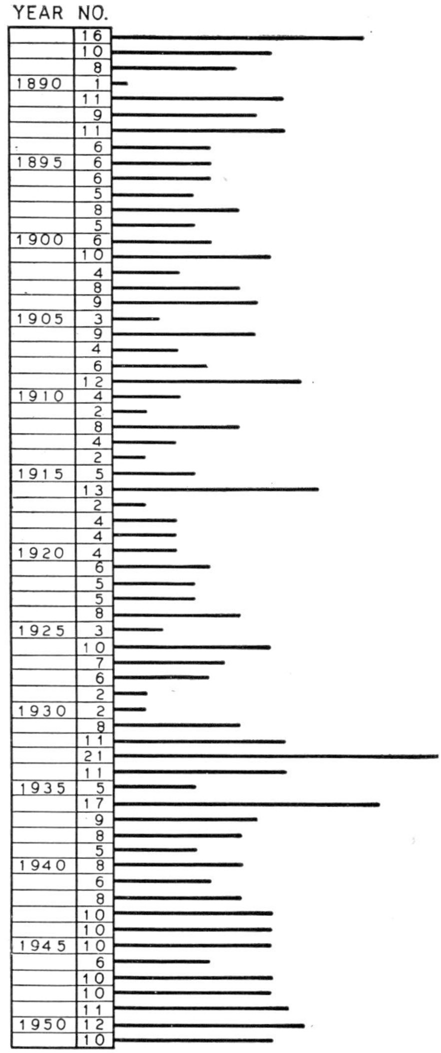

FIG. 25.3 This graph shows the total number of tropical storms by years which occurred in the North Atlantic for the 65-year period from 1887 to 1951. The grand total was 485.

There are never very many ship reports or upper air observations available in the large general area where a hurricane is first noticed and therefore it has been difficult to study the causes and early development of this type storm. Several theories have been offered as explanations, but none has been accepted as fully explaining the cause. Some believe the convectional theory best describes the cause, others rely on the triple point theory, and still others believe that hurricanes are formed on the Waves in the Easterlies. Since no agreement has been reached, no attempt will be made here to explain the theories mentioned.

Many tropical cyclones develop but die out before ever reaching very great strength. Furthermore it should be pointed out that only slightly more than one half of the storms which are classified as tropical cyclones have winds which ever attain hurricane force.

25.5. General Nature. The fully developed hurricane is circular or elliptical in shape. Although hurricanes vary in size, the area covered by them is roughly 300 miles in diameter. At the center is a small area about 10 to 20 miles in diameter known as the *Eye of the Storm.* Here the pressure is low, there is little or no wind, and the waves are confused and mountainous. The skies are sometimes clear or at least not completely overcast. Surrounding the calm center are found the strongest winds of the storm, which may reach a velocity of 125 knots. Gusts are likely to be higher in velocity. No one knows what the maximum wind speeds in tropical storms might be. Farther from the storm's center the winds are strong, but not as violent. At the outer edges of the storm the wind blows intermittently and is gentle to light, but gusty. Heavy, even torrential rain and clouds are present in all quadrants, and there may be lightning and thunder. In the northern hemisphere the wind blows counterclockwise around the storm and is inclined inward. The direction of wind is clockwise for storms of the southern hemisphere.

The average life of a tropical cyclone in the North Atlantic is about 10 days. Soon after its formation it begins to move at the rate of about 12 miles an hour, usually toward the northwest. Upon reaching the horse latitudes it tends to slow and it recurves toward the northeast. The records, however, show many exceptions, and many unusual tracks have been followed. There is then a progressive movement by the storm as a whole, as well as a circulation of the wind within the storm. After recurving, the storm usually becomes larger but less violent, and it speeds up along its path.

For those who have never experienced a hurricane at sea, its fury is difficult to imagine. Torrential rains are accompanied by deafening, violent winds which create tremendous waves whose tops are blown off into sheets of water and spray. At times it is difficult to distinguish ocean from air. As a hurricane approaches a coast line, storm tides often add to the destruction caused by the storm proper. Ships which may be present in harbors batter to pieces piers, bridges, and buildings along the waterfront. The rain driven by the hurricane winds and the waves undermine the foundations of buildings, and the wind and waves complete their destruction. Parts of the broken buildings may be picked up by the winds and, together with poles, uprooted trees, and other wreckage, act as battering rams to demolish other structures.

25.6. Hurricane Advisories and Storm Signals. During the hurricane season, the U.S. Navy, U.S. Weather Bureau, and the U.S. Air Force work together to provide the coastal areas of the United States and shipping interests with timely, tropical storm warnings. By joint agreement these agencies operate a hurricane warning center from which coordinated warnings are issued. Once the position and forecast are agreed upon, each agency disseminates the warning to its respective interests. The Weather Bureau is primarily concerned with civil interests and merchant shipping, the Navy is concerned with its coastal activities and ships at sea, and the Air Force has the responsibility of warning Army installations along with its own interests in the critical area.

The Navy and Air Force are assigned areas of reconnaissance responsibility and are frequently called upon to investigate areas of possible hurricane formation. Once a tropical storm is located, daily reconnaissance is made of its position and reports are radioed to the hurricane forecast center giving the exact location, size, speed of forward movement, wind information and other data. Thus the Atlantic tropical storms are closely watched and reported.

A similar arrangement is in effect in the Pacific; thus shipping interests and the military forces are constantly aware of any threat to life and property by destructive winds and seas emanating from tropical storms.

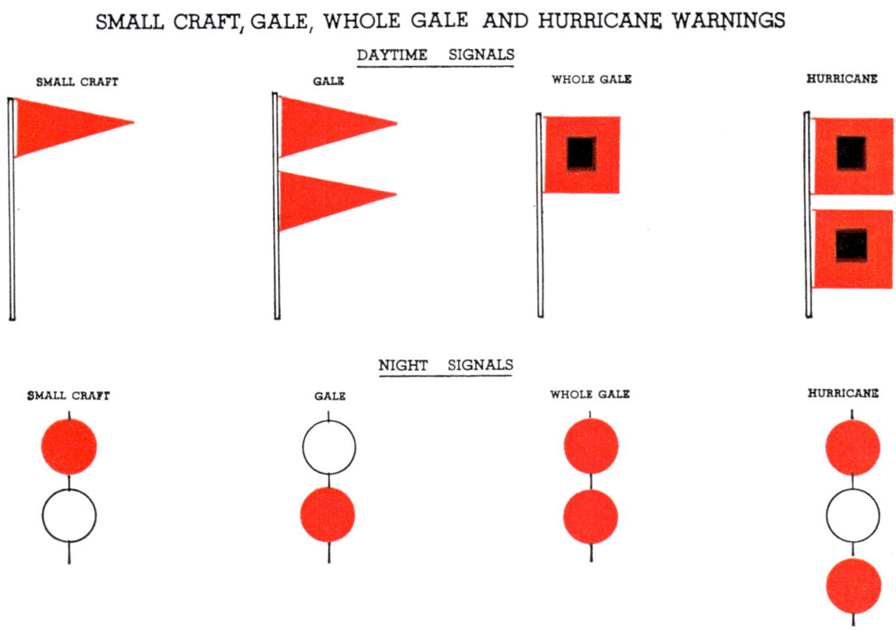

FIG. 25.4 STORM AND HURRICANE WIND DISPLAYS

The hurricane warning system in operation furnishes timely warnings to most interested parties. Its development was made possible by recent advances in tropical meteorology, reconnaissance techniques, suitable aircraft, and highly developed navigation and communication equipment.

In addition to hurricane and storm communiqués by radio, a system of flags and lights (Fig. 25.4) is displayed at many coastal points along the United States seacoasts when winds dangerous to navigation are forecast for any coastal section. An explanation of the various warnings follow:

Small Craft Warning: One red pennant displayed by day and a red light over a white light at night indicate that winds up to 38 miles an hour (33 knots) and/or sea conditions dangerous to small craft are forecast for the area.

Gale Warning: Two red pennants displayed by day and a white light above a red light at night indicate that winds ranging from 39 to 54 miles an hour (34 to 48 knots) are forecast for the area.

Whole Gale Warning: A single square red flag with a black center displayed during daytime and two red lights at night indicate that winds ranging from 55 to 73 miles an hour (48 to 63 knots) are forecast for the area.

Hurricane Warning: Two square red flags with black centers displayed by day and a white light between two red lights at night indicate that winds 74 miles an hour (64 knots) and above are forecast for the area.

25.7. Locating a Hurricane by Local Signs. Radio has made it less essential to depend upon one's own observations in locating and avoiding hurricanes. However, no good navigator would want to rely entirely upon radio reports but should understand how to locate by his own observations the direction and distance of these storms and how to avoid their violence.

First Indications. Hurricanes are usually preceded by a day having good visibility, atmospheric pressure somewhat higher than usual, and a change in the strength or direction of the usual breezes. Day temperatures are likely to be higher than normal, whereas night temperatures may be lower. Other early indications of the storm are the appearance of cirrus clouds and long period sea swells. A long period swell [1] in the Caribbean Sea or Gulf of Mexico during the summer or fall is a reliable indication of the existence of a tropical cyclone, and the bearing of the storm is the direction from which the swell moves. Outside those areas over the ocean proper the direction of the swell is not a direct indication of the bearing of the storm.

Convincing Signs. A drop in barometric pressure of 0.10 inch or more within a 3-hour period, exclusive of the usual daily change,[2] is significant of the approach of a hurricane. This rule does not hold outside the tropics because in higher latitudes storms of other types may similarly affect the barometer. With the approach of the storm the cirrus clouds are replaced by a veil of cirrostratus which often produce solar and lunar halos; brilliant ruby and crimson skies at sunrise and sunset are common. Presently a bank of clouds appears on the horizon in the direction of the storm. The wind then is light but gusty and blows from your left as you face the storm. The clouds become darker and gradually cover the entire sky. A heavy cross sea develops and, as the storm grows nearer, rain and wind increase, and the barometer consistently falls, though at times it is unsteady.

25.8. Buys Ballot's Law. About 1850 Buys Ballot, a Dutch physicist, stated an important principle with reference to the relation of wind and pressure in storm areas. It is in effect that in the northern hemisphere, if one stands with his back to the wind, the pressure is low to the left and high to the right. The

[1] Waves are caused by the winds of the storm, and their motion is faster than the progress of the storm itself. They therefore outrun the storm and give notice of its existence. The size of a wave depends on the force of the wind which causes it and the length of time and stretch of water over which it blows. After the wave outruns the storm which produced it, the wave decreases in height and increases in length and becomes a swell. The period of a swell is the time interval in seconds between the passage of successive swell crests.

[2] In the tropics atmospheric pressure readings have very pronounced regular daily rises and falls with two maximum and two minimum values within a 24-hour period.

reverse is true in the southern hemisphere. This law at once had wide practical application and was useful to sea captains in determining the directions of storms. Actually the direction of the lowest pressure might be anywhere from directly to the left to 30 or 40 degrees ahead of left.

25.9. The Law of Storms. The term *Law of Storms* is frequently encountered, particularly in the earlier writings on weather when investigators were making important definite discoveries regarding the relation of storm areas, pressure values, and wind directions. This law may be illustrated by referring

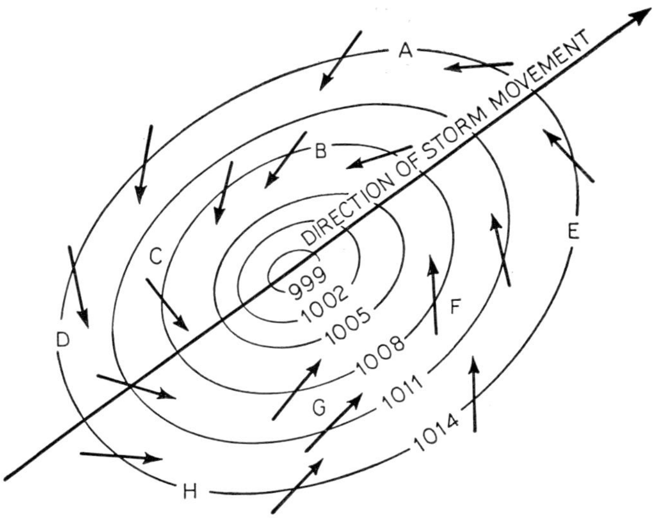

FIG. 25.5 The figure shows changes of wind direction, barometer, and direction of storm center when a storm passes to the south and to the north of an observer.

to Fig. 25.5. Imagine that you are at point *A* and that the storm is moving toward the northeast. On a slip of paper write the pressure values, wind direction, and direction of the storm center from you when you occupy the positions *B*, *C*, and *D*. Then imagine that you are at position *E*. Record the values of the pressure, wind direction, and direction of the storm center from you when you occupy the positions *F*, *G*, and *H*. In the first case, the storm on its journey from west to east passed south of you. In the second case, the storm center passed north of your location. As cloudiness, precipitation, temperature, and humidity vary within different portions of a storm area, it is readily apparent that the law of storms was at once important to an understanding of weather changes.

25.10. Dangerous and Navigable Semicircles. Place yourself on the storm track and face the direction toward which the storm is moving. That portion of the storm to your right is known as the dangerous semicircle, whereas that to your left is the navigable semicircle. Refer to Fig. 25.6. It is evident that a vessel in the dangerous semicircle would tend to be carried by wind and ocean currents into the path of the storm; a vessel in the navigable semicircle would most likely be forced to a safer position to the rear of the cyclone.

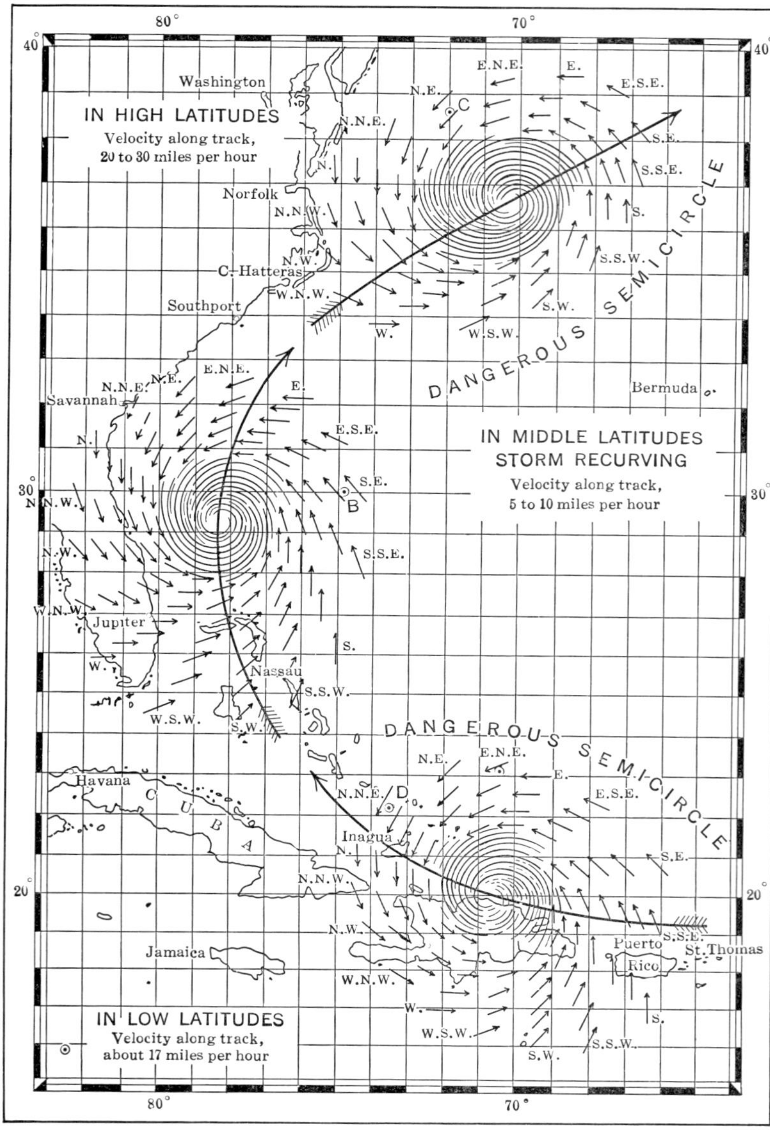

FIG. 25.6 CHARACTERISTIC PATH FOLLOWED BY TROPICAL CYCLONES OF THE NORTH ATLANTIC. (*From Pilot Chart of the Central American Waters, October, 1946*)

There is another reason why the right-hand semicircle is the most dangerous. It is that portion of the storm where the winds blow in the same general direction that the storm as a whole is moving. Therefore the force of the wind is increased by the rate of forward motion of the storm. In the navigable semi-

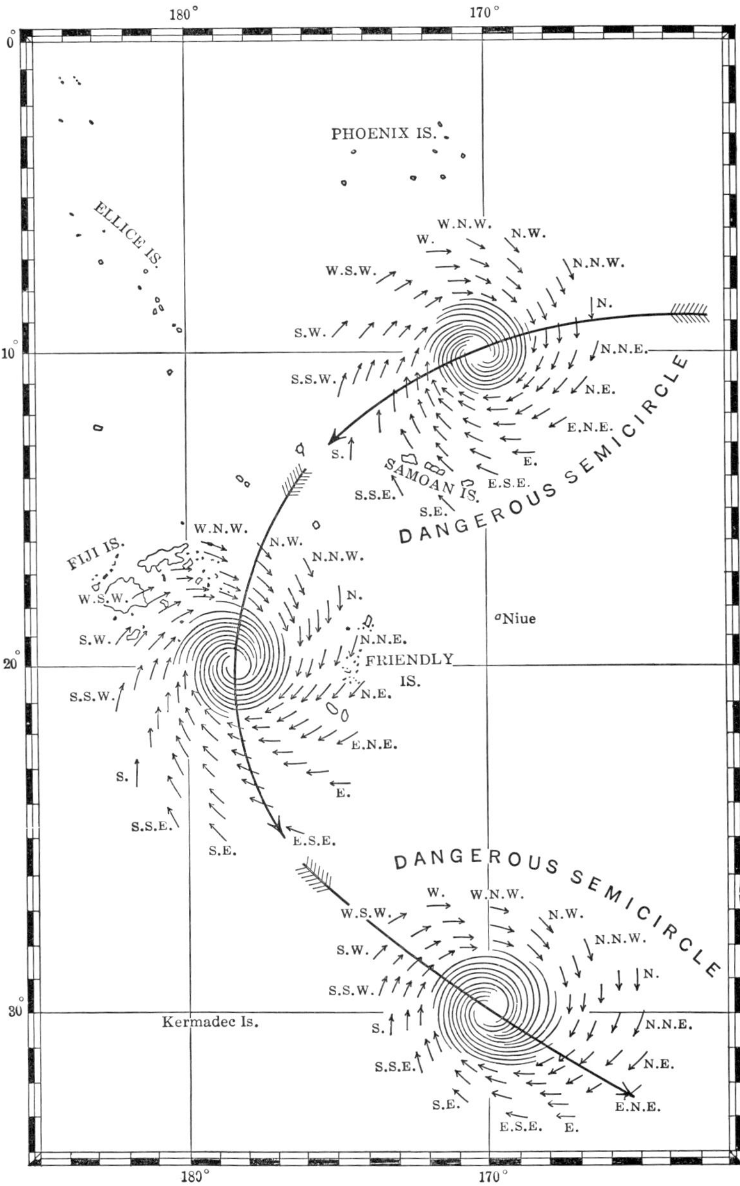

FIG. 25.7 CHARACTERISTIC PATH FOLLOWED BY TROPICAL CYCLONES OF THE SOUTH PACIFIC. *(From Pilot Chart of the Central American Waters, October, 1946)*

circle the wind blows in a direction roughly opposite to the direction of storm motion. It is apparent, then, that the relative motion between wind and water is greater in the dangerous semicircle than in the navigable semicircle.

25.11. Handling Ships in a Hurricane. The navigator is faced with three problems:

1. He must determine the bearing, distance, and the track along which the storm is moving.
2. The storm center and dangerous semicircle must be avoided, and escape from the dangerous semicircle must be made if the ship becomes involved in it.
3. The ship must ride out the storm in safety if unable to escape from it.

In the absence of radio information, it is well to heave to and note the shift in wind direction and the trend of the barometer. Buys Ballot's Law is useful in determining the direction of the storm center. The distance to the storm center may be only roughly estimated. Generally, the lower the barometer, the heavier the seas, and the stronger the wind, the closer will be the storm center.

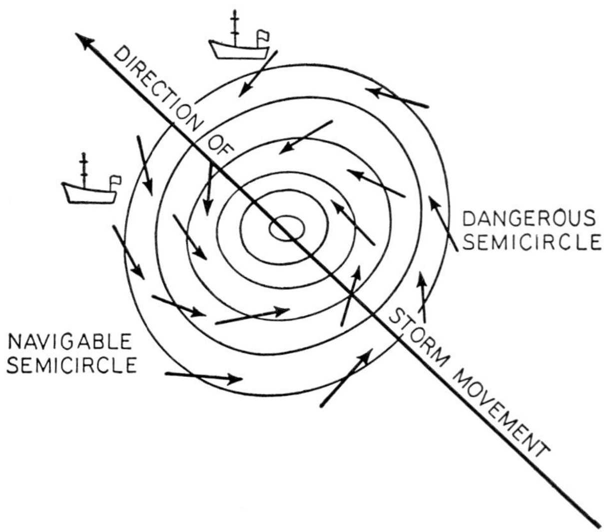

FIG. 25.8 A ship hove-to in the dangerous semicircle will note the wind to shift to the right (clockwise); the wind will shift to the left (counterclockwise) if the ship is in the navigable semicircle.

Refer to Fig. 25.8. In the northern hemisphere if the wind hauls to the right, the ship is in the dangerous semicircle; if it hauls to the left, it is in the navigable semicircle. If the wind continues from the same direction with a steadily falling barometer, the ship is likely to be in the direct path of the storm center.

25.12. Steamships (Northern Hemisphere). If the ship is believed to be in the direct path of the storm, and plenty of sea room is available, bring the wind on the starboard quarter, note the course, and run. In this way the distance from the storm center will be constantly increased. The wind will draw more forward, but the vessel will be in the navigable semicircle, making good her escape from the storm center. This rule also applies if it is believed that the vessel is anywhere in the navigable semicircle. For vessels caught in the dangerous semicircle bring the wind on the starboard bow and make as much headway as possible. If obliged to heave to, do so head to sea. If a particular vessel behaves better hove to with the sea astern or on the quarter, it is safe to do so in the navigable semicircle, but it should not be attempted in the dangerous semicircle because of the possibility of being drawn into the center of the storm.

25.13. Sailing Ships (Northern Hemisphere). While making a preliminary study of the storm, sailing ships should be hove to on the starboard tack. If the wind shifts to the right, the ship is in the dangerous semicircle but on the proper tack. It may then be attempted to work away from the track of the storm center, close-hauled on the starboard tack. If necessary to heave to, do so on the same tack.

If the wind hauls to the left while hove to on the starboard tack, the ship is in the navigable semicircle and heading away from the track of the storm center. In this case the wind should be brought on the starboard quarter and then run as long as possible. If it is necessary to heave to, do so on the port tack; try to make as little headway as possible.

If the wind direction remains the same while hove to on the starboard tack and the barometer falls steadily, it is likely that the ship is ahead and in the path of the hurricane. This is true if we assume that the storm is circular in shape rather than elliptical. The ship should run with the wind on the starboard quarter and hold the compass course thus noted until the barometer begins to rise.

25.14. Illustration of Rules for Sailing Vessels. If in a sailing vessel you find it necessary to heave to, you should always do it on whatever tack permits the shift of wind to draw aft. Figure 25.9 shows the rules for lying to for sailing vessels in the northern hemisphere. The storm is circular with the storm center ten points to the right of the wind direction. The storm is moving toward the north-northwest. The ship D has the wind at ENE; she is to the right of the track, or in the dangerous semicircle. The ship N has the wind at NNE and is in the navigable semicircle. As the storm progresses, these ships if lying to on the tacks indicated successively take the positions indicated by 1, 2, 3 and 4. It will be noted that the wind shifts to the left in the case of ship N, whereas it shifts to the right for ship D. In both cases the wind draws aft and diminishes the probability of either ship being struck aback with possible serious damage to spars and rigging. If the ships were lying to on the opposite tacks, the wind would tend to draw forward and expose the vessels to the possibility of being struck aback.

25.15. Storm Card. It is advisable to construct a storm card similar to Fig. 25.9 so that the position of the ship with relation to the storm may be readily visualized. This may be done on a thin sheet of paper or tracing cloth. The outline of a ship's hull may then be drawn on another sheet of paper and the storm card moved over it in the direction in which it is believed the storm is moving.

25.16. Proximity of Land—Anchorages—Latitude—Ship Characteristics. Adjacent land areas are always to be taken into consideration when arriving at decisions for maneuvering in a storm. If winds are expected to be off

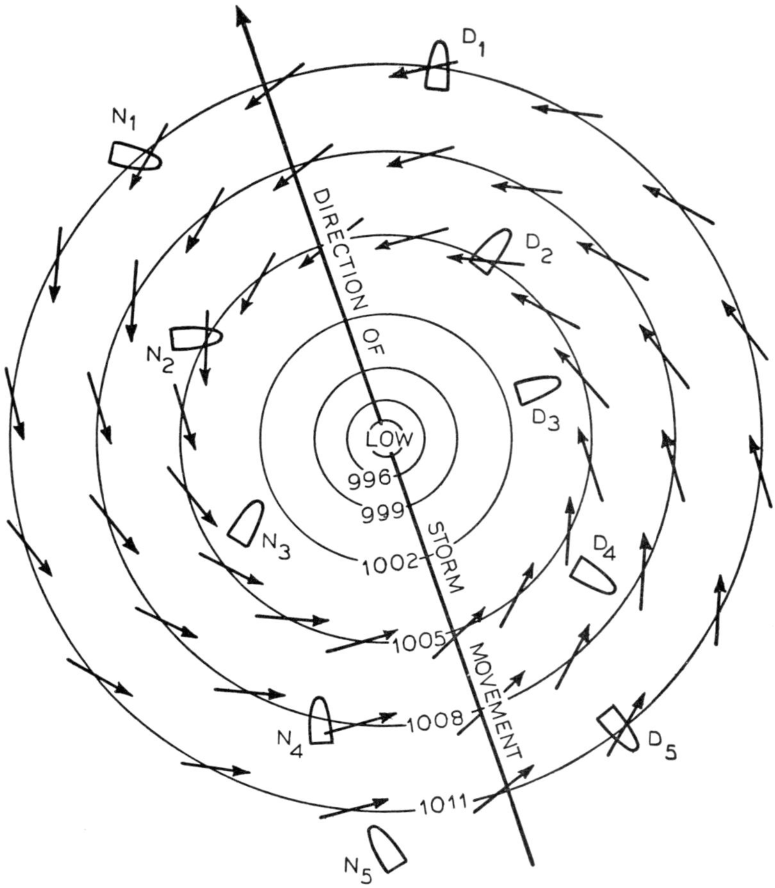

FIG. 25.9 This figure illustrates the rules for lying to by sailing ships in the northern hemisphere. Note that the wind draws aft in the cases of both vessels N and D.

shore, there is, of course, less danger of closing with a coast. Also it may be safe to anchor in an open roadstead if the wind shifts are certain to give a lee.

It should be kept in mind that hurricanes in the tropics and trade-wind belt often move toward the northwest and that they often recurve toward the northeast after reaching the horse latitudes. Consequently ships headed north have been known to escape from a storm only to encounter it again in a higher latitude.

The wind from a given direction in one latitude may mean that the ship is in the dangerous semicircle; in another latitude it may mean that the ship is in the navigable semicircle. The seaworthiness of the ship, her speed, and many other conditions are factors in the problem of maneuvering.

25.17. Other Sources of Information. After the storm reaches higher latitudes, it becomes larger in diameter but loses some of its violence. The problem then becomes one of handling a ship in rough weather rather than one of how to escape from the storm. Chapter 24 includes information applicable to ship-handling in storms of the higher latitudes. The Pilot Charts of the oceans issued by the Hydrographic Office of the Navy Department should be consulted. These are climatic charts of the oceans by months that contain a wealth of information concerning winds, weather, and ocean currents.

25.18. Summary of Maneuvering Rules—Northern Hemisphere. *Right or Dangerous Semicircles.* Steamers: Bring the wind on the starboard bow, make as much way as possible, and if obliged to heave to, do so head to sea. Sailing vessels: Keep close-hauled on the starboard tack, make as much way as possible, and if obliged to heave to, do so on the starboard tack.

Left or Navigable Semicircle. Steam and sailing vessels: Bring the wind on the starboard quarter, note the course, and hold it. If obliged to heave to, steamers may do so stern to sea; sailing vessels on the port tack.

On the Storm Track in Front of Center. Steam and sailing vessels: Bring the wind two points on the starboard quarter, note the course, hold it, and run for the left semicircle, and when in that semicircle maneuver as above.

On the Storm Track in Rear of Center. Avoid the center by the best practicable route with due regard to the tendency of cyclones to recurve to the northward and eastward.

25.19. Southern Hemisphere. *Left or Dangerous Semicircle.* Steamers: Bring the wind on the port bow, make as much way as possible, and, if obliged to heave to, do so head to sea. Sailing vessels: Keep close-hauled on the port tack, make as much way as possible, and, if obliged to heave to, do so on the port tack.

Right or Navigable Semicircle. Steam and sailing vessels: Bring the wind on the port quarter, note the course, and hold it. If obliged to heave to, steamers may do so stern to sea; sailing vessels, on the starboard tack.

On the Storm Track, in Front of Center. Steam and sailing vessels: Bring the wind two points on the port quarter, note the course, hold it, and run for the right semicircle, and when in that semicircle maneuver as above.

On the Storm Track, in Rear of Center. Avoid the center by the best practicable route, having due regard to the tendency of cyclones to recurve to the southward and eastward.

The above rules depend, of course, upon having sea room. In case land interferes, a vessel should heave to as recommended for the semicircle in which she finds herself.

25.20. Hurricanes of the Southern Hemisphere. The foregoing treatment of the hurricane has been confined for the most part to those storms which form and move in the northern hemisphere. The weather elements of hurricanes in the southern hemisphere are quite similar to those of the northern oceans, but there are several differences regarding wind, storm motion, and rules for maneuvering. South of the equator hurricanes at first move toward the southwest and then usually recurve and speed up toward the southeast. The winds around the storm center blow in a clockwise direction. Because of these differences ships must be maneuvered under rules different from those which apply in northern latitudes. The same principles apply, however, and the rules for maneuvering in southern latitude storms have been summarized.

The Thunderstorm

Thunderstorms are essentially intense local storms associated with the cumulonimbus cloud and are characterized by squalls, gustiness, turbulence, showers, thunder and lightning, and often hail. They are, therefore, of interest in the

FIG. 25.10 ICING OF AIRPLANE'S WING. *Official U.S. Navy Photograph*

operation of small surface craft and aircraft. The strong and gusty surface winds are of serious concern to surface craft and to aircraft on the ground; the strong vertical currents in and below the cumulonimbus cloud offer great trouble in flight. Upward and downward currents may be so great as to put dangerous strains and stresses on a plane's structure. Visibility is invariably poor in a thunderstorm, ceilings are low, and landings difficult to make. Airplanes may also be seriously damaged by hail; lightning may affect radio antennae, Pitot tubes, and other protruding equipment; icing of the plane's surfaces and carburetor is likely (Fig. 25.10).

Thunderstorms form in a number of ways, all of which may be classed under two general headings: the air mass and frontal types. In general, all methods of formation depend upon a volume of air being given an initial upward impulse which is sometimes spoken of as the *trigger action;* if the air is unstable, it will continue upward many thousands of feet often to the base of the stratosphere. This means, of course, that as long as the air continues to rise, it is finding itself surrounded by an environment colder than its own temperature.

25.21. Air Mass Thunderstorms. These form within an air mass and are not the result of frontal activity. They are commonly caused by the heating of the air at the earth's surface and may occur at any season in the tropics; in the middle latitudes they are confined mostly to the warmer portions of the year. They are of rare occurrence in the high latitudes. Air mass thunder-

FIG. 25.11 CUMULONIMBUS. (*USWB—F. Ellerman*)

storms over land are most often encountered in the mid-afternoon of hot, sultry days when wind velocities are light. If the wind is strong, particularly at high levels, the convectional air currents which would form the clouds are broken up. The heat and moisture furnish the energy needed to build up the great tops known as thunderheads. During a hot summer afternoon temperatures at an altitude of 4,000 feet are likely to be 20 degrees or more cooler than at the surface. Over the ocean thunderstorms are more likely to be encountered at night than during the day, which is the reverse of conditions over land areas. The reason is that the sea surface temperature remains almost as warm at night as it is during the day. However, temperatures at the upper levels over the ocean drop a considerable amount throughout the night because of radiation. This results in the same type of temperature contrast between the warm surface air and the cooler upper air as is found over the land during summer afternoons. Many persons who have gone to sea may recall clear days and evenings which have been followed by showers prior to daybreak on the following mornings.

While thunderstorms are most common during the warmer seasons over land, they occur at sea with greater frequency during colder months.

Another way in which air mass thunderstorms form is known as *orographic* (Fig. 25.11). Thunderstorms occur quite commonly on the windward slopes of mountains. The upward deflection of the air by the mountains furnishes the trigger action which may result in thunderstorm activity. Even in relatively flat country, such as the middle western states, thunderstorm activity may break out when air ascends only a few hundred feet in the vicinity of low rolling hills.

25.22. Cold-Front Thunderstorms. These may occur during the day or night over both land and water surfaces. A series of thunderstorms several hundred miles in length may form along the narrow banded cold front of an extra-tropical cyclone where the wedge of cold air furnishes the trigger action which results in the ascent of warm air ahead of the front. This type is confined to the storm tracks of the middle latitudes. In the southern portions of the United States they may occur at any time throughout the year, whereas in the northern states they occur only rarely in the winter. Cold-front thunderstorms are followed by definitely cooler weather because of the advance of the cooler air mass behind the front. Aviators often find it impractical to combat this type of thunderstorm activity because individual storms may be spaced so closely together along the front that it is impossible to find an opening to go through. Severe turbulence often exists upward to thousands of feet. When conditions warrant flying through a front, sufficient elevation should be gained and the passage made at right angles to the front to ensure the shortest possible time of exposure to the turbulence. In general, it is advisable to suspend flying; the line squall, as the cold front is sometimes called, soon passes, and flights may then go through safely and without the loss of much time. Thunderstorms also may occur along warm fronts.

25.23. Forecasting Thunderstorms. Meteorologists prepare graphs and charts based on pilot balloon and radiosonde data. These graphs, when considered with the surface map, reveal the places and times where thunderstorm activity is likely. The vast closely knit network of airport weather stations throughout this and other countries which report conditions hourly and oftener make it possible now to know where thunderstorms are in progress. The direction and rate of movement are also apparent by comparing the reports received.

In the absence of reports it is not practicable to attempt to forecast the approach of cold-front thunderstorms until the clouds and winds incident to the front manifest themselves. However, the air mass type which is due to surface heating may often be anticipated. One should become familiar with the altocumulus castellatus and mammato cumulus clouds (Figs. 25.12 and 25.13). At some localities in the tropics thunderstorms are virtually a daily occurrence during the middle or late afternoon, followed by cloudless evenings and nights. In middle latitudes they may be expected during the afternoon of a warm, sultry day when winds are light and when cumulus clouds build to great heights during the morning hours. On such days the sky toward the west should be watched,

FIG. 25.12 ALTOCUMULUS CASTELLATUS. (*USWB—L. E. Johnson*)

FIG. 25.13 MAMMATO CUMULUS. (*USWB—Harold Photo Studio, Sioux Falls*)

because thunderstorms in middle latitudes move in general from west to east. The rate of movement in the United States is from 30 to 40 miles per hour; in Europe, from 20 to 30. Prior to the appearance of a thunderhead there is often a gentle southerly wind. With the appearance of the thunderheads the surface wind tends to die away or to blow directly toward the storm. When one views a cumulonimbus cloud from the side, the top is seen to be spread out with the upper leading portion resembling an anvil; the parts of the cloud at high levels are frozen. There are also cirrus clouds present. As the anvil moves overhead, the boiling, churning squall cloud directly below can be seen. Beyond the squall cloud may be seen the dark rain cloud with its curtain of heavy showers. The surface wind next shifts to one which is strong and gusty from the west, and lightning and thunder occur. Precipitation begins in the form of heavy showers and often hail (Fig. 25.14). Large hailstones indicate very strong vertical currents. As the storm continues its journey eastward, the wind dies down but continues from the west, and the rain diminishes in intensity. Presently it becomes lighter to the west, and the clouds break through. The duration of the storm depends on its size and rate of movement. With the retreat of the storm the skies clear, the wind settles back to the south, and the thunderheads and lightning may be observed to the east.

FIG. 25.14 The size of these hailstones may be compared with a baseball. (*USWB—C. L. Gillespie*)

Tornadoes

Although the tornado is the least extensive of all storm types, it is the most violent and sharply defined. Tornadoes may occur in other parts of the world but are most common in the United States, particularly in the plains and southern areas. They average only a few hundred yards in diameter and move along a path averaging about 25 miles in length. In the northern hemisphere the wind of the tornado whirls counterclockwise, and it is possible that in many cases velocities exceed 300 miles per hour. The tornado may be recognized by its funnel-shaped cloud (Fig. 25.15) which builds downward from a cumulonimbus cloud. This funnel may or may not reach the ground and does little damage unless it does touch the ground. The pressure within a tornado is very low, and a building literally explodes when struck. The explanation is that the normal pressure within a building pushes the walls outward toward the region of the tornado's low pressure. The violent winds then complete the structure's destruc-

FIG. 25.15 THE FUNNEL-SHAPED CLOUD OF A TORNADO. *Official U.S. Navy Photograph*

FIG. 25.16 WATERSPOUT OFF GUAM. *Official U.S. Navy Photograph*

tion and scatter parts in every direction. The cloud mass is one of condensed vapor, though it also contains much dust and debris picked up along its journey.

Tornadoes are most common in spring and early summer when air masses from the Gulf of Mexico which are quite warm encounter masses from the northern part of the continent which are still cold at that season. At that time of year the temperature contrast is greatest between masses of air of southern and northern origin. Regions along and just ahead of surface cold fronts at that time favor tornado development. In general, tornadoes move toward the northeast at a rate of from 20 to 40 miles per hour. They may therefore be avoided easily.

25.24. Waterspouts. Tornadoes at sea are known as waterspouts; they are less violent than those on land. When they touch the sea surface, spray is drawn up into the funnel. They are common off the east coast of North America, in the Gulf of Mexico, and off the coast of Japan and China. The winds in some of the less violent waterspouts rotate clockwise in the northern hemisphere. This type more nearly resembles the common dust whirlwind of land than it does the tornado and is often encountered off the west coasts of Africa, Mexico, and Central America (Fig. 25.16).

APPENDICES

APPENDIX

1

Rope and Cordage

Ropes in common use are of two general classifications—fiber rope and wire rope.

A1.1. Fiber Rope. The following raw materials are used in the manufacture of fiber rope: manila, hemp, cotton, linen, sisal, henequen, jute, and man-made fibers, such as nylon (Saran, Orlon, Dynel, Dacron), and other similar synthetics. The ropes commonly used on shipboard are:

1. *Manila, Made from Abacá.* Manila rope has almost completely replaced hemp for general purposes on shipboard. It is made from the fiber of the abacá plant. The abacá, which resembles the banana plant, is not a hemp as is commonly believed. It is a perennial plant, growing to a height of from 10 to 20 feet and found principally in the Philippine Islands, Central America, Sumatra, and Borneo.

When ready for harvesting, the matured abacá plants are felled and the stalks are then cut into thin slabs called "tuxies." Fiber extraction is then done by stripping the "tuxies" with a knife. Complete mechanical decortication methods are now being used for this, although, previously, all stripping was done by hand. The perfection and strength of the finished rope depend upon the thorough cleaning of the fiber during the stripping process. When the fiber is cleaned, it is spread out and thoroughly dried by sun and air. The fiber is then carefully sorted, classified and baled, and is ready for the rope factory.

Taken from the bales, the fibers are combed smooth and straight by machines. At this stage, an emulsion of oils is added to soften and lubricate the fibers. From 10 to 15 percent by weight of the finished rope consists of oil, making for preservation against both excessive dampness and excessive heat.

The combed fibers are then spun into twisted yarns, wound on bobbins. In this connection it should be noted that in the U.S. Naval Service the terms "thread" and "yarn" are used interchangeably. The yarns are then twisted into strands by means of the bobbins and capstans. Three or four strands, each strand composed of a number of yarns twisted together, go to make the finished rope. This is accomplished in the laying machine. To counteract the tendency to unlay, the successive twists are taken in opposite directions, yarns being usually twisted to the right.

Rope is commonly made with three strands, but four are sometimes used.

Also left-laid rope may be produced by reversing the direction of the twists. For special uses three ropes are sometimes twisted into a cable to provide greater elasticity.

2. *Hemp, Made from True Hemp.* Hemp rope is made of fibers from the stalk of the hemp plant which is cultivated extensively in many parts of the world, but especially in Italy, Russia, Central Europe, Chile, and the United States. American Hemp, grown in Wisconsin, Illinois, and Kentucky, is generally used for hemp cordage manufactured in the United States.

Hemp rope is now seldom used, but when used it is almost invariably tarred. In this form it was used on shipboard for standing rigging which is now made of wire. The tar preserves the rope from deterioration due to dampness but reduces its strength and flexibility. In present United States Naval practice tarred hemp is used only for "small stuff": ratline, marline, houseline, spun yarn, roundline, and seizing stuff.

3. *Cotton, Made from Cotton.* Cotton ropes are used principally for lead lines, taffrail log lines, and signal halyards.

4. *Linen, Made from Flax.* Linen rope is now used mostly for yachting purposes as sheets and halyards. Formerly signal halyards also were made from the unfinished flax twine.

5. *Sisal, Henequen, and Jute.* Other fiber ropes, such as sisal, henequen, and jute are used on shipboard and for other marine uses in times of great fiber shortage.

Sisal is made from the leaves of the *Agave sisalana* plant, grown chiefly in Haiti, Africa, and Java. Its most common use now is in agriculture as baler twine for hay baling. With the supply of manila hemp cut off in wartime, it finds increasing use for rope and cordage. It is also used largely for the cores or centers of wire rope. Sisal resembles manila in appearance and has about 80 percent of its tensile strength.

Henequen, erroneously called Mexican sisal, is made from the *Agave fourcroydes* plant grown chiefly in Yucatan and Cuba, in the latter case being known as Cuban sisal. Its principal use is for binder twine and agricultural and hardware ropes. It is generally of coarser texture than true sisal.

Jute rope is seldom used for marine purposes, although in times of fiber shortage it may be substituted for "small stuff," marlines, spun yarn, etc. It is not as strong as manila or sisal and does not withstand the rotting action of water.

6. *Synthetic Rope* (Fig. A1.1). Nylon is a synthetic fiber made from mineral products. The Bureau of Ships, Navy Department, has authorized the use of nylon rope for mooring lines for submarines and certain surface ships, and nylon is being introduced aboard merchant vessels. Its tensile strength is approximately twice that of manila. Nylon lines are lighter, more flexible, less bulky, and easier to handle and stow. They resist rot, decay, and marine fungus growth. Nylon's inherent properties provide the ability to stretch, absorb shocks, and resume normal length when strain is removed. Nylon is no cure-all for the hazards involved in any line-handling situation. Dependence upon its additional

strength may cause the seaman to ignore ordinary precautions. Nylon lines will part, as will manila or wire rope, when they are weakened, subjected to greater strains than they can withstand, or when given improper handling. Nylon's terrific backlash can cause serious injury or death.

The use of nylon lines presents hazards unique to nylon's physical properties. This is particularly true under conditions conducive to heavy strain, such as

FIG. A1.1 EIGHT-INCH NYLON HAWSER FLEMISHED DOWN. *Official U.S. Coast Guard Photograph*

while under way with headway while mooring or maneuvering into locks. When nylon is stretched more than 40 percent, it is likely to part. The stretch is immediately recovered with a snapback that will sound like a pistol shot. The snapback can also be as deadly as a bullet. It is therefore imperative that no one stand in the direct line of pull when a heavy strain is applied. This is also true for other types of lines, but overconfidence in nylon's strength may lead one to underestimate its sting.

A1.2. Fiber Rope. Fiber rope is designated as to size by its circumference and runs from about 5/8 inch to 16 inches and even more. However, the largest sizes are never seen on shipboard, 12 inches being about the maximum that even the largest ship would carry. The length is measured in fathoms; for shipment, it is made up in coils. (Fig. A1.2.)

(A) Manila Fiber, Baled

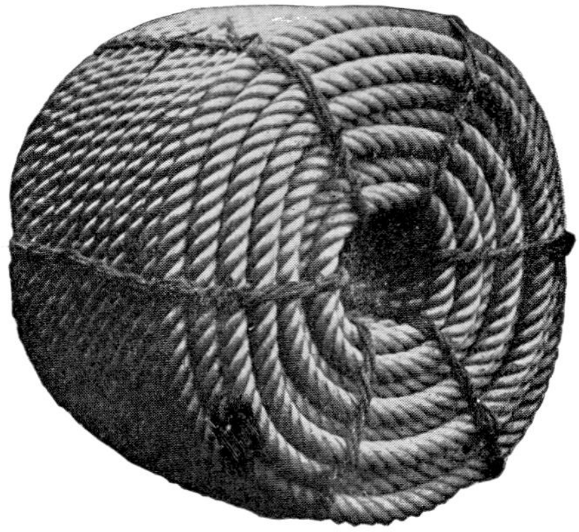

(B) Rope in Coil

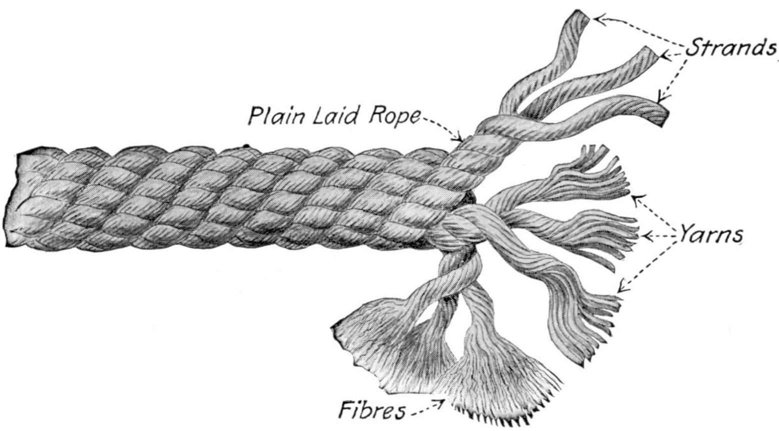

(C) Cable Laid Rope

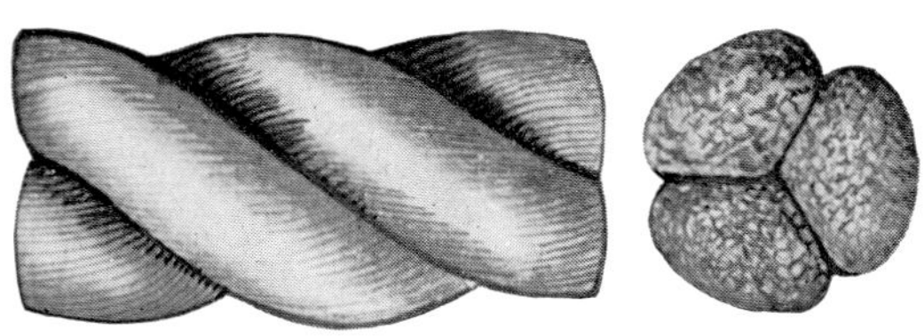

(D) Cross Section, 3-Strand Rope

FIG. A1.2 MANILA ROPE

A1.3. Care of Fiber Rope. Unlike metal, fiber has not a permanent elastic limit within which it can be worked indefinitely. Therefore, no attempt should be made to put a maximum strain on a rope which has seen continuous service under a moderate strain or on one which has once been close to the breaking point. The safety of a rope decreases comparatively rapidly with use, dependent to some extent upon the amount of strain. This is due to the fact that the fibers slip a small amount under each strain in spite of the twisting.

Rope tends to contract when wet, and, unless allowed to do so freely, may be injuriously strained. It is for this reason that running gear is slacked in damp weather. On the other hand, advantage may be taken of this tendency for tautening lashings, etc., by wetting the rope.

A1.4. Coiling, Uncoiling, and Stowage. *How to Coil and Uncoil Properly.* Right-hand rope should be coiled clockwise. Lay out the rope (perhaps a boat fall) along the deck. Begin coiling it down, close to where it is made fast. Turn the coil over. It is now ready for use. To uncoil, reach down and pull the end of line up through the inside of the coil. The line should uncoil counterclockwise as it pays out.

A1.5. What Not to Do. 1. Don't put strains on kinked lines with buckled strands and don't pull the kink through a block. Coax the strands back into place.

2. Don't drag the line around and get it full of grit and sand.

3. Don't let wear become localized in one spot. Use chafing gear on the line, reverse the line end for end, or cut off the end so that wear is at a different spot.

4. Don't let a weak or damaged spot ruin the whole line. Cut it out and splice the line together again.

5. Don't let the lay of a line become unbalanced by continued use on a winch in the same direction. Reverse the turns periodically, and keep the kinks out.

6. Don't use chain stoppers—use rope stoppers.

7. Don't let the line become fouled in machinery and gears.

8. Don't surge lines on running capstans.

9. Don't put sudden strains on the lines. Surge smoothly.

10. Don't let lines tighten up when wet. Slack off wet lines and halyards.

11. Don't permit rat guards and sharp edges to wear mooring lines. Use chafing gear. Lash well.

12. Don't use the wrong size block for falls. As an approximate rule, blocks should be three times the circumference and sheaves twice the circumference of the fall.

13. Don't neglect blocks. Inspect and lubricate blocks frequently. Repair or replace them when necessary.

14. Don't lubricate lines. They already are properly lubricated, and over-lubrication can cause loss of strength and difficulty in handling on capstan.

15. Don't throw away worn-out lines. Turn in large, surveyed manila lines for salvage of the sound inner yarns. Use ends and yarns from old lines for

lashings and whipping new lines and, finally, turn in all not used for scrap and salvage.

A1.6. Wire Rope. Manufacturers commonly designate the size of wire rope by the diameter, and this designation is used in the U.S. Naval specifications. Wire runs from about ¼ to 2¾ inches in diameter.

Manufacture of Wire Rope. Wire ropes are made almost exclusively from steel. The first requisite in the manufacture of steel wire is the selection and blending of the different iron ores which determine the chemical composition of the steel ingots. After selecting the proper steel (iron, cast steel, extra-strong cast steel, plow steel, or high-grade plow steel) from the hearth furnace, the large ingots are reheated and rolled to billets about 4 by 4 inches, cut in small lengths, again reheated and rolled to small bars, and again into small rods of ¼ to ½ inch in diameter. Next, the wire is drawn cold, through dies, to the desired size. The action of drawing has the effect of hardening the wire and making it brittle, so that annealing must be done, at intervals, to soften it again for further drawing. During wire drawing, various lubricants, such as oil, tallow, and soapy water, are employed to facilitate the operation.

The fundamental unit in wire rope construction is the strand. Strands are formed by winding individual wires in a spiral direction around a core. This is done by a stranding machine. A vast number of geometrical combinations of wires is possible, but, for ordinary work, the practice is to use one wire in the center of the strand, surrounding this with a layer of 6 wires, then successively with layers of 12 and 18 wires. The addition of one layer of 6 wires around a center will produce a strand for a haulage rope. A supplementary layer of 12 wires makes a 19-wire strand for a hoisting rope, etc. (Fig. A1.3.)

The finished strands are wound upon large bobbins and twisted into rope by means of a rope-closing machine. The standard wire rope is made of 6 wire strands and a fiber core. This arrangement affords the most convenient and compact form, as the strands and the core are practically all of the same size.

The use of fiber for the center of the rope not only contributes to flexibility, but has the further advantage that the fiber forms a cushion upon which the strands close in as the rope contracts under heavy pull, thus acting with the elasticity of the wire and the "give" which results from the spiral lay, to reduce the effect of sudden stresses. When the fiber center has absorbed its share of the lubricant which should be used upon the rope from time to time, this cushioning will result in a lubrication of the interior wires and greatly reduce the interior friction.

Each wire, in the wire rope, is carefully tested and inspected throughout its length. This thorough inspection and the careful methods of manufacture are largely responsible for the reputation of "reliability" which has been given universally to wire rope.

The purposes for which the types of rope are applicable are as follows:

Material. The strength of wire rope increases with the strength of the material from which it is made. The strength increases in the following order: Phosphor bronze, cast steel, extra-strong cast steel, high-grade plow steel. The weight and the diameter of a wire rope for a particular purpose can be reduced by using a stronger material, as, for example, by using a plow steel instead of an extra-

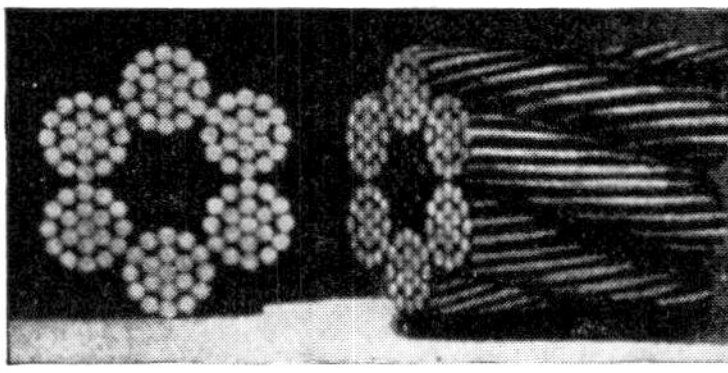

Strands of 19 Wires (6 × 19)

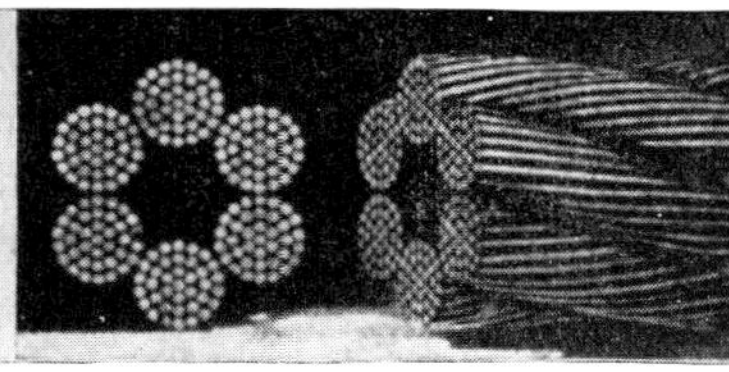

6 Strands of 37 Wires (6 × 37)

Strands of 12 Wires (6 × 12)

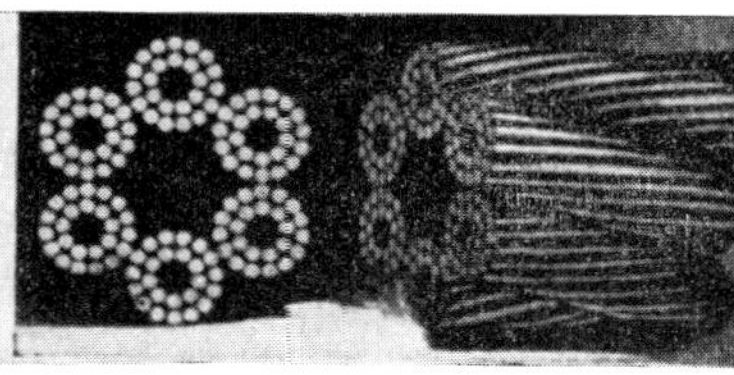

6 Strands of 24 Wires (6 × 24)

Wire Rope Unlaid, Showing Hemp Core

FIG. A1.3 TYPES OF WIRE ROPE

strong cast-steel rope; or by using a stronger material, the strength and, therefore, the safe working load can be increased for a given weight and diameter.

Galvanized-Wire Rope. Galvanized-wire rope should be used if the rope is likely to corrode because of the presence of moisture, as for the standing rigging of a ship. Because the zinc coating is rapidly removed by wear, it should not, in general, be used for hoisting. It may, however, be used for the running rigging and for wheel (steering) ropes on ships, as these ropes do not wear so rapidly.

Uncoated-Wire Rope. Uncoated-wire rope should be used where it is protected from moisture, as in a building, and for more or less continuous hoisting. It may be used instead of galvanized-wire rope, where it is exposed to moisture, as for derrick guys, if a protective coating is applied to the rope at regular intervals.

Phosphor-Bronze Wire Rope. Phosphor-bronze wire rope has lower strength than steel-wire rope; therefore the working loads should be lower. The sheaves should also be larger than those for steel rope. It is nonmagnetic and can be used for conditions under which galvanized steel rope does not give satisfaction. Because of these properties it is used on small vessels.

6 by 19 Ungalvanized Steel Wire Rope. This rope is principally used for heavy hoisting and is of very great strength, particularly useful on derricks and dredges. It is the stiffest and strongest construction of the types listed, which is suitable for general hoisting purposes. Sheaves for this type of rope should be larger, if possible, than those used for the other more flexible types, to obtain the best results. This rope may be applied to a great variety of uses on elevators, derricks, dredges, cableways, coal hoists, conveyers, and similar equipment.

6 by 19 Galvanized Steel Wire Rope. Galvanized rope of this construction may be used for standing rigging, guys, boat slings, topping lifts for coaling booms, running rigging 7⁄16 inch and under, and wheel rope 7⁄16 inch and under. This construction, both ungalvanized and galvanized, is furnished in two grades of steel: "extra strong cast steel" and "high-grade plow steel," of which the latter is the stronger.

6 by 19 Phosphor-Bronze Wire Rope. This rope may be used for lifelines, clearing lines, wheel ropes, or rigging, mainly on submarines and destroyers, where either noncorrosive or nonmagnetic properties are desired.

6 by 37 Ungalvanized Steel Wire Rope. This is a very flexible rope, suitable for cranes and similar machinery where sheaves are, of necessity, smaller than desirable. It may be used for heavy hoisting, especially where conditions are unusually severe. Hoisting ropes larger than 1¾-inch diameter are usually of this type, as are also ropes for clamshell buckets on colliers. Its wires are smaller than in the 6-strand, 19-wire rope, and consequently will not stand as much abrasive wear. It is a very efficient rope, because a little over 50 percent of the wires, and consequently over 50 percent of the strength, are in the inner layers of the strand, protected from abrasion. This explains its particular advantage in addition to its flexibility.

6 by 37 Galvanized Steel Wire Rope. Galvanized rope of this construction may be used for steering-gear transmission rope, boat crane falls (afloat and ashore), hawsers, where great strength is required, relieving tackle, towing hawsers, bridles (large and small), tiller ropes, tiller ropes on ships' boats, cat and fish pendants, clear hawse pendants, dip ropes, torpedo slings, slings for general hoisting, heavy running rigging, and wheel ropes over 7⁄16 inch. This construction, both ungalvanized and galvanized, is furnished in two grades of steel: "extra-strong cast steel" and "high-grade plow steel," of which the latter is stronger.

6 by 12 Galvanized Steel Wire Rope. This construction is more flexible than either the 6 by 19 or 6 by 37 construction, but is not as strong. This construction may be used for lifelines, guys, ridge ropes, and boat ladders, Jacob's ladders, gun port lanyards, boom pendants, and running rigging. It is desired for running rigging service where extreme flexibility is required and exposure to moisture is frequent. This rope is also used as a wire mooring line and, when so used, shall have a large eye splice at one end and a thimble in the opposite end to attach to reel. The large eye splice shall be 7 feet long, and the smaller one shall be 10 inches, with a thimble. This construction is furnished in two grades of steel: "cast steel" and "high-grade plow steel," of which the latter is the stronger.

6 by 12 Phosphor-Bronze Wire Rope. This rope may be used for lifelines, wheel ropes, or rigging, mainly on submarines and destroyers, where either noncorrosive or nonmagnetic properties are required.

6 by 24 Galvanized Steel Wire Rope. This construction has almost the same flexibility as the 6 by 12 construction, but is stronger. It is used primarily in the larger sizes, where the strength of a 6 by 12 rope of same size will not be satisfactory and where extreme flexibility is the major consideration.

6 by 7 Galvanized Steel Wire Rope. This construction is the stiffest of the types listed herein. This construction is not suitable for general hoisting, but is applicable mainly for permanent standing guys. This construction is also applicable for haulage and gravity systems, coal-haulage systems, operating grip cars, drilling, and, in general, where abrasion is severe and flexibility a minimum requisite.

Marline-Covered Wire Rope, 5 by 19. Marline-covered wire rope is stronger and more durable than manila rope. The marline covering prevents wearing of the wires and supplies lubricant to them. As the marline wears to a smooth surface, the rope is easily handled or laid in a flat coil. Compared with uncovered wire rope, the marline-covered rope is more easily handled, has greater friction, which is an advantage if it is used on a smooth drum, and is very durable, particularly if it is exposed to gases, grit, or moisture.

Fiber Cores. Cotton may be spun into cores of small diameter more satisfactorily than other fibers. It is therefore required for the cores of small wire ropes. For wire rope of large diameter, hard fibers (manila, sisal, or java, African, Mexican, or Yucatan) are used. Jute is not used for the cores of any wire rope. Compared with the hard fibers, jute has less cellulose for its volume and is therefore weaker mechanically; it is more hydroscopic and therefore deteriorates more rapidly. In the event of war, jute would probably be unobtainable.

Seizing. The end of a wire rope should be seized to prevent unlaying, which, if it occurs, makes the rope useless. Annealed iron wire should be wound tightly in a close helix around the rope. The seizings may be replaced by fittings if they prevent unlaying of the rope.

Care of Wire Rope. Wire rope needs in some ways better care than hemp or manila and far better care than it generally receives on shipboard. It should be kept on a reel when not in use. A single kink in the finest wire rope practically ruins it at once.

In receiving a line and transferring it from one reel to another, care should be taken to unreel it, instead of slipping off the successive bights over the end of the reel, as is sometimes done.

Fig. A1.4 illustrates the right way and the wrong way of dealing with wire rope under various conditions.

A wire hawser should be gone over thoroughly every month or two with linseed oil; or better, if it is not to be used for some time, with crude petroleum or some heavy lubricating oil to which a small quantity of graphite has been added. If the lubricant used is inclined to be stiff, it is better used hot. Neither oils nor tars can be regarded as perfect preservatives, as they all contain more or less acid, which attacks the steel. The principal reliance must therefore be upon the galvanizing. If the rope is to remain under water for some time, the best preservative is made by adding, to one part of Stockholm tar, one part fresh slaked lime. Boil well and use while hot for saturating the rope. After using wire rope towing hawsers in salt water, the wire rope should be washed off with fresh water and then gone over with linseed oil.

It is important that the lubricant, whatever its composition, should be thin enough to penetrate into interstices of the rope and yet that it should have consistency enough to adhere to the wire for a reasonable length of time, after which it should be renewed. Care must be taken to ensure covering the rope all around. A hawser used for towing should be relubricated after use while being reeled up.

Wherever wire rope is to be worked over a sheave, the diameter of the sheave and the speed of running become very important factors.

The larger the sheave and the lower the speed, the better. All manufacturers of wire rope prescribe a minimum diameter for sheaves, and their guaranteed breaking strains and estimated safe-working loads are for these minimum diameters and for moderate speed. A high speed increases the wear upon the rope, not only by the friction on the sheaves but still more by the friction of the wires upon each other—a point which is often overlooked.

The importance of this interior friction will be realized if we consider the "play" which necessarily goes on between the fibers of a rope which is being alternately bent and straightened in running over a sheave. This play, of course, increases with the speed and is greater with a small sheave than with a large one. The same consideration enters in where a rope is alternately stretched and relaxed under a straight but varying pull, as, for example, in towing.

The diameter of the sheave over which the rope is worked should never be less than twenty times that of the rope itself; and the less flexible the rope, the larger should be the sheave.

It is important that the score of the sheave should be of such size as to carry

Right Way

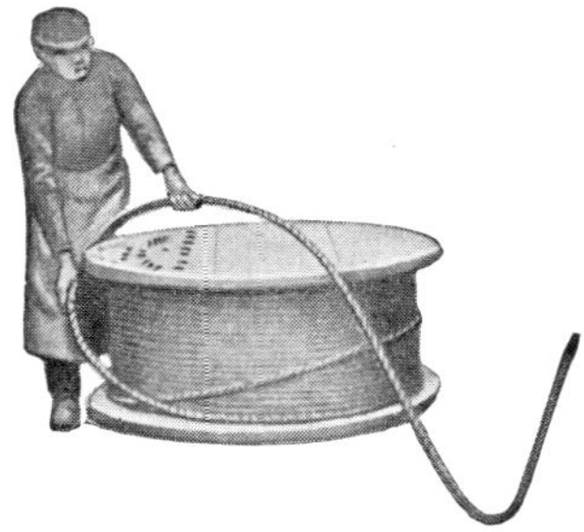

Wrong Way

To Take Rope From Reel

Right Way

Wrong Way

To Take Rope From Coil

An Overworked Rope

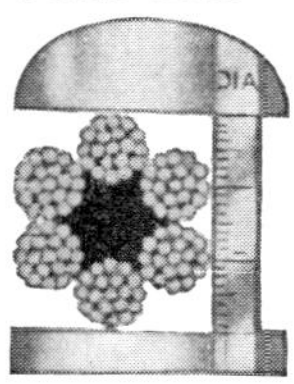

Right Way

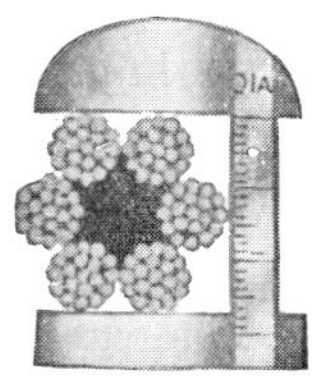

Wrong Way

To Measure Diameter

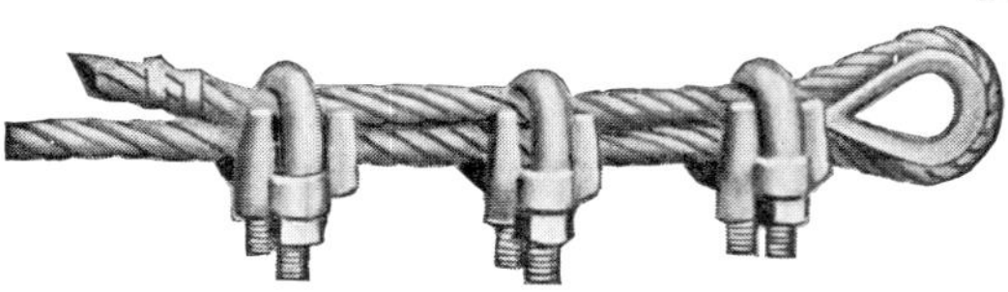

Right Way (U-bolt on Dead End)

Wrong Way (U-bolt on Tension End)

Wire Rope Clips

FIG. A1.4 HANDLING WIRE ROPE

the rope without excessive play and, above all, without friction against the sides of the score. Metal sheaves are not required to be lined with wood or leather.

The turns of the rope should never be allowed to overlap on the drum of the winch.

As the wear of the rope over a sheave increases more rapidly with the speed than with the load, it is better, when an increased output is demanded, to increase the load rather than the speed.

In addition to the question of friction in running over a sheave, the distortion of the rope wherever it passes around a relatively sharp bend, whether on a moving sheave or a stationary chock or bollard, is a factor of great importance. Those fibers which lie farthest from the center of the curve are stretched, while those which hug the round of the bend are more or less compressed. Thus the outer fibers give way before the inner ones begin to feel the strain.

By far the most unfavorable conditions to which wire rope can be subjected are those wherein it runs over sheaves that give it a reverse bend like the letter S; hence it passes over one sheave with a bend to the right and immediately afterwards swings around another with a bend to the left.

While the strength, lightness, and durability of wire rope are important factors in its favor, its most valuable characteristic, as compared with hemp, manila, and chain, is its reliability. Within proper working limits, it almost never fails. Hemp and manila may be rotten at the core and show no sign, or they may have been weakened by excessive strains and give no indication of it except, perhaps, that they are a little "long-jawed." A chain may be made of worthless material, or, if of the best material and made with every care, it may have flaws which no inspection can reveal.

Hemp, sisal, and manila ropes are made up of a great number of fibers from a few inches to several feet in length. Wire rope is made of small numbers of wires of the full length of the rope, each of which is manufactured, inspected, and tested individually throughout its full length before it goes into the rope. The inspection is so simple that a flaw can hardly be overlooked, and it is most improbable that any number of wires can have flaws which in the end appear at the same point of a strand. Thus it is almost impossible that any serious flaw should exist in a wire rope as manufactured.

A flaw due to kinking can always be seen.

An accident with wire rope is almost necessarily due to carelessness.

Assuming the rope to have a well-lubricated hemp core and to be used over properly proportioned sheaves, the outside strands will be the first to wear out, and the reduction in their diameter becomes the measure of the wear of the rope as a whole.

Wire rope should be condemned when the outside wires are worn down to one half their original diameter, or when it is apparent from broken wires or other abnormal indications that it has been subjected to danger by excessive strains, or to a sharp bend resulting in a pronounced kink.

When wire rope is to be cut, a whipping of small soft (annealed) iron wire should be wrapped on each side of the point where the cut is to be made, to prevent the rope from unlaying. If suitable annealed wire is not available, a fairly heavy seizing stuff may be used, the turns being passed very taut.

A1.7. Guide to Using Nylon: *Maintenance.*

1. Nylon rope will hold a load even though a considerable number of the yarns are abraided. Ordinarily, when abrasion is localized, the rope may be made satisfactory for reuse by cutting away the chafed section and splicing the ends. Chafing and stretching do not necessarily indicate the load-carrying ability of nylon rope.

2. Splice nylon rope as you would manila rope except that tape instead of seizing stuff should be used for whipping the strands and rope. Also nylon rope, because of its smoothness and elasticity, requires at least one extra tuck over that for manila rope. For heavy load applications such as towing, take an additional backtuck with each strand.

3. Should nylon rope become iced over, thaw it carefully at moderate temperature and drain before stowing.

4. Should nylon rope become slippery because of the accumulation of oil or grease, scrub it down. Isolated spots may be removed by the use of light burning oils.

5. *General Use:* Do not uncoil new nylon rope by pulling the end up through the eye of the coil. Unreel it as you would wire rope.

6. New cable-laid nylon hawsers tend to be stiff and difficult to handle. To alleviate this condition, tension the cables for 20 minutes at 30 percent extension (100 feet when tensioned would measure 130 feet).

7. When the stretch of nylon becomes excessive, double up the lines by passing the bight, thereby halving the elongation under load. This reduces the hazard of snapback, since the rope will usually part near the eye. For drydocking and other close control work, stretch can be reduced to one half by doubling the lines.

8. When new cable-laid nylon hawsers are strained, sharp cracking noises will be heard. The noises are associated with readjustment of the rope strands in the stretched cable. Under normal safe-working loads, the rope will stretch one third of its length.

9. Wet nylon hawsers under strain emit steamlike water vapor. This phenomenon is normal under safe-working loads.

10. Nylon rope can withstand repeated stretching with no serious effect. When under load it thins out; but when free of tension, it returns to its normal size. The critical point of loading is 40 percent extension, i.e., a 10-foot length would stretch to 14 feet when under load. Should the stretch exceed 40 percent, the rope is in danger of parting.

11. When sets of ropes are to be used in parallel, as are boatfalls, do not pair nylon rope with low elongation rope, such as wire of manila.

12. Use nylon rope stopper for holding nylon hawsers under load. Do not use manila or chain.

13. When handling nylon rope without a powered reel, avoid coiling it in the same direction all the time, since this will tend to unbalance the lay.

14. Bitts, chocks, and other holding devices used with nylon rope should have smooth surfaces to reduce abrasion and minimize surging of nylon ropes under working conditions. Use chafing gear where there are sharp metal edges. During reeling or heaving-in operations, take care that thimbles and connecting links do not chafe or cut the nylon hawsers.

15. Since, normally, plain-laid nylon rope is right-laid, coil it on bitts, capstans, or reels in a clockwise direction.

16. Do not use wire or spring lay rope on the same chock or bitt with nylon rope.

17. Plain-laid nylon hawsers tend to elongate around bitts when loaded. To minimize excessive lengthening, take a turn under the horn and cross the line on itself before taking more turns.

18. When nylon hawsers are used on capstans for heavy towing or impact loading, take six turns on the capstan and two turns overlaying the last four turns. This procedure reduces the hazard of sudden surges on rendering cut.

19. For mooring purposes with low freeboard vessels where the tide differential is average, make up at half tide. No further handling should be required.

20. Nylon rope under heavy strain may develop glazed areas where it has worked against bitt and chock surfaces. This condition may be caused by the removal of paint from metal surfaces or the fusing of nylon fibers. In either case, the effect on the rope strength is negligible.

Alongside Towing:

21. Make up forward and backing tow lines as close as possible without regard to sharp bends.

22. Take up slack in relaxed line while the other line is under heavy load.

23. When easing pull, the tug may have to reverse engines slightly to counteract the elastic property of nylon and thus avoid "snapback" action.

Precautions:

24. Nylon rope on parting is stretched 50 percent. The stretch is recovered instantaneously with resulting snapback. In view of this, it is imperative that no one stand in direct line of pull when heavy loads are applied.

25. Do not use a single part of plain-laid rope for hauling or hoisting any load that is free to rotate. If one part of rope is essential, use cable-laid nylon hawsers.

26. Do not stow nylon rope in strong sunlight for long periods. Cover it with tarpaulins. During stowage, keep it away from heat and strong chemicals.

27. Be extremely careful when easing out nylon rope around bitts and cleats under heavy load. Because its coefficient of friction is lower than that of manila, the nylon rope may slip when eased out and cause injury to personnel unfamiliar with its oddities.

28. For control in easing out, take two or three round turns on the bitt before figure-eighting the line. Use of the round turns provides a means for

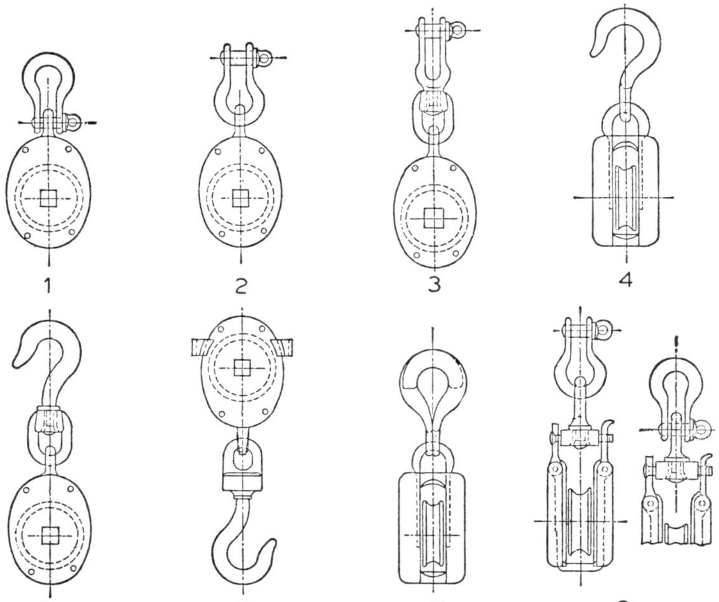

BLOCKS—FOR MANILA ROPE

1—Shackle, Regular, Front; Loose.
2—Shackle, Upset, Front; Loose.
3—Shackle, Upset, Swivel Link; Loose.
4—Hook, Single, Side; Loose.
5—Hook, Single, Swivel Link; Loose.
6—Hook, Single, Ball Bearing Swivel Link; Loose.
7—Hooks, Sister, Side; Loose.
8—Shackle, Upset, or Regular Swivel Eye; Loose.

GENERAL NOTES

Size of Block Refers to Approximate Length of Cheek.
Mortise is the Clear Distance between Cheeks.
Swallow is the Opening over the Sheave Through which the Rope Passes.
Breech is the End of the Block Opposite to the Swallow.
Becket is the Fitting which Takes the Standing Part of the Rope, the Thimble is that Part of the Becket over which the Rope Eye is Spliced.
A Fitting the Outer End of which is in the Plane of the Sheave is a Front Fitting; One at Right Angles to the Plane of the Sheave is a Side Fitting.
An Upset Shackle is One the Bow of which Passes Through the Strap Bail or Intermediate Fitting whereas in the Regular Shackle the Shackle Pin Passes Through the Adjacent Part.
When Ordering Blocks Observe the Following.
Procedure:
1st. State Whether Wood or Metal.
2nd. State Size, also Whether Single, Double, Etc.
3rd. State the Rig No. as shown on Buships Standard Plans.
4th. State Whether with or without Becket.
Example:
Block, Wood, 8 Inch Double, Rig No. 4 with Becket.
Details of Fittings for Both Standard and Non-Standard Rigs Appear on Standard Plans of Wood or Metal Block Rigs for the Respective Sizes.

FIG. A1.5

closer control in easing out or surging. Always stand well clear of the bitts during these operations.

Life Expectancy: Nylon rope properly handled and maintained should remain serviceable more than five times longer than manila rope subjected to the same use. Adherence to the foregoing instructions combined with the usual safe practices followed for manila rope will give all the advantages of nylon rope plus savings in cordage allowances.

Harmful Chemicals: The chemicals listed here have a permanent effect on nylon yarn. Action takes place at various temperatures and concentrations according to the chemical involved.

Concentrated formic acid
Benzyl alcohol (at a boil)
Phenol
Cresols
Xylenols
Chlorinated phenols
Concentrated nitric acid
Concentrated sulfuric acid
Calcium chloride (in methanol)
Calcium chloride (in glacial acetic acid; ethylene chlorohydrin; ethylene glycol)
Zinc chloride in methanol
Concentrated hydrochloric acid

Fiber Ropes, Kinds, Sizes, Strength

Table A1.1—Strength Rating, Type Designations, and Fiber Content

Type	Strength rating (percent)	Type designation marker tape		Fiber
		Consecutive printing on tape	Color of printing	
M.......	100	Type M—Manila; type M—Manila	Red	Abaca (Musa textilis)
S........	80	Type S—Sisal; type S—Sisal	Purple	Agave sisalana
SX......	75	Type SX—Sisal mixed; type SX—Sisal mixed	Blue	Agave sisalana and not more than 20 percent by weight of agave fourcroydes and other hard fiber
SH......	70	Type SH—Sisal hemp; type SH—Sisal hemp	Green	Agave sisalana blended with canabis sativa
A........	60	Type A—Agave; type A—Agave	Yellow	Agave fourcroydes (Henequen) and other hard fiber
J........	60	Type J—Jute; type J—Jute	Orange	Corchorus capsularis and/or corchorus olitorius
CMS.....	90	Type CMS—Composite; type CMS—Composite	Maroon	Outer yarns manila; inner yarns sisal
	*	....	....	....

* See Table A1.2.

TABLE A1.2—MINIMUM STRENGTH (POUNDS)

Nominal size (inches)		Manila, type M (100 percent)	Substitute for Manila *				
Circumference	Diameter		Type CMS (90 percent)	Type S (80 percent)	Type SX (75 percent)	Type SH (70 percent)	Type A or J (60 percent)
5/8	3/16	450	......	360	340	310	270
3/4	1/4	600	......	480	450	420	360
1	5/16	1,000	......	800	750	700	600
1 1/8	3/8	1,350	......	1,080	1,010	950	810
1 1/4	7/16	1,750	......	1,400	1,310	1,230	1,050
1 1/2	1/2	2,650	......	2,120	1,990	1,850	1,590
1 3/4	9/16	3,450	......	2,760	2,590	2,410	2,070
2	5/8	4,400	......	3,520	3,300	3,080	2,640
2 1/4	3/4	5,400	......	4,320	4,050	3,780	3,240
2 1/2	13/16	6,500	......	5,200	4,880	4,550	3,900
2 3/4	15/16	7,700	......	6,160	5,780	5,390	4,620
3	1	9,000	......	7,200	6,750	6,300	5,400
3 1/4	1 1/16	10,500	......	8,400	7,870	7,350	6,300
3 1/2	1 1/8	12,000	......	9,600	9,000	8,400	7,200
3 3/4	1 1/4	13,500	......	10,800	10,120	9,450	8,100
4	1 5/16	15,000	......	12,000	11,250	10,500	9,000
4 1/2	1 1/2	18,500	16,600	14,800	13,900	12,950	11,100
5	1 5/8	22,500	20,300	18,000	16,900	15,800	13,500
5 1/2	1 3/4	26,500	23,800	21,200	19,900	18,500	15,900
6	2	31,000	27,900	24,800	23,200	21,700	18,600
7	2 1/4	41,000	36,900	32,800	30,800	28,700	......
8	2 5/8	52,000	46,800	41,600	39,000	36,400	......
9	3	64,000	57,500	51,200	48,000	44,800	......
10	3 1/4	77,000	69,300	61,600	57,800	53,900	......
11	3 5/8	91,000	81,900	72,800	68,200	63,700	......
12	4	105,000	94,500	84,000	78,800	73,500	......

* See Table A1.1.

TABLE A1.3—COIL LENGTH AND COIL WEIGHT

Nominal size (circumference) (inches)	Minimum length (feet)	Approximate coil weight (gross) (pounds)
5/8	3,000	45
3/4	2,750	55
1	2,250	65
1 1/8	1,620	66
1 1/4	1,200	63
1 1/2	1,200	90
1 3/4	1,200	125
2	1,200	160
2 1/4	1,200	200
2 1/2	1,200	234
2 3/4	1,200	270
3	1,200	324
3 1/4	1,200	375
3 1/2	1,200	432
3 3/4	1,200	502
4	900	432
4 1/2	900	540
5	900	670
5 1/2	900	805
6	600	645
7	600	875
8	600	1,145
9	600	1,450
10	600	1,795
11	600	2,200
12	600	2,610

TABLE A1.4—RATLINE STUFF AND SIMILAR MATERIAL MADE UP ON SPECIAL ORDERS OF TARRED AMERICAN HEMP

Name and circumference	Weight per fathom (pound)	Length of coil (fms)	Weight of coil (pounds)	Breaking strain (pounds)
6-thread, 3/4-inch	0.18	100	18	650
9-thread, 1-inch	0.25	100	25	1,000
12-thread, 1 1/8-inch	0.35	100	35	1,400
15-thread, 1 1/4-inch	0.45	100	45	1,800
18-thread, 1 3/8-inch	0.53	100	53	2,100
21-thread, 1 1/2-inch	0.60	100	60	2,400

TABLE A1.5—SEIZING STUFF—TARRED AMERICAN HEMP

Use: For serving manila and wire where material more pliable than ratline is required.

Name and circumference	Weight per fathom (pound)	Length of coil (fms)	Breaking strain (pounds)	Remarks
4-thread, $\frac{1}{2}$-inch.....	0.095	100	365	5 lengths to package
6-thread, $\frac{5}{8}$-inch.....	0.12	100	560	4 lengths to package
9-thread, $\frac{7}{8}$-inch......	0.165	100	700	3 lengths to package
12-thread, 1-inch.....	0.21	100	955	3 lengths to package

Halyards should be of the following types:

Type I. Cotton thread.
Type II. Cotton twine.
Type III. Flax.

Halyards should conform to the requirements of Table A1.6.

TABLE A1.6—PHYSICAL PROPERTIES

	Type I		Type II		Type III	
Nominal circumference (inches).....	$\frac{3}{4}$	1	$\frac{3}{4}$	1	$\frac{3}{4}$	1
Minimum feet per pound...........	34	23	35	23	38	25
Minimum strength (pounds)........	500	750	450	650	700	1,000

SYNTHETIC ROPES, CHARACTERISTICS

1. *Materials.* The material used in the rope should be bright, virgin, continuous-filament nylon fiber of at least six-denier size having at least 6.5 grams per denier strength. The nylon should be a long chain polymer made of hexamethylene diamine and adipic acid. The following are government specifications for synthetic rope.

2. *Construction.* The ropes shall be of three strands to conform to the requirements specified herein. Each strand shall be made of one size of balanced plied nylon yarns. The plied yarns shall be made of three single nylon yarns. The single yarns shall be made from grouped filaments conforming to the sizes and twists specified in Table A1.7 for the respective size ropes. Heat setting of the rope or any of its twisted components will not be permitted. The lay of the rope shall be right-hand or "Z" lay.

3. *Physical Properties.* The physical properties of the finished rope shall be as specified in Table A1.8. The elongation at the breaking point shall not exceed 55 percent. The load-elongation curve, drawn autographically, shall not

TABLE A1.7—CONSTRUCTION

Rope size (circumference)	Turns per foot minimum	Denier of single yarns
Inches		
5/8 to 1 1/2, incl.	22	2,500 to 8,000
1 3/4 to 2 1/2, incl.	22	7,500 to 8,000
2 3/4 to 3 incl.	18	10,000 to 12,000
3 1/2 to 6 1/2, incl.	15	15,000 to 16,000
7 to 10, incl.	15	15,000 minimum

TABLE A1.8—PHYSICAL PROPERTIES

Circumference at load P (inches)	Tolerance plus or minus (inch)	Approximate diameter (inches)	Load P ($200 \times D^2$) (pounds)	Feet per pound (at load P) minimum (feet)	Hardness		Breaking strength minimum (pounds)
					Minimum (pounds)	Maximum (pounds)	
5/8	1/16	3/16	7	100.0	5	25	1,000
3/4	1/16	1/4	12	66.0	5	25	1,500
1	1/16	5/16	20	36.0	5	25	2,500
1 1/8	1/16	3/8	28	28.5	5	25	3,000
1 3/8	1/16	7/16	38	20.0	5	25	4,500
1 1/2	1/16	1/2	50	16.5	5	25	5,500
1 3/4	1/16	9/16	65	12.5	5	25	7,000
2	1/16	5/8	80	9.7	5	25	8,400
2 1/4	1/16	3/4	110	7.2	5	25	11,500
2 1/2	1/16	13/16	130	6.2	5	25	14,000
2 3/4	1/16	15/16	175	5.0	5	25	16,000
3	1/8	1	200	4.1	20	100	20,000
3 1/2	1/8	1 1/8	250	3.0	20	100	26,000
4	1/8	1 5/16	345	2.3	20	100	34,000
4 1/2	1/8	1 1/2	450	1.8	20	100	42,000
5	1/8	1 5/8	530	1.5	20	100	52,000
5 1/2	1/8	1 3/4	610	1.25	20	100	62,000
6	1/8	2	800	1.00	20	100	74,000
6 1/2	1/8	2 1/8	900	0.90	20	100	82,000
7	1/4	2 1/4	1,000	0.71	20	100	100,000
8	1/4	3 5/8	1,400	0.55	20	100	125,000
9	1/4	3	1,800	0.43	20	100	155,000
10	1/4	3 1/4	2,100	0.34	20	100	185,000

exhibit evidence of changes in load applications greater than 5 percent of the mean weighed load due to strand readjustments. Changes due to splice slippage shall not be considered in this determination.

4. *Hardness.* The rope lay shall be sufficiently tight to meet the minimum hardness value when tested as received, but shall not exceed the maximum value specified in Table A1.8 when tested after being immersed in tap water for 16 hours and drained for 2 hours.

5. *Identification.* The manufacturer shall identify his product by inserting a kraft paper or water repellent cotton marker within one strand and completely enveloped by the cover yarns of that strand in all ropes larger than 1½ inches circumference. Unless otherwise specified in the contract or order, the manufacturer's name, year of manufacture and type of fiber (NYLON) shall be printed on the marker in bold, easily read type which shall be suitably permanent upon exposure to water or mineral oil.

6. *Finish.* No extraneous substance shall be added for the purpose of weighing the rope. Finishing materials will be permitted up to 3.0 percent of the weight of the rope. The finishing materials shall be resistant to aging and shall not cause rope strength loss greater than 10 percent when tested.

7. *Spliceability.* The finished ropes shall be spliceable and shall not develop yarn displacement or strand cockles in the splicing test.

8. *Lengths on Reels.* The ropes shall be furnished on nonreturnable wooden reels not larger than 6 feet in diameter in the amounts specified in Table A1.9.

TABLE A1.9—NET LENGTH AND WEIGHT ON REELS

Circumference (inches)	Minimum length (feet)	Approximate weight (pounds)
1	2,250	65
1⅛	1,620	59
1⅜	1,200	64
1½	1,200	78
1¾	1,200	103
2	1,200	133
2¼	1,200	179
2½	1,200	214
2¾	1,200	262
3	1,200	320
3½	1,200	430
4	600	280
4½	600	355
5	600	425
5½	600	510
6	600	640
6½	600	720
7	600	900
8	600	1,200
9	600	1,500
10	600	1,900

Unless otherwise specified, broken lengths will be permitted but no length shall be shorter than 600 feet when measured in the relaxed condition. For ropes of ⅝- and ¾-inch circumference, the reels shall contain a minimum of 30 pounds of rope.

Wire Rope, Kinds, Sizes, Strength

Types, Navy Rope, Wire. Wire rope shall be of the following types, as specified:

A. 6 by 19 wire rope, extra-strong cast steel, uncoated or galvanized, fiber core.

B. 6 by 19 wire rope, high-grade plow steel, uncoated or galvanized, fiber core.

C. 6 by 19 wire rope, phosphor-bronze, fiber core.

D. 6 by 37 wire rope, extra-strong cast steel, uncoated or galvanized, fiber core.

E. 6 by 37 wire rope, high-grade plow steel, uncoated or galvanized, fiber core.

F. 6 by 12 wire rope, cast steel, galvanized, fiber core.

G. 6 by 12 wire rope, high-grade plow steel, galvanized, fiber core.

H. 6 by 12 wire rope, phosphor-bronze, fiber core.

I. 6 by 24 wire rope, cast steel, galvanized, fiber core.

J. 6 by 24 wire rope, high-grade plow steel, galvanized, fiber core.

K. 6 by 7 wire rope, extra-strong cast steel, galvanized, fiber core.

L. 5 by 19 marline-covered wire rope, extra-strong cast steel, uncoated, fiber core.

M. 1 by 7 seizing strand, galvanized.

N. 1 by 19 seizing strand, galvanized.

O. Special types.

TABLE A1.10—TYPE A, 6 BY 19 WIRE ROPE, EXTRA-STRONG CAST STEEL, UNCOATED OR GALVANIZED, FIBER CORE

Nominal diameter, minimum diameter (inches)		Approximate circumference (inches)	Approximate weight per foot (pounds)	Approximate weight per fathom (6 feet) (pounds)	Uncoated		Galvanized	
					Nominal breaking strength (tons—2,000 pounds)	Minimum breaking strength (pounds)	Nominal breaking strength (tons—2,000 pounds)	Minimum breaking strength (pounds)
1/4	0.250	3/4	0.10	0.60	2.3	4,480	2.07	4,040
5/16	0.312	1	0.16	0.96	3.5	6,820	3.15	6,140
3/8	0.375	1 1/8	0.23	1.38	5.0	9,800	4.50	8,780
7/16	0.438	1 1/4	0.31	1.86	6.6	12,900	5.94	11,690
1/2	0.500	1 1/2	0.40	2.40	8.5	16,600	7.65	14,900
9/16	0.562	1 3/4	0.51	3.06	10.6	20,700	9.54	18,600
5/8	0.625	2	0.63	3.78	13.1	25,500	11.8	23,000
3/4	0.750	2 1/4	0.90	5.40	18.7	36,500	16.8	32,800
7/8	0.875	2 3/4	1.23	7.38	25.4	49,500	22.9	44,700
1	1.000	3	1.60	9.60	33.0	64,400	29.7	57,900
1 1/8	1.125	3 1/2	2.03	12.18	41.5	80,900	37.4	72,900
1 1/4	1.250	4	2.50	15.00	51.0	99,400	45.9	89,500
1 3/8	1.375	4 1/4	3.03	18.18	61.5	112,000	55.4	108,000
1 1/2	1.500	4 3/4	3.60	21.60	72.5	141,000	65.2	127,000
1 5/8	1.625	5	4.23	25.38	85	166,000	76.5	149,000
1 3/4	1.750	5 1/2	4.90	29.40	98	191,000	88.2	172,000
1 7/8	1.875	5 3/4	5.63	33.78	112	218,000	101	197,000
2	2.000	6 1/4	6.40	38.40	127	248,000	114	223,000
2 1/4	2.250	7 1/8	8.10	48.60	160	312,000	144	281,000
2 1/2	2.500	7 7/8	10.00	60.00	195	380,000	176	342,000
2 3/4	2.750	8 5/8	12.10	72.60	234	456,000	211	411,000

TABLE A1.11—TYPE B, 6 BY 19 WIRE ROPE, HIGH-GRADE PLOW STEEL, UNCOATED OR GALVANIZED, FIBER CORE

Nominal diameter, minimum diameter (inches)		Approximate circumference (inches)	Approximate weight per foot (pounds)	Approximate weight per fathom (6 feet) (pounds)	Uncoated		Galvanized	
					Nominal breaking strength (tons—2,000 pounds)	Minimum breaking strength (pounds)	Nominal breaking strength (tons—2,000 pounds)	Minimum breaking strength (pounds)
1/4	0.250	3/4	0.10	0.60	2.9	5,660	2.61	5,090
5/16	0.312	1	0.16	0.96	4.5	8,780	4.05	7,900
3/8	0.375	1 1/8	0.23	1.38	6.3	12,300	5.67	11,100
7/16	0.438	1 1/4	0.31	1.86	8.4	16,400	7.56	14,700
1/2	0.500	1 1/2	0.40	2.40	10.8	21,100	9.72	19,000
9/16	0.562	1 3/4	0.51	3.06	13.5	26,300	12.2	23,800
5/8	0.625	2	0.63	3.78	16.6	32,400	14.9	29,100
3/4	0.750	2 1/4	0.90	5.40	23.7	46,200	21.3	41,500
7/8	0.875	2 3/4	1.23	7.38	32.2	62,800	29.0	56,600
1	1.000	3	1.60	9.60	42.0	81,900	37.8	73,700
1 1/8	1.125	3 1/2	2.03	12.18	53.0	103,000	47.7	93,000
1 1/4	1.250	4	2.50	15.00	65.0	127,000	58.5	114,000
1 3/8	1.375	4 1/4	3.03	18.18	78.5	153,000	70.6	138,000
1 1/2	1.500	4 3/4	3.60	21.60	92.5	180,000	83.2	162,000
1 5/8	1.625	5	4.23	25.38	108	211,000	97.2	190,000
1 3/4	1.750	5 1/2	4.90	29.40	124	242,000	112	218,000
1 7/8	1.875	5 3/4	5.63	33.78	142	277,000	128	249,000
2	2.000	6 1/4	6.40	38.40	161	314,000	145	283,000
2 1/4	2.250	7 1/8	8.10	48.60	202	394,000	182	355,000
2 1/2	2.500	7 7/8	10.00	60.00	246	480,000	221	432,000
2 3/4	2.750	8 5/8	12.10	72.60	294	573,000	265	516,000

TABLE A1.12—TYPE C, 6 BY 19 WIRE ROPE, PHOSPHOR-BRONZE, FIBER CORE

Nominal diameter, minimum diameter (inches)		Approximate circumference (inches)	Approximate weight per foot (pounds)	Approximate weight per fathom (6 feet) (pounds)	Nominal breaking strength (tons—2,000 pounds)	Minimum breaking strength (pounds)
1/4	0.250	3/4	0.11	0.66	1.19	2,140
5/16	0.312	1	0.16	0.96	1.84	3,310
3/8	0.375	1 1/8	0.23	1.38	2.62	4,720
7/16	0.438	1 1/4	0.32	1.92	3.54	6,370
1/2	0.500	1 1/2	0.41	2.46	4.61	8,300
9/16	0.562	1 3/4	0.52	3.12	5.75	10,400
5/8	0.625	2	0.64	3.84	7.08	12,700
3/4	0.750	2 1/4	0.92	5.52	9.98	18,000
13/16	0.812	2 1/2	1.08	6.48	11.6	20,900
7/8	0.875	2 3/4	1.26	7.56	13.4	24,100

TABLE A1.13—TYPE D, 6 BY 37 WIRE ROPE, EXTRA-STRONG CAST STEEL, UNCOATED OR GALVANIZED, FIBER CORE

Nominal diameter, minimum diameter (inches)		Approximate circumference (inches)	Approximate weight per foot (pounds)	Approximate weight per fathom (6 feet) (pounds)	Uncoated: Nominal breaking strength (tons—2,000 pounds)	Uncoated: Minimum breaking strength (pounds)	Galvanized: Nominal breaking strength (tons—2,000 pounds)	Galvanized: Minimum breaking strength (pounds)
3/8	0.375	1 1/8	0.22	1.32	4.90	9,560	4.41	8,600
7/16	0.438	1 1/4	0.30	1.80	6.60	12,900	5.94	11,600
1/2	0.500	1 1/2	0.39	2.34	8.40	16,400	7.56	14,700
9/16	0.562	1 3/4	0.49	2.94	10.5	20,500	9.45	18,400
5/8	0.625	2	0.61	3.66	12.8	25,000	11.5	22,400
3/4	0.750	2 1/4	0.87	5.22	18.1	35,300	16.3	31,800
7/8	0.875	2 3/4	1.19	7.14	24.2	47,200	21.8	42,500
1	1.000	3	1.55	9.30	31.5	61,400	28.4	55,400
1 1/8	1.125	3 1/2	1.96	11.76	39.7	77,400	35.7	69,600
1 1/4	1.250	4	2.42	14.52	48.9	95,400	44.0	85,800
1 3/8	1.375	4 1/4	2.93	17.58	59.0	115,000	53.1	104,000
1 1/2	1.500	4 3/4	3.49	20.94	70.0	136,000	63.0	123,000
1 5/8	1.625	5	4.09	24.54	82.0	160,000	73.8	144,000
1 3/4	1.750	5 1/2	4.75	28.50	95.0	185,000	85.5	167,000
1 7/8	1.875	5 3/4	5.45	32.70	108.5	212,000	97.6	190,000
2	2.000	6 1/4	6.20	37.20	123.0	240,000	111	216,000
2 1/4	2.250	7 1/8	7.85	47.10	153.0	298,000	138	269,000
2 1/2	2.500	7 7/8	9.69	58.14	188.0	367,000	169	330,000
2 3/4	2.750	8 5/8	11.72	70.32	226.0	441,000	203	397,000

Table A1.14—Type E, 6 by 37 Wire Rope, High-Grade Plow Steel, Uncoated or Galvanized, Fiber Core

Nominal diameter, minimum diameter (inches)		Approximate circumference (inches)	Approximate weight per foot (pounds)	Approximate weight per fathom (6 feet) (pounds)	Uncoated		Galvanized	
					Nominal breaking strength (tons—2,000 pounds)	Minimum breaking strength (pounds)	Nominal breaking strength (tons—2,000 pounds)	Minimum breaking strength (pounds)
1/4	0.250	3/4	0.10	0.60	2.4	4,620	2.10	4,180
5/16	0.312	1	0.15	0.90	4.1	7,900	3.84	7,500
3/8	0.375	1 1/8	0.22	1.32	6.1	11,900	5.49	10,700
7/16	0.438	1 1/4	0.30	1.80	8.3	16,200	7.47	14,600
1/2	0.500	1 1/2	0.39	2.34	10.6	20,700	9.54	18,600
9/16	0.562	1 3/4	0.49	2.94	13.2	25,700	11.9	23,200
5/8	0.625	2	0.61	3.66	16.1	31,400	14.5	28,200
3/4	0.750	2 1/2	0.87	5.22	22.8	44,500	20.5	40,000
7/8	0.875	2 3/4	1.19	7.14	30.5	59,500	27.4	53,400
1	1.000	3	1.55	9.30	39.5	77,000	35.6	69,400
1 1/8	1.125	3 1/2	1.96	11.76	49.9	97,300	44.9	87,600
1 1/4	1.250	4	2.42	14.52	61.5	120,000	55.4	108,000
1 3/8	1.375	4 1/4	2.93	17.58	74.3	145,000	66.9	130,000
1 1/2	1.500	4 3/4	3.49	20.94	88.2	172,000	79.4	155,000
1 5/8	1.625	5	4.09	24.54	103.3	201,000	93	181,000
1 3/4	1.750	5 1/2	4.75	28.50	119.5	233,000	108	210,000
1 7/8	1.875	5 3/4	5.45	32.70	137	267,000	123	240,000
2	2.000	6 1/4	6.20	37.20	155	302,000	140	272,000
2 1/4	2.250	7 1/8	7.85	47.10	194	378,000	175	340,000
2 1/2	2.500	7 7/8	9.69	58.14	237	462,000	213	416,000
2 3/4	2.750	8 5/8	11.72	70.32	285	556,000	256	500,000

TABLE A1.15—TYPE F, 6 BY 12 WIRE ROPE, CAST STEEL, GALVANIZED, FIBER CORE

Nominal diameter, minimum diameter (inches)		Approximate circumference (inches)	Approximate weight per foot (pounds)	Approximate weight per fathom (6 feet) (pounds)	Nominal breaking strength (tons—2,000 pounds)	Minimum breaking strength (pounds)
5/16	0.312	1	0.10	0.60	1.76	3,430
3/8	0.375	1 1/8	0.15	0.90	2.47	4,820
7/16	0.438	1 1/4	0.20	1.20	3.29	6,420
1/2	0.500	1 1/2	0.26	1.56	4.23	8,250
9/16	0.562	1 3/4	0.33	1.98	5.26	10,300
5/8	0.625	2	0.41	2.46	6.46	12,600
3/4	0.750	2 1/4	0.59	3.54	9.26	18,100
13/16	0.812	2 1/2	0.69	4.14	10.8	21,100
7/8	0.875	2 3/4	0.80	4.80	12.5	24,400
1	1.000	3	1.05	6.30	16.3	31,800
1 1/16	1.062	3 1/4	1.19	7.14	18.3	35,700
1 1/8	1.125	3 1/2	1.33	7.98	20.5	40,000
1 3/16	1.188	3 3/4	1.48	8.88	22.8	44,500
1 1/4	1.250	4	1.64	9.84	25.1	48,900
1 3/8	1.375	4 1/4	1.99	11.94	30.2	58,900
1 7/16	1.438	4 1/2	2.17	13.02	33.0	64,400
1 1/2	1.500	4 3/4	2.36	14.16	35.8	69,800
1 5/8	1.625	5	2.77	16.62	41.8	81,500
1 11/16	1.688	5 1/4	2.99	17.94	45.0	87,800
1 3/4	1.750	5 1/2	3.22	19.32	48.3	94,200
1 13/16	1.812	5 3/4	3.45	20.70	51.7	101,000
1 15/16	1.938	6	3.94	23.64	59.0	115,000
2	2.000	6 1/4	4.20	25.20	62.7	122,000
2 1/16	2.062	6 1/2	4.47	26.82	66.6	130,000

Table A1.16—Type G, 6 by 12 Wire Rope, High-Grade Plow Steel, Galvanized, Fiber Core

Nominal diameter, minimum diameter (inches)	Approximate circumference (inches)	Approximate weight per foot (pounds)	Approximate weight per fathom (6 feet) (pounds)	Nominal breaking strength (tons—2,000 pounds)	Minimum breaking strength (pounds)
1/4 0.250	3/4	0.07	0.42	1.59	3,100
5/16 0.312	1	0.10	0.60	2.46	4,800
3/8 0.375	1 1/8	0.14	0.84	3.46	6,750
7/16 0.438	1 1/4	0.20	1.20	4.61	8,990
1/2 0.500	1 1/2	0.26	1.56	5.9	11,500
9/16 0.562	1 3/4	0.33	1.98	7.4	14,400
5/8 0.625	2	0.42	2.52	9.1	17,700
3/4 0.750	2 1/4	0.59	3.54	13.0	25,400
13/16 0.812	2 1/2	0.68	4.08	15.5	30,200
7/8 0.875	2 3/4	0.80	4.80	17.7	34,500
1 1.000	3	1.05	6.30	23.0	44,800
1 1/16 1.062	3 1/4	1.18	7.08	26.0	50,700
1 1/8 1.125	3 1/2	1.33	7.98	29.1	56,700
1 3/16 1.188	3 3/4	1.47	8.82	32.3	63,000
1 1/4 1.250	4	1.63	9.78	35.6	69,400
1 3/8 1.375	4 1/4	2.00	12.00	43.0	83,800
1 7/16 1.438	4 1/2	2.16	12.96	46.8	91,300
1 1/2 1.500	4 3/4	2.36	14.16	50.9	99,300
1 5/8 1.625	5	2.76	16.56	59.2	115,000
1 11/16 1.688	5 1/4	2.94	17.64	63.6	124,000
1 3/4 1.750	5 1/2	3.23	19.38	68.2	133,000
1 13/16 1.812	5 3/4	3.42	20.52	72.8	142,000
1 15/16 1.938	6	3.83	22.98	83.0	162,000
2 2.000	6 1/4	4.20	25.20	88.3	172,000
2 1/16 2.062	6 1/2	4.43	26.58	93.4	182,000

Table A1.17—Type H, 6 by 12 Wire Rope, Phosphor-Bronze, Fiber Core

Nominal diameter, minimum diameter (inches)	Maximum diameter (inches)	Approximate circumference (inches)	Approximate weight per foot (pounds)	Approximate weight per fathom (6 feet) (pounds)	Nominal breaking strength (tons—2,000 pounds)	Minimum breaking strength (pounds)
1/4 0.250	9/32 0.281	3/4	0.075	0.45	0.71	1,270
5/16 0.312	11/32 0.344	1	0.115	0.69	1.11	1,990
3/8 0.375	13/32 0.406	1 1/8	0.168	1.01	1.59	2,860
7/16 0.438	15/32 0.469	1 1/4	0.227	1.36	2.16	3,900
1/2 0.500	17/32 0.531	1 1/2	0.295	1.77	2.83	5,090
9/16 0.562	19/32 0.594	1 3/4	0.370	2.22	3.58	6,440
5/8 0.625	21/32 0.656	2	0.460	2.76	4.42	7,950
3/4 0.750	25/32 0.781	2 1/4	0.660	3.96	6.36	11,400
13/16 0.812	55/64 0.859	2 1/2	0.766	4.60	7.46	13,400
7/8 0.875	59/64 0.922	2 3/4	0.895	5.37	8.65	15,600

TABLE A1.18—TYPE I, 6 BY 24 WIRE ROPE, CAST STEEL, GALVANIZED, FIBER CORE

Nominal diameter, minimum diameter (inches)		Approximate circumference (inches)	Approximate weight per foot (pounds)	Approximate weight per fathom (6 feet) (pounds)	Nominal breaking strength (tons—2,000 pounds)	Minimum breaking strength (pounds)
3/4	0.750	2 1/4	0.78	4.68	13.6	26,500
13/16	0.812	2 1/2	0.91	5.46	15.8	30,800
7/8	0.875	2 3/4	1.06	6.36	18.3	35,700
1	1.000	3	1.38	8.28	23.7	46,200
1 1/16	1.062	3 1/4	1.56	9.36	26.7	52,100
1 1/8	1.125	3 1/2	1.75	10.50	29.9	58,300
1 3/16	1.188	3 3/4	1.95	11.70	33.2	64,700
1 1/4	1.250	4	2.16	12.96	36.7	71,600
1 3/8	1.375	4 1/4	2.61	15.66	44.4	86,600
1 7/16	1.438	4 1/2	2.85	17.10	48.4	94,400
1 1/2	1.500	4 3/4	3.11	18.66	52.6	103,000
1 5/8	1.625	5	3.64	21.84	61.6	120,000
1 11/16	1.688	5 1/4	3.93	23.58	66.3	129,000
1 3/4	1.750	5 1/2	4.23	25.38	71.2	139,000
1 13/16	1.812	5 3/4	4.53	27.18	76.2	149,000
1 15/16	1.938	6	5.18	31.08	86.6	169,000
2	2.000	6 1/4	5.52	33.12	92.0	179,000
2 1/16	2.062	6 1/2	5.87	35.22	98.0	191,000

TABLE A1.19—TYPE J, 6 BY 24 WIRE ROPE, HIGH-GRADE PLOW STEEL, GALVANIZED, FIBER CORE

Nominal diameter, minimum diameter (inches)		Approximate circumference (inches)	Approximate weight per foot (pounds)	Approximate weight per fathom (6 feet) (pounds)	Nominal breaking strength (tons—2,000 pounds)	Minimum breaking strength (pounds)
3/4	0.750	2 1/4	0.78	4.68	18.9	36,900
13/16	0.812	2 1/2	0.90	5.40	22.0	42,900
7/8	0.875	2 3/4	1.05	6.30	25.3	49,300
1	1.000	3	1.38	8.28	32.8	64,000
1 1/16	1.062	3 1/4	1.56	9.36	37.0	72,200
1 1/8	1.125	3 1/2	1.75	10.50	41.4	80,700
1 3/16	1.188	3 3/4	1.96	11.76	46.1	89,900
1 1/4	1.250	4	2.17	13.02	51.0	99,400
1 3/8	1.375	4 1/4	2.65	15.90	61.6	120,000
1 7/16	1.438	4 1/2	2.90	17.40	67.2	131,000
1 1/2	1.500	4 3/4	3.15	18.90	73.1	143,000
1 5/8	1.625	5	3.68	22.08	85.6	167,000
1 11/16	1.688	5 1/4	3.97	23.82	92.1	180,000
1 3/4	1.750	5 1/2	4.27	25.62	99.0	193,000
1 13/16	1.812	5 3/4	4.57	27.42	106.0	207,000
1 15/16	1.938	6	5.22	31.32	121.0	235,000
2	2.000	6 1/4	5.55	33.30	129.0	251,000
2 1/16	2.062	6 1/2	5.77	34.62	136.0	264,000

Table A1.20—Type K, 6 by 7 Wire Rope, Extra-Strong Cast Steel, Galvanized, Fiber Core

Nominal diameter, minimum diameter (inches)		Approximate circumference (inches)	Approximate weight per foot (pounds)	Approximate weight per fathom (6 feet) (pounds)	Nominal breaking strength (tons—2,000 pounds)	Minimum breaking strength (pounds)
9/32	0.281	7/8	0.12	0.72	2.45	4,780
5/16	0.312	1	0.15	0.90	3.02	5,890
3/8	0.375	1 1/8	0.21	1.26	4.23	8,250
7/16	0.438	1 1/4	0.29	1.74	5.67	11,100
1/2	0.500	1 1/2	0.38	2.28	7.38	14,400
9/16	0.562	1 3/4	0.48	2.88	9.27	18,100
5/8	0.625	2	0.59	3.54	11.3	22,000
11/16	0.688	2 1/8	0.72	4.32	13.7	26,700
3/4	0.750	2 1/4	0.84	5.04	16.3	31,800
7/8	0.875	2 3/4	1.15	6.90	22.1	43,100
1	1.000	3	1.50	9.00	28.7	56,000
1 1/8	1.125	3 1/2	1.90	11.40	36.0	70,200
1 1/4	1.250	4	2.34	14.04	43.8	85,400
1 3/8	1.375	4 1/4	2.84	17.04	52.4	102,000
1 1/2	1.500	4 3/4	3.38	20.28	61.8	121,000

Table A1.21—Type L, 5 by 19 Wire Rope, Marline-Covered, Extra-Strong Cast Steel, Uncoated, Fiber Core

Approximate diameter before serving (inches)		Nominal diameter after serving (inches)		Approximate weight per foot (pounds)	Approximate weight per fathom (6 feet) (pounds)	Nominal breaking strength (tons—2,000 pounds)	Minimum breaking strength (pounds)
1/4	0.250	9/16	0.562	0.21	1.26	2.1	4,100
3/8	0.375	11/16	0.688	0.36	2.16	4.6	8,970
1/2	0.500	13/16	0.812	0.51	3.06	7.8	15,200
9/16	0.562	7/8	0.875	0.62	3.72	9.7	18,900
5/8	0.625	1	1.000	0.81	4.86	12.0	23,400
3/4	0.750	1 1/8	1.125	1.10	6.60	17.1	33,300
7/8	0.875	1 1/4	1.250	1.32	7.92	23.3	54,400
1	1.000	1 3/8	1.375	1.70	10.20	30.2	58,900
1 1/8	1.125	1 1/2	1.500	2.12	12.72	38.0	74,100
1 1/4	1.250	1 5/8	1.625	2.58	15.48	46.8	91,300
1 3/8	1.375	1 3/4	1.750	3.14	18.84	56.4	110,000
1 1/2	1.500	1 7/8	1.875	3.69	22.14	66.5	130,000
1 5/8	1.625	2	2.000	4.29	25.74	77.9	152,000
1 3/4	1.750	2 1/8	2.125	5.00	30.00	89.8	175,000

TABLE A1.22—TYPE M, 1 BY 7 SEIZING STRAND, GALVANIZED

Nominal diameter, minimum diameter (inch)		Approximate diameter of wire (inch)	Approximate weight per foot (pound)	Approximate weight per fathom (6 feet) (pound)	Approximate weight of reel (pounds)	Approximate length of strand on reel		Minimum breaking strength (pounds)
						Feet	Fathoms	
1/16	0.062	0.022	0.010	0.06	50	5,000	834	140
3/32	0.094	0.032	0.020	0.12	50	2,500	416	300
1/8	0.125	0.042	0.033	0.20	50	1,500	250	530
5/32	0.156	0.052	0.050	0.30	50	1,000	166	810

TABLE A1.23—TYPE N, 1 BY 19 SEIZING STRAND, GALVANIZED

Nominal diameter, minimum diameter (inch)		Approximate diameter of wire (inch)	Approximate weight per foot (pound)	Approximate weight per fathom (6 feet) (pound)	Approximate weight of reel (pounds)	Approximate length of strand on reel		Minimum breaking strength (pounds)
						Feet	Fathoms	
3/16	0.188	0.039	0.080	0.048	100	1,200	200	1,100
1/4	0.250	0.052	0.133	0.80	100	750	125	1,900

TABLE A1.24—SIZE OF WIRE ROPES TO REPLACE MANILA LINES

Manila rope		6 x 12 wire rope type "G"		6 x 24 wire rope type "J"		6 x 37 wire rope type "E"	
Circumference (inches)	Strength (pounds)	Diameter (inches)	Strength (pounds)	Diameter (inches)	Strength (pounds)	Diameter (inches)	Strength (pounds)
4	15,000	5/8	17,700	.....	......	1/2	18,600
5	22,500	3/4	25,400	.....	......	9/16	23,200
6	31,000	7/8	34,500	3/4	36,900	3/4	40,000
7	41,000	1	44,800	7/8	49,300	7/8	53,400
8	52,000	1 1/8	56,700	1	64,000	1	69,400
9	64,000	1 1/4	69,400	1 1/16	72,200	(1)	(69,400)
10	77,000	1 3/8	83,800	1 3/16	89,900	1 1/8	87,600
11	91,000	1 1/2	99,300	1 1/4	99,400	1 1/4	108,000
12	105,000	1 5/8	115,000	1 3/8	120,000	1 3/8	130,000

APPENDIX

2

Knotting and Splicing

A2.1. Working in Hemp and Manila—Knots (Fig. A2.1). This plate shows a number of simple knots, all of which are made with the end of a single rope. (The word *line* is properly used in the Navy instead of *rope*.)

Overhand Knot. A. A simple knot formed by passing the end of a rope over the standing part and through the bight. Seldom used.

Bowline. B. One of the most common and useful knots. It forms an eye which may be of any length and which cannot slip. It is used for lowering men over the side, to form an eye in the end of a hawser to be thrown over a bollard in handling a ship alongside of a dock, and for a great variety of similar purposes. Place the end part on the standing part and form a loop with the end through; pass the end under the standing part and bring it back through the loop.

Running Bowline. C. A convenient form of running an eye. Formed by making a bowline over its own standing part.

Bowline on a Bight. D and E. The bowline made with bight of a line. Dip the open single bight (b) over both double bights (a) and up above the loop and haul taut. Used where a greater strength than given by a single bowline is necessary, or when the end of the rope is not accessible.

Cat's Paw. F. A double loop formed by twisting two bights of a rope. The hook of a tackle is passed through them. Convenient and secure.

Sheep Shank. G. Used to shorten a rope. Lay the bight of the rope in three parts, and half hitch each part around the bight of the other two parts.

Figure of Eight. H. A hitch resembling the figure eight and used to prevent the end of the rope from unreeving when rove through blocks. Pass the end of the rope around the bight over its own part and through the loop.

Blackwall Hitch. I and J. A hitch either single or double around the back of a hook, the end on one side of the hook and the standing part on top, on the other side of the hook.

French Bowline (*Fig. A2.2*). When sending a man over the side on hazardous work, into a smoke-filled hold, or anywhere where he may have to use both hands, make use of the French bowline. The bowline is formed in the same fashion as the regular bowline, except that the end (D), instead of going about the standard part (E) at once, is given a round turn about the loop of the eye

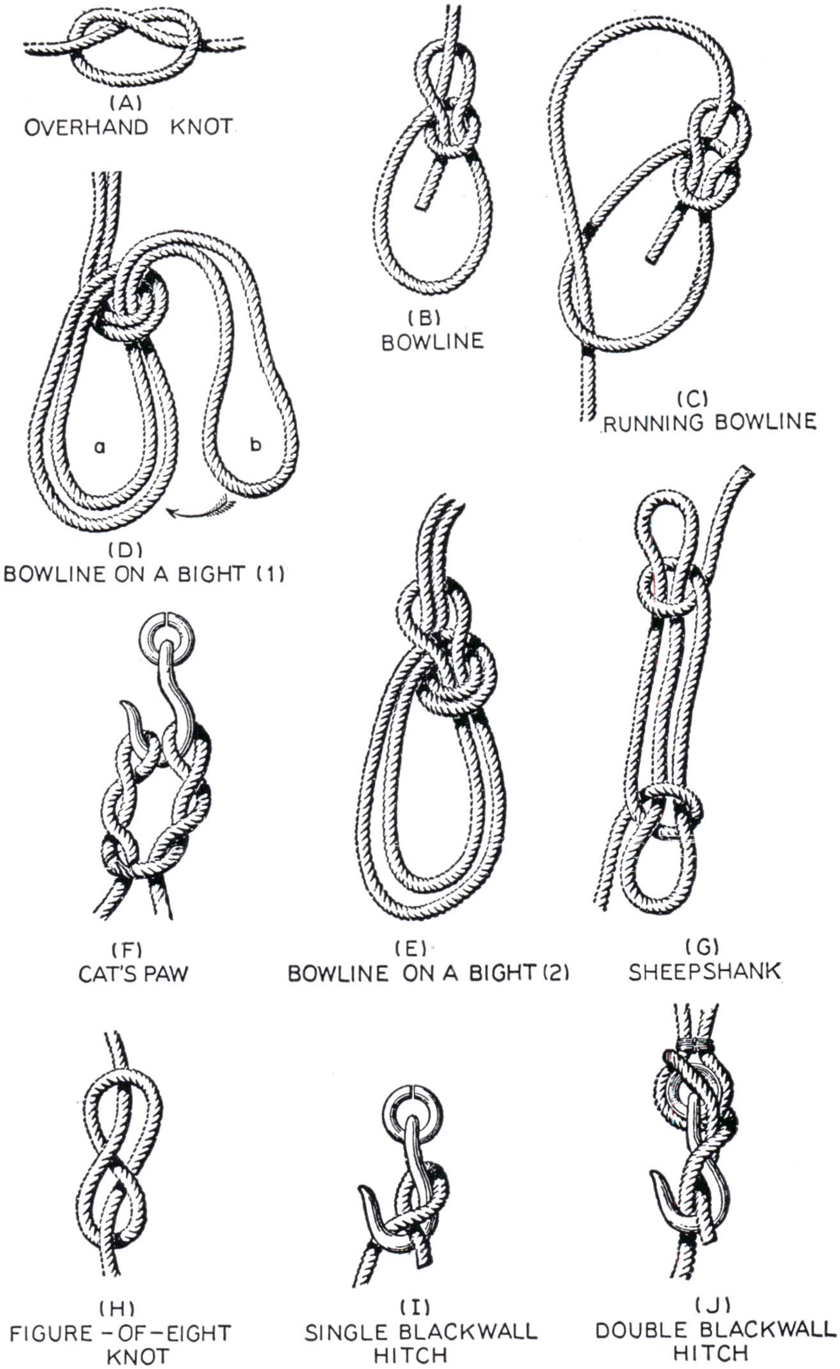

FIG. A2.1 KNOTS

(A), and then the knot is finished off as before. This leaves two eyes that are loosely connected through the loop neck. The eyes are made so that a man can sit in one—(b)—while the other—(c)—goes under his armpits, the knot being at his breast. The weight of the man hauls the armpit eye taut, he is safe against falling out and is held upright if unconscious. Never use manila if there is a chance of it catching fire.

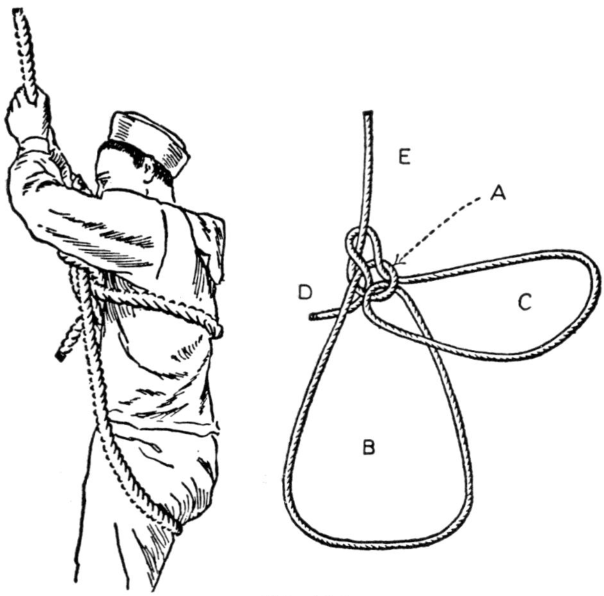

FIG. A2.2

A2.2. Bends (Fig. A2.3). This plate shows a series of knots for joining the ends of the two ropes or the two ends of the same rope. This is perhaps the most important group of knots with which we have to deal.

Square or Reef Knot. A. A most important knot and used for a great variety of purposes. Formed from an overhand knot by crossing the ends and bringing one end up through the bight alongside its own part.

Granny Knot. B. Frequently confused with the square knot. Unseamanlike knot made, instead of a square knot, by a green hand.

Sheet or Becket Bend, single. C. Used in bending on flags not fitted with snap hooks and very efficient for the purpose of uniting two ends of a rope or a rope's end to an eye. Made by passing the end of a rope up through the bight (eye) of another, around both parts and under its own part.

Sheet Bend, double. D. Here the end of the bending line is passed twice around the standing line and through its own part, giving added security.

Single Carrick Bend. E and F. A knot made by first crossing the end of one rope and then passing the end of the other down through the bight, under the

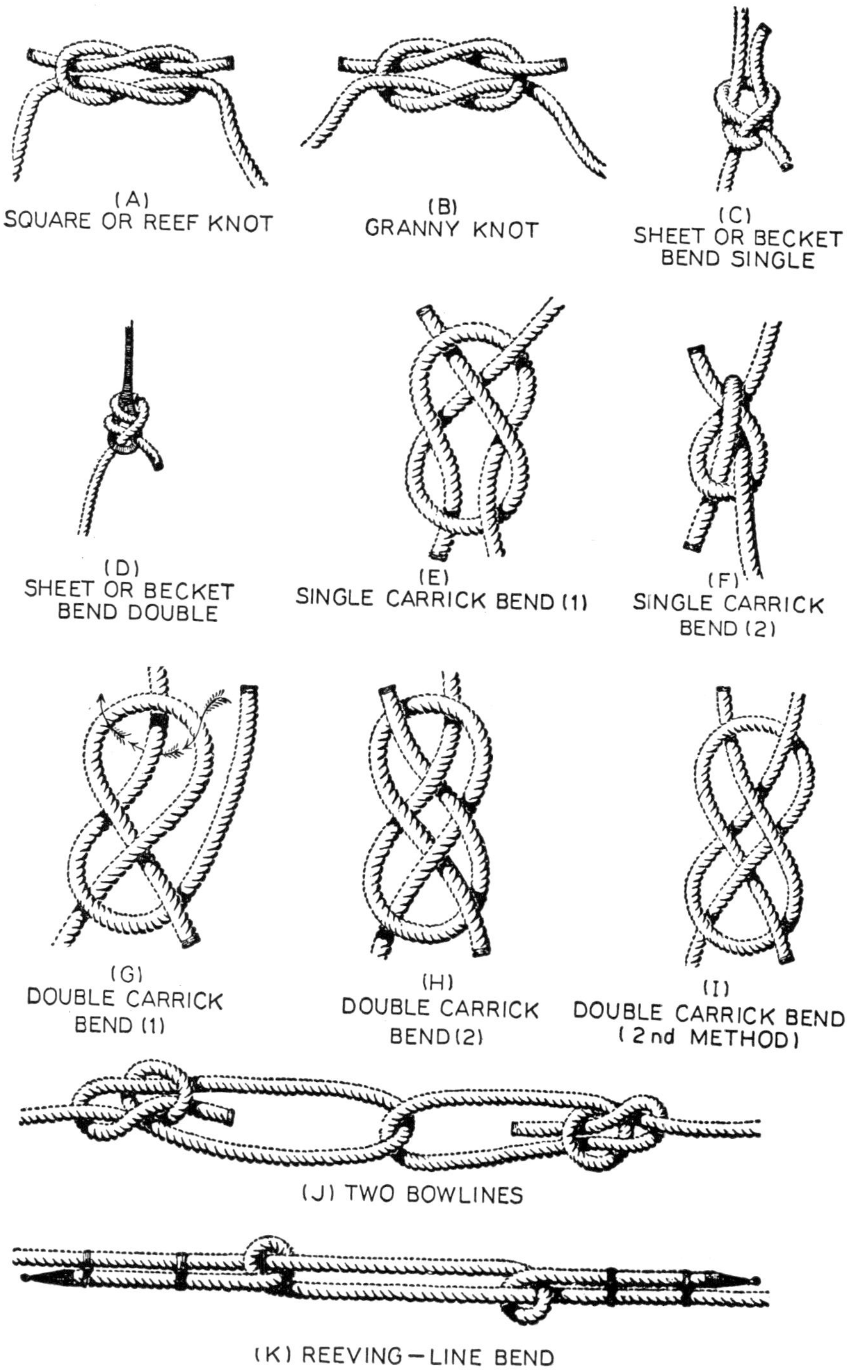

FIG. A2.3 BENDING TWO ROPES TOGETHER

standing part, over the end and down through the bight again. Used for joining two hawsers together.

Double Carrick Bend. G, H, and I. More secure than the single Carrick (both ends coming out on different sides). With both, it is well to seize the ends to their own parts.

Two Bowlines. J. A safe and convenient way of bending two hawsers together. Will not slip. Is somewhat bulky for use where the lines are to be veered out through a chock.

Reeving-Line Bend. K. A method of connecting two hawsers in such a way that they will reeve through an opening, offering as little obstruction as possible. Made by taking a half hitch with each end around the other hawser and seizing the ends.

A2.3. Bends (Fig. A2.4). This plate shows various ways of securing lines to spars, posts, rings, etc.

Studding Sail Tack Bend. A. A useful bend for a variety of purposes where it is important to have no danger of coming adrift through the flapping of a sail.

Studding Sail Halyard Bend. B. The greater the pull on the halyards, the more tightly the parts of the bend are jammed against the spar. Formed by taking a round turn with the end coming around the standing part, under both turns and tucked over and under the turns.

Fisherman's Bend. C. Used for securing a rope to a buoy, or a hawser to the ring or jew's-harp of an anchor. Formed by passing the end twice round the ring and under the turns. Seize the end back.

Timber Hitch. D. Useful when towing spars. Formed by passing the end around the spar and its own standing part; then pass several turns around its own part.

Timber Hitch and Half Hitch. E. The half hitch is taken first and the timber hitch afterwards formed with the end.

Rolling Hitch. F and H. Very useful where a rope is to be bent to a spar or to the standing part (not the end) of another rope, or to a chain (as in clearing hawse). Pass the end twice round the spar or rope, crossing the standing part on the top side each time; then hitch the end round the spar or rope, on the opposite side of the two turns.

Round Turn and Two Half Hitches. G. For making a hawser fast to a bollard, spar, or another rope. Most useful and popular way to make a line fast.

Two Half Hitches. I. Another method of bending a rope's end to a spar, stanchion, bollard, or ring. Formed by leading the end over and under and up through the standing part and repeating the process.

Clove Hitch. J. One of the most common hitches for attaching a rope to a spar for fastening ratlines to shrouds or to the standing part of a rope. Pass the end around the spar, crossing the standing part, then around the spar again, bringing the end through between the end part and standing part under its own part.

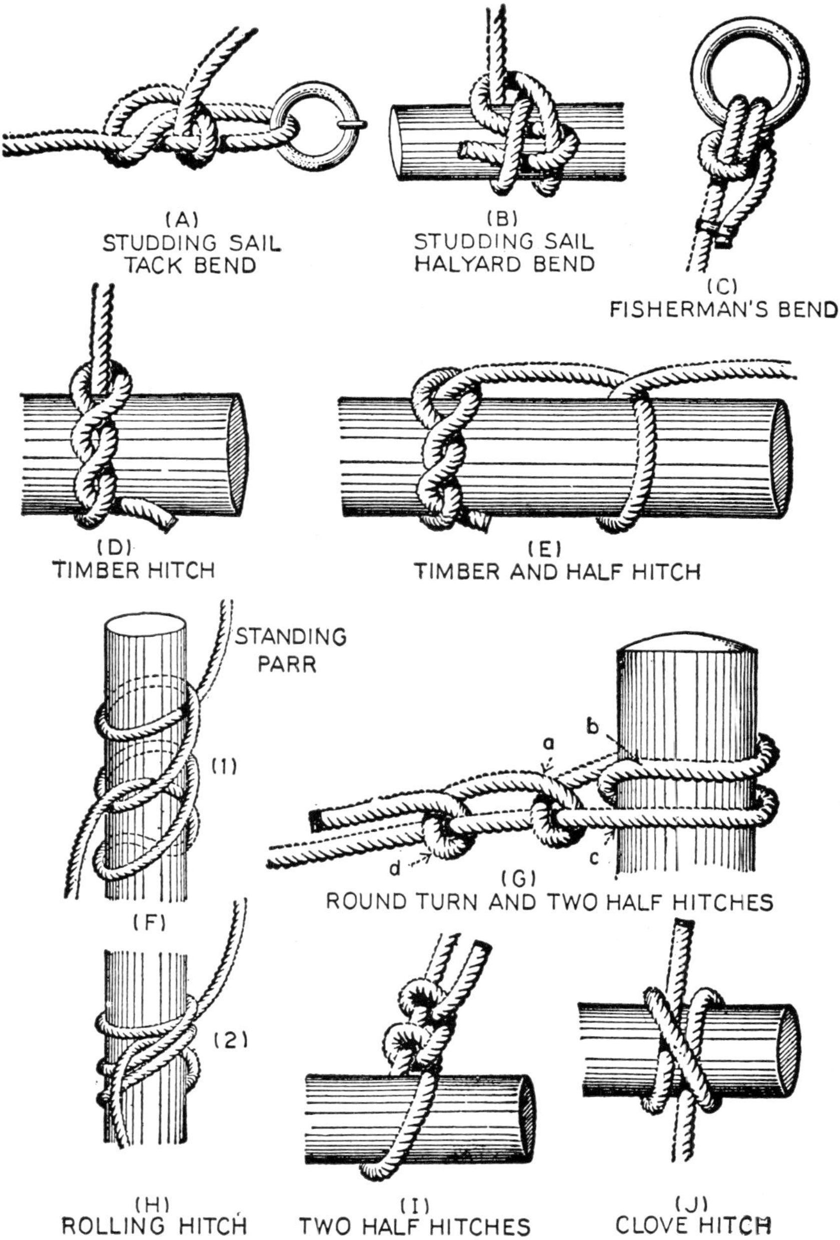

FIG. A2.4 BENDING A ROPE TO A POST, SPAR OR RING

Knotting a Rope Yarn (*Fig. A2.5A*). Used in making a selvagee strap where rope yarn is used. The parts shown in the drawing are hauled taut. The rope yarn will then be secured without give.

Stopper on a Rope (*Fig. A2.5B*). A short length of rope secured at one end, and used in securing or checking a running rope, e.g., deck stopper, boat fall stopper, etc.

Marline Hitch. Used in lashing hammocks, securing sails to spars.

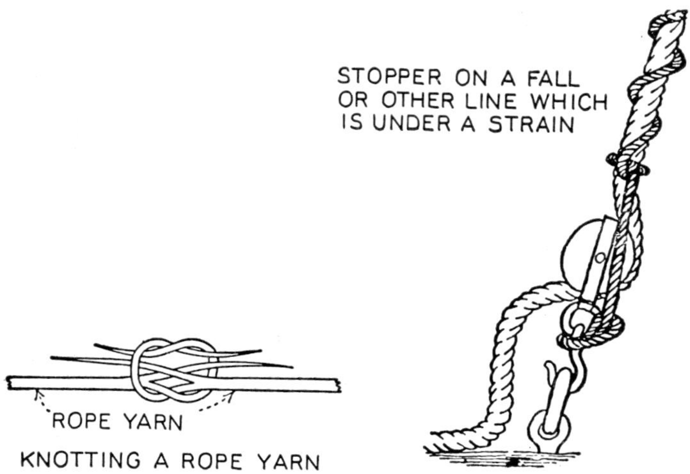

(A) (B)

FIG. A2.5

A2.4. Passing a Strap. A selvagee or a single strap is used to attach a block to a part of the rigging or to a spar. It is twisted around the rigging and the block is hooked into the bight as follows:

First Method (*Fig. A2.6, top*). A. Take the loop of one end of strap and hold same on hawser; now pass other end of strap close around hawser until it is expended, allowing sufficient length to bring the two ends together.

Second Method (*Fig. A2.6, top*). B. Take strap and hold center against hawser and pass both (crossing same) ends around hawser in opposite directions. Bring ends together and hook block.

Third Method (*Fig. A2.6, top*). C. Spread loop of one end of strap at right angles to hawser, now pass other end of strap around hawser inside of loop until strap is expended. Bring ends together to hook block.

Signal Halyard Splice (*Fig. A2.6, bottom*). A signal halyard splice is an eye turned in the end of signal halyard for snap hooks and rings.

1. About 1 inch or 1½ inches from end of halyard cut out three strands. Insert sail needle and twine through halyard about 1½ inches from end and make marline hitch; make another marline hitch with needle and twine near end of halyard (this tapers end of halyard for tucking).

2. Six inches from end of halyard insert a pricker down through center of halyard for about 2 inches.
3. Haul sail needle, twine, and tapered end through halyard ring or snap hook to be used. Withdraw pricker and insert sail needle, twine, and

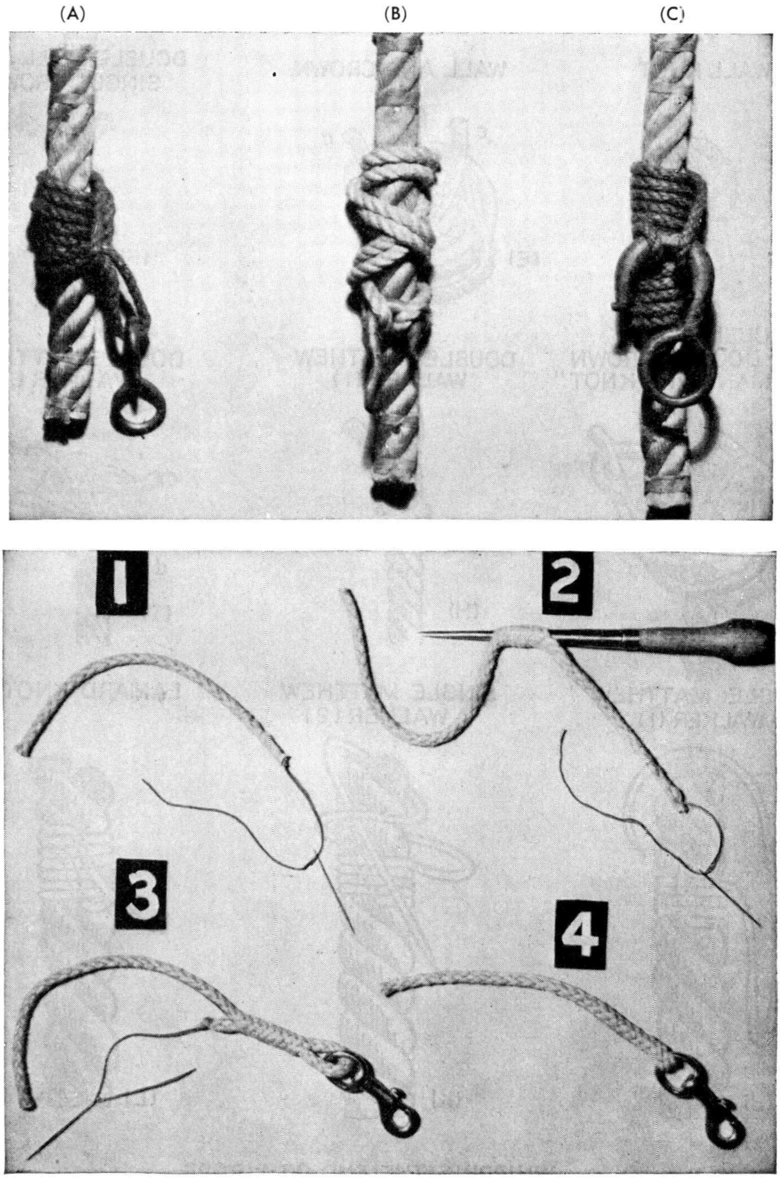

FIG. A2.6 (*Top*) THREE METHODS OF PASSING A STRAP
(*Bottom*) SIGNAL HALYARD SPLICE

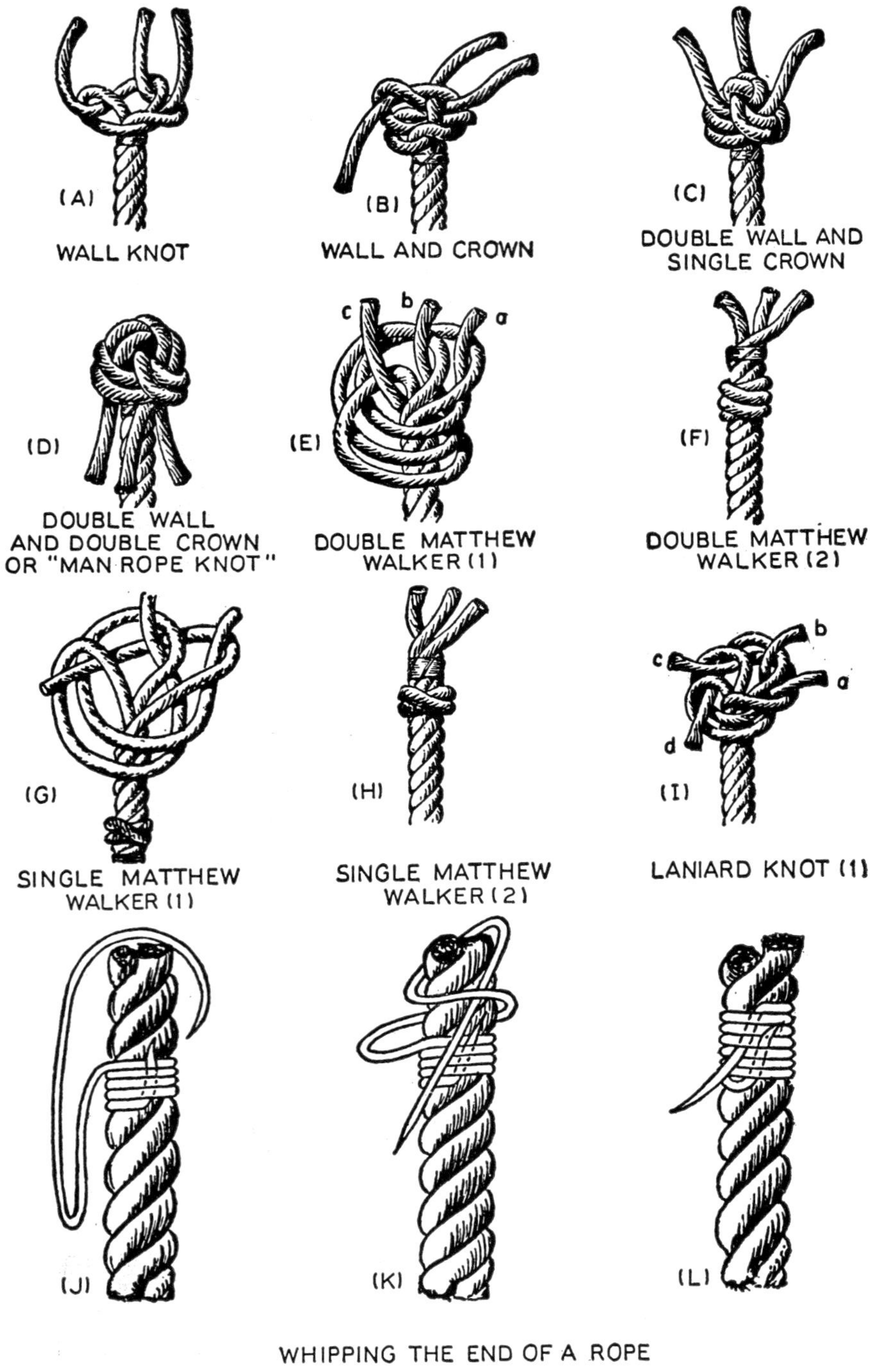

FIG. A2.7 KNOTS WORKED IN THE END OF A ROPE

tapered end through center of halyard. Cut off tapered end, leaving about ½ inch of end.

4. Haul out and smooth splice.

Note: For preservation, splice and fittings may be dipped in a solution of boiling beeswax.

A2.5. Knots Worked in the End of a Rope (Fig. A2.7). This plate shows a series of knots which are worked in the end or the body of the rope by unlaying the rope and using its own strands. A whipping is usually put on below the point where the knot is to be. Knots of this kind are sometimes used to give a finish to the end of the rope, sometimes to prevent unreeving, and sometimes merely for ornamental purposes.

Wall Knot. A. Unlay the strands sufficiently to form the knot. Form a bight with strand 1 and pass strand 2 around the end of it, also strand 3 around the end of 2 and through the bight of 1, hauling the ends taut.

Wall and Crown. B. Lay strand 1 over the center of the wall, strand 2 over 1, and strand 3 over 2 and under 1. Haul the strands taut and form the knot.

Double Wall and Single Crown. C.

Double Wall and Double Crown or "Man-Rope Knot." D. The ends come out underneath and are hidden. Used at the end of a man-rope.

Matthew Walker Knot, Double and Single. E, F, G, and H.

Laniard Knot. I.

Whipping the End of a Rope. J, K, and L. To prevent unlaying or fraying out. Lay the end of a length of stout twine along the rope and pass a number of turns around the rope and over the end of the whipping, hauling each turn taut. J. Then lay the other end of the whipping along the rope in the direction from the first end, and pass a number of turns, on the bight, over this end. K. Haul the end through and trim off.

Sailmaker's Whipping. Put on with a needle and twine. The rope is first struck through between two strands with the needle and the twine is drawn to the end. Several turns are passed around the rope and the rope struck through at each end of the whipping between the strands. Crossing turns are passed between strands and secured with two half hitches and hauled close down.

A2.6. Splicing (Fig. A2.8). There are various forms of splices for joining the ends of two ropes permanently, or for bending the end of a rope upon itself to form a permanent eye. In making these, when the strands of a rope are to be tucked, the lay of the rope where they go through is opened out by means of a fid, and the strand tucked through once in its full size, then reduced in size by cutting away a certain number of threads, tucked a second time, reduced once more in size, and tucked again. It is thus tucked once full size, once two-thirds, and once one-third size. This produces a tapered and much neater splice than if it were tucked three times in full size.

Length of Splices. For eye splices and short splices whether in manila or wire, from 12 to 24 inches should be allowed for forming the splice; for long splices, ten times the circumference of the rope.

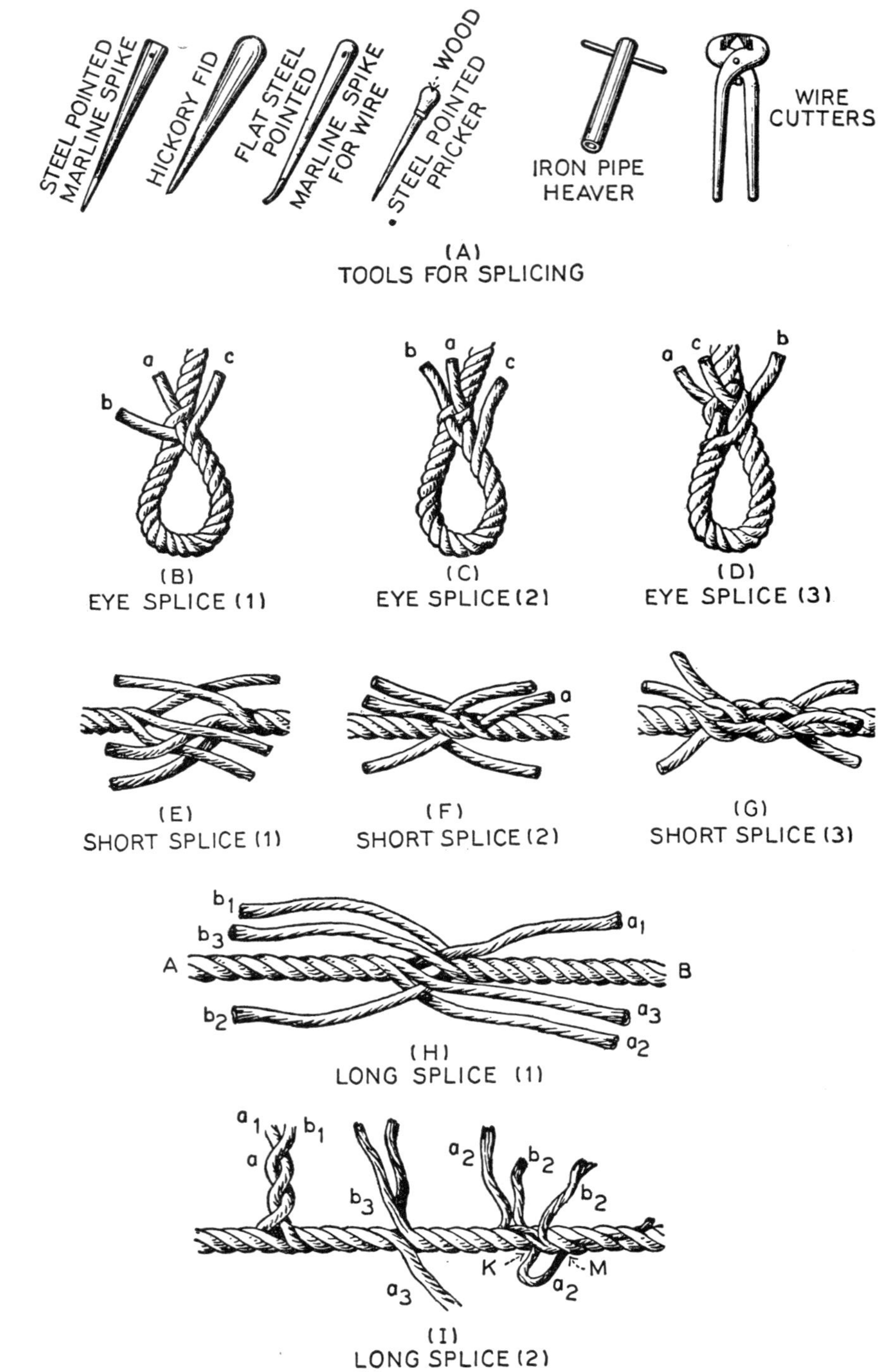

FIG. A2.8 SPLICES, HEMP AND MANILA

A Three-Strand Eye Splice. B, C, and D. The rope is unlaid for perhaps a foot from the end, and the strands brought back upon the body of the rope at a point which will form an eye of the size that is wanted.

Middle strand (a) up and a strand laying on either side have eye toward you. Tuck as follows: (B). Middle strand (a) under strand below it, left hand strand (b) over the strand under which the first strand was tucked and under the next (C). Now turn the splice over (D). Give the last strand (c) an extra twist with the lay and tuck it under the remaining strand. All strands are tucked from right to left. After first full tuck with the three strands, tuck each over and under twice more. If splice is to be finished and served, cut out a third of the yarns of the strand.

A Four-Strand Eye Splice. Here the first strand (left strand) is tucked under two (but this is for the first tuck only); remaining strands each under one. All tucks from right to left.

A Short Splice. E, F, and G. Two ropes are unlaid for a short distance and married together with strands interlacing (E). The strands of each rope are then tucked through the lay of the other rope exactly as has been described in the case of an eye splice.

A Long Splice. H and I. Here the ropes are unlaid for a greater distance than for a short splice and the ends brought together as before, with strands interlacing. Instead now of tucking at once, we proceed as follows: Unlay a_1, one of the strands of A, for a considerable distance, and in place of it lay up b_1, the adjoining strand of B, thus working a strand of B into A, for, say, a foot and a half to two feet. For convenience now trist up a_1 and b_1 together temporarily, as in I. Turn the rope end for end, unlay b_2, one of the strands of B, and in place of it lay up a_2, the adjoining strand of A. a_3 and b_3 are left lying beside each other without being unlaid. We now have three pairs of strands at different points of the rope. Beginning with a_2 and b_2 (for example) separate each of the strands into two parts, and taking one half of each strand, overhand knot these together (K, I) and tuck them as in a short splice, over one and under one of the full remaining strands in the rope (M, I). The other pairs of strands (a_1, b_1) (a_2, b_2) are similarly reduced, knotted, and tucked. The spare half of each strand is trimmed off smooth, as are the ends of the other halves after they have been tucked.

A2.7. Working in Wire Rope. As already stated, wire rope is usually six-stranded, with a hemp center. In splicing, we may work with the strands separately or in pairs. The work calls for special appliances and for a degree of skill such as can be acquired only by long practice under expert instruction. Something may be learned from careful description and much more from an occasional visit to a rigging loft; but the facilities which are available on shipboard do not admit of doing such work as is possible with a rigger's bench, a turning-in machine, etc. Where a heavy rope is to be bent around a thimble or the parts otherwise brought together for splicing or seizing, a rigger's screw is needed. In the absence of this, a vise may be used, but less conveniently.

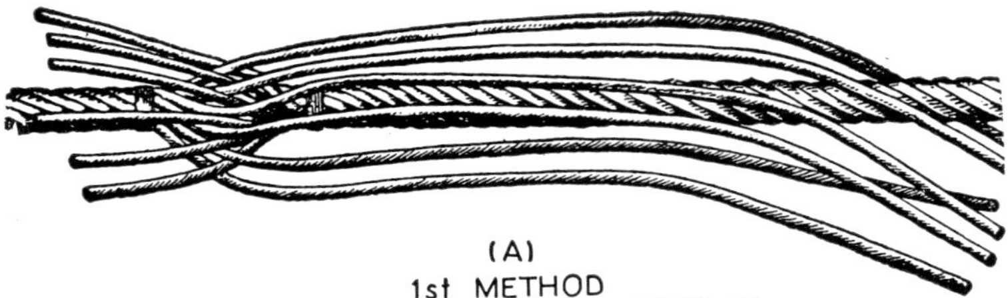

(A)
1st METHOD
LONG SPLICE IN WIRE ROPE (1)

(B)
1st METHOD
LONG SPLICE IN WIRE ROPE (2)

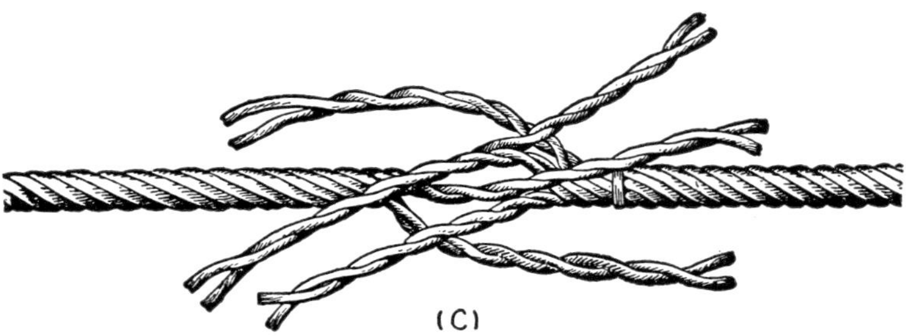

(C)
2nd METHOD
LONG SPLICE IN WIRE

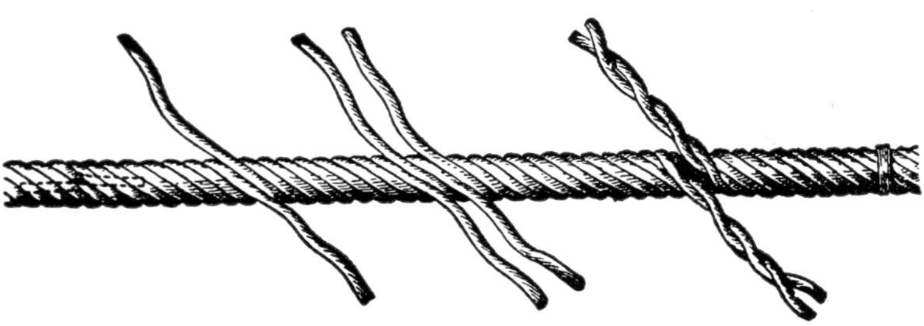

(D)
2nd METHOD
LONG SPLICE IN WIRE

FIG. A2.9 LONG SPLICE IN WIRE

In tucking the strands of a splice, the lay of the rope is opened out and the spike left in, holding the strands apart until the tuck has been made. For hauling the strands through, a jigger is used on each one, the body of the rope being held by another jigger or a lashing. After a tuck, the parts of the ropes are hammered down tightly upon each other. Wire-cutters are used for cutting off ends.

A Long Splice in Wire (Fig. A2.9A and B). Put on a good seizing 6 to 10 feet—according to the size of the rope—from the end of one of the ropes to be spliced, and a similar seizing 1 to 2 feet from the end of the other rope. Unlay, open out the strands, cut out the center, and marry the ends together with strands interlacing. Cut the seizing on the short end. Unlay one of the short strands, following it up in the same lay with the opposite long strand, leaving end enough to tuck. Continue in the same manner with the remaining strands, except as to the distance to which they are laid up, this distance being varied in such a way as to leave the successive pairs an equal distance apart, as shown in Fig. A2.9B. Commencing with any two strands, half knot them together (full size), then divide each into three parts and tuck these parts separately as shown; or, cut out a few inches of the center and insert the ends of the strands in its place in the center of the rope. When a splice is to be served, the latter way of finishing it off answers very well.

Second Method (C and D). Will be clear from Fig. A2.9.

A Short Splice in Wire (Fig. A2.10). A. Put on a good seizing 2 or 3 feet, according to size of the rope, from the end of the ropes to be spliced, and a similar seizing 1 or 2 feet from the end of the other rope. Unlay the ends and open out the strands, cutting out the heart close to the seizings. Marry them together and clap on a temporary seizing around the short ends and the body of the rope, to hold the parts close together. Commencing with any one of the long strands, tuck each in succession over one and under two strands, opening out the lay with a spike. Tuck the remaining strands in the same manner; twice whole strands, once one half, and once one quarter, hauling through with a jigger each time. Then turn the splice around, cutting the temporary seizing on the short ends, and tuck the short strands once one half and once one quarter, heaving them through with a jigger. Hammer down all parts and trim off the ends.

An Eye Splice in Wire. B. Put the rope on a stretch, allow from 18 to 24 inches from the end for splicing, and put on a mark with a couple of turns of twine. Measure along the rope from this mark the length of the eye (one and one half the round of the thimble) and put on another similar mark. Paint with red lead, worm, parcel, paint again, and double-serve between the marks. Now come up the stretch and seize the thimble in, breaking the wire into the shape of the thimble, heaving both parts together with a rigging screw and putting a good racking seizing around both parts. Come up the screw, unlay the end of the wire, and cut out the heart close to the service. Now, with the thimble toward you, counting from left to right 1, 2, 3, etc., stick No. 4 (Fig. A2.10B)

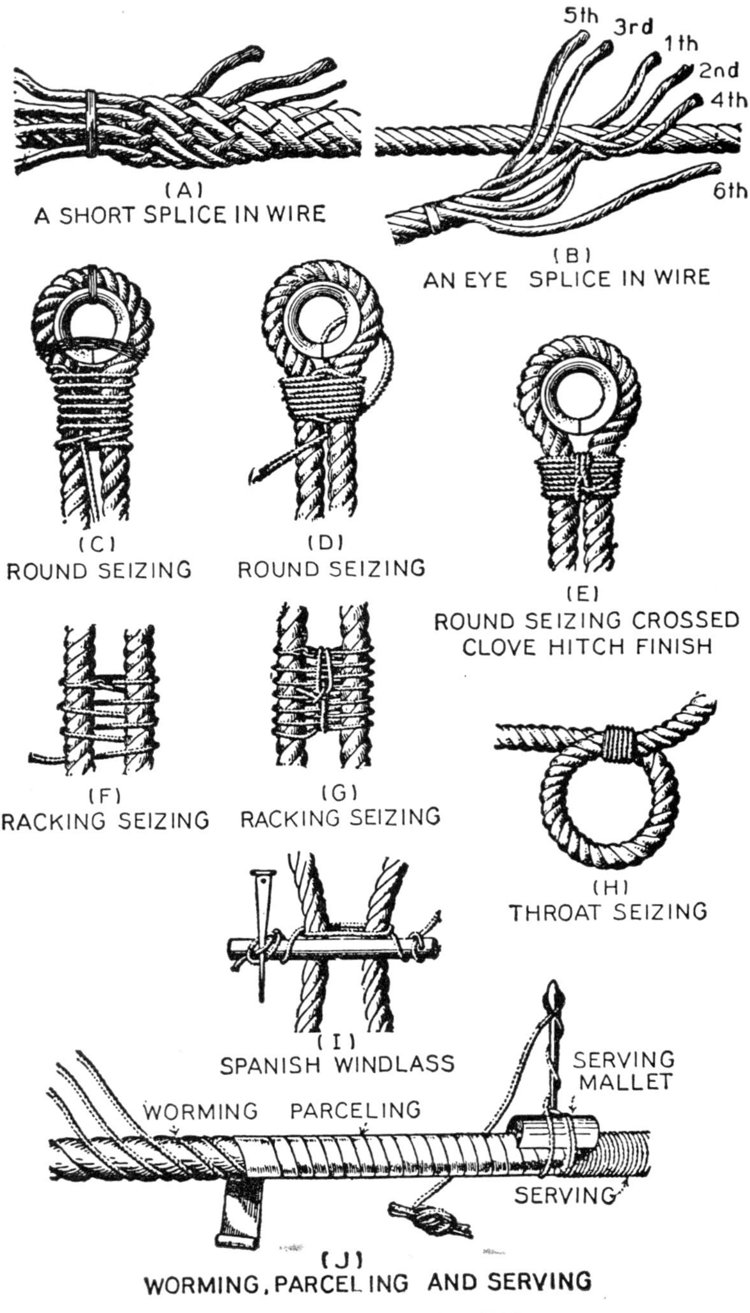

FIG. A2.10 SPLICES—SEIZINGS

strand from right to left under the two upper strands of the rope just clear of the service, opening the strands by a spike. Haul through by hand. In the same manner—under two strands—tuck the remaining strands, in the following order: 3, 5, 2, 6, 1. Now, commencing with any strand, tuck again whole over one and under one and haul through by means of a jigger. Hammer the strands down in place, cut each strand down to two-thirds size and tuck again, hauling through with a jigger as before. Cut the strands down to one third and tuck again. Hammer down all strands and cut off the wire with a wire-cutter.

The eye splice is most often used on board ship. Expert riggers favor the following method of turning in this splice.

1st. Clap on a stout whipping from 1 to 4 feet from end of wire rope, depending upon its size.

2nd. Whip the end of each strand with strong sail twine. Unlay the strands and cut out heart (core) of rope (not of strands).

3rd. Break the eye around the thimble place in riggers screw and seize it in place, put the wire on a stretch at about waist height.

4th. With the strands lying about parallel to the part of the wire through which they are to be tucked, stand with the thimble away from you, bight of rope under your right arm.

5th. Open a way through the middle of the bight, spike horizontal, pointing away from you. This is easy when enough turn has been taken out of the bight by a heaver.

6th. Take top one of strands to be tucked, and shove it through the middle of the rope, following the spike which may be withdrawn as tucking strand goes through. When through, tuck this strand around the strand of bight lying above it.

7th. Take the next strand down through middle having opened the way again, but only under two strands, and around the strand lying just above it.

8th. Take next strand down through middle opening, but only under and around one strand.

9th. Now take the next strand (fourth), and tuck it over and around the next strand to the right.

10th. Take next one over and around the next.

11th. Take last strand over and around the last untouched strand on the bight.

Note: All strands are tucked around in the same direction that the wires run in the strand. Strands are then tucked once more, around and around, sailmaker fashion, then heart is taken out of strands and half of the wires are cut out and the splice is tucked twice more.

Finish the splice by parceling with tarred canvas and serve over all with marline.

A2.8. Wire-Rope Grommets. A popular method of laying up a grommet of wire is to use two strands, carefully unlaying them from the rope and keeping them in their normal position in relation to one another. The length of the two strands used is something more than three times the circumference of the grommet to be made. Form the grommet of the bight of a left-handed overhand knot. Lay each double end around in the lay of the bight until they meet. This will bring four ends together, with a pair pointing each way, and six strands around the bight or grommet. Now cast off the seizing that holds the ends of the pair of strands together. Unlay one strand's end in the lay vacated as the strand was unlaid. This will separate the points of opposing parts, cross and tuck the ends in the same manner as in tucking the ends for a long splice.

Figure A2.10C, D, E, F, G, and H. Seizings for binding two ropes or two parts of the same rope. The manner of passing them is made clear by the figures. With heavy ropes, the parts must be hove together by power of some kind, such as a Spanish Windlass, a rigger's screw, or a turning-in machine.

A Spanish Windlass for Drawing Two Taut Ropes Together. I.

Worming, Parceling, and Serving. J. Rope which is to be exposed to the weather or to exceptionally hard usage is protected by worming, parceling, and serving.

Worming consists in following the "lay" of the rope, between the strands, with small-stuff, tarred, which keeps moisture from penetrating to the interior of the rope, and at the same time fills out the round of the rope, giving a smooth surface for the parceling and serving.

Parceling consists in wrapping the rope spirally with long strips of canvas, following the lay of the rope, and overlapping like shingles on a roof to shed moisture.

Serving consists in wrapping small-stuff snugly over the parceling and is done against the lay of the rope. A "serving mallet" is used for passing the turns, each turn being hove taut by the leverage of the handle so that the whole forms a stiff protecting cover for the rope.

Mechanical Eye Splice. A method of making eye splices in wire rope by swaging a metal sleeve around the rope with a hydraulic press is now available. This mechanical splice is superior in appearance, is easier to make, and is stronger than a manual splice. No seizing is necessary for this splice and splices can be made in all sizes of wire rope from one-eighth inch to one and one-half inches in diameter.

A2.9. Appliances for Use with Wire Rope (Figs. A2.11 and A2.12). Most of the appliances for use with wire rope are designed to provide some sort of an eye on the end of the rope, by which it can be connected with another rope or with a tackle, or otherwise secured. Figure A2.11 illustrates a number of these. It will be seen that any two of the ends shown can be joined together, either directly or by the aid of a shackle. B shows a handy clip and the manner of applying it. These clips, when made of drop-forged steel and properly applied, are little if at all inferior to a splice, and they can be applied in a few moments where a splice would take as many hours. In the emergencies which sometimes arise on shipboard, as in handling anchors, taking a vessel in tow, etc., when an eye is needed in a hawser and there is not time to make a splice,

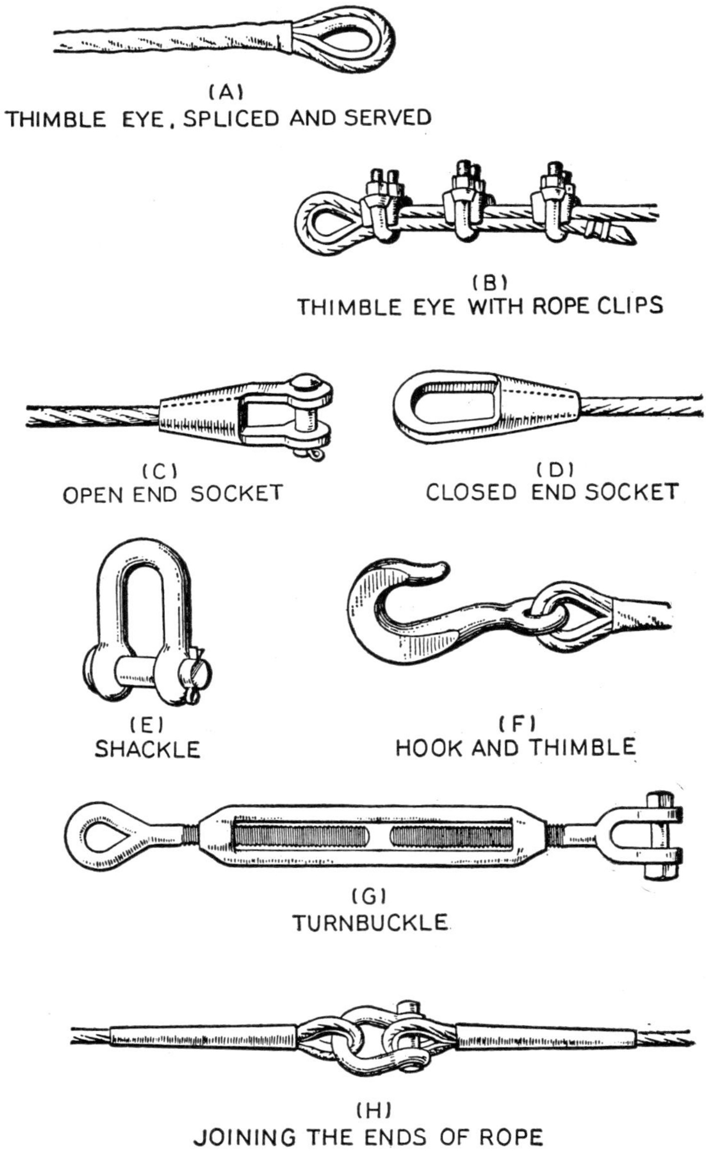

FIG. A2.11 APPLIANCES FOR USE WITH WIRE ROPE

these clips would be invaluable. They can be removed as quickly as they are applied, breaking down the eye at once.

A set of these in sizes to fit any hawser on board might well be issued to all ships. Note that the U-bolt is always applied to the "dead" end of the rope.

Number of wire rope clips required to make a fastening having 80 percent of the strength of 6 x 19 plow steel wire rope.

Diameter of the rope (inches)	Number of clips
¾	5
⅞	5
1	5
1⅛	5
1¼	6

The distance between clips should not be less than six times the diameter of the rope.

For wire rope smaller than ¾-inch diameter, at least four clips should be used.

For wire rope larger than 1¼ inches, it is preferable to socket the rope and avoid the use of clips.

Clips should be inspected daily and the bolts tightened, if they become loose as the rope stretches.

Sockets. Sockets are devices secured to the ends of wire rope to provide ready means for bending ropes together or to attach a load. Sockets are classed as open or closed, depending on whether they are the jaw and pin or loop types.

The rope end should be whipped (seized) near the end. Put on an additional seizing at a distance from the end of the rope equal to the length of the basket of the socket. It is very important that the seizings be secure to prevent untwisting of the wires and strands and a resultant unequal tension between the several wires after the socket is attached. Place the rope end upright in a vise. Remove the seizing at the bitter end. Cut out the hemp heart down to remaining seizing. If the heart is wire, allow it to remain. Untwist the strands and broom out the wire. The wires should be separated from each other but should not be straightened. Clean the wires carefully with benzine, naphtha, or gasoline, for the distance they will be taken into the socket; then dip them in a bath of commercial muriatic acid for 30 seconds to 1 minute, or until the acid has thoroughly cleaned each wire. To remove the acid, dip into boiling water containing a small amount of soda. Draw the wires together again (a piece of seizing wire will be suitable to do this) so that the socket can be forced down over them. Force the socket down over the rope end until it reaches the seizing. Free the wires within the socket basket and allow them to spread evenly and naturally. The ends of the wires should be level with the large end of the socket basket. Care should be taken to see that the centerline of the basket is lined up exactly with that of the rope, i.e., that the socket is in a true straight line with the rope, so that when under load each element will sustain its due share. Seal the small end of the socket around the rope with putty, clay, or similar substance. Fill the socket basket with molten zinc. The zinc must not be too hot, particularly on small ropes. From 800° to 850°F is the correct temperature. Allow to cool in air or by plunging into cool, fresh water. Remove the seizing and the socket is ready for service.

The socket shown in Fig. A2.11C and D is the strongest attachment known for use on the end of a wire rope. The interior of the base of the socket is coned and the bare end of the rope is bedded in and sealed with molten zinc, which forms a head that holds without distortion of the wires.

Figure A2.12 shows a wire clamp that is very handy and of great strength. The rigger's screw is invaluable in splicing and working fittings on wire rope.

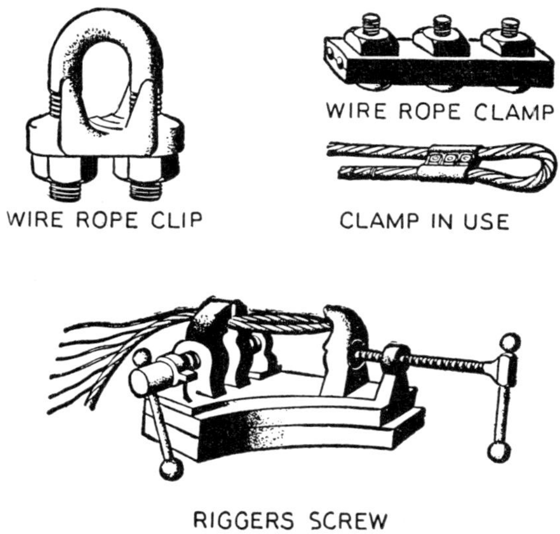

FIG. A2.12 RIGGER'S SCREW

APPENDIX

3

Mechanical Appliances Aboard Ship

The advent of steam power, steel ships, and other modern innovations used by present-day seamen has not relieved these seamen of the necessity of understanding the same fundamental knowledge of ropes, tackles, and weight handling that was required of their predecessors. True, the weights are heavier, are moved greater distances, and wire or wire rope is used more often than manila, but the same principles of mechanical advantage and friction still apply.

Tasks requiring the use of booms, tackles, topping lifts, and other mechanical appliances are the daily lot of seamen. Merchant vessels and naval cargo types require the rigging of many different purchases for positioning and steadying the cargo booms and for handling the weights after the booms have been rigged. Naval vessels of all types employ chain hoists and small tackles to load, strike below, and break out stores and ammunition.

Anyone who is concerned with rigging or operating cargo booms should be thoroughly familiar with the forces which are set up during the operation. To this end, he should consult the manufacturer's tables to determine the strength of the wire or rope used in the standing and running rigging, and he should never exceed these loads. He should inspect all the rigging periodically, replacing worn parts and performing such preventative maintenance as may be specified by the manufacturer. Finally, he should be familiar with the fundamentals of applied mechanics and the effects of acceleration and deceleration on the rigging.

A3.1. Moving Weights. Most heavy weight handling on cargo vessels is done by means of one or more cargo booms. With the use of one boom the weight is lifted from its initial position, the boom swung until the weight is over its intended position, and the weight is lowered. This method is satisfactory only when light loads are being moved because the boom guys must be readjusted as the boom swings.

A more common method is the Yard and Stay rig in which two booms are used. One boom called the *hatch boom* is rigged to plumb the cargo hatch being worked, while the other, or outboard, boom is rigged over the side to plumb a lighter or the dock. Two winches and two cargo runners or whips are used, one on each boom. Each runner is attached to a common hook which engages

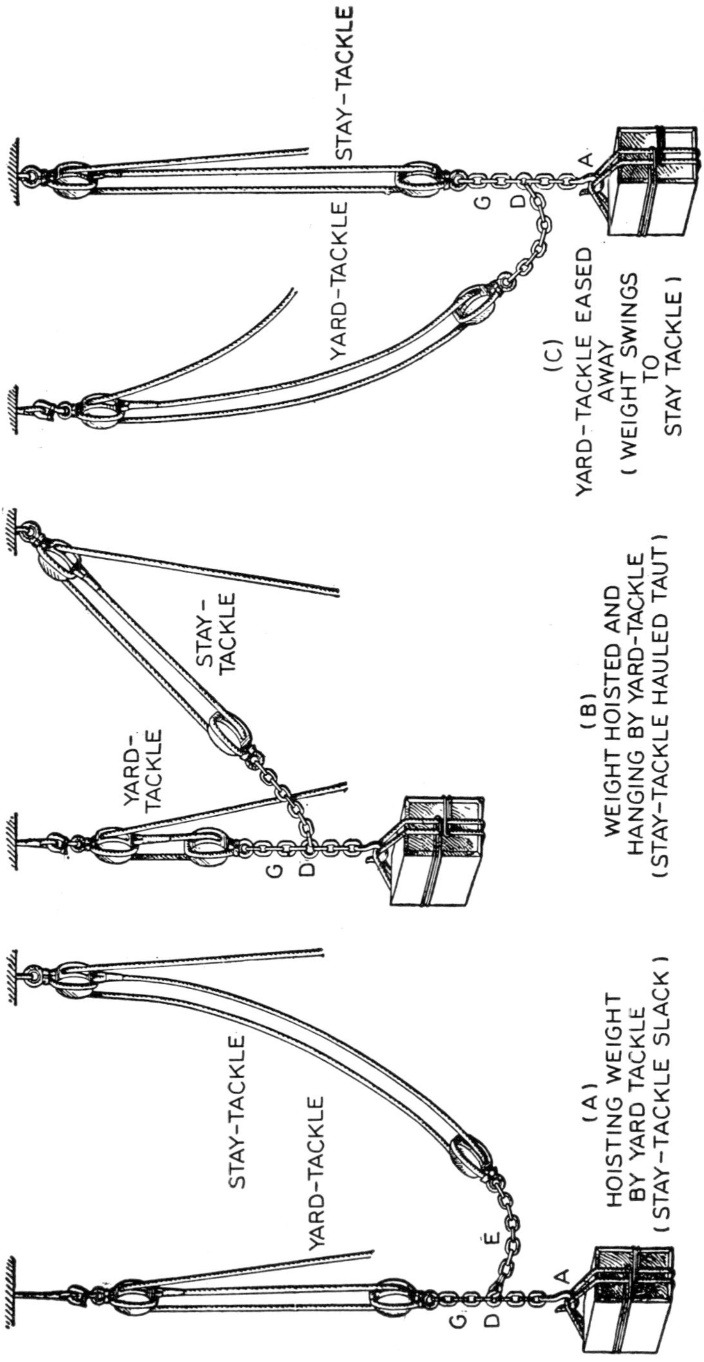

FIG. A3.1 YARD AND STAY TACKLES

the load. For loading the winch of the outboard boom hoists the load, or draft, high enough to clear the ship's side or other obstruction. The draft is then "racked" inboard with the hatch boom winch heaving in and the other winch slacking off slowly until the load is entirely supported by the hatch boom. Finally, the hatch boom winch lowers the draft into the hold, with the outboard cargo runner being kept slack. For unloading, the cycle is reversed. See Fig. A3.1.

In any weight-handling operation speed should be subordinated to safety and smoothness of operation. Jerky movements caused by too rapid acceleration and deceleration put enormous strains on the standing and running rigging. This could result in the parting of one or more lines and the collapse of the entire rig with resulting damage to the load, the ship, and the operating personnel.

A3.2. Blocks. A "block," in the nautical sense, consists of a frame of wood or steel, within which is fitted one or more sheaves (pulleys).

Blocks take their names from the purpose for which they are used, the places they occupy, or from some peculiarity in their shape or construction. They are designated further as single, double, or triple blocks, according to the number of sheaves they have.

The size of the block is determined by the circumference of the rope to be used with it. For manila rope the size of the block in inches is three times the circumference of the manila rope; the diameter of the sheave is twice the circumference of the manila rope. Thus a block for use with a three-inch manila rope would be (3 times 3) a nine-inch block, measured as the length of the cheek, and its sheave would be (2 times 3) six inches in diameter.

Blocks for use with wire rope are not so well standardized. Wire rope can now be made to conform to widely varying specifications. When specifications for blocks are drawn, the advice of the manufacturer should be followed as to diameter of sheave or drum over which the wire is to be rove and the speed of operation (linear speed of the wire). If such advice is not followed, the life of the rope will be materially shortened due to alternate bending and straightening as the wire passes over the sheave (drum).

The usual procedure is to base the diameter of sheaves and drums on the diameter of the rope which is to be used thereon. This procedure is technically correct because the diameter of the rope is a linear function of the diameter of the wires in a rope. The bending stress in a wire rope is a function of the wires and diameter of the sheave or drum. A preferable procedure from a practicable standpoint would be to base the sheave or drum diameter on the breaking strength of the rope. If a sheave or drum diameter is chosen such that the bending stress is 10 percent of the breaking strength, a reasonable load stress may be applied with a good factor of safety. This is a proper criterion of minimum diameter to follow in the design of sheaves or drums; the diameters of sheaves associated with wire ropes should be as large as practical considerations permit. The values given here are the minimum values of the ratio of diameter of the sheave at the bottom of the score (tread diameter) to the diameter of the wire rope used.

Construction	Minimum diameter ratio
6 x 7	28
5 x 19	20
6 x 19	20
6 x 12	14
6 x 24	14
6 x 37	14

A3.3. Tackle. An assemblage of ropes (falls) and blocks for the purpose of multiplying force is a tackle (Figs. A3.2 and A3.3).

The seaman speaks of "reeving" when he passes ropes around the sheaves of the blocks. These ropes are called "falls." The "standing part" is that part of the fall made fast to one of the blocks. The hauling part is the end of the falls to which force is applied to handle the weight. To "overhaul" the falls is to separate the blocks. To "round in" is to bring the blocks together. The blocks are said to be "chock a block" or "two-blocked" when they are tight together.

Tackles are designated either according to the number of sheaves in the blocks that are used to make the tackle, e.g., single, two-fold, three-fold purchase; or according to the purpose for which the tackle is used, e.g., yard-tackles, stay-tackles, fore-and-aft tackles. Other designations handed down from the past still persist, as luff-tackles, gun-tackles, Spanish-burtons.

A single-whip. (A, Fig. A3.3.) A single block fixed.

A runner. (B, Fig. A3.3.) A single block movable.

A whip and runner would be a whip hooking to the hauling part of a runner.

A gun-tackle purchase. (C, Fig. A3.3.)

A luff-tackle. (D, Fig. A3.3.) A single and double block.

A luff upon luff. The double block of one luff-tackle hooked to the hauling part of another, thus multiplying the power. (See Fig. A3.6.)

A two-fold purchase. (Fig. 4, Fig. A3.2.) Two double blocks.

A double luff. A double and treble block. (Fig. 5, Fig. A3.2.)

A three-fold purchase. Two treble blocks. (See Fig. A3.4, in which two purchases are shown.)

A three-fold purchase is the heaviest purchase commonly used, designated by terms descriptive of their use without reference to the blocks or sheaves involved:

Thwartship-tackles are used on the heads of boat-davits for rigging in. In a more general sense the term is applied to any tackle leading across the deck. Similarly, a tackle for hauling out the backbone of an awning or for any other purpose where it has a fore-and-aft lead is a fore-and-aft tackle.

Hatch-tackles are used at hatches for hoisting, lowering stores, etc.

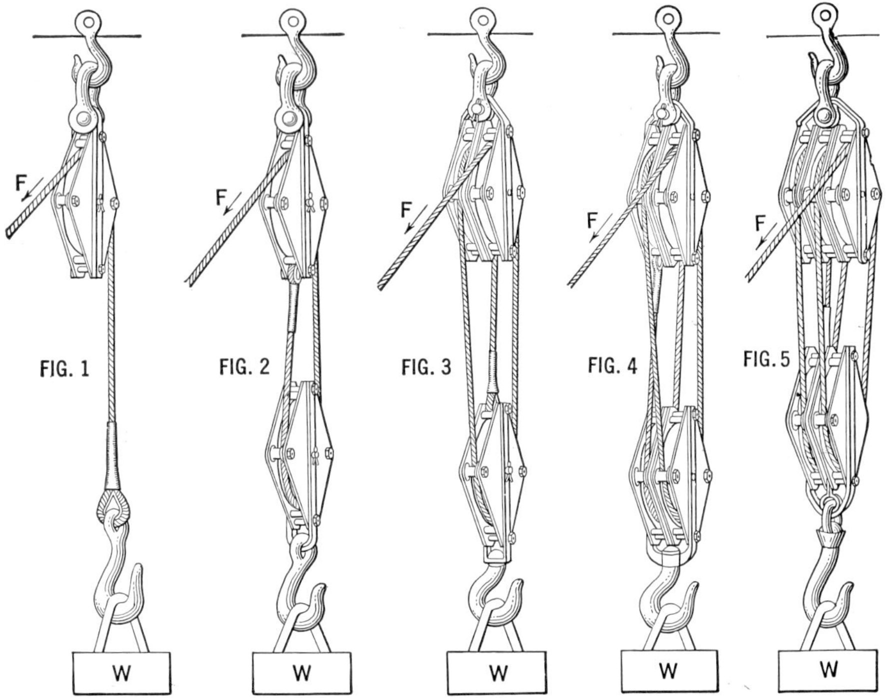

Ratio of Weight, W to Force F Necessary to Raise Weight.

	Disregarding Friction.	Allowing for Friction.
Fig. 1	$F = W$	$\frac{F}{W} = \frac{11}{10}$
Fig. 2	$\frac{F}{W} = \frac{10}{20}$	$\frac{F}{W} = \frac{12}{20}$
Fig. 3	$\frac{F}{W} = \frac{10}{30}$	$\frac{F}{W} = \frac{13}{30}$
Fig. 4	$\frac{F}{W} = \frac{10}{40}$	$\frac{F}{W} = \frac{14}{40}$
Fig. 5	$\frac{F}{W} = \frac{10}{50}$	$\frac{F}{W} = \frac{15}{50}$
In a three-fold purchase	$\frac{P}{W} = \frac{10}{60}$	$\frac{F}{W} = \frac{16}{60}$
Luff on Luff	$\frac{P}{W} = \frac{10}{120}$	$\frac{F}{W} = \frac{16}{120}$

Note: In this plate the hauling part leads from the fixed block. The mechanical efficiency can always be increased if the hauling part is led from the movable block.

FIG. A3.2 TYPES OF TACKLES

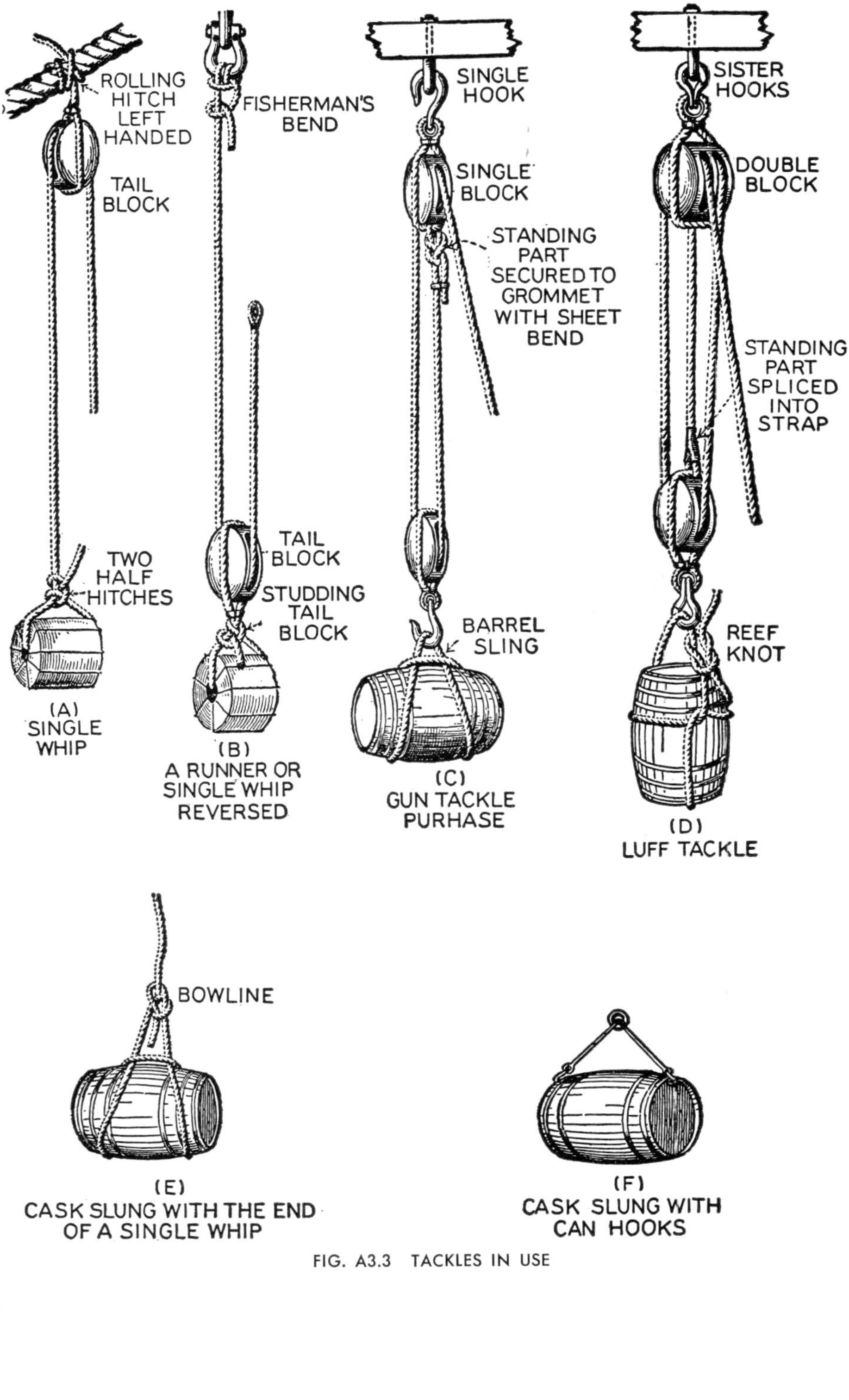

FIG. A3.3 TACKLES IN USE

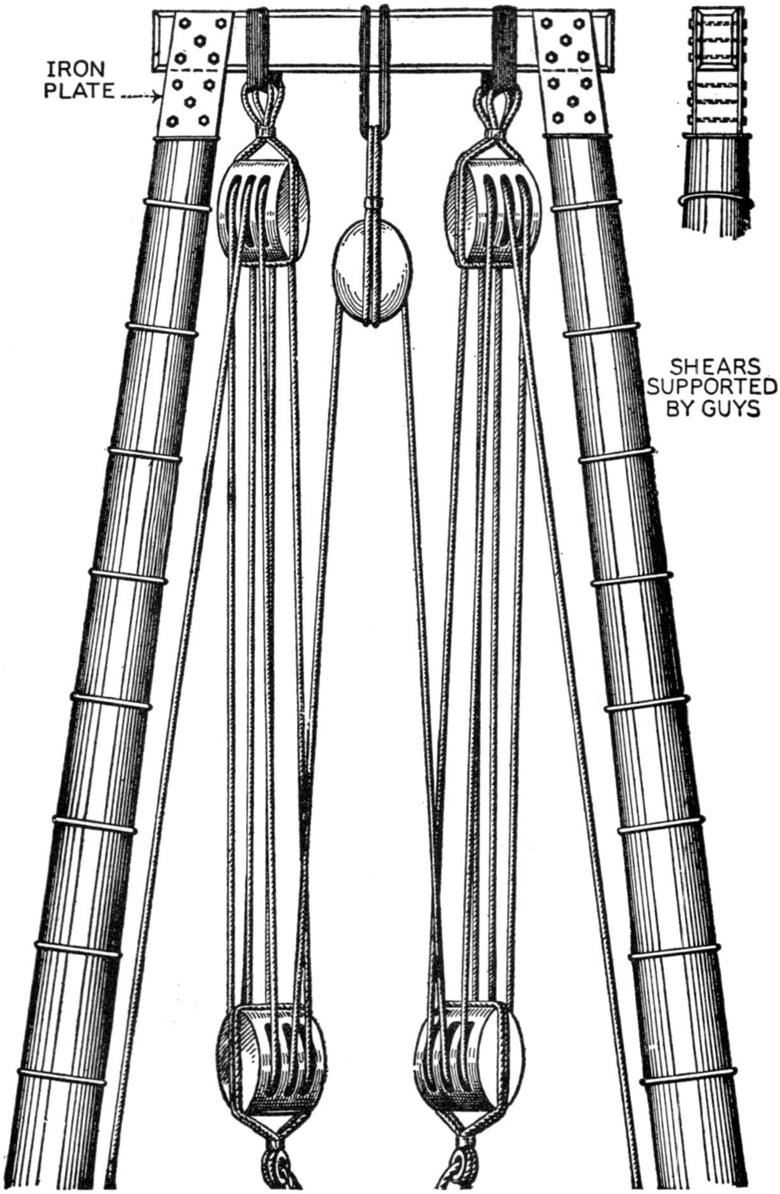

FIG. A3.4 SHEARS WITH DOUBLE THREE-FOLD PURCHASE FOR HANDLING HEAVY WEIGHT

Note: The rope straps shown in use for very heavy work should preferably be of wire. Even where iron-strapped blocks are available they are less reliable than if strapped with wire rope, though they are, of course, more convenient.

Jiggers are small light tackles used for miscellaneous work about the ship.

A deck-tackle is a heavy purchase, usually two-fold, used in handling ground-tackle, mooring ship, and generally for heavy work of any kind about the deck.

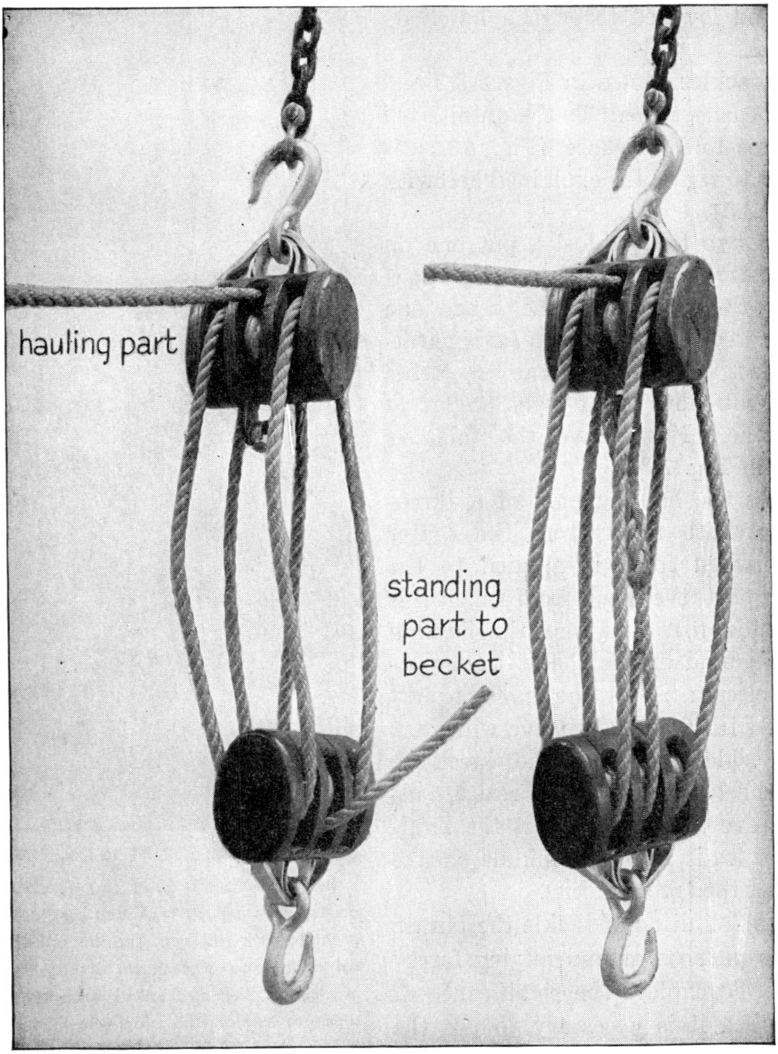

FIG. A3.5 RIGGING THE THREE-FOLD PURCHASE

Fig. A3.4 shows a special purchase which may be used for very heavy weights. The tension on the various parts of the fall is partially equalized, by applying the power to both ends.

Yard and stay tackles take their names from their application on ships with masts and yards, where they were used together for transferring stores from a boat alongside to the deck or hatch of the ship. The general principle involved in the "yard and stay" is of wide application on merchant ships where a weight is to be lifted from a dock and lowered through a hatch on a vessel.

The tackles shown in Figs. A3.2 and 3.3 are easily rigged. To the uninitiated the three-fold purchase (Fig. A3.4) is difficult to rig and is explained herewith (Fig. A3.5).

Take two treble blocks, put one on deck resting on its cheek (axis perpendicular to the deck); the other one rests on deck on its sheaves (axis parallel to deck). The axes of the two treble blocks will now be at right angles to each other. Now reeve the falls as shown in Fig. A3.5.

Unless the hauling part of a three-fold purchase leads from the center sheave when force is applied to the hauling part, the block from which this part leads will cant and the hauling part will bind on the cheek. Under extreme circumstances the hauling part when led from an end sheave will exert enough side thrust on the cheek to break the block. For these reasons always reeve the falls so that the hauling part leads from the middle sheave of one of the treble blocks.

In the definition of tackle the phrase "for the purpose of multiplying force" is used. To explain the significance of this phrase it is necessary to use the basic formulae of work.

$$fS = Fs$$

This formula may be expressed as "a force times the distance through which it moves equals the weight (or resistance) times the distance through which

FIG. A3.6 LUFF UPON LUFF

It is sometimes a good plan to clap a tackle on the hauling part of another tackle where great force is required. The theoretical mechanical advantage of such an arrangement is the product of the mechanical advantages of the separate tackles. The proper method of rigging a luff upon luff is to bend the double block of one luff to the hauling part of the other luff. Such a method of rigging would have a mechanical advantage of 3×4 to 1 or 12 to 1.

The simplest manner of attaching a tackle to the hauling part of another is to form a cat's-paw or a blackwall hitch in the hauling part and slip it over the hook of the block. *Official U.S. Navy Photograph* it moves. In the formula, therefore, when used in connection with tackles, f represents a smaller force than F which is normally the weight being lifted and s is a smaller vertical height than S. Thus in a two-fold purchase if f acting through four feet (S) will raise a weight F one foot (s), we can write the fundamental equation:

$$f \times 4 = F \times 1$$

or $f = F/4$, the force (f) has been multiplied four times. The mechanical advantage of this purchase is four.

If friction is neglected, there are two thumb rules that are useful in determining the multiplication of the force:

1. Pass a plane above the block to which the weight is attached; the number of parts of the falls cut by this plane will be the ratio of force lifted to the force applied.

In Fig. A3.2 the application of Rule 1 will be readily seen. Observe though, that if the two blocks of any purchase were interchanged, that is the movable block made fast so as to become the fixed block and the weight attached to the other, which would then be the movable block, the ratio of Force to Weight is changed to

Fig. 1	Fig. 2	Fig. 3	Fig. 4	Fig. 5
$\frac{1}{2}$	$\frac{1}{3}$	$\frac{1}{4}$	$\frac{1}{5}$	$\frac{1}{6}$

Therefore, when practicable make the block having the greatest number of sheaves the movable block. When the two blocks have the same number of sheaves, rig the tackle so that the hauling part leads from the movable block.

2. Rig a tackle similar to the tackle in question. Place sufficient weight on the movable block to take all the slack out of the falls and let it just touch the deck; make a reference mark on the hauling part of the falls. Pull on the hauling part until the weight has been raised one foot, measure the amount of line taken on the hauling part. This amount of line (in feet) will be the ratio of force lifted to force applied.

Friction is allowed for, practically, by adding 10 percent of the weight for each sheave passed over by the falls.

To illustrate: a two-fold purchase with its hauling part led from the fixed block and to the source of power through a fair lead has a theoretical advantage of 1 to 4. There are four sheaves in the two blocks plus one in the fair lead. Hence, $50/100W$ is added to W and since the theoretical advantage is 1 to 4, we have:

$$\left(W + \frac{50W}{100}\right) \times \frac{1}{4} = \frac{15W}{40} = \frac{3W}{8} \quad \text{(See A3.2)}$$

When hoisting, the maximum tension will be on the hauling part; in lowering the maximum tension will be on the standing part. These effects are caused by friction which always acts to oppose motion. As an illustration consider the luff-tackle. If a weight of W pounds is suspended motionless from the purchase (D, Fig. A3.3), the strain on all falls is $W/3$. If the weight is hoisted, the strain on the hauling part is increased by friction from $W/3$ to $13W/30$, an increase of $3W/30$ but the strain on the standing part is still $W/3$ because friction has not acted on it. These conditions are just reversed when lowering.

When applying power to the hauling part of one tackle by means of another tackle, the mechanical advantage of the system will be the product of the mechanical advantage of each system. To illustrate this, suppose you make a gun-tackle purchase fast to the hauling part of a two-fold purchase. There are four combinations of these tackles, viz., the two-fold purchase can be rigged so as to have a mechanical advantage of $\frac{1}{4}$ or $\frac{1}{5}$ and the gun-tackle purchase can be rigged so as to have mechanical advantage of $\frac{1}{2}$ or $\frac{1}{3}$. The following combinations could be used.

$$\left(\frac{1}{4} \times \frac{1}{2}\right) = \frac{1}{8}\text{; allowing for friction } \left(\frac{14}{40} \times \frac{12}{20}\right) = \frac{168}{800}\text{, approximately } \frac{1}{5}$$

$$\left(\frac{1}{4} \times \frac{1}{3}\right) = \frac{1}{12}\text{; allowing for friction } \left(\frac{14}{40} \times \frac{12}{30}\right) = \frac{168}{1{,}200}\text{, approximately } \frac{1}{7}$$

$$\left(\frac{1}{5} \times \frac{1}{2}\right) = \frac{1}{10}\text{; allowing for friction } \left(\frac{14}{50} \times \frac{12}{20}\right) = \frac{168}{1{,}000}\text{, approximately } \frac{1}{6}$$

$$\left(\frac{1}{5} \times \frac{1}{3}\right) = \frac{1}{15}\text{; allowing for friction } \left(\frac{14}{50} \times \frac{12}{30}\right) = \frac{168}{1{,}500}\text{, approximately } \frac{1}{9}$$

A3.4. Chain Hoists (Fig. A3.7). Various types of chain hoists (mechanical purchases) are used on board ship. Inasmuch as the chain hoists are classified according to the weight they are built to handle, and their mechanical advantage is so designed that one man can pull enough force to lift the weight for which they are designed, no discussion of the mechanical principles of their design is given here. Suffice to say that if the weight for which a given chain hoist is designed is exceeded, disaster is invited. Overloading usually makes itself apparent by requiring an unusually heavy pull to raise the weight, by the weight hook spreading, or by the lifting chain parting. Overloading is to be carefully guarded against when using chain hoists as yard and stay tackles.

In a *differential purchase* (Fig. 6) an endless chain is taken over two sheaves of slightly different diameters which are keyed to the same shaft and revolve together. A movable block to which a weight may be attached is hung in the bight of the endless rope or chain. If force is applied to one of the parts leading from the larger sheave, the rope is unwound from the sheave but is at the same time wound up on the slightly smaller sheave alongside. Thus the change in the length of the bight which carries the movable block is very slight for a great distance moved by the hauling part. By a simple mathematical demon-

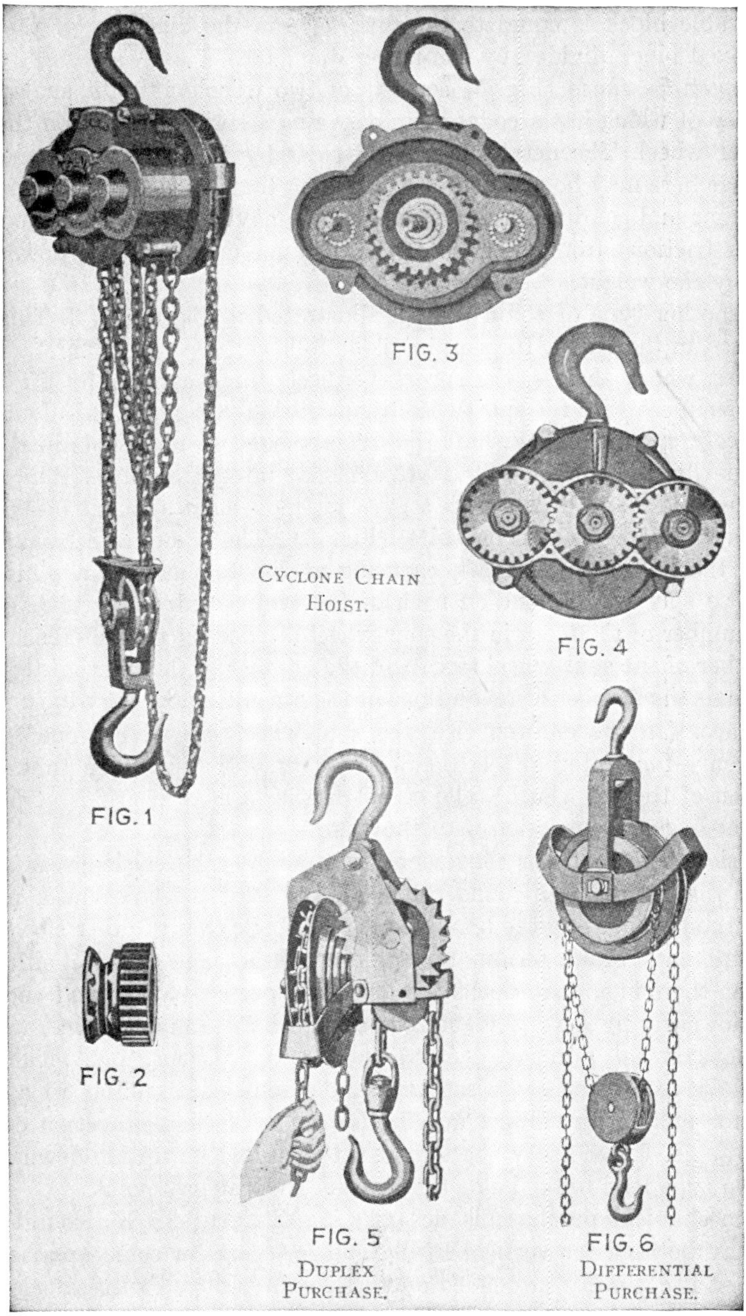

FIG. A3.7 MECHANICAL PURCHASES

stration it can be shown that the ratio of the power applied to the power on the movable block is equal to the difference in the diameter of the sheaves of the fixed block divided by the larger diameter.

A *Duplex Purchase* (Fig. 5) consists of two wheels at right angles to each other, one of which has a cogged rim engaging a series of cams on the face of the other wheel. The details are made clear by the figure. The mechanical advantage here may be made almost anything that is desired by proper design of the cams and gearing. In any given case it may be determined theoretically (without friction) from the ratio of the distance moved by the power to that moved by the weight.

Still another type of chain hoist is illustrated in Figs. 1, 2, 3. This type is much used in the Navy. Its working is as follows:

The lift wheel, that is, the sprocket wheel which carries the lift chain, is cast in one piece with the spur-wheel that drives it (Fig. 2). This double wheel turns freely upon a hollow shaft rigidly supported at both ends in the frame. The spur-wheel is encircled by a yoke having internal teeth meshing into the spur-wheel teeth and driven with a gyrating movement about it by two eccentrics placed diametrically opposite (Fig. 3). The hand wheel shaft passes through the hollow main shaft, carrying at the further end a pinion which drives two spur-wheels, one on each of the two eccentric shafts (Fig. 4).

The number of the teeth in the spur-wheel divided by the difference between the number of the spur-wheel teeth and the number of the internal teeth of the yoke equals the number of revolutions of the eccentric necessary to turn the lift wheel once. (In the one-ton size, the spur-wheel has twenty-one teeth, the yoke twenty-four internal teeth, and the eccentrics turn seven times to each revolution of the lift wheel.) The eccentric shafts have bearings at both ends and roller-bushed connection with the yoke.

The friction loss of this movement is so slight (the efficiency is about 80 percent) that it has been found practicable to gear the hoists to a very high speed without increasing the hand wheel pull above that of other slower hoists.

The automatic brake permits the spinning of the hand wheel in either direction when there is no load, locks the load with perfect safety, and yet permits its free lowering by a very slight reverse pull on the hand chain.

The mechanical purchase (A3.8) is a very efficient, lightweight, spur-geared device operated by a ratchet handle somewhat similar to a jack. It is used for pulling any object in a horizontal or vertical direction or at any possible angle. If necessary the unit can be used in an upside-down position. A friction type load brake holds the load securely at any point.

This mechanical purchase is designed particularly for horizontal pulling, but it may be used for vertical lifting and lowering for some applications. A chain hoist is generally suspended overhead and equipped with sufficient chain for a lift of eight feet or more.

The method of operating this mechanical purchase requires that the handle be within easy reach of the operator. The top swivel hook is attached to some stationary object or "anchor." Buttons "A" and "B" are pulled out and given ¼ turn to a neutral position. This provides "free wheeling" as the lifting wheel is free, and the chain can then be pulled with very little effort. Attach the bottom swivel hook to the load or weight to be lifted. Then turn button B back to its active position and turn button A to the UP position (button is plainly marked UP and DOWN). Operate the handle up and down, and the object

Horizontal Pulling

Compact for Tool Box Storage

Vertical Lifting

FIG. A3.8 RATCHET HANDLE TYPE MECHANICAL PURCHASE

will be lifted or pulled. Turn button A to the DOWN position, and the object will be lowered by the same motion of the handle.

A3.5. Strength of Ropes, Blocks, Tackles, Etc. When definite information as to the strength of the wire or rope in question is at hand, it should be utilized. In the absence of such definite information the following rules are convenient and safe. Those given below for the strength of manila or hemp are generally very accurate. Those rules given below for the strength of wire are applicable only to the type of wire mentioned and do not necessarily apply to all types though they may be used as rough thumb rules as long as maximum loads are not approached. In the interests of operating safety, manufacturer's tables should be consulted prior to using any wire.

B = Breaking strength, lbs. or tons (2,000 lbs.).
P = Safe working load; viz., safe tension for single part of rope.
C = Circumference (inches).
D = Diameter (inches) of wire.

Rule 1. Strength of manila or hemp.

$$B = \frac{C^2}{2.5} \text{ tons} = C^2 \times (750 \text{ to } 1{,}000) \text{ lbs. (large to small sizes)}$$

In the above formula, the constant of 900 has been found to give a close approximation for all sizes.

Rule 2. Strength of wire. Type D (Navy Standard).

$$B = D^2 \times 27 \text{ tons} = D^2 \times 54{,}000 \text{ lbs.}$$

The breaking strength being known from either the above formulae, from other sources (see Appendix 1) or from the manufacturer's tables, the working load adopted as safe for a single part of the rope will depend upon the factor of safety which we adopt after consideration of the condition of the rope itself and the use to be made of it.

The factor of safety may be taken as follows:

(a) Under average conditions,

Working load $P = 1/6$ breaking strength B.

(b) Under best conditions—new rope to be used occasionally,

Working load $P = 1/4$ breaking strength B.

(c) Under unfavorable conditions, where rope is used frequently and for an indefinite period, as in the case of running rigging and boat falls,

Working load $P = 1/8$ breaking strength B.

If, without knowing the breaking strength in a given case, we wish to derive the safe working load directly from the diameter or circumference of the rope, we have the following rules.

Rule 3. Safe working load for manila or hemp.

Under average conditions:

$$P = \frac{C^2}{15} \text{ tons} = C^2 \times (120 \text{ to } 160) \text{ lbs.}$$

Rule 4. Safe working load for wire. Type D (Navy Standard).

Under average conditions:

$$P = D^2 \times 4.5 \text{ tons} = D^2 \times 9{,}000 \text{ lbs.}$$

Under best conditions, add 30 percent to above values of P.

Under very unfavorable conditions, subtract 30 percent from above values of P.

Rule 5. To find the size of manila rope to lift a given load in tons.

From rule 3 above, we have

$$C \text{ (inches)} = \sqrt{15 \times P} \text{ (tons)}$$

Hence, multiply the load *in tons* by 15, then take the square root of the product for the circumference in inches.

Notes:

Rule 5 above gives an unnecessarily large factor of safety for the newer and stronger types of wire, but it is convenient and safe.

A well-made splice weakens either manila or wire by from five to ten percent.

A sharp nip may weaken manila or wire by from 25 to 50 percent.

Manila deteriorates rapidly if stowed away wet, or if exposed, either wet or dry, to continued high temperature.

Wire rope should be discarded when its outer wires are worn down to one-half their original diameter.

The strength of two ropes of different sizes but similar construction is proportional to the *squares* of the circumferences. Thus a 2-inch rope is to a 3-inch as 4 to 9.

As a working rule, wire rope is six times as strong as manila of the same size.

In cases where a load is applied suddenly, with a blow or a jerk, its effect is doubled, and this should be allowed for in calculating the size of rope required.

Rule 6. To find the size of wire rope to lift a given load (Type D).

From Rule 4, above, we have

$$D = \sqrt{\frac{P \text{ tons}}{4.5}} \text{ inches}$$

Rule 7. To find the size of manila rope when rove as a tackle to lift a given weight.

Add to the weight one-tenth of its value for every sheave to be used in hoisting. This gives the total resistance, including friction. Divide this by the number of parts at the movable block, for the maximum tension on the fall. Reeve the fall of a size to stand this tension as a safe working load.

Example: To lift 10 tons with a three-fold purchase, the fall of which, coming from the upper block, is taken through an extra sheave on deck for a fair lead. Determine the size of fall needed.

Total resistance, including friction $= 10 + \frac{7}{10} \times 10 = 17$ tons

Maximum tension on fall $= \frac{17}{6} = 2.8$ tons

Size of fall (Rule 5) $= \sqrt{15 \times 2.8} = 6\frac{1}{2}$ inches (nearly)

Rule 8. To find the weight which a given purchase will lift with safety.

Find the safe-working load for the rope to be used (Rule 1). Multiply this by the number of parts at the movable block. This gives the total resistance including friction.

Multiply the total resistance by 10 and divide by (10 + the number of sheaves used). The result is the weight that may be lifted.

Example: To find the weight which may be lifted by a fall of 6½-inch manila rope as a three-fold purchase, the fall of which leads from the upper block through an extra leader on deck. (Problem is reverse of one above.)

$$\text{Safe-working load, } \frac{6.5^2}{15} = 2.8 \text{ tons} = P$$

Total resistance, including friction, $6 \times 2.8 = 16.8$ tons

$$\text{Weight to be lifted, } \frac{16.8 \times 10}{10 + 7} = \frac{168}{17} = 10 \text{ tons approximately}$$

Rule 9. Strength of hooks and shackles.

It may generally be assumed that the safe load for a well-made *block* is in excess of that of any hemp or manila rope that it will reeve. This, however, is not always true of the *hook*, which is almost invariably the weakest part, and often gives way under strains for which the block is otherwise amply strong. The strength of the hook is therefore the measure of the strength of the block. The difficulty here comes from the tendency of the hook to open out—a tendency which should be guarded against, in heavy work, by careful "mousing" of the hook—preferably by an iron link.

For heavy work, *shackles* are fitted to blocks in place of hooks and are very much to be preferred, as will be apparent from the rules and tables which follow.

(a) Hooks.

$D =$ Diameter at back of hook

$P = \frac{2}{3}D^2$ tons

(b) Shackles.

$D =$ Diameter at sides

$P = 3D^2$ tons

Note: For a given diameter of material, a shackle is approximately 5 times as strong as a hook.

The following table (A3.1) gives the results of practical tests made at the Watertown Arsenal. The hooks and shackles tested were of the ordinary commercial form. The diameter given is that of the metal at the back of the hook and the sides of the shackle.

TABLE A3.1—TEST OF HOOKS

Diameter of metal (inches)	Broke at (pounds)	Remarks
½	2,385	Hook partly straightened, then fractured across the back.
¾	4,130	
1	10,315	
1¼	14,510	
1½	20,940	
1¾	27,420	
2	38,100	
2½	55,380	

TEST OF SHACKLES

Diameter of metal (inches)	Broke at (pounds)	Remarks
¾	20,700	Eye of shackle parted.
⅞	38,100	Eye of shackle parted.
1	51,900	Eye of shackle parted.
1¼	75,200	Eye of shackle parted.
1½	119,980	Eye of shackle parted.
1¾	146,400	Sheared shackle-pin.
2	196,600	Eye of shackle parted.
2½	210,400	Eye of shackle parted.

Rule 10. To find the strength of a spar to resist compression (derrick or shears).

$$T = \text{Safe thrust in tons}$$

$$R = \text{Radius of spar in inches}$$

$$L = \text{Length of spar in feet}$$

$$T = \frac{4R^4}{L^2}$$

Note: The multiplier 4 in this formula is safe for all ordinary kinds of wood. For very strong woods, like oak, mahogany, etc., it could be increased 50 percent without danger.

Example: To find safe thrust for a spar 10 inches in diameter, of fir, 17 feet long.

$$T = \frac{4 \times 5^4}{17^2} = 8.6 \text{ tons}$$

Modern American cargo type merchantmen are rigged with cargo booms constructed to handle loads somewhat in excess of 5 tons, but to allow a safety factor, they are designated as 5-ton booms. The other three factors deciding the load limit for a given boom are:

1. The winch capacity, or the strength of the winch, which usually runs from 4 to 7 tons.
2. The strength of the cargo runner, or the line actually hoisting the load. Cargo runners are usually of $\frac{5}{8}$ inch wire, with a strength of about 10 to 12 tons.
3. The strength of the mast, guys, stays, and topping lift. This is a variable factor, which depends on how the gear is rigged.

In the event that a given load exceeds the capacity of the boom, topping lift, cargo runner or winch, various methods of doubling up are used to divide the load between two rigs.

In addition to this, there is usually a pair of heavy duty booms, usually located over the largest hatch of modern merchantmen. On naval merchant types, these booms have a capacity of as much as 50 tons, and are fitted with special heavy standing and running rigging. Civilian merchantmen do not carry gear as heavy as this, since most of their heavy lifting is done by dock-side cranes, but their large booms may run as high as 10-15 tons.

General Precautions. 1. Remember that the giving way of one part breaks and destroys other parts, frequently to an extent not readily repaired, and, furthermore, endangers the men.

2. Heavy weights must never be allowed to drop, even for the shortest distances, but must be lowered to rest with a gentle motion, and at the same time chocked to prevent rolling or sliding.

3. In raising or lowering heavy weights always, whenever possible, closely follow up with blocks or chocks to guard against any possible giving way of jacks or tackle.

4. All motions with heavy weights must be slow, so as not to generate momentum.

5. Supports must have a firm base and cribbing, a level foundation, and be built up vertically.

6. All fittings or appliances used for securing lines must be strong and secure beyond any possibility of carrying away.

7. Be careful at all times to avoid any sudden shocks or strains.

8. Every operation should be done with spirit but without bustle or confusion.

9. Vigilance on the part of the person in charge should be unceasing to see all gear rigged, handled, and operated correctly.

10. Do not permit men to step on a taut fall, or to get in positions of danger, such as under weights, in the bight of a running rope, or at the end of a taut rope or cable which might give way.

11. Special precautions must be taken in wet weather when the material is slippery.

12. Overhaul tackle as frequently as necessary, keep free from rust, corrosion and dirt and keep well oiled and operating freely.

13. Always insure that decks are adequately shored to withstand the additional stresses of handling heavy weights.

14. Overloading decreases the strength of rope materially. Rope should never be loaded beyond one-third of its breaking load.

15. Keep wire rope well coated with a preservative lubricant. Keep wire free of kinks or sharp bends.

16. Do not reeve a tackle with old and worn rope.

17. Rope exposed to moisture and weather deteriorates very rapidly.

18. Always use a shackle or a mousing on hooks when handling heavy weights.

Index